Cosmology

DANIEL BAUMANN

Universiteit van Amsterdam

Shaftesbury Road, Cambridge CB2 8EA, United Kingdom

One Liberty Plaza, 20th Floor, New York, NY 10006, USA

477 Williamstown Road, Port Melbourne, VIC 3207, Australia

314–321, 3rd Floor, Plot 3, Splendor Forum, Jasola District Centre, New Delhi – 110025, India

103 Penang Road, #05–06/07, Visioncrest Commercial, Singapore 238467

Cambridge University Press is part of Cambridge University Press & Assessment,
a department of the University of Cambridge.

We share the University's mission to contribute to society through the pursuit of
education, learning and research at the highest international levels of excellence.

www.cambridge.org
Information on this title: www.cambridge.org/highereducation/isbn/9781108838078
DOI: 10.1017/9781108937092

First published 2022 (version 3, April 2023)

Printed in the United Kingdom by TJ Books Limited, Padstow, Cornwall, 2022

A catalogue record for this publication is available from the British Library.

Library of Congress Cataloging-in-Publication Data
Names: Baumann, Daniel, 1978– author.
Title: Cosmology / Daniel Baumann, Universiteit van Amsterdam.
Description: Cambridge, United Kingdom ; New York, NY : Cambridge
University Press, 2022. | Includes bibliographical references and index.
Identifiers: LCCN 2022000140 | ISBN 9781108838078 (hardback)
Subjects: LCSH: Cosmology. | BISAC: SCIENCE / Space Science / Cosmology
Classification: LCC QB981 .B3645 2022 | DDC 523.1–dc23/eng20220315
LC record available at https://lccn.loc.gov/2022000140

ISBN 978-1-108-83807-8 Hardback

Additional resources for this publication at www.cambridge.org/baumann

Cosmology

Progress in modern cosmology has been remarkable, leading to a detailed understanding of the structure and evolution of the universe. In this approachable and self-contained textbook, the author—a leading theoretical cosmologist—expands on his widely acclaimed lecture notes, presenting the theoretical foundations of cosmology and describing the observations that have turned the subject into a precision science.

Notable features of the book are:

- many worked examples and exercises
- numerous end-of-chapter problems
- high-quality figures to illustrate the concepts introduced in the text
- detailed appendices to provide the necessary background material
- online resources—including an annotated list of important cosmology papers, Mathematica notebooks and updates on observational results—to keep students abreast of the latest research.

The step-by-step approach and explicit presentation of key calculations make this book a definitive resource for advanced undergraduate or beginning graduate students in physics, astronomy and applied mathematics, as well as for self-study.

Daniel Baumann is a world-leading theoretical cosmologist. He obtained his doctorate from Princeton University in 2008, after which he was a postdoctoral researcher at Harvard University and the Institute for Advanced Study in Princeton. From 2011, he was a faculty member at Cambridge University, until he was appointed Professor of Theoretical Cosmology at the University of Amsterdam in 2015. Baumann has received numerous awards, including an ERC Starting Grant, an NWO VIDI Grant and a Yushan Professorship at National Taiwan University. His cosmology lecture notes are used worldwide. He is the author of the book *Inflation and String Theory* (with Liam McAllister).

For Celia and Kosmo

Contents

Part I The Homogeneous Universe

Part II The Inhomogeneous Universe

Appendices

Preface

Cosmology is both an old and a new subject. While questions about the origin and structure of the universe have been asked for a long time, concrete answers have emerged only relatively recently. A century ago, we didn't know that there are other galaxies beyond our own, we didn't know why stars shine, and we didn't know that the universe is expanding. Even just twenty-five years ago, we didn't know what most of the universe is made of.

Progress in cosmology has been nothing short of astonishing. We have learned that the light elements were created in the first few minutes of the hot Big Bang and that the heavy elements were made inside of stars. We have taken a picture of the universe when it was just 370 000 years old, discovering small fluctuations in the first light that eventually grew into all of the structures we see around us. We have developed powerful numerical simulations to predict the growth of this large-scale structure and measured its statistical properties in galaxy surveys. We have discovered that the universe is dominated by dark matter and dark energy, although their origin is still a mystery. Finally, we have found evidence that the primordial density perturbations originated from microscopic quantum fluctuations, stretched to cosmic scales during a period of inflationary expansion.

This book is about these developments in modern cosmology. I will present the theoretical foundations of the subject and describe the observations that have turned cosmology into a precision science.

About this book

This book grew out of lectures given in Part III of the Mathematical Tripos at the University of Cambridge. Elements of the book were also developed in courses taught at numerous graduate schools. Although the material has been significantly expanded and reworked, the book retains some of the flavor and tone of the original lecture notes. In particular, many calculations are presented in more detail than is common in the traditional textbooks on the subject. A large number of worked examples and problems has been included and coordinated closely with the material. The book is therefore particularly well-suited for self-study or as a course companion.

Although the material covered in the book has mostly been tested on Masters students, I have tried to be very pedagogical, so that large parts of the book should also be accessible to advanced undergraduates. Familiarity with basic relativity, quantum mechanics and statistical mechanics will be helpful, but the book includes all necessary background material and is therefore self-contained.

Structure of the book

Chapter 1 is a brief invitation to the subject. I describe the enormous range of scales involved in cosmology, introduce the forms of matter and energy that fill the universe and sketch key events in the history of the universe.

The rest of the book is organized in two parts. Part I deals with the dynamics of the homogeneous universe, while Part II discusses the evolution of small fluctuations.

- Chapter 2 introduces the expanding universe. I first derive the spacetime geometry of the universe and then determine its evolution in the different stages of the universe's history. The chapter also includes a summary of the key observations that have helped to establish the ΛCDM concordance model.
- Chapter 3 describes the hot Big Bang. I first explain how the thermal history of the universe is shaped by a competition between the rate of particle interactions and the expansion rate. I discuss the concept of local thermal equilibrium and show how the principles of statistical mechanics are applied in the cosmological context. Finally, I introduce the Boltzmann equation as the key tool to go beyond thermal equilibrium. This is then applied to three examples: dark matter freeze-out, Big Bang nucleosynthesis and recombination.
- Chapter 4 explores cosmological inflation. The chapter begins with a careful examination of the horizon and flatness problems of the hot Big Bang. I then explain how inflation solves these problems and discuss the physics of inflation from a modern perspective. The chapter closes with a discussion of several open problems in the field of inflationary cosmology.
- Chapter 5 introduces structure formation in Newtonian gravity. I derive the linearized equations of motion for a non-relativistic fluid and apply them to the evolution of dark matter fluctuations. I present the key statistical properties of these fluctuations in the linear regime. I then give a brief discussion of the nonlinear clustering of dark matter fluctuations and review current observations of the large-scale structure of the universe.
- Chapter 6 develops cosmological perturbation theory in general relativity. This allows us to extend the Newtonian treatment of the previous chapter to relativistic fluids and superhorizon scales. I derive the linearized evolution equations for the coupled matter and metric perturbations, and show how the initial conditions are imposed on superhorizon scales. I then use the relativistic framework to discuss the evolution of dark matter, photons and baryons. The chapter closes with a heuristic description of the CMB anisotropies.
- Chapter 7 presents the physics of the CMB anisotropies in more detail. I derive the propagation of sound waves in the photon–baryon fluid before recombination and show how it leads to a characteristic pattern of fluctuations at the moment

of photon decoupling. I also describe how these fluctuations are modified as the photons travel through the inhomogeneous universe, compute the angular power spectrum of the CMB anisotropies and show how the result depends on the cosmological parameters. Finally, the chapter ends with a discussion of CMB polarization, including the E/B decomposition of the polarization field. I explain why B-modes are a clean probe of primordial gravitational waves.

- Chapter 8 derives the initial conditions predicted by inflation. I first show that each Fourier mode of the inflationary perturbations satisfies the equation of a harmonic oscillator. This implies that the perturbations experience the same zero-point fluctuations as a harmonic oscillator in quantum mechanics. I devote some time to explaining the natural choice of the vacuum state. Finally, I derive the two-point functions of scalar and tensor fluctuations created by inflation. The chapter also includes a discussion of the most promising observational tests of the inflationary paradigm.

- Chapter 9 provides a brief outlook on the bright future of modern cosmology.

The book contains four appendices: Appendix A is a review of the fundamentals of general relativity. Appendix B presents a detailed analysis of the CMB anisotropies based on the Boltzmann equation. This complements the more approximate hydrodynamic treatment of Chapter 7. Appendix C collects parameters and relations that are frequently used in cosmological calculations. Finally, Appendix D reviews the properties of the special functions that make an appearance in the book.

Sections marked with an asterisk (*) contain more advanced material that can be skipped without loss of continuity.

Teaching from this book

My goal in writing this book has been to produce a text that is ideally suited for teaching a class on cosmology (or for self-study). I have therefore limited my choice of topics to those that I think a first introduction to the subject should include. I tried to resist the temptation to include more advanced and specialized topics. In my experience, it takes about one and a half semesters to teach all of the material covered in this book. However, it is also possible to use the book for a one-semester course if some of the details in Chapters 6, 7 and 8 are sacrificed.

Although the book presents the material in the order that I find most logical, it is possible to change that order in some places and I have experimented with this in my own teaching. For example, in my lectures, I have sometimes interchanged the order of Chapters 3 and 4. My students at Cambridge were rather mathematically minded and easily put off by the more dirty aspects of physics. It therefore helped to first seduce them to the subject, before revealing to them the real complexities of the hot Big Bang. Physics and astronomy students might be less put off by the use of nuclear and atomic physics in Chapter 3 and so don't mind seeing it earlier in the course. I have opted to start with the Newtonian treatment of structure formation in the book because it provides a quick and intuitive path to the relevant equations of motion. However, in my lectures, I have often cut down the discussion of Newtonian structure formation in Chapter 5 and merged it with the treatment of relativistic perturbation theory in Chapter 6. Getting to the relativistic treatment directly works for students with a strong math background, but sacrifices the physically intuitive aspects of the Newtonian analysis. In fact, if the book is used in an undergraduate course then presenting only the Newtonian analysis is probably the best option. Finally, I have often interchanged the order of Chapters 7 and 8. Since Chapter 7 is quite detailed, I suggest teaching this material separately in a second part of the course. The first part of the course could instead contain a more heuristic description of the CMB anisotropies, like that in Chapter 6.

Web page for the book

Together with Cambridge University Press, I will maintain a web page for the book (www.cambridge.org/baumann). This web page includes the following content:

- Color versions of selected figures.
- Updates on observational results.
- Notes on historically important cosmology papers.
- Mathematica notebooks for calculations in cosmological perturbation theory.
- For instructors: solutions to the problems and exercises.

Notation and Conventions

This book uses the $(-+++)$ signature of the metric, so that the line element in Minkowski space is $ds^2 = -c^2 dt^2 + d\mathbf{x}^2$. We will often employ natural units with $c = \hbar \equiv 1$ and define the reduced Planck mass as $M_{\mathrm{Pl}} \equiv 1/\sqrt{8\pi G}$. The conversion between units is explained in Appendix C.

We use Greek letters for spacetime indices, $\mu = 0, 1, 2, 3$, and Latin letters for spatial indices, $i = 1, 2, 3$. Spatial vectors are denoted by $\mathbf{x}$ and their components by x^i. The corresponding three-dimensional wavevectors are $\mathbf{k}$, with magnitudes $k \equiv |\mathbf{k}|$ and unit vectors written as $\hat{\mathbf{k}} \equiv \mathbf{k}/|\mathbf{k}|$. Our convention for the Fourier transform of a function $f(\mathbf{x})$ is

$$f(\mathbf{k}) = \int d^3x \, f(\mathbf{x}) \, e^{-i\mathbf{k}\cdot\mathbf{x}} \, .$$

Derivatives with respect to physical time t are denoted by overdots, while those with respect to conformal time η are given by primes, $f' = a\dot{f}$, where $a(t)$ is the scale factor.

Most of the notation in this book will be introduced as we go along. Here, I just list some of the most commonly used variables, especially if they do not have a uniform notation in the literature:

η	conformal time (or baryon-to-photon ratio!)
τ	proper time (or optical depth!)
H	Hubble parameter, $H \equiv \dot{a}/a$
$\mathcal{H}$	conformal Hubble parameter, $\mathcal{H} \equiv a'/a = Ha$
q	deceleration parameter, $q \equiv -\ddot{a}/(\dot{a}H)$
ρ	energy density
P	pressure
w	equation of state, $w \equiv P/\rho$
δ	density contrast, $\delta \equiv \delta\rho/\rho$
Δ	density contrast in comoving gauge
δ_{D}	Dirac delta function
v_i	bulk velocity, $v_i \equiv \partial_i v + \hat{v}_i$
θ	velocity divergence, $\theta \equiv \partial_i v^i$
q_i	momentum density, $q_i \equiv (\bar{\rho} + \bar{P})v_i$
P^μ	four-momentum
U^μ	four-velocity, $U^\mu \equiv \mathrm{d}x^\mu/\mathrm{d}\tau$
Π_{ij}	anisotropic stress
Ψ	gravitational potential in Newtonian gauge, $\delta g_{00} \equiv -2a^2\Psi$
Φ	curvature perturbation in Newtonian gauge, $\delta g_{ij} \equiv -2a^2\Phi\,\delta_{ij}$
ζ	curvature perturbation in uniform density gauge
$\mathcal{R}$	curvature perturbation in comoving gauge
h_{ij}	tensor metric perturbation, $\delta g_{ij} \equiv a^2 h_{ij}$
A_{s}	amplitude of scalar fluctuations
A_{t}	amplitude of tensor fluctuations
n_{s}	scalar spectral index
n_{t}	tensor spectral index
r	tensor-to-scalar ratio
ε	Hubble slow-roll parameter, $\varepsilon \equiv -\dot{H}/H^2$
κ	Hubble slow-roll parameter, $\kappa \equiv \dot{\varepsilon}/(\varepsilon H)$
ϕ	inflaton field
$V(\phi)$	inflaton potential
δ	dimensionless acceleration, $\delta \equiv -\ddot{\phi}/(H\dot{\phi})$
ε_V	potential slow-roll parameter, $\varepsilon_V \equiv (M_{\mathrm{Pl}}^2/2)(V_{,\phi}/V)^2$
η_V	potential slow-roll parameter, $\eta_V \equiv M_{\mathrm{Pl}}^2(V_{,\phi\phi}/V)$
$\mathcal{P}_f(k)$	power spectrum, $\langle f(\mathbf{k})f(\mathbf{k}')\rangle \equiv (2\pi)^3\delta_{\mathrm{D}}(\mathbf{k}+\mathbf{k}')\,\mathcal{P}_f(k)$
$\Delta_f^2(k)$	dimensionless power spectrum, $\Delta_f^2(k) \equiv (k^3/2\pi^2)\mathcal{P}_f(k)$
C_l	CMB power spectrum, $\langle a_{lm}a_{l'm'}^*\rangle \equiv C_l\,\delta_{ll'}\delta_{mm'}$
Δ_T^2	rescaled CMB power spectrum, $\Delta_T^2 \equiv [l(l+1)/2\pi]C_l\,T_0^2$
g_*	effective number of relativistic degrees of freedom
N_{eff}	effective number of neutrino species
μ	chemical potential

Acknowledgments

This book would not exist without the many emails I have received from students around the world expressing their appreciation of my lecture notes. I am very grateful for their encouragement and I hope that this book will be a useful resource for them and others.

The core material of the book was developed for my cosmology course at the University of Cambridge. I inherited the course from Anthony Challinor and his detailed lecture notes have been an extremely valuable resource. I am grateful to my colleagues at Cambridge for encouraging my efforts in developing the course, especially Paul Shellard, Anne Davies and Eugene Lim. Thanks also to David Tong for sharing his passion for teaching with me and for his advice in the preparation of my classes at Cambridge.

Many people have contributed to my own understanding of cosmology: I am especially grateful to my friends and long-term collaborators Daniel Green and Liam McAllister for countless discussions on all aspects of physics and cosmology. I have learned so much from them. Thanks also to my PhD advisor Paul Steinhardt for guiding my early years as a researcher and for encouraging my teaching activities at Princeton. Finally, discussions with Nima Arkani-Hamed and Matias Zaldarriaga have had a great impact on my view of cosmology and fundamental physics.

Thanks to the members of my research group—Peter Adshead, Soner Albayrak, Valentin Assassi, Matteo Biagetti, Pieter Braat, Wei-Ming Chen, Horng Sheng Chia, Anatoly Dymarsky, Garrett Goon, Austin Joyce, Hayden Lee, Manuel Loparco, David Marsh, Guilherme Pimentel, Carlos Duaso Pueyo, John Stout, Lotte ter Haar, Gimmy Tomaselli, Benjamin Wallisch and Yi Wang—for sharing with me the excitement about this wonderful subject. Thanks also to my collaborators for teaching me so much and for their understanding when the writing of this book took up too much of my time. Thanks to my colleagues in Amsterdam—especially Gianfranco Bertone, Alejandra Castro, Miranda Cheng, Jan de Boer, Ben Freivogel, Diego Hofman, Samaya Nissanke, Jan Pieter van der Schaar, Erik Verlinde, Christoph Weniger and Jasper van Wezel—for their support of the project.

Many friends and colleagues have kindly provided their time and expertise to help improve the content of this book: First of all, I cannot thank Eiichiro Komatsu enough for his careful reading of the book. He gave remarkably detailed comments on the entire manuscript. Similarly, Anthony Challinor suggested important corrections to Chapter 7 and Appendix B. I am immensely grateful to both Eiichiro and Anthony for their efforts. Any remaining mistakes are of course my own. John Stout introduced me to drawing figures in Tikz and I often consulted his help in producing the figures for this book. Benjamin Wallisch provided Python scripts for some of the figures. Cyril Pitrou's Mathematica code PRIMAT was used to compute the BBN predictions in Chapter 3. Thomas Tram patiently answered questions on the Boltzmann code CLASS and Pieter Braat helped to verify some of the numerical calculations in Chapters 6 and 7. Marius Millea's Python notebook was used to produce the matter power spectrum in Chapter 5. Mustafa Amin gave very helpful feedback on the content of Chapter 6. Jens Chluba and Matteo Biagetti suggested references for Chapters 3 and 5, respectively. Discussions with Eiichiro Komatsu inspired some of the presentation in Chapter 7. Parts of Chapters 4 and 8 were discussed with Daniel Green, Liam McAllister and Eva Silverstein. Gimmy Tomaselli provided important corrections to Appendix A, and I received many comments on a draft version of the book from Carlos Duaso Pueyo, Amanda van Hembert, Pim Herbschleb and Ethan van Woerkom. Finally, a number of anonymous referees also made helpful suggestions.

Thanks to my editors at Cambridge University Press, Vince Higgs and Stefanie Seaton, for their support of the project and their expert guidance during the editorial process, and to Rachel Norridge for overseeing the production of the book.

I am grateful to the physics department at the National Taiwan University—especially Professors Yu-tin Huang and Jiunn-Wei Chen—for its kind hospitality during the COVID-19 pandemic and to the Taiwanese Ministry of Science and Technology for supporting my stay with a Yushan Professorship.

Finally, I wish to thank my family for its support and encouragement as this book was being written.

Introduction

This book is about 13.8 billion years of cosmic evolution. We will trace the history of the universe from fractions of a second after the Big Bang until today. Before we begin our journey, however, it will be useful to have a bird's eye view of the subject. In this chapter, I will therefore set the stage by giving a qualitative account of the structure and evolution of the universe, as we now understand it. In the rest of the book, I will then show you where this knowledge comes from.

1.1 Scales of the Universe

The universe is big. In fact, the length scales in cosmology are so enormous that they are hard to grasp (see Table 1.1). Evolution simply hasn't equipped the human brain with the ability to have any intuition for the vastness of the cosmos. Apparently, being able to visualize the size of the universe has not increased the survival chances of our species.

Let us start with the **Solar System**. We live on the third planet from the Sun. Our nearest neighbour—at a distance of 384 400 km (about sixty times the size of the Earth)—is the Moon. Compared to the other objects of the Solar System, the Moon is very close. Although the Apollo spacecraft took three days to travel to the Moon, light makes the trip in just 1.3 seconds. For comparison, the light from the Sun takes over eight minutes to reach us. That's a distance of 150 million km (or 11 780 Earths lined up side to side).

Table 1.1	Important length scales of the universe (in different units)		
Object	Size [km]	Size [ly]	Size [Mpc]
Earth	6371	6.7×10^{-10}	2.1×10^{-16}
Distance to Sun	1.5×10^{8}	1.6×10^{-5}	4.8×10^{-12}
Solar System	4.5×10^{9}	4.7×10^{-4}	1.5×10^{-10}
Milky Way Galaxy	1.0×10^{18}	105 700	0.032
Local Group	9×10^{19}	9×10^{6}	3
Local Supercluster	5×10^{21}	5×10^{8}	150
Universe	4.4×10^{23}	46.5 billion	14 000

While it takes light only a few hours to cross the Solar System, the light from even the nearest stars takes several years to reach us. To measure the distances of objects that lie beyond our Solar System, it is therefore convenient to introduce the distance traveled by light in one year. A *light-year* (ly) is

$$1 \text{ ly} \approx 9.5 \times 10^{15} \text{ m} . \tag{1.1}$$

Besides light-years, astronomers also use *parsecs* (pc). As the Earth moves around the Sun, the positions of nearby stars shift ever so slightly, because they are being viewed from different directions. Given the size of an observed shift and the known radius of the Earth's orbit, the distance to a star follows from simple trigonometry. A parsec is defined as the distance at which the radius of the Earth's orbit around the Sun subtends an angle of one arcsecond (i.e. 1/3600 of a degree). Expressed in terms of light-years, this is

$$1 \text{ pc} \approx 3.26 \text{ ly} . \tag{1.2}$$

The nearest star, Proxima Centauri, is 4.2 light-years (or 1.3 pc) away.

Our Galaxy, the **Milky Way**, contains about 100 billion stars arranged in a flattened disc. It is about 30 kpc across, which means that it takes light about 100 000 years to cross our Galaxy. The Sun is located at the edge of a spiral arm, 30 000 light-years from the center. It takes 250 million years for the Sun to complete one orbit around the center of the Galaxy.

There are about 100 billion galaxies in the observable universe, each with about 100 billion stars. Our nearest neighbour is the Andromeda Galaxy (M31), a spiral galaxy about 2 million light-years away. The distances to galaxies are therefore measured in Mpc, which will be the unit of choice in cosmology. Andromeda is one of about fifty nearby galaxies that are gravitationally bound together. This arrangement of galaxies, called the **Local Group**, spans about 10 million light-years.

On even larger scales, galaxies arrange themselves into clusters and superclusters, with filamentary structures and giant voids in between them. Although the clusters within the superclusters are gravitationally bound, they are spreading apart as the universe is expanding. The largest such structures, like our **Local Supercluster** (Laniakea), are about 500 million light-years across.

Because the universe has a finite age, there is a maximum distance that we can see. This radius of the **observable universe** is

$$46.5 \text{ billion ly} \approx 14 \text{ Gpc} \approx 4.4 \times 10^{26} \text{ m} . \tag{1.3}$$

Notice that this is larger than the naive distance that light traveled in the age of the universe (which is 13.8 billion years). This discrepancy arises because the universe is expanding and the light is carried along with the expansion.

Having developed some sense for the vastness of the universe, let us now discuss what it is made of.

1.2 The Invisible Universe

Most of the matter and energy in the universe is invisible. The stuff that we can see—ordinary atoms—accounts for less than 5% of the total. The rest is in the form of dark matter and dark energy.

The majority of mass in the universe is composed of **dark matter**. This invisible matter is required to explain the stability and growth of structure in the universe. Early evidence for the existence of dark matter was collected by Fritz Zwicky in 1933. When studying galaxies in the Coma cluster, he found that they were moving faster than allowed by the gravity of the visible matter alone and an extra "dunkle materie" seemed to be required to hold the cluster together [1]. It took a long time, however, for the existence of dark matter to become accepted by the astronomical community. Important further evidence for the existence of dark matter came in the 1970s when Vera Rubin and collaborators measured the rotation speeds of hydrogen gas in the outer reaches of galaxies [2]. The large speeds that they found could only be explained if these galaxies were embedded in halos of dark matter.[1]

Today, some of the most striking evidence for dark matter comes from the gravitational lensing of the cosmic microwave background (CMB). As the CMB photons travel through the universe, they get deflected by the intervening large-scale structure, which distorts the hot and cold spots of the CMB. The effect depends on the total amount of matter in the universe and has been measured by the Planck satellite [5]. At the same time, the observed abundances of the light elements (D, He, Li), which were created by Big Bang nucleosynthesis (see below), imply a smaller amount of ordinary baryonic matter. The mismatch between the two measurements points to the existence of non-baryonic dark matter. The same amount of dark matter also explains the pattern of CMB anisotropies and the rate of gravitational clustering. In particular, the small density variations observed in the early universe only grow fast enough if assisted by the gravitational pull of the dark matter. Although the influence of dark matter is now seen over a wide range of scales, the true nature of the putative dark matter remains a pressing open problem in astrophysics and cosmology.

In the 1980s, cosmology was in crisis; the age of a matter-only universe seemed to be shorter than the ages of the oldest stars within it. In addition, observations of the large-scale clustering of galaxies implied that the total matter density was only around 30% of the critical density required for a spatially flat universe [6, 7], in conflict with the theoretical expectation from inflationary cosmology [8]. To account for this "missing energy," some theorists suggested the existence of a new form of **dark energy** [7, 9, 10], but direct evidence for it was lacking. Both problems were resolved with the discovery of the accelerating universe [11, 12]. By observing distant

---

[1] The history of the dark matter problem is considerably more complex and nuanced than this brief description might suggest. More on the fascinating history of dark matter can be found in [3, 4].

supernova explosions, cosmologists were able to measure both the distances and recession speeds of far-away galaxies. The results showed that the rate of expansion was decelerating at early times (as expected for a universe dominated by ordinary matter), but started to *accelerate* a few billion years ago. As we will see, such an accelerated expansion is possible in Einstein's theory of gravity if the universe is filled with an energy density that doesn't dilute and exerts a negative pressure. The accelerated expansion increases the estimate for the age of the universe, reconciling it with the ages of the oldest stars. Moreover, shortly after the supernova observations, balloon-borne CMB experiments [13, 14] provided strong observational evidence for the spatial flatness of the universe and hence solidified the case that dark energy was needed to explain 70% of the total energy in the universe today.

However, while dark energy solves the age problem, it has led to a new crisis in cosmology. We have given dark energy a name, but we don't know what it is. A natural candidate for dark energy is the energy density of empty space itself, since it doesn't dilute with the expansion of space. In fact, such a "vacuum energy" is predicted by quantum field theory, but its estimated size is many orders of magnitude larger than the observed dark energy density. Explaining this discrepancy remains one of the biggest open problems in cosmology and fundamental physics [15].

1.3 The Hot Big Bang

The universe is expanding [16]. It was therefore denser and hotter in the past. Particles were colliding frequently and the universe was in a state of thermal equilibrium with an associated temperature T. It is convenient to set Boltzmann's constant to unity, $k_B = 1$, and measure temperature in units of energy. Moreover, we will often use the particle physicists' convention of measuring energies in *electron volt* (eV):

$$\begin{aligned} 1\,\text{eV} &\approx 1.60 \times 10^{-19}\,\text{J} \\ &\approx 1.16 \times 10^4\,\text{K}\,. \end{aligned} \tag{1.4}$$

For reference, typical atomic processes are measured in eV, while the characteristic scale of nuclear reactions is MeV. A useful relation between the temperature of the early universe and its age is

$$\frac{T}{1\,\text{MeV}} \approx \left(\frac{t}{1\,\text{s}}\right)^{-1/2}\,. \tag{1.5}$$

One second after the Big Bang the temperature of the universe was therefore about $1\,\text{MeV}$ (or $10^{10}\,\text{K}$). While there was very little time available in the early universe, the rates of reactions were extremely high, so that many interesting things happened in a short amount of time (see Table 1.2).

Above $100\,\text{GeV}$ (or a trillionth of a second after the Big Bang), all particles of the Standard Model were in equilibrium and were therefore present in roughly equal

Table 1.2	Key events in the history of the universe		
Event	**Temperature**	**Energy**	**Time**
Inflation	$< 10^{29}$ K	$< 10^{16}$ GeV	$> 10^{-34}$ s
Dark matter decouples	?	?	?
Baryons form	?	?	?
EW phase transition	10^{15} K	100 GeV	10^{-11} s
Hadrons form	10^{12} K	150 MeV	10^{-5} s
Neutrinos decouple	10^{10} K	1 MeV	1 s
Nuclei form	10^9 K	100 keV	200 s
Atoms form	3460 K	0.29 eV	290 000 yrs
Photons decouple	2970 K	0.25 eV	370 000 yrs
First stars	50 K	4 meV	100 million yrs
First galaxies	20 K	1.7 meV	1 billion yrs
Dark energy	3.8 K	0.33 meV	9 billion yrs
Einstein born	2.7 K	0.24 meV	13.8 billion yrs

abundances. This state can be viewed as the initial condition for the hot Big Bang. The density at that time was a staggering 10^{36} kg cm^{-3}, which is what you would get if you compressed the mass of the Sun to the size of a marble. In a billionth of a second, the universe expanded by a factor of 10^4. During this expansion, the temperature dropped and the universe went through different evolutionary stages.

At around 100 GeV (or 10^{15} K), the electroweak (EW) symmetry of the Standard Model was broken during the **EW phase transition**. The electromagnetic and weak nuclear forces became distinct entities and the matter particles received their masses. Although the basics of EW symmetry breaking are well understood—and have been experimentally verified by the discovery of the Higgs boson [17, 18]—the detailed dynamics of the EW phase transition and its observational consequences are still a topic of active research.

Once the temperature drops below the mass of a particle species, particles and antiparticles start to annihilate, while the reverse process—the spontaneous creation of particle–antiparticle pairs—becomes inefficient. The first particles to disappear from the universe in this way were the top quarks (the heaviest particles of the Standard Model). W and Z bosons followed. Then the Higgs, the bottom and charm quarks and the tau lepton. Around 150 MeV, the **QCD phase transition** occurred and the remaining quarks condensed into hadrons (mostly protons, neutrons and pions).

Particles fall out of thermal equilibrium when their interaction rate drops below the expansion rate of the universe. At that moment, the particles stop interacting with the rest of the thermal bath and a relic abundance will be created. It is likely that the dark matter was created in this way, but the details are still unknown. A known decoupling event is the decoupling of neutrinos about *one second* after the Big Bang, which produced the **cosmic neutrino background** (CνB). During the early radiation-dominated period, these cosmic neutrinos carried about 40% of the total energy density and had a significant effect on the expansion of the universe.

Imprints of the cosmic neutrino background have recently been detected in both the CMB and the clustering of galaxies. These observations provide a window into the universe when it was just one second old.

After the QCD phase transition, the universe was a plasma of mostly free electrons, protons and neutrons, as well as very energetic photons that prevented any heavier nuclei from forming. About *one minute* after the Big Bang, the temperature of the universe had dropped enough for the synthesis of helium-4 and lithium-7 to become efficient [19–21]. The predicted amounts of these elements are consistent with the abundances found in early gas clouds, where very little post-processing of the primordial abundances has taken place. **Big Bang nucleosynthesis** (BBN) produced very few nuclei that are heavier than lithium, because there are no stable nuclei with 5 or 8 nucleons that would be required to sustain the nuclear chain reactions. Heavier elements were instead produced in the interior of stars, where the high densities and long timescales involved allow for three helium-4 nuclei to fuse. These heavier elements include carbon and oxygen which were spread throughout the universe after the stars exploded.

About *370 000 years* after the Big Bang, the universe had cooled enough for the first stable atoms to form in a process called **recombination**. At that moment, light stopped scattering off the free electrons in the plasma and started to propagate freely through the universe. These free-streaming photons are still seen today as an afterglow of the Big Bang. Stretched by 13.8 billion years of cosmic expansion, the universe's first light is observed as a faint microwave radiation [22]—the **cosmic microwave background**. The CMB contains tiny variations in its intensity (as a function of direction on the sky) that reflect perturbations in the density of matter in the early universe. The pattern of these fluctuations contains critical information about the primordial universe. More than any other cosmological probe, the study of the CMB anisotropies has transformed cosmology into a precision science.

What I have described so far are either facts (like BBN and recombination) or theoretical extrapolations that we can make with extremely high confidence (like the EW and QCD phase transitions). However, there are two important events in the history of the early universe that we know must have occurred, but whose details we are much less certain of. The first is **dark matter production**. Some process in the early universe must have led to the abundance of dark matter that we observe in the universe today. There are many ways in which this could have occurred, depending on the precise nature of the dark matter. For example, weakly interacting massive particles (WIMPs) could have decoupled from the primordial plasma at high energies, producing a cosmological abundance of dark matter. Or, a massive boson (maybe an axion) could have started to oscillate around the minimum of its potential when the expansion rate dropped below the mass of the particle, producing a bosonic condensate that acts like dark matter. The nature of the dark matter and its production in the early universe remain important open problems.

Another event that we believe must have occurred in the early universe, but whose details are unknown is **baryogenesis**. This refers to the mechanism by which an asymmetry was created between the amount of matter and antimatter in the universe. The required asymmetry is very small: for every 10^{10} particles of antimatter there must have been one extra matter particle. In this way, the annihilation between matter and antimatter into photons produces the observed matter-to-photon ratio in the universe. It isn't that we have no idea how this asymmetry might have been generated. In fact, there are many models of baryogenesis, but no way, so far, to decide which of these (if any) is the correct one.

1.4 Growth of Structure

The density fluctuations in the early universe eventually grew into all of the structures we see around us. On large scales, the gravitational clustering of matter can be described analytically, while, on small scales, the process becomes highly nonlinear and can only be captured by numerical simulations. The dark matter formed a web-like structure with high-density nodes connected by filaments. The baryonic gas collected in the regions of high dark matter density where it collapsed into stars which then congregated into galaxies.

The first stars—called **Population III stars**[2]—formed when the universe was about *100 million years* old. Computer simulations suggest that these stars were very massive, about a few hundred times more massive than the Sun. They were also very luminous and, hence, burned up their fuel rapidly. Although the first stars were short-lived, their impact on the universe was significant. They emitted ultraviolet light which heated and ionized the surrounding gas. The dynamics of this process of **reionization** are still not completely known and are actively investigated through numerical simulations. The first stars also may have provided the seeds for the growth of supermassive black holes which are found at the centers of most galaxies. And, finally, they created the first heavy elements in their interiors which were dispersed throughout the cosmos when the stars exploded. Enriched with these heavy elements, the baryonic gas cooled more efficiently, allowing smaller and more long-lived stars to be formed.

The first **galaxies** started to appear about *one billion years* after the Big Bang. Over time, these galaxies formed clusters and superclusters, a process that is still ongoing today. In the future, however, the growth of structure will stop as dark energy starts to dominate the universe. The details of galaxy formation are intricate and still an active area of research. In this book, we will be more interested in galaxies as tracers of the underlying distribution of dark matter, which in turn is determined by the seed fluctuations in the early universe. This distribution isn't random but has interesting spatial correlations which have been measured in large galaxy surveys.

---

[2] This terminology displays the full range of the weirdness of astronomical nomenclature. All elements heavier then helium are called "metals." The youngest metal-rich stars are called Population I stars, while older metal-poor stars are called Population II. By extension, the first stars, containing no metals, are called Population III stars.

1.5 Cosmic Palaeontology

Cosmology is famously an observational rather than an experimental science. No experimentalists were present in the early universe, and the birth and subsequent evolution of the universe cannot be repeated. Instead, we can only measure the spatial correlations between cosmological structures at late times. A central challenge of modern cosmology is to construct a consistent "history" of the universe that explains these correlations. This cosmological history is a narrative, a story we tell to give a rational accounting of the patterns that we see in the cosmological correlations.

This parallels the way palaeontologists infer the history of the Earth by studying the pattern of fossilized bones in the ground today. Like astronomical objects, these fossils are not randomly spread throughout space, but display interesting correlations which we try to explain by invoking past events. In much the same way, cosmologists study the pattern of cosmological structures observed today to infer the history of the early universe.

A remarkable feature of the observed correlations in the CMB is that they span scales that are larger than the distance traveled by light between the beginning of the hot Big Bang and the time when the CMB was created. This is in conflict with causality, unless the correlations were generated *before* the hot Big Bang. Indeed, there is now growing evidence that the hot Big Bang was not the beginning of time, but that the primordial density fluctuations were produced during an earlier period of accelerated expansion called **inflation** [8, 23, 24]. Small quantum fluctuations during inflation were amplified by the rapid expansion of the space and became the seeds for the large-scale structure of the universe [25–29].

If inflation really occurred, it was a rather dramatic event in the history of the universe. In just a billionth of a trillionth of a trillionth of a second, the universe doubled in size about 80 times. A region of space the size of a mosquito got stretched to the size of an entire galaxy. The entire observable universe then originated from a microscopic, causally-connected region of space. The correlations observed in the afterglow of the Big Bang were inherited from the correlations of the quantum fluctuations during inflation. While this picture provides an elegant explanation for the initial conditions of the primordial universe, it must be emphasized that inflation is *not* yet a fact—at the same level that, for example, BBN is a fact. The theoretical framework for inflation, however, is sufficiently well developed to justify including it in an introductory textbook on standard cosmology. Moreover, many new observations of the primordial correlations are being carried out— or are in the planning stages—that will subject the inflationary paradigm to further tests.

Further Reading

In writing this book, I have drawn on many excellent earlier treatments of cosmology. In the following, I will list some of the sources that I have found particularly useful. At the end of each chapter, I will give additional reading suggestions, including pointers to the research literature.

The following textbooks are all excellent and may be consulted for further details or alternative points of view:

- Dodelson, *Modern Cosmology*
  This book is already a modern classic. It is fantastically written and all explanations are exceptionally clear.

- Mukhanov, *Physical Foundations of Cosmology*
  Written by one of the architects of inflation, this book is especially good for early universe cosmology.

- Weinberg, *Cosmology*
  Sadly Steven Weinberg died during the writing of this book. He was a giant of theoretical physics and a hero of so many of us. His cosmology textbook is a great resource, especially for the more subtle aspects of the subject.

- Peter and Uzan, *Primordial Cosmology*
  This is a very comprehensive book, which despite its title also covers important parts of late-time cosmology.

- Kolb and Turner, *The Early Universe*
  Despite its advanced age, this book is still a very useful resource. Kolb and Turner's treatment of the thermal history of the universe remains unparalleled.

- Ryden, *Introduction to Cosmology*
  This is one of the few good cosmology books for undergraduates. Its level is lower than that of this book, but it is beautifully written and contains many very clear explanations.

Problems

1.1 The range of length scales involved in cosmology is hard to grasp. The best we can do is to consider relative distances and compare them to something more familiar. In this exercise, we will make some attempt at obtaining a more intuitive understanding of the vastness of the cosmos.

1. Consider shrinking the Earth to the size of a basketball. What would then be the size of the Moon and its orbit around the Earth?

2. Now imagine scaling the Earth down to the size of a peppercorn. What would then be the size of the Sun and the Earth's orbit? How far away would the most distant planet in the Solar System be?

3. The "Solar Neighbourhood" is a collection of about fifty nearby stars, spread across about 65 light-years, that travel together with the Sun. Scaling this region down, so that it fits inside a basketball court, what would be the size of the Solar System?

4. Shrinking our Galaxy to the size of the basketball court, what would now be the size of the Solar Neighbourhood?

5. The "Local Group" comprises about fifty nearby galaxies, spread across about 10 million light-years. If we squeezed this region into the size of the basketball court, what would be the size of our Milky Way galaxy?

6. The largest structures in the universe, like our "Local Supercluster," are about 500 million light-years across. Scaling these superclusters down to the dimensions of the basketball court, what would be the size of our Local Group?

7. The radius of the observable universe is 46.5 billion light-years. Compressing the observable universe to the size of the basketball court, what would be the size of the largest superclusters?

1.2 A key parameter in cosmology is the Hubble constant

$$H_0 \approx 70 \, \text{km s}^{-1} \text{Mpc}^{-1} \, .$$

In the following, you will use the measured value of the Hubble constant to estimate a few fundamental scales of our universe.

1. What is the Hubble time $t_{H_0} \equiv H_0^{-1}$ in years? This is a rough estimate of the age of the universe.

2. What is the Hubble distance $d_{H_0} \equiv c H_0^{-1}$ in meters? This is a rough estimate of the size of the observable universe.

3. The average density of the universe today is

$$\rho_0 = \frac{3 H_0^2}{8 \pi G} \, ,$$

where $G = 6.67 \times 10^{-11} \, \text{m}^3 \, \text{kg}^{-1} \, \text{s}^{-2}$ is Newton's constant. What is ρ_0 in g cm^{-3}? How does this compare to the density of water?

4. Let us assume that the universe is filled with only hydrogen atoms. What is then the total number of atoms in the universe? How does this compare to the number of hydrogen nuclei in your brain?

Hint: Assume that the brain is mostly water. Use $m_{\mathrm{H}} \approx 2 \times 10^{-24}$ g and $m_{\mathrm{H_2O}} \approx 3 \times 10^{-23}$ g.

5. The maximal energy scale probed by the Large Hadron Collider (LHC) is $E_{\max} \sim$ TeV. What length scale $\ell_{\min}$ does this correspond to? How does this compare to the size of the universe d_{H_0}?

PART I

THE HOMOGENEOUS UNIVERSE

2 The Expanding Universe

Our goal in this chapter is to derive, and then solve, the equations governing the evolution of the entire universe. This may seem like a daunting task. How can we hope to describe the long-term evolution of the cosmos when we have such a hard time just predicting the weather or the stability of the Solar System?

Fortunately, the coarse-grained properties of the universe are remarkably simple. While the distribution of galaxies is clumpy on small scales, it becomes more and more uniform on large scales. In particular, when averaged over sufficiently large distances (say larger than 100 Mpc), the universe looks *isotropic* (the same in all directions). Assuming that we don't live at a special point in space—and that nobody else does either—the observed isotropy then implies that the universe is also *homogeneous* (the same at every point in space). This leads to a simple mathematical description of the universe because the spacetime geometry takes a very simple form.

Since a static universe filled with matter and energy is unstable, we expect the spacetime to be dynamical. Indeed, observations of the light from distant galaxies have shown that the universe is expanding. Running this expansion backwards in time, we predict that nearly 14 billion years ago our whole universe was in a hot dense state. The Big Bang theory describes what happened in this fireball, and how it evolved into the universe we see around us today. In Part I of this book, I will describe our modern understanding of this theory. In this chapter, we will study the geometry and dynamics of the homogeneous universe, while in the next chapter, we will discuss the many interesting events that occurred in the hot Big Bang.

I will assume some familiarity with the basics of special relativity (at the level of manipulating spacetime tensors), but will introduce the necessary elements of general relativity (GR) as we go along. I will mostly state results in GR without derivation, which are then relatively easy to apply in the cosmological context. Although this plug-and-play approach loses some of the geometrical beauty of Einstein's theory, it gets the job done and provides the fastest route to our explorations of cosmology. Further background on GR is given in Appendix A.

2.1 Geometry

Gravity is a manifestation of the geometry of spacetime. To describe the evolution of our universe, we therefore start by determining its spacetime geometry. This geometry is characterized by a spacetime metric, whose form we will derive in this section. The large degree of symmetry of the homogeneous universe means that its metric will take a rather simple form.

2.1.1 Spacetime and Relativity

I will assume that you have been introduced to the concept of a **metric** before. Just to remind you, the metric is an object that turns coordinate distances into physical distances. For example, in three-dimensional Euclidean space, the physical distance between two points separated by the infinitesimal coordinate distances $\mathrm{d}x$, $\mathrm{d}y$ and $\mathrm{d}z$ is

$$\mathrm{d}\ell^2 = \mathrm{d}x^2 + \mathrm{d}y^2 + \mathrm{d}z^2 = \sum_{i,j=1}^{3} \delta_{ij}\,\mathrm{d}x^i\mathrm{d}x^j \,, \tag{2.1}$$

where I have introduced the notation $(x^1, x^2, x^3) = (x, y, z)$. In this example, the metric is simply the Kronecker delta $\delta_{ij} = \mathrm{diag}(1,1,1)$. However, you also know that if we were to use spherical polar coordinates, the square of the physical distance would no longer be the sum of the squares of the coordinate distances. Instead, we would get

$$\mathrm{d}\ell^2 = \mathrm{d}r^2 + r^2\mathrm{d}\theta^2 + r^2\sin^2\theta\,\mathrm{d}\phi^2 \equiv \sum_{i,j=1}^{3} g_{ij}\,\mathrm{d}x^i\mathrm{d}x^j \,, \tag{2.2}$$

where $(x^1, x^2, x^3) = (r, \theta, \phi)$. In this case, the metric has taken a less trivial form, namely $g_{ij} = \mathrm{diag}(1, r^2, r^2\sin^2\theta)$. Observers using different coordinate systems won't necessarily agree on the coordinate distances between two points, but they will always agree on the physical distance, $\mathrm{d}\ell$. We say that $\mathrm{d}\ell$ is an *invariant*. Hence, the metric turns observer-dependent coordinates into invariants.

A fundamental object in relativity is the **spacetime metric**. It turns observer-dependent spacetime coordinates $x^\mu = (ct, x^i)$ into the invariant line element[1]

$$\mathrm{d}s^2 = \sum_{\mu,\nu=0}^{3} g_{\mu\nu}\,\mathrm{d}x^\mu\mathrm{d}x^\nu \equiv g_{\mu\nu}\,\mathrm{d}x^\mu\mathrm{d}x^\nu \,. \tag{2.3}$$

In special relativity, the spacetime is **Minkowski space**, $\mathbb{R}^{1,3}$, whose line element is

$$\mathrm{d}s^2 = -c^2\mathrm{d}t^2 + \delta_{ij}\mathrm{d}x^i\mathrm{d}x^j \,, \tag{2.4}$$

---

[1] Throughout this book, I will use Einstein's summation convention where repeated indices are summed over. Our metric signature will be $(-, +, +, +)$. In this chapter, I will keep the speed of light explicit, but in the rest of the book I will use natural units with $c \equiv 1$.

so that the metric is simply $g_{\mu\nu} = \text{diag}(-1,1,1,1)$. An important feature of this metric is that the associated spacetime curvature vanishes. In general relativity, on the other hand, the metric will depend on the position in spacetime, $g_{\mu\nu}(t, \mathbf{x})$, in such a way that the spacetime curvature is nontrivial. This spacetime dependence of the metric incorporates the effects of gravity. How the metric varies throughout spacetime is determined by the distribution of matter and energy in the universe.

Indices You should be familiar with the manipulation of indices from special relativity. In particular, you should know that the metric is used to raise and lower indices on four-vectors (and on general tensors),

$$A_\mu = g_{\mu\nu} A^\nu \quad \text{and} \quad A^\mu = g^{\mu\nu} A_\nu , \tag{2.5}$$

where $g^{\mu\nu}$ is the inverse metric defined by

$$g^{\mu\lambda} g_{\lambda\nu} = \delta^\mu_\nu = \begin{cases} 1 & \mu = \nu , \\ 0 & \mu \neq \nu . \end{cases} \tag{2.6}$$

Four-vectors with a lower index, A_μ, are sometimes called *covariant* to distinguish them from the *contravariant* four-vectors, A^μ, with an upper index. Covariant and contravariant vectors can be *contracted* to produce an invariant *scalar*,

$$S \equiv A \cdot B = A^\mu B_\mu = g_{\mu\nu} A^\mu B^\nu . \tag{2.7}$$

If any of this isn't old news to you, please have a look at Appendix A.

The spatial homogeneity and isotropy of the universe mean that it can be represented by a time-ordered sequence of three-dimensional spatial slices, each of which is homogeneous and isotropic (see Fig. 2.1). The four-dimensional line element can then be written as[2]

$$ds^2 = -c^2 dt^2 + a^2(t) d\ell^2 , \tag{2.8}$$

where $d\ell^2 \equiv \gamma_{ij}(x^k) dx^i dx^j$ is the line element on the spatial slices and $a(t)$ is the **scale factor**, which describes the expansion of the universe. We will first determine the allowed forms of the spatial metric γ_{ij} and then discuss how the evolution of the scale factor is related to the matter content of the universe.

2.1.2 Symmetric Three-Spaces

Homogeneous and isotropic three-spaces must have constant intrinsic curvature. There are then only three options: the curvature of the spatial slices can be zero, positive or negative. Let us determine the metric for each case:

---

[2] Skeptics might worry about uniqueness. Why didn't we include a g_{0i} component? Because it would introduce a preferred direction and therefore break isotropy. Why didn't we allow for a nontrivial g_{00} component? Because it can be absorbed into a redefinition of the time coordinate, $dt' \equiv \sqrt{-g_{00}}\, dt$.

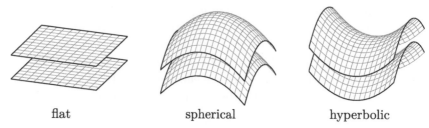

flat   spherical   hyperbolic

Fig. 2.1 The spacetime of the universe can be foliated into flat, spherical (positively curved) or hyperbolic (negatively curved) spatial hypersurfaces.

- **Flat space**: The simplest possibility is three-dimensional Euclidean space E^3. This is the space in which parallel lines don't intersect. Its line element is

$$\mathrm{d}\ell^2 = \mathrm{d}\mathbf{x}^2 = \delta_{ij}\mathrm{d}x^i\mathrm{d}x^j\,, \tag{2.9}$$

which is clearly invariant under spatial translations $x^i \mapsto x^i + a^i$ and rotations $x^i \mapsto R^i{}_k x^k$, with $\delta_{ij}R^i{}_k R^j{}_l = \delta_{kl}$.

- **Spherical space**: The next possibility is a three-space with constant positive curvature. On such a space, parallel lines will eventually meet. This geometry can be represented as a three-sphere S^3 embedded in four-dimensional Euclidean space E^4:

$$\mathrm{d}\ell^2 = \mathrm{d}\mathbf{x}^2 + \mathrm{d}u^2\,, \qquad \mathbf{x}^2 + u^2 = R_0^2\,, \tag{2.10}$$

where R_0 is the radius of the sphere. Homogeneity and isotropy on the surface of the three-sphere are inherited from the symmetry of the line element under four-dimensional rotations.

- **Hyperbolic space**: Finally, we can have a three-space with constant negative curvature. On such a space, parallel lines diverge. This geometry can be represented as a hyperboloid H^3 embedded in four-dimensional Lorentzian space $\mathbb{R}^{1,3}$:

$$\mathrm{d}\ell^2 = \mathrm{d}\mathbf{x}^2 - \mathrm{d}u^2\,, \qquad \mathbf{x}^2 - u^2 = -R_0^2\,, \tag{2.11}$$

where $R_0^2 > 0$ is a constant determining the curvature of the hyperboloid. Homogeneity and isotropy of the induced geometry on the hyperboloid are inherited from the symmetry of the line element under four-dimensional pseudo-rotations—i.e. Lorentz transformations, with u playing the role of time.

The line elements of the spherical and hyperbolic cases can be combined as

$$\mathrm{d}\ell^2 = \mathrm{d}\mathbf{x}^2 \pm \mathrm{d}u^2\,, \qquad \mathbf{x}^2 \pm u^2 = \pm R_0^2\,. \tag{2.12}$$

The differential of the embedding condition gives $u\,\mathrm{d}u = \mp\mathbf{x}\cdot\mathrm{d}\mathbf{x}$, so that we can eliminate the dependence on the auxiliary coordinate u from the line element

$$\mathrm{d}\ell^2 = \mathrm{d}\mathbf{x}^2 \pm \frac{(\mathbf{x}\cdot\mathrm{d}\mathbf{x})^2}{R_0^2 \mp \mathbf{x}^2}\,. \tag{2.13}$$

Finally, we can unify (2.13) with the Euclidean line element (2.9) by writing

$$d\ell^2 = d\mathbf{x}^2 + k\,\frac{(\mathbf{x}\cdot d\mathbf{x})^2}{R_0^2 - k\mathbf{x}^2}\,, \qquad \text{for} \qquad k \equiv \begin{cases} 0 & E^3\,, \\ +1 & S^3\,, \\ -1 & H^3\,. \end{cases} \tag{2.14}$$

To make the symmetries of the space more manifest, it is convenient to write the metric in spherical polar coordinates, (r, θ, ϕ). Using

$$d\mathbf{x}^2 = dr^2 + r^2(d\theta^2 + \sin^2\theta\,d\phi^2)\,,$$
$$\mathbf{x}\cdot d\mathbf{x} = r\,dr\,, \tag{2.15}$$
$$\mathbf{x}^2 = r^2\,,$$

the metric in (2.14) becomes

$$d\ell^2 = \frac{dr^2}{1 - kr^2/R_0^2} + r^2 d\Omega^2\,, \tag{2.16}$$

where $d\Omega^2 \equiv d\theta^2 + \sin^2\theta\,d\phi^2$ is the metric on the unit two-sphere. In Problem 2.1 at the end of this chapter, you are invited to show that this form of the metric can also be derived by imposing that the space has constant positive, negative or zero Ricci curvature at every point. Note that despite appearance $r = 0$ is not a special point in (2.16); there is no center in the universe. Every point in space is equivalent to every other point.

2.1.3 Robertson–Walker Metric

Substituting (2.16) into (2.8), we obtain the **Robertson–Walker metric** in polar coordinates:

$$ds^2 = -c^2 dt^2 + a^2(t)\left[\frac{dr^2}{1 - kr^2/R_0^2} + r^2 d\Omega^2\right]\,, \tag{2.17}$$

which is sometimes also called the Friedmann–Robertson–Walker (**FRW**) metric. Notice that the symmetries of the universe have reduced the ten independent components of the spacetime metric $g_{\mu\nu}$ to a single function of time, the scale factor $a(t)$, and a constant, the curvature scale R_0.

A few important features of this result are worth highlighting:

- First of all, the line element (2.17) has a rescaling symmetry

$$a \to \lambda a\,, \quad r \to r/\lambda\,, \quad R_0 \to R_0/\lambda\,. \tag{2.18}$$

This means that the geometry of the spacetime stays the same if we simultaneously rescale a, r and R_0 by a constant λ as in (2.18). We can use this freedom to set the scale factor today, at $t = t_0$, to be unity, $a(t_0) \equiv 1$. The scale R_0 is then the physical curvature scale today, justifying the use of the subscript.

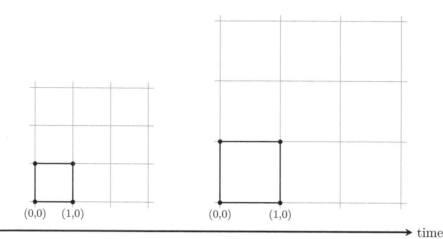

$(0,0)$ $(1,0)$ $(0,0)$ $(1,0)$

$\longrightarrow$ time

Fig. 2.2 Expansion of the universe. The comoving distance between points on an imaginary coordinate grid remains constant as the universe expands. The physical distance, on the other hand, grows in proportion to the scale factor $a(t)$.

- The coordinate r is called a **comoving coordinate**, which can be changed using the rescaling in (2.18) and hence is not a physical observable. Instead, physical results can only depend on the **physical coordinate**, $r_{\rm phys} = a(t)\,r$ (see Fig. 2.2).

  Consider a galaxy with a trajectory $\mathbf{r}(t)$ in comoving coordinates and $\mathbf{r}_{\rm phys} = a(t)\mathbf{r}$ in physical coordinates. The physical velocity of the galaxy is

$$\mathbf{v}_{\rm phys} \equiv \frac{d\mathbf{r}_{\rm phys}}{dt} = \frac{da}{dt}\mathbf{r} + a(t)\frac{d\mathbf{r}}{dt} \equiv H\mathbf{r}_{\rm phys} + \mathbf{v}_{\rm pec}\,, \qquad (2.19)$$

where we have introduced the **Hubble parameter**

$$\boxed{H \equiv \frac{\dot{a}}{a}}\,. \qquad (2.20)$$

Here, and in the following, an overdot denotes a time derivative, $\dot{a} \equiv da/dt$. We see that (2.19) has two contributions: The first term, $H\mathbf{r}_{\rm phys}$, is the **Hubble flow**, which is the velocity of the galaxy resulting from the expansion of the space between the origin and $\mathbf{r}_{\rm phys}(t)$. The second term, $\mathbf{v}_{\rm pec} \equiv a(t)\dot{\mathbf{r}}$, is the **peculiar velocity**, which is the velocity measured by a "comoving observer" (i.e. an observer who follows the Hubble flow). It describes the motion of the galaxy relative to the cosmological rest frame, typically due to the gravitational attraction of other nearby galaxies.

- The complicated g_{rr} component of (2.17) can sometimes be inconvenient. In that case, we may redefine the radial coordinate, $d\chi \equiv dr/\sqrt{1 - kr^2/R_0^2}$, such that

$$ds^2 = -c^2 dt^2 + a^2(t)\left[d\chi^2 + S_k^2(\chi)\,d\Omega^2\right], \qquad (2.21)$$

where

$$S_k(\chi) \equiv R_0 \begin{cases} \sinh(\chi/R_0) & k = -1\,, \\ \chi/R_0 & k = 0\,, \\ \sin(\chi/R_0) & k = +1\,. \end{cases} \tag{2.22}$$

Note that for a flat universe ($k = 0$) there is no distinction between r and χ. The form of the metric in (2.21) will be useful when we define observable distances in Section 2.2.3.

- Finally, it is often helpful to introduce **conformal time**,

$$\mathrm{d}\eta = \frac{\mathrm{d}t}{a(t)}\,, \tag{2.23}$$

so that (2.21) becomes

$$\mathrm{d}s^2 = a^2(\eta)\left[-c^2\mathrm{d}\eta^2 + \left(\mathrm{d}\chi^2 + S_k^2(\chi)\mathrm{d}\Omega^2\right)\right]. \tag{2.24}$$

We see that the metric has now factorized into a static part and a time-dependent conformal factor $a(\eta)$. This form of the metric is particularly convenient for studying the propagation of light, for which $\mathrm{d}s^2 = 0$. As we will see in Section 2.3.6, going to conformal time will also be a useful change of variables to obtain certain exact solutions to the Einstein equations.

2.2 Kinematics

Having found the metric of the expanding universe, we now want to determine how particles evolve in this spacetime. It is an essential feature of general relativity that freely falling particles in a curved spacetime move along **geodesics**. I will first introduce some basic facts about geodesic motion in GR and then apply it to the FRW spacetime. More details are given in Appendix A.

2.2.1 Geodesics

Before we study how particles evolve in the FRW spacetime, let us first look at the simpler problem of a free particle in two-dimensional Euclidean space.

Curvilinear coordinates

In the absence of any forces, Newton's law simply reads $\mathrm{d}^2\mathbf{x}/\mathrm{d}t^2 = 0$, which in Cartesian coordinates $x^i = (x, y)$ becomes

$$\frac{\mathrm{d}^2 x^i}{\mathrm{d}t^2} = 0\,. \tag{2.25}$$

In a general coordinate system, however, $\ddot{\mathbf{x}} = 0$ does *not* have to imply $\ddot{x}^i = 0$. Let me illustrate this explicitly for the polar coordinates (r, ϕ). Using $x = r \cos \phi$ and $y = r \sin \phi$, the equations of motion in (2.25) imply

$$
\begin{aligned}
0 &= \ddot{x} = \ddot{r} \cos \phi - 2 \sin \phi \, \dot{r} \dot{\phi} - r \cos \phi \, \dot{\phi}^2 - r \sin \phi \, \ddot{\phi} \,, \\
0 &= \ddot{y} = \ddot{r} \sin \phi + 2 \cos \phi \, \dot{r} \dot{\phi} - r \sin \phi \, \dot{\phi}^2 + r \cos \phi \, \ddot{\phi} \,.
\end{aligned}
\tag{2.26}
$$

Solving this for $\ddot{r}$ and $\ddot{\phi}$, we find

$$
\begin{aligned}
\ddot{r} &= r \dot{\phi}^2 \,, \\
\ddot{\phi} &= -\frac{2}{r} \dot{r} \dot{\phi} \,.
\end{aligned}
\tag{2.27}
$$

We see that $\ddot{r} \neq 0$ and $\ddot{\phi} \neq 0$, except in the special case $\dot{\phi} = 0$. The reason is simply that in polar coordinates the basis vectors $\hat{\mathbf{r}}$ and $\hat{\boldsymbol{\phi}}$ vary in the plane. To keep $\ddot{\mathbf{x}} = 0$, the coordinates must then satisfy (2.27).

Exercise 2.1 Show that the equations of motion (2.27) can also be derived from the Lagrangian of the free particle

$$
L = \frac{m}{2} \left(\dot{r}^2 + r^2 \dot{\phi}^2 \right),
\tag{2.28}
$$

where m is the mass of the particle.

To derive the equations of motion in an arbitrary coordinate system, with metric $g_{ij} \neq \delta_{ij}$, we start from the Lagrangian

$$
L = \frac{m}{2} g_{ij}(x^k) \dot{x}^i \dot{x}^j \,.
\tag{2.29}
$$

Substituting this into the Euler–Lagrange equation (see below), we find

$$
\boxed{\frac{\mathrm{d}^2 x^i}{\mathrm{d} t^2} = -\Gamma^i_{ab} \frac{\mathrm{d} x^a}{\mathrm{d} t} \frac{\mathrm{d} x^b}{\mathrm{d} t}} \,,
\tag{2.30}
$$

where we have introduced the **Christoffel symbol**

$$
\Gamma^i_{ab} \equiv \frac{1}{2} g^{ij} \left(\partial_a g_{jb} + \partial_b g_{ja} - \partial_j g_{ab} \right), \quad \text{with} \quad \partial_j \equiv \partial / \partial x^j \,.
\tag{2.31}
$$

In the special case $g_{ij} = \mathrm{diag}(1, r^2)$, the non-vanishing components of the Christoffel symbol are $\Gamma^r_{\phi\phi} = -r$ and $\Gamma^\phi_{r\phi} = \Gamma^\phi_{\phi r} = r^{-1}$, and (2.30) leads to (2.27).

Derivation The Euler–Lagrange equation is

$$
\frac{\mathrm{d}}{\mathrm{d} t} \left(\frac{\partial L}{\partial \dot{x}^k} \right) = \frac{\partial L}{\partial x^k} \,.
\tag{2.32}
$$

Since the mass m will cancel on both sides, we will not write it explicitly. The left-hand side of (2.32) then becomes

$$\frac{d}{dt}\left(\frac{\partial L}{\partial \dot{x}^k}\right) = \frac{d}{dt}\left(g_{ik}\dot{x}^i\right) = g_{ik}\ddot{x}^i + \frac{dx^j}{dt}\frac{\partial g_{ik}}{\partial x^j}\dot{x}^i$$

$$= g_{ik}\ddot{x}^i + \partial_j g_{ik}\dot{x}^i\dot{x}^j$$

$$= g_{ik}\ddot{x}^i + \frac{1}{2}\left(\partial_i g_{jk} + \partial_j g_{ik}\right)\dot{x}^i\dot{x}^j \,, \qquad (2.33)$$

while the right-hand side is

$$\frac{\partial L}{\partial x^k} = \frac{1}{2}\partial_k g_{ij}\dot{x}^i\dot{x}^j \,. \qquad (2.34)$$

Combining (2.33) and (2.34) then gives

$$g_{ki}\ddot{x}^i = -\frac{1}{2}\left(\partial_i g_{jk} + \partial_j g_{ik} - \partial_k g_{ij}\right)\dot{x}^i\dot{x}^j \,. \qquad (2.35)$$

Multiplying both sides by g^{lk}, and using $g^{lk}g_{ki} = \delta_i^l$, we get

$$\ddot{x}^l = -\frac{1}{2}g^{lk}\left(\partial_i g_{jk} + \partial_j g_{ik} - \partial_k g_{ij}\right)\dot{x}^i\dot{x}^j \equiv -\Gamma_{ij}^l \dot{x}^i\dot{x}^j \,, \qquad (2.36)$$

which is the desired result (2.30).

The equation of motion of a massive particle in general relativity will take a similar form as (2.30). However, in this case, the term involving the Christoffel symbol cannot be removed by going to Cartesian coordinates, but is a physical manifestation of the spacetime curvature.

Curved spacetime

For massive particles, a geodesic is the timelike curve $x^\mu(\tau)$ which extremizes the proper time $\Delta\tau$ between two points in the spacetime.[3] In Appendix A, I show that this extremal path satisfies the following **geodesic equation**

$$\boxed{\frac{d^2 x^\mu}{d\tau^2} = -\Gamma^\mu_{\alpha\beta}\frac{dx^\alpha}{d\tau}\frac{dx^\beta}{d\tau}} \,, \qquad (2.37)$$

where the Christoffel symbol is

$$\Gamma^\mu_{\alpha\beta} \equiv \frac{1}{2}g^{\mu\lambda}(\partial_\alpha g_{\beta\lambda} + \partial_\beta g_{\alpha\lambda} - \partial_\lambda g_{\alpha\beta})\,, \quad \text{with} \quad \partial_\alpha \equiv \partial/\partial x^\alpha \,. \qquad (2.38)$$

Notice the similarity between (2.37) and (2.30).

---

[3] In our conventions, proper time is defined by the relation $c^2 d\tau^2 = -ds^2$. The relativistic action of a massive particle is $S = -mc^2 \int d\tau$, so extremizing the proper time between two events is equivalent to extremizing the action. See Section A.3.1 for more details.

It will be convenient to write the geodesic equation in terms of the **four-momentum** of the particle

$$P^\mu \equiv m \frac{\mathrm{d}x^\mu}{\mathrm{d}\tau} \, . \tag{2.39}$$

Using the chain rule, we have

$$\frac{\mathrm{d}}{\mathrm{d}\tau} P^\mu(x^\alpha(\tau)) = \frac{\mathrm{d}x^\alpha}{\mathrm{d}\tau} \frac{\partial P^\mu}{\partial x^\alpha} = \frac{P^\alpha}{m} \frac{\partial P^\mu}{\partial x^\alpha} \, , \tag{2.40}$$

so that (2.37) becomes

$$\boxed{ P^\alpha \frac{\partial P^\mu}{\partial x^\alpha} = -\Gamma^\mu_{\alpha\beta} P^\alpha P^\beta } \, . \tag{2.41}$$

Rearranging this expression, we can also write

$$P^\alpha \left(\partial_\alpha P^\mu + \Gamma^\mu_{\alpha\beta} P^\beta \right) = 0 \, . \tag{2.42}$$

The term in brackets is the so-called **covariant derivative** of the four-vector P^μ, which we denote by $\nabla_\alpha P^\mu \equiv \partial_\alpha P^\mu + \Gamma^\mu_{\alpha\beta} P^\beta$. This allows us to write the geodesic equation in the following slick way:

$$P^\alpha \nabla_\alpha P^\mu = 0 \, . \tag{2.43}$$

In a GR course, you would have derived this form of the geodesic equation directly by thinking about the "parallel transport" of the vector P^μ (see Section A.3.3). Such a course would also have told you more about the geometrical meaning of the covariant derivative. We sacrificed beauty for speed.

The form of the geodesic equation in (2.43) is particularly convenient because it also applies to massless particles. In that case, our derivation of the geodesic equation breaks down because the proper time $\Delta\tau$ between any null separated points vanishes identically. However, the more subtle derivation of the geodesic equation for massless particles gives exactly the same result as in (2.43) if we interpret $P^\mu \equiv \mathrm{d}x^\mu/\mathrm{d}\lambda$ as the four-momentum of a massless particle, where λ now parameterizes the curve $x^\mu(\lambda)$. Accepting that the geodesic equation (2.43) is valid for both massive and massless particles, we will move on and apply it to particles in the FRW spacetime.

Free particles in FRW

To evaluate the right-hand side of (2.41), we need the Christoffel symbols for the FRW metric,

$$\mathrm{d}s^2 = -c^2\mathrm{d}t^2 + a^2(t)\gamma_{ij}\mathrm{d}x^i\mathrm{d}x^j \, , \tag{2.44}$$

where the form of the spatial metric γ_{ij} depends on our choice of spatial coordinates (e.g. Cartesian versus polar coordinates) and on the curvature of the spatial slices. Substituting $g_{\mu\nu} = \mathrm{diag}(-1, a^2\gamma_{ij})$ into the definition (2.38), it is straightforward to compute the Christoffel symbols. I will derive $\Gamma^0_{\alpha\beta}$ as an example (see box below) and leave the rest as an exercise. All Christoffel symbols with two time indices vanish, i.e. $\Gamma^{\mu}_{00} = \Gamma^{0}_{0\beta} = 0$. The only nonzero components are

$$\Gamma^{0}_{ij} = c^{-1} a \dot{a} \gamma_{ij} ,$$
$$\Gamma^{i}_{0j} = c^{-1} \frac{\dot{a}}{a} \delta^{i}_{j} , \tag{2.45}$$
$$\Gamma^{i}_{jk} = \frac{1}{2} \gamma^{il} (\partial_j \gamma_{kl} + \partial_k \gamma_{jl} - \partial_l \gamma_{jk}) ,$$

or are related to these by symmetry (note that $\Gamma^{\mu}_{\alpha\beta} = \Gamma^{\mu}_{\beta\alpha}$).

Example Let us derive $\Gamma^{0}_{\alpha\beta}$ for the metric (2.44). The Christoffel symbol with upper index equal to zero is

$$\Gamma^{0}_{\alpha\beta} = \frac{1}{2} g^{0\lambda} (\partial_\alpha g_{\beta\lambda} + \partial_\beta g_{\alpha\lambda} - \partial_\lambda g_{\alpha\beta}) . \tag{2.46}$$

The factor $g^{0\lambda}$ vanishes unless $\lambda = 0$, in which case it is equal to -1. Hence, we have

$$\Gamma^{0}_{\alpha\beta} = -\frac{1}{2} (\partial_\alpha g_{\beta 0} + \partial_\beta g_{\alpha 0} - \partial_0 g_{\alpha\beta}) . \tag{2.47}$$

The first two terms reduce to derivatives of g_{00} (since $g_{i0} = 0$). The FRW metric has constant g_{00}, so these terms vanish and we are left with

$$\Gamma^{0}_{\alpha\beta} = \frac{1}{2} \partial_0 g_{\alpha\beta} . \tag{2.48}$$

The derivative is only nonzero if α and β are spatial indices, $g_{ij} = a^2 \gamma_{ij}$. In that case, we find

$$\Gamma^{0}_{ij} = c^{-1} a \dot{a} \gamma_{ij} , \tag{2.49}$$

which confirms the result in (2.45). The factor of the speed of light arises because $x^0 \equiv ct$ and hence $\partial_0 = c^{-1} \partial_t$.

Exercise 2.2 Derive Γ^{i}_{0j} and Γ^{i}_{jk} for the metric (2.44).

We are now ready to discuss the physical content of the geodesic equation (2.41). I will specialize to the case of massless particles (like photons), leaving you to work out the result for massive particles as an exercise. Using that $\lambda = \lambda(t)$, the left-hand side of (2.41) can be written as

$$\frac{dP^{\mu}}{d\lambda} = \frac{dt}{d\lambda} \frac{dP^{\mu}}{dt} = \frac{P^0}{c} \frac{dP^{\mu}}{dt} , \tag{2.50}$$

and we have

$$\frac{P^0}{c} \frac{dP^{\mu}}{dt} = -\Gamma^{\mu}_{\alpha\beta} P^{\alpha} P^{\beta} . \tag{2.51}$$

Let us consider the $\mu = 0$ component of this equation. Since all Christoffel symbols with two time indices vanish, the right-hand side is $-\Gamma^0_{ij} P^i P^j$. Using $P^0 = E/c$ and $\Gamma^0_{ij} = c^{-1} a\dot{a}\gamma_{ij}$, we then find

$$\frac{E}{c^3} \frac{\mathrm{d}E}{\mathrm{d}t} = -\frac{1}{c} a\dot{a}\gamma_{ij} P^i P^j \, . \tag{2.52}$$

For massless particles, the four-momentum $P^\mu = (E/c, P^i)$ obeys the constraint

$$g_{\mu\nu} P^\mu P^\nu = -c^{-2} E^2 + a^2 \gamma_{ij} P^i P^j = 0 \, . \tag{2.53}$$

This allows us to write (2.52) as

$$\frac{1}{E} \frac{\mathrm{d}E}{\mathrm{d}t} = -\frac{\dot{a}}{a} \, , \tag{2.54}$$

which implies that the energy of massless particles decays with the expansion of the universe, $E \propto a^{-1}$.

Exercise 2.3 Repeating the analysis for massive particles, with

$$g_{\mu\nu} P^\mu P^\nu = -m^2 c^2 \, , \tag{2.55}$$

show that the *physical* three-momentum, defined as $p^2 \equiv g_{ij} P^i P^j$, satisfies $p \propto a^{-1}$. Starting from the definition of the four-momentum, show that the momentum can be written as

$$p = \frac{mv}{\sqrt{1 - v^2/c^2}} \, , \tag{2.56}$$

where $v^2 \equiv g_{ij} \dot{x}^i \dot{x}^j$ is the *physical* peculiar velocity. Since $p \propto a^{-1}$, freely falling particles will therefore converge onto the Hubble flow.

2.2.2 Redshift

Most of what we know about the universe is inferred from the light we receive from distant objects. The light emitted by a distant galaxy can be interpreted either quantum mechanically as freely propagating photons or classically as propagating electromagnetic waves. To analyze the observations correctly, we need to take into account that the wavelength of the light gets stretched (or, equivalently, the photons lose energy) by the expansion of the universe.

Recall that the wavelength of light is inversely proportional to the photon energy, $\lambda = h/E$, where h is Planck's constant. Since the energy of a photon evolves as $E \propto a^{-1}$, the wavelength therefore scales as $\lambda \propto a(t)$. Light emitted at a time t_1 with wavelength λ_1 will therefore be observed at a later time t_0 with a larger wavelength

$$\lambda_0 = \frac{a(t_0)}{a(t_1)} \lambda_1 \, . \tag{2.57}$$

This increase of the observed wavelength is called **redshift**, since red light has a longer wavelength than blue light.

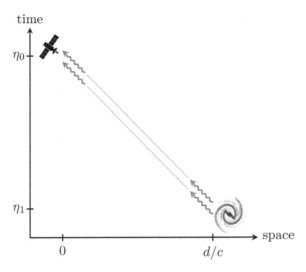

Fig. 2.3 In conformal time, the period of a light wave ($\delta\eta$) is equal at the times of emission (η_1) and detection (η_0). However, measured in physical time, $\delta t = a(\eta)\delta\eta$, the period is longer when it reaches us, $\delta t_0 > \delta t_1$, because the scale factor has increased. We say that the light has redshifted since its wavelength is now longer, $\lambda_0 > \lambda_1$.

The same result can also be derived by treating light as a classical electromagnetic wave, without the heavy machinery of the geodesic equation. Consider a galaxy at a fixed comoving distance d. Since the spacetime is isotropic, we can always choose coordinates for which the light travels purely in the radial direction, with $\theta = \phi = \text{const}$. The evolution is then determined by a two-dimensional line element[4]

$$\mathrm{d}s^2 = a^2(\eta)\left[-c^2\mathrm{d}\eta^2 + \mathrm{d}\chi^2\right], \tag{2.58}$$

where η is conformal time. Since photons travel along null geodesics, with $\mathrm{d}s^2 = 0$, their path is defined by

$$\Delta\chi(\eta) = \pm c\Delta\eta, \tag{2.59}$$

where the plus sign corresponds to outgoing photons and the minus sign to incoming photons. This shows the main benefit of working with conformal time: light rays correspond to straight lines in the χ–η coordinates. If we had used the physical time t instead, then the path of the light would be curved. Moreover, the comoving distance to a source is simply equal to the difference in conformal time between the moments when the light was emitted and when it was received.

At a time η_1, the galaxy emits a signal of short conformal duration $\delta\eta$ (see Fig. 2.3). The light arrives at our telescopes at the time $\eta_0 = \eta_1 + d/c$. The conformal

---

[4] We have used the parameterization (2.24) for the radial coordinate χ, so that (2.58) is conformal to a two-dimensional Minkowski space and the curvature scale R_0 of the three-dimensional spatial slices is absorbed into the definition of the coordinate χ. Had we used the regular polar coordinate r, the two-dimensional line element would have retained an explicit dependence on R_0. For flat slices, χ and r are of course the same.

duration of the signal measured by the detector is the same as at the source, but the physical time intervals are different at the points of emission and detection,

$$\delta t_1 = a(\eta_1)\delta\eta \quad \text{and} \quad \delta t_0 = a(\eta_0)\delta\eta. \tag{2.60}$$

If δt is the period of the light wave, then the light is emitted with wavelength $\lambda_1 = c\delta t_1$, but is observed with wavelength $\lambda_0 = c\delta t_0$, so that

$$\frac{\lambda_0}{\lambda_1} = \frac{a(\eta_0)}{a(\eta_1)}, \tag{2.61}$$

which agrees with the result in (2.57).

It is conventional to define the redshift as the fractional shift in the wavelength:

$$z \equiv \frac{\lambda_0 - \lambda_1}{\lambda_1}. \tag{2.62}$$

This shift in the wavelength is measured by observing spectral lines in the light of galaxies and stars. By comparing the observed wavelengths of these lines with those measured in a laboratory on Earth, we determine the redshift. Using (2.61), and setting $a(t_0) \equiv 1$, we can write the redshift parameter as

$$\boxed{1 + z = \frac{1}{a(t_1)}}. \tag{2.63}$$

A galaxy at redshift $z = 1$ therefore emitted the observed light when the universe was half its current size, a galaxy at redshift $z = 2$ when it was one-third its current size, and so on. It is conventional to label events in the history of the universe by their redshift. For example, the CMB photons were released from the cosmic plasma at redshift $z = 1100$ and the first galaxies formed around redshift $z \sim 10$.

For nearby sources ($z < 1$), we can expand the scale factor in a power series around $t = t_0$:

$$a(t_1) = 1 + (t_1 - t_0)H_0 + \cdots, \tag{2.64}$$

where $t_0 - t_1$ is the **look-back time** and H_0 is the **Hubble constant**

$$H_0 \equiv \frac{\dot{a}(t_0)}{a(t_0)}. \tag{2.65}$$

Equation (2.63) then gives $z = H_0(t_0 - t_1) + \cdots$. For close objects, $t_0 - t_1$ is simply equal to d/c, so that the redshift increases linearly with distance. This linear relation is called the **Hubble–Lemaître law**. In terms of the recession speed of the object, $v \equiv cz$, it reads

$$v = cz \approx H_0 d. \tag{2.66}$$

Hubble's original measurements of the velocity–distance relations of galaxies are shown in Fig. 2.4. For distant objects ($z > 1$), we have to be more careful about the meaning of "distance."

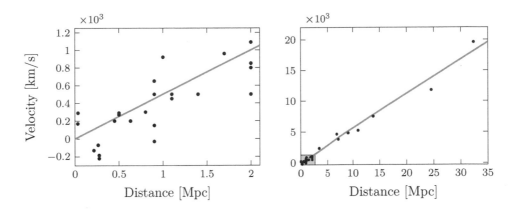

Fig. 2.4 Historical measurements of the velocity–distance relationship in an expanding universe. The left plot shows Hubble's original data from 1929, including an optimistic linear fit to Hubble's law [1]. The right plot includes additional data from 1931 collected together with Humason [2]. Both measurements contain systematic errors and overestimate Hubble's constant significantly, $H_0 \approx 500 \, \mathrm{km \, s^{-1} \, Mpc^{-1}}$.

2.2.3 Distances*

Measuring distances in cosmology is notoriously difficult. Note that the distances appearing in the metric are *not* observable. Even the physical distance $d_{\mathrm{phys}} = a(t)\chi$ cannot be observed because it is the distance between separated events at a *fixed* time. A more practical definition of "distance" must take into account that the universe is expanding and that it takes light a finite amount of time to reach us.

Luminosity distance

An important way to measure distances in cosmology uses so-called **standard candles**. These are objects of known intrinsic brightnesses, so that their observed brightnesses can be used to determine their distances. Essentially, an object of a given brightness will appear dimmer if it is further away.

Hubble used **Cepheids** to discover the expansion of the universe. These are stars whose brightnesses vary periodically. The observed periods were found to be correlated with the intrinsic brightness of the stars. By measuring the time variation of Cepheids, astronomers can infer their absolute brightnesses and then use their observed brightnesses to infer their distances. However, the Cepheid method only works for relatively small distances. To measure larger distances we need brighter sources. These are provided by **type Ia supernovae**—stellar explosions that arise when a white dwarf accretes too much matter from a companion star. Supernovae are rare (roughly a few per century in a typical galaxy), but outshine all stars in the host galaxy and can therefore be seen out to enormous distances. By observing many galaxies, astronomers can then measure a large number of supernovae. The

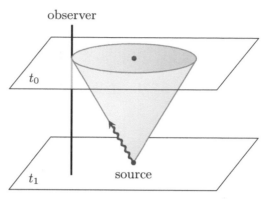

observer

t_0

source

t_1

Fig. 2.5 Spacetime diagram illustrating the concept of the luminosity distance. The light from the source spreads over a spherical surface, part of which gets captured by our detectors. In addition, the energy and arrival frequency of the photons is reduced by the expansion of the universe.

supernova explosions occur at a relatively precise moment—when the mass of the white dwarf exceeds the Chandrasekhar limit—and therefore have a fixed brightness. Residual variations of the supernova brightnesses can be corrected for phenomenologically. The use of supernovae as standard candles has been instrumental in the discovery of the acceleration of the universe and they continue to play an important role in observational cosmology.[5]

Let us assume that we have identified an astronomical object with a known luminosity L (= energy emitted per unit time). The observed flux F (= energy per unit time per receiving area) can then be used to infer its (luminosity) distance. Our task is to determine the relation between L and F in an expanding universe.

Consider a source at a redshift z. The comoving distance to the object is

$$\chi(z) = c \int_{t_1}^{t_0} \frac{\mathrm{d}t}{a(t)} = c \int_0^z \frac{\mathrm{d}z}{H(z)} \,, \tag{2.67}$$

where the redshift evolution of the Hubble parameter, $H(z)$, depends on the matter content of the universe (see Section 2.3). We assume that the source emits radiation isotropically (see Fig. 2.5). In a static Euclidean space, the energy would then spread uniformly over a sphere of area $4\pi\chi^2$, and the fraction going through an area A

---

[5] In the future, gravitational waves (GWs) will allow for robust measurements of luminosity distances using so-called **standard sirens**. The observed waveforms of the GW signals determine the parameters of the sources (like the masses of the objects in a binary system) and general relativity then predicts the emitted GW power. Comparing this to the observed GW amplitude provides a clean measurement of the luminosity distance. Moreover, if the signal comes with an electromagnetic counterpart (e.g. in the merger of two neutron stars) then the redshift to the source can be measured independently and the Hubble constant can be extracted [3].

(e.g. the collecting area of a telescope) is $A/4\pi\chi^2$. The relation between the absolute luminosity and the observed flux would then be

$$F = \frac{L}{4\pi\chi^2} \quad \text{(static space)}. \tag{2.68}$$

In an expanding spacetime, this result is modified for three reasons:

1. When the light reaches the Earth, at the time t_0, a sphere with radius χ has an area

$$4\pi a^2(t_0) d_M^2 , \tag{2.69}$$

where $d_M \equiv S_k(\chi)$ is the "metric distance" defined in (2.22). In a curved space, the metric distance d_M differs from the radius χ.

2. The arrive rate of the photons is smaller than the rate at which they are emitted at the source by a factor of $a(t_1)/a(t_0) = 1/(1 + z)$. This reduces the observed flux by the same factor.

3. The energy of the observed photons is less than the energy with which they were emitted by the same redshift factor $1/(1 + z)$. This lowers the observed flux by another factor of $1/(1 + z)$.

Hence, the correct formula for the observed flux of a source with luminosity L at coordinate distance χ and redshift z is

$$F = \frac{L}{4\pi d_M^2 (1 + z)^2} \equiv \frac{L}{4\pi d_L^2} \quad \text{(expanding space)}. \tag{2.70}$$

In the second equality, we have defined the **luminosity distance**, d_L, so that the relation between luminosity, flux and luminosity distance is the same as in (2.68). Hence, we find

$$\boxed{d_L(z) = (1 + z) d_M(z)} , \tag{2.71}$$

where the metric distance for an object with redshift z depends on the cosmological parameters. Figure 2.6 shows the luminosity distance as a function of redshift in a universe with and without dark energy. We see that the luminosity distance out to a fixed redshift is larger in a dark energy-dominated universe than in a matter-only universe.

For objects at low redshifts ($z < 1$), we can define perturbative corrections to Hubble's law (2.66). We first extend the Taylor expansion of $a(t)$ in (2.64) to higher order in the look-back time

$$a(t) = 1 + H_0(t - t_0) - \frac{1}{2} q_0 H_0^2 (t - t_0)^2 + \cdots , \tag{2.72}$$

where we have defined

$$q_0 \equiv -\frac{\ddot{a}}{aH^2}\bigg|_{t=t_0} . \tag{2.73}$$

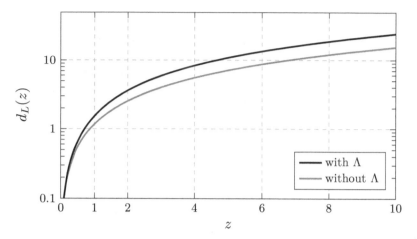

Fig. 2.6 Luminosity distance (in units of c/H_0) as a function of redshift in a flat universe, with matter only (*gray*) and with 70% dark energy (*black*).

The parameter q_0 goes by the unfortunate name of the **deceleration parameter**. Today, we know that it is negative, $q_0 \approx -0.5$, and hence measures the *acceleration* of the universe. Substituting (2.72) into (2.63), we obtain the redshift as a function of the look-back time

$$z = \frac{1}{a(t_1)} - 1 = H_0(t_0 - t_1) + \frac{1}{2}(2 + q_0)H_0^2(t_0 - t_1)^2 + \cdots . \tag{2.74}$$

This can be inverted to give

$$H_0(t_0 - t_1) = z - \frac{1}{2}(2 + q_0)z^2 + \cdots , \tag{2.75}$$

where the higher-order terms can be ignored as long as $z < 1$. Using (2.72) and (2.75), we can also write the comoving distance in terms of the look-back time and the redshift

$$\chi = c \int_{t_1}^{t_0} \frac{dt}{a(t)} = c \int_{t_1}^{t_0} dt \left[1 - H_0(t - t_0) + \cdots \right]$$

$$= c(t_0 - t_1) + \frac{1}{2} \frac{H_0}{c} c^2(t_1 - t_0)^2 + \cdots$$

$$= \frac{c}{H_0} \left(z - \frac{1}{2}(1 + q_0)z^2 + \cdots \right). \tag{2.76}$$

Through (2.71) and (2.22), this determines the luminosity distance as a function of the redshift, $d_L(z)$. For a flat universe, with $S_k(\chi) = \chi$, the modified Hubble–Lemaître law reads

$$d_L = \frac{c}{H_0} \left(z + \frac{1}{2}(1 - q_0)z^2 + \cdots \right). \tag{2.77}$$

The values of H_0 and q_0 can be extracted by fitting this functional form to the observed $d_L(z)$. The value of q_0 will depend on the energy content of the universe. I will present the results of such a measurement in Section 2.4.

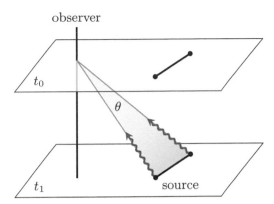

Fig. 2.7 Spacetime diagram illustrating the concept of the angular diameter distance. The observed angular size depends on the transverse physical size of the object and the physical distance to the object at the time when the light was *emitted*.

Measurements of the Hubble constant used to come with very large uncertainties. It therefore became conventional to define

$$H_0 \equiv 100\,h\,\mathrm{km\,s^{-1}Mpc^{-1}}, \tag{2.78}$$

where the parameter h is used to keep track of how uncertainties in H_0 propagate to the inferred values of other cosmological parameters. Today, measurements of H_0 have become much more precise. The latest supernovae measurements have found [4]

$$h = 0.730 \pm 0.010 \quad \text{(supernovae)}. \tag{2.79}$$

The Hubble constant can also be extracted from the CMB anisotropy spectrum (see Chapter 7). These observations give [5]

$$h = 0.674 \pm 0.005 \quad \text{(CMB)}. \tag{2.80}$$

As you can see, there is currently a statistically significant discrepancy between the two measurements. It is unclear whether this "Hubble tension" is due to an unidentified observational systematic in either of the measurements or signals a breakdown in the standard cosmological model.

Angular diameter distance

An alternative way to measure distances is to use **standard rulers**, which are objects of known physical size. The observed angular sizes of such objects then depends on their distances. As we will see in Chapter 7, the typical size of hot and cold spots in the CMB can be predicted theoretically and is therefore a standard ruler. The observed angular size of these spots then determines the distance to the CMB's surface of last-scattering.[6]

---

[6] We observe the CMB at a fixed moment in time when the first neutral atoms were formed and the photons scattered for the last time off the free electrons in the primordial plasma

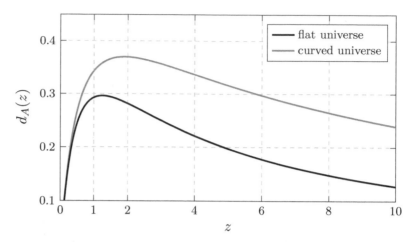

Fig. 2.8 Angular diameter distance (in units of c/H_0) as a function of redshift in a flat matter-only universe (*black*) and a universe with negative spatial curvature (*gray*).

Let us assume again that an object is at a comoving distance χ and that the photons which we observe today were emitted at a time t_1 (see Fig. 2.7). The transverse physical size of the object is D. In static Euclidean space, we would expect its angular size to be

$$\delta\theta = \frac{D}{\chi} \quad \text{(static space)}, \tag{2.81}$$

where we have assumed that $\delta\theta \ll 1$ (in radians), which is true for objects at cosmological distances. In an expanding universe, this formula instead becomes

$$\delta\theta = \frac{D}{a(t_1)\,d_M} \equiv \frac{D}{d_A} \quad \text{(expanding space)}, \tag{2.82}$$

where $d_M \equiv S_k(\chi)$. Notice that the observed angular size depends on the distance at the time t_1 when the light was *emitted*. The second equality in (2.82) has defined the **angular diameter distance** by analogy with the Euclidean formula (2.81). In terms of the metric distance d_M and the redshift z, the angular diameter distance therefore is

$$\boxed{d_A(z) = \frac{d_M(z)}{1+z}}. \tag{2.83}$$

Figure 2.8 shows the angular diameter distance as a function of redshift in a flat matter-only universe and in a universe with negative spatial curvature. We see that the angular diameter distance starts to *decrease* around $z_m \sim 1.5$. At redshifts larger than z_m, objects of a given proper size will therefore appear bigger on the sky with increasing redshift. This effect is due to the expansion of the universe. The spacetime was compressed when the light was emitted and the objects were closer to us than they are today. The observed angular size is therefore larger.

(see Chapter 3). The observed radiation therefore comes from a spherical surface around the observer called the "surface of last-scattering."

2.3 Dynamics

So far, we have used the symmetries of the universe to determine its geometry and studied the propagation of particles in the expanding spacetime. The scale factor $a(t)$ has remained an unspecified function of time. The evolution of the scale factor follows from the **Einstein equation**

$$G_{\mu\nu} = \frac{8\pi G}{c^4} T_{\mu\nu} \,, \tag{2.84}$$

where $G = 6.67 \times 10^{-11}\,\mathrm{m^3\,kg^{-1}\,s^{-2}}$ is Newton's constant. This relates the **Einstein tensor** $G_{\mu\nu}$ (a measure of the "spacetime curvature" of the universe) to the **energy-momentum tensor** $T_{\mu\nu}$ (a measure of the "matter content" of the universe). We will first discuss the possible forms of cosmological energy-momentum tensors, then compute the Einstein tensor for the FRW background, and finally put them together to solve for the evolution of the scale factor $a(t)$ as a function of the matter content.

2.3.1 Perfect Fluids

We have seen how the spacetime geometry of the universe is constrained by homogeneity and isotropy. Now, we will determine which types of matter are consistent with these symmetries. We will find that the coarse-grained energy-momentum tensor is required to be that of a **perfect fluid**,

$$T_{\mu\nu} = \left(\rho + \frac{P}{c^2} \right) U_\mu U_\nu + P\, g_{\mu\nu} \,, \tag{2.85}$$

where ρc^2 and P are the energy density and the pressure in the rest frame of the fluid, and U^μ is its four-velocity relative to a comoving observer.

Number density

Before we get to the energy-momentum tensor, let us study a slightly simpler object: the number current four-vector N^μ. The $\mu = 0$ component, N^0, measures the number density of particles, where a "particle" in cosmology may be an entire galaxy. The $\mu = i$ component, N^i, is the flux of the particles in the direction x^i. We call N^0 a three-scalar because its value doesn't change under a purely spatial coordinate transformation. In contrast, N^i transforms like a three-vector under such a transformation. We want to determine which form of the number current is consistent with the homogeneity and isotropy seen by a comoving observer.

Isotropy requires that the mean value of any three-vector, such as N^i, must vanish, and homogeneity requires that the mean value of any three-scalar, such as N^0, is only a function of time. Hence, the number current measured by a comoving observer has the following components

$$N^0 = cn(t), \qquad N^i = 0, \tag{2.86}$$

where $n(t)$ is the number of galaxies per proper volume. A general observer (i.e. an observer in motion relative to the mean rest frame of the particles), would measure the following number current four-vector

$$N^\mu = nU^\mu, \tag{2.87}$$

where $U^\mu \equiv \mathrm{d}x^\mu/\mathrm{d}\tau$ is the relative four-velocity between the particles and the observer. Of course, we recover (2.86) for a comoving observer, with $U^\mu = (c, 0, 0, 0)$. Moreover, for $U^\mu = \gamma(c, v^i)$, the expression (2.87) gives the correctly boosted results. For instance, you may recall that the boosted number density is γn, where the Lorentz factor γ accounts for the fact that one of the dimensions of the volume is Lorentz contracted.

How does the number density evolve with time? If the number of particles is conserved, then the rate of change of the number density must equal the divergence of the flux of the particles. In Minkowski space, this implies the following **continuity equation**, $\partial_0 N^0 = -\partial_i N^i$, or, in relativistic notation,

$$\partial_\mu N^\mu = 0. \tag{2.88}$$

Equation (2.88) is generalized to curved spacetimes by replacing the partial derivative with a **covariant derivative** (see Appendix A):

$$\nabla_\mu N^\mu = 0. \tag{2.89}$$

This covariant form of the continuity equation is valid independent of the choice of coordinates.

Covariant derivative   The covariant derivative is an important object in differential geometry and it is of fundamental importance in general relativity. The origin and properties of the covariant derivative are discussed in Appendix A (and much more can be found in any textbook on general relativity). Since we don't have time to waste, I will not get into these details here, but simply tell you how the covariant derivative acts on four-vectors with upstairs and downstairs indices:

$$\nabla_\mu A^\nu = \partial_\mu A^\nu + \Gamma^\nu_{\mu\lambda} A^\lambda, \tag{2.90}$$
$$\nabla_\mu B_\nu = \partial_\mu B_\nu - \Gamma^\lambda_{\mu\nu} B_\lambda. \tag{2.91}$$

Please take a moment to study the signs and index placements in these expressions. We already encountered (2.90) in our discussion of the geodesic equation, while (2.91) can be derived from this by requiring the covariant derivative of the scalar function $f \equiv A^\nu B_\nu$ to reduce to a partial derivative, $\nabla_\mu f = \partial_\mu f$. The action on a general tensor is a straightforward generalization of these two results. For example, the covariant derivative of a rank-2 tensor with mixed indices is

$$\nabla_\mu T^\sigma{}_\nu = \partial_\mu T^\sigma{}_\nu + \Gamma^\sigma_{\mu\lambda} T^\lambda{}_\nu - \Gamma^\lambda_{\mu\nu} T^\sigma{}_\lambda . \tag{2.92}$$

The expressions above are all that we will need for the computations in this chapter.

Using (2.90) in (2.89), we get

$$\partial_\mu N^\mu = -\Gamma^\mu_{\mu\lambda} N^\lambda . \tag{2.93}$$

Let us evaluate this in the rest frame of the fluid. Substituting the components of the number current in (2.86), we obtain

$$\frac{1}{c} \frac{dn}{dt} = -\Gamma^\mu_{\mu 0} n$$
$$= -\frac{3}{c} \frac{\dot{a}}{a} n , \tag{2.94}$$

where we have used that $\Gamma^0_{00} = 0$ and $\Gamma^i_{i0} = 3c^{-1}\dot{a}/a$. Hence, we find that

$$\frac{\dot{n}}{n} = -3\frac{\dot{a}}{a} \quad \Rightarrow \quad \boxed{n(t) \propto a^{-3}} . \tag{2.95}$$

As expected, the number density decreases in proportion to the increase of the proper volume, $V \propto a^3$.

Energy-momentum tensor

We will now use a similar logic to determine which form of the energy-momentum tensor $T_{\mu\nu}$ is consistent with the requirements of homogeneity and isotropy. First, we decompose $T_{\mu\nu}$ into a three-scalar, T_{00}, two three-vectors, T_{i0} and T_{0j}, and a three-tensor, T_{ij}. The physical meaning of these components is

$$T_{\mu\nu} = \left(\begin{array}{c|c} T_{00} & T_{0j} \\ \hline T_{i0} & T_{ij} \end{array} \right) = \left(\begin{array}{c|c} \text{energy density} & \text{momentum density} \\ \hline \text{energy flux} & \text{stress tensor} \end{array} \right) . \tag{2.96}$$

In a homogeneous universe, the energy density must be independent of position, but can depend on time, $T_{00} = \rho(t)c^2$. Moreover, isotropy again requires the mean values of the three-vectors to vanish in the comoving frame, $T_{i0} = T_{0j} = 0$. Finally, isotropy around a point $\mathbf{x} = 0$ constrains the mean value of any three-tensor, such as T_{ij}, to be proportional to δ_{ij}. Since the metric g_{ij} equals $a^2 \delta_{ij}$ at $\mathbf{x} = 0$, we have

$$T_{ij}(\mathbf{x} = 0) \propto \delta_{ij} \propto g_{ij}(\mathbf{x} = 0). \tag{2.97}$$

Homogeneity requires the proportionality coefficient to be only a function of time. Moreover, since this is a proportionality between two tensors, T_{ij} and g_{ij}, it remains unaffected by transformations of the spatial coordinates, including those transformations that preserve the form of g_{ij} while taking the origin into any other point. Hence, homogeneity and isotropy require the components of the energy-momentum tensor everywhere to take the form

$$T_{00} \equiv \rho(t)c^2, \qquad T_{i0} \equiv c\pi_i = 0, \qquad T_{ij} \equiv P(t)\,g_{ij}(t, \mathbf{x}). \tag{2.98}$$

Raising one of the indices, we find

$$T^{\mu}{}_{\nu} = g^{\mu\lambda} T_{\lambda\nu} = \begin{pmatrix} -\rho c^2 & 0 & 0 & 0 \\ 0 & P & 0 & 0 \\ 0 & 0 & P & 0 \\ 0 & 0 & 0 & P \end{pmatrix}, \tag{2.99}$$

which we recognize as the energy-momentum tensor of a perfect fluid in the frame of a comoving observer. More generally, the energy-momentum tensor can be written in the following, explicitly covariant, form

$$T_{\mu\nu} = \left(\rho + \frac{P}{c^2} \right) U_\mu U_\nu + P\, g_{\mu\nu}, \tag{2.100}$$

where U^μ is the relative four-velocity between the fluid and the observer, while ρc^2 and P are the energy density and pressure in the rest frame of the fluid. Of course, we recover (2.98) for a comoving observer, $U^\mu = (c, 0, 0, 0)$.

Continuity equation

How do the density and pressure evolve with time? In Minkowski space, energy and momentum are conserved. The energy density therefore satisfies the **continuity equation**, $\dot{\rho} = -\partial_i \pi^i$, i.e. the rate of change of the density equals the divergence of the energy flux. Similarly, the evolution of the momentum density satisfies the **Euler equation**, $\dot{\pi}_i = \partial_i P$. These conservation laws can be combined into a four-component conservation equation for the energy-momentum tensor

$$\partial_\mu T^{\mu}{}_{\nu} = 0. \tag{2.101}$$

In general relativity, this is promoted to the covariant conservation equation

$$\boxed{\nabla_\mu T^{\mu}{}_{\nu} = 0}. \tag{2.102}$$

To work with (2.102), we need to unpack the covariant derivative. Using (2.92), we get

$$\nabla_\mu T^{\mu}{}_{\nu} = \partial_\mu T^{\mu}{}_{\nu} + \Gamma^{\mu}{}_{\mu\lambda} T^{\lambda}{}_{\nu} - \Gamma^{\lambda}{}_{\mu\nu} T^{\mu}{}_{\lambda} = 0. \tag{2.103}$$

This corresponds to four separate equations (one for each value of ν). The evolution of the energy density is determined by the $\nu = 0$ equation

$$\partial_\mu T^\mu{}_0 + \Gamma^\mu_{\mu\lambda} T^\lambda{}_0 - \Gamma^\lambda_{\mu 0} T^\mu{}_\lambda = 0 \,. \tag{2.104}$$

Since $T^i{}_0$ vanishes by isotropy, this reduces to

$$\frac{1}{c}\frac{\mathrm{d}(\rho c^2)}{\mathrm{d}t} + \Gamma^\mu_{\mu 0}(\rho c^2) - \Gamma^\lambda_{\mu 0} T^\mu{}_\lambda = 0 \,, \tag{2.105}$$

where $\Gamma^\lambda_{\mu 0}$ vanishes unless λ and μ are spatial indices and equal to each other, in which case we have $\Gamma^i_{i0} = 3c^{-1}\dot{a}/a$. The continuity equation (2.105) then becomes

$$\boxed{\dot{\rho} + 3\frac{\dot{a}}{a}\left(\rho + \frac{P}{c^2}\right) = 0} \,. \tag{2.106}$$

This important equation describes "energy conservation" in the cosmological context. Notice that the usual notion of energy conservation in flat space—as derived from Noether's theorem—relies on a symmetry under time translations. This symmetry is broken in an expanding space, so the familiar energy conservation does not have to hold and is replaced by (2.106).

Exercise 2.4 Show that (2.106) can also be obtained from the thermodynamic relation $\mathrm{d}U = -P\mathrm{d}V$, where $U = (\rho c^2)V$ and $V \propto a^3$.

Most cosmological fluids can be parameterized in terms of a constant **equation of state**, $w = P/(\rho c^2)$. In that case, we get

$$\frac{\dot{\rho}}{\rho} = -3(1+w)\frac{\dot{a}}{a} \quad \Rightarrow \quad \boxed{\rho \propto a^{-3(1+w)}} \,, \tag{2.107}$$

which shows how the dilution of the energy density depends on the equation of state. In the following, I will describe three types of fluids—matter, radiation and dark energy—that play important roles in the evolution of our universe.

2.3.2 Matter and Radiation

I will use the term **matter** to refer to a fluid whose pressure is much smaller than its energy density, $|P| \ll \rho c^2$. As we will see in Chapter 3, this is the case for a gas of *non-relativistic particles*, for which the energy density is dominated by their rest mass. Setting $w = 0$ in (2.107) gives

$$\rho \propto a^{-3} \,. \tag{2.108}$$

This dilution of the energy density simply reflects the fact that the volume of a region of space increases as $V \propto a^3$, while the energy within that region stays constant.

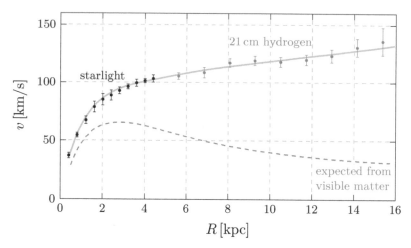

Fig. 2.9 Rotation curve of M33 (figure adapted from [6]). The dashed line is the expected rotation curve accounting only for the visible matter in the stellar disc. The data clearly shows the need for an additional dark matter component.

- **Baryons** Cosmologists refer to ordinary matter (nuclei and electrons) as baryons. Of course, this is technically incorrect (electrons are *leptons*), but nuclei are so much heavier than electrons that most of the mass is in the baryons. If this terminology upsets you, you should ask your astronomer friends what they mean by "metals."

- **Dark matter** Most of the matter in the universe is in the form of dark matter. Figure 2.9 shows the orbital speeds of visible stars or gas in the galaxy M33 versus their radial distance from the center of the galaxy. We see that the rotation speeds stay constant far beyond the extent of the visible disc. This suggests that there is an invisible halo of dark matter that holds the galaxy together. On larger scales, dark matter has been detected by gravitational lensing. As the CMB photons travel through the universe they get deflected by the total matter in the universe. This effect can be measured statistically and provides one of the best pieces of evidence for dark matter on a cosmic scale.

The precise nature of the dark matter is unknown. We usually think of it as a new particle species, but we don't know what it really is. There are many popular dark matter candidates such as WIMPs, axions, MACHOs, and primordial black holes (see [7] for a comprehensive review). While finding a microscopic explanation for dark matter is a very active area of research, the precise details don't affect the large-scale evolution of the universe. On large scales, all dark matter models are described by the same pressureless fluid. This *cold dark matter* (CDM) is an integral part of the standard model of cosmology.

I will use the term **radiation** to denote anything for which the pressure is one-third of the energy density, $P = \frac{1}{3}\rho c^2$. This is the case for a gas of *relativistic* *particles*, for which the energy density is dominated by the kinetic energy. Setting $w = 1/3$ in (2.107) gives

$$\rho \propto a^{-4}. \tag{2.109}$$

The dilution now includes not just the expansion, $V \propto a^3$, but also the redshifting of the energy of the particles, $E \propto a^{-1}$.

- **Light particles** At early times, all particles of the Standard Model acted as radiation, because the temperature of the universe was larger than the masses of the particles. As the temperature of the universe dropped, the masses of many particles became relevant and they started to behave like matter.

- **Photons** Being massless, photons are always relativistic. Together with neutrinos, they were the dominant energy density during Big Bang nucleosynthesis. Today, we observe these photons in the form of the cosmic microwave background.

- **Neutrinos** For most of the history of the universe, neutrinos behaved like radiation. Only recently have their small masses become relevant, making them act like matter.

- **Gravitons** The early universe may also have produced a background of gravitons. The density of these gravitons is predicted to be very small, so they have a negligible effect on the expansion of the universe. Nevertheless, experimental efforts are underway to detect them (see Section 8.4.2).

2.3.3 Dark Energy

We have recently learned that matter and radiation aren't enough to describe the evolution of the universe. Instead, the universe today is dominated by a mysterious form of **dark energy** with negative pressure, $P = -\rho c^2$, and hence constant energy density,

$$\rho \propto a^0. \tag{2.110}$$

Since the energy density doesn't dilute, energy has to be created as the universe expands. As described above, this doesn't violate the conservation of energy, as long as equation (2.106) is satisfied.

- **Vacuum energy** A natural candidate for a constant energy density is the energy associated to empty space itself. As the universe expands, more space is being created and this energy therefore increases in proportion to the volume. In quantum field theory, this so-called "vacuum energy" is actually predicted, leading to an energy-momentum tensor of the form

$$T_{\mu\nu}^{\text{vac}} = -\rho_{\text{vac}} c^2 g_{\mu\nu}. \tag{2.111}$$

As expected from Lorentz symmetry, the energy-momentum tensor associated to the vacuum is proportional to the spacetime metric. Comparison with (2.100) shows that this indeed gives $P_{vac} = -\rho_{vac}c^2$. Unfortunately, as I will describe below, quantum field theory also predicts the size of the vacuum energy ρ_{vac} to be much larger than the value inferred from cosmological observations.

- **Cosmological constant** The observation of dark energy has also revived the old concept of a "cosmological constant," originally introduced by Einstein to make the universe static. To understand the origin of the cosmological constant, let us first note that the left-hand side of the Einstein equation (2.84) isn't uniquely defined. We can add to it the term $\Lambda g_{\mu\nu}$, for some constant Λ, without changing the conservation of the energy-momentum tensor, $\nabla^\mu T_{\mu\nu} = 0$. To see this, you need to convince yourself that $\nabla^\mu g_{\mu\nu} = 0$. In other words, we could have written the Einstein equation as

$$G_{\mu\nu} + \Lambda g_{\mu\nu} = \frac{8\pi G}{c^4}\, T_{\mu\nu}\,. \tag{2.112}$$

However, it has become standard practice to move this extra term to the right-hand side and treat it as a contribution to the energy-momentum tensor

$$T_{\mu\nu}^{(\Lambda)} = -\frac{\Lambda c^4}{8\pi G}\, g_{\mu\nu} \equiv -\rho_\Lambda c^2\, g_{\mu\nu}\,, \tag{2.113}$$

which is of the same form as (2.111).

- **Dark energy** The term "dark energy" is often used to describe a more general fluid whose equation of state is not exactly that of a cosmological constant, $w \approx -1$, or may even be varying in time. It is the failure of quantum field theory to explain the size of the observed vacuum energy that has led theorists to consider these more exotic possibilities. Desperate times call for desperate measures.

Since there are no indications for a deviation from $w = -1$, I will typically use the terms vacuum energy, cosmological constant and dark energy interchangeably. Whatever it is, dark energy plays a key role in the standard cosmological model—the ΛCDM model (see Section 2.4).

The cosmological constant problem*

When the cosmological constant was discovered, it came as a relief to observers, but was a shock to theorists. While the cosmological constant reconciled the age of the universe with the ages of the oldest stars within it, its observed value is much smaller than all particle physics scales, making it hard to understand from a more fundamental perspective. This **cosmological constant problem** is the biggest crisis in modern theoretical physics. Let me digress briefly to describe the problem.

In Section 2.4, we will see that the observed value of the vacuum energy is

$$\rho_\Lambda c^2 \approx 6 \times 10^{-10}\,\mathrm{J\,m^{-3}}\,. \tag{2.114}$$

To compare this to our expectation from particle physics, we need to write this in units that are more natural from the perspective of high-energy physics. These units are electron volts (eV). Using $1\,\mathrm{J} \approx 6.2 \times 10^{18}\,\mathrm{eV}$ and $\hbar c \approx 2.0 \times 10^{-7}\,\mathrm{eV\,m}$, we get $(\hbar c)^3 \rho_\Lambda c^2 \approx (10^{-3}\,\mathrm{eV})^4$. Written in natural units, with $\hbar = c \equiv 1$, we then have

$$\rho_\Lambda \approx (10^{-3}\,\mathrm{eV})^4 . \tag{2.115}$$

We see that the scale appearing in the vacuum energy, $M_\Lambda \sim 10^{-3}\,\mathrm{eV}$, is much smaller than the typical scales relevant to particle physics. In the following, we will dig a bit deeper into the issue.

I mentioned that quantum field theory (QFT) predicts the existence of vacuum energy. However, I will now show that the expected value of this vacuum energy is vastly too large. To start, recall that in quantum field theory every particle arises as an excitation of a fundamental field. Moreover, each field can be represented by an infinite number of Fourier modes with frequencies ω_k. These Fourier modes satisfy the equation of a harmonic oscillator and therefore experience the same **zero-point fluctuations** as a quantum harmonic oscillator (see Section 8.2.1). In other words, the vacuum energy receives contributions of the form $\frac{1}{2}\hbar\omega_k$ for each mode. The sum over all Fourier modes actually diverges. This divergence arises because we extrapolated our QFT beyond its regime of validity. Let us therefore assume that our theory is valid only up to a cutoff scale M_* and only include contributions to the vacuum energy from modes with frequencies below the cutoff. The vacuum energy will now be finite, but depend on the arbitrary cutoff scale. This dependence on the cutoff can be removed by adding a counterterm in the form of a bare cosmological constant Λ_0. After this process of renormalization, the remaining vacuum energy is

$$\rho_{\mathrm{QFT}} \sim \sum_i m_i^4 , \tag{2.116}$$

where the sum is over all particles with masses below the cutoff. In the absence of gravity this constant vacuum energy wouldn't affect any dynamics and is therefore usually discarded. The equivalence principle, however, implies that all forms of energy gravitate and hence affect the curvature of the spacetime. The vacuum energy is no exception.

The first to worry about the gravitational effects of zero-point energies was Wolfgang Pauli in the 1920s. Considering the contribution from the electron in (2.116), he estimated that the induced curvature radius of the universe would be $10^6\,\mathrm{km}$, so that "the world would not even reach to the moon" [8]. Including the rest of the Standard Model only makes the problem worse:

$$\rho_{\mathrm{QFT}} \gtrsim (1\,\mathrm{TeV})^4 = 10^{60}\rho_\Lambda , \tag{2.117}$$

where we have added contributions up to the TeV scale. The enormous discrepancy between this estimate and the observed value in (2.114) is the cosmological constant problem.

Now, you could argue that we are always free to add an arbitrary constant to the QFT estimate to match the observed value of the vacuum energy

$$\rho_\Lambda = \rho_{\text{QFT}} + \rho_0 \,. \tag{2.118}$$

In fact, precisely such a constant was added in the process of renormalization. To explain the small size of the observed vacuum energy would require two large numbers, ρ_{QFT} and ρ_0, have 60 digits in common, but differ at the 61st digit, so that the sum of the two is smaller by 60 orders of magnitude. While it is true that you can fine-tune the cosmological constant in this way, "nobody will applaud you when you do it" [9]. Moreover, each time the universe goes through a phase transition, the value of ρ_{QFT} changes, while the value of ρ_0 presumably doesn't. For example, we expect the electroweak and QCD phase transitions to induce jumps in the vacuum energy of the order of $\Delta\rho_{\text{QFT}} \sim (200\,\text{GeV})^4$ and $(0.3\,\text{GeV})^4$, respectively. Any fine-tuning must explain the small value of the cosmological constant today and not at the beginning.

One way to explain the apparent fine-tuning to the vacuum energy is to appeal to the **anthropic principle**. Suppose that our universe is part of a much larger **multiverse** and that the fundamental physical parameters exhibit different values in different regions of space. In particular, imagine that the value of the vacuum energy varies across the multiverse. As pointed out by Weinberg [10], in such a multiverse, we shouldn't be surprised to find ourselves in a universe with the observed value for the cosmological constant. If the vacuum energy were much bigger, dark energy would start to dominate before galaxies would have formed and complex structures could not have evolved. In other words, we shouldn't be shocked by the small value of the vacuum energy, since if it was any bigger we would not be around to ask the question. "We live where we can live" in a vast landscape of possibilities, and the observed value of the cosmological constant is simply an accident of "environmental selection."

2.3.4 Spacetime Curvature

Having introduced the relevant types of matter and energy in the universe, we now want to see how they source the dynamics of the spacetime. To do this, we have to compute the Einstein tensor on the left-hand side of the Einstein equation (2.84),

$$\boxed{G_{\mu\nu} = R_{\mu\nu} - \frac{1}{2}R g_{\mu\nu}}\,, \tag{2.119}$$

where $R_{\mu\nu}$ is the **Ricci tensor** and $R = R^\mu{}_\mu = g^{\mu\nu}R_{\mu\nu}$ is its trace, the **Ricci scalar**. In Appendix A, we give the following definition of the Ricci tensor in terms of the Christoffel symbols

$$R_{\mu\nu} \equiv \partial_\lambda \Gamma^\lambda_{\mu\nu} - \partial_\nu \Gamma^\lambda_{\mu\lambda} + \Gamma^\lambda_{\lambda\rho}\Gamma^\rho_{\mu\nu} - \Gamma^\rho_{\mu\lambda}\Gamma^\lambda_{\nu\rho}\,. \tag{2.120}$$

Again, there is a lot of beautiful geometry behind these ideas, which I encourage you to learn. We will simply keep plugging-and-playing: given the Christoffel symbols (2.45) nothing stops us from computing (2.120) and (2.119) for the FRW spacetime.

We don't need to calculate $R_{i0} = R_{0i}$, because it is a three-vector and therefore must vanish due to the isotropy of the Robertson–Walker metric. The non-vanishing components of the Ricci tensor are

$$R_{00} = -\frac{3}{c^2}\frac{\ddot{a}}{a}, \tag{2.121}$$

$$R_{ij} = \frac{1}{c^2}\left[\frac{\ddot{a}}{a} + 2\left(\frac{\dot{a}}{a}\right)^2 + 2\frac{kc^2}{a^2 R_0^2}\right]g_{ij}. \tag{2.122}$$

These results are derived in the boxes below. Note that we had to find that $R_{ij} \propto g_{ij}$ to be consistent with homogeneity and isotropy.

Example Setting $\mu = \nu = 0$ in (2.120), we have

$$R_{00} = \partial_\lambda \Gamma^\lambda_{00} - \partial_0 \Gamma^\lambda_{0\lambda} + \Gamma^\lambda_{\lambda\rho}\Gamma^\rho_{00} - \Gamma^\rho_{0\lambda}\Gamma^\lambda_{0\rho}. \tag{2.123}$$

Since Christoffel symbols with two time indices vanish, this reduces to

$$R_{00} = -\partial_0 \Gamma^i_{0i} - \Gamma^i_{0j}\Gamma^j_{0i}. \tag{2.124}$$

Using $\Gamma^i_{0j} = c^{-1}(\dot{a}/a)\delta^i_j$, we find

$$R_{00} = -\frac{1}{c^2}\frac{d}{dt}\left(3\frac{\dot{a}}{a}\right) - \frac{3}{c^2}\left(\frac{\dot{a}}{a}\right)^2 = -\frac{3}{c^2}\frac{\ddot{a}}{a}, \tag{2.125}$$

which is the result cited in (2.121).

Example Computing R_{ij} by brute force would be very tedious (especially, for $k \neq 0$). However, the analysis can be simplified greatly using the following trick: we will compute R_{ij} at $\mathbf{x} = 0$ where the spatial metric is $\gamma_{ij} = \delta_{ij}$ and then argue that the result for general $\mathbf{x}$ must have the form $R_{ij} \propto g_{ij}$.

It is convenient to work in Cartesian coordinates, where the spatial metric is

$$\gamma_{ij} = \delta_{ij} + \frac{kx_ix_j}{R_0^2 - k(x_kx^k)}. \tag{2.126}$$

The key point is to think ahead and anticipate that we will set $\mathbf{x} = 0$ at the end. This allows us to drop many terms. You may be tempted to use $\gamma_{ij}(\mathbf{x} = 0) = \delta_{ij}$ straight away. However, the Christoffel symbols contain a derivative of the metric and the Ricci tensor has another derivative, so there will be terms in the final answer with two derivatives acting on the metric. These terms get a contribution from the second term in (2.126). Nevertheless, we only need to keep terms to quadratic order in x:

$$\gamma_{ij} = \delta_{ij} + kx_ix_j/R_0^2 + O(x^4). \tag{2.127}$$

Plugging this into the definition of the Christoffel symbol gives

$$\Gamma^i_{jk} = \frac{k}{R_0^2}x^i\delta_{jk} + O(x^3). \tag{2.128}$$

This vanishes at $\mathbf{x} = 0$, but its derivative does not. From the definition of the Ricci tensor, we then have

$$R_{ij}(\mathbf{x} = 0) \equiv \underbrace{\partial_\lambda \Gamma^\lambda_{ij} - \partial_j \Gamma^\lambda_{i\lambda}}_{(A)} + \underbrace{\Gamma^\lambda_{\lambda\rho}\Gamma^\rho_{ij} - \Gamma^\rho_{i\lambda}\Gamma^\lambda_{j\rho}}_{(B)} . \tag{2.129}$$

Let us first look at the two terms labeled (B). Dropping terms that are zero at $\mathbf{x} = 0$, we find

$$\begin{aligned}
(B) &= \Gamma^l_{l0}\Gamma^0_{ij} - \Gamma^0_{il}\Gamma^l_{j0} - \Gamma^l_{i0}\Gamma^0_{jl} \\
&= \frac{3}{c^2}\frac{\dot{a}}{a}a\dot{a}\delta_{ij} - \frac{1}{c^2}a\dot{a}\delta_{il}\frac{\dot{a}}{a}\delta^l_j - \frac{1}{c^2}\frac{\dot{a}}{a}\delta^l_i a\dot{a}\delta_{jl} \\
&= \frac{\dot{a}^2}{c^2}\delta_{ij} .
\end{aligned} \tag{2.130}$$

The two terms labeled (A) in (2.129) can be evaluated by using (2.128),

$$\begin{aligned}
(A) &= \partial_0 \Gamma^0_{ij} + \partial_l \Gamma^l_{ij} - \partial_j \Gamma^l_{il} \\
&= c^{-2}\partial_t(a\dot{a})\delta_{ij} + \frac{k}{R_0^2}\delta^l_l\delta_{ij} - \frac{k}{R_0^2}\delta^l_j\delta_{il} \\
&= c^{-2}\left(a\ddot{a} + \dot{a}^2 + 2\frac{kc^2}{R_0^2}\right)\delta_{ij} .
\end{aligned} \tag{2.131}$$

Hence, we get

$$\begin{aligned}
R_{ij}(\mathbf{x} = 0) &= (A) + (B) \\
&= \frac{1}{c^2}\left[a\ddot{a} + 2\dot{a}^2 + 2\frac{kc^2}{R_0^2}\right]\delta_{ij} \\
&= \frac{1}{c^2}\left[\frac{\ddot{a}}{a} + 2\left(\frac{\dot{a}}{a}\right)^2 + 2\frac{kc^2}{a^2 R_0^2}\right]g_{ij}(\mathbf{x} = 0) .
\end{aligned} \tag{2.132}$$

Since this is a relation between tensors, it holds for general $\mathbf{x}$, so we get the result quoted in (2.122).

To be clear, these days nobody in their right mind would do a computation like this by hand. However, it is good to have seen it once, if only to gain an appreciation for modern computer algebra.

Given the components of the Ricci tensor, it is now straightforward to complete the calculation. The Ricci scalar is

$$\begin{aligned}
R &= g^{\mu\nu}R_{\mu\nu} \\
&= -R_{00} + \frac{1}{a^2}\gamma^{ij}R_{ij} = \frac{3}{c^2}\frac{\ddot{a}}{a} + \frac{\delta^i_i}{c^2}\left[\frac{\ddot{a}}{a} + 2\left(\frac{\dot{a}}{a}\right)^2 + 2\frac{kc^2}{a^2 R_0^2}\right] \\
&= \frac{6}{c^2}\left[\frac{\ddot{a}}{a} + \left(\frac{\dot{a}}{a}\right)^2 + \frac{kc^2}{a^2 R_0^2}\right] ,
\end{aligned} \tag{2.133}$$

and the nonzero components of the Einstein tensor, $G^\mu{}_\nu \equiv g^{\mu\lambda} G_{\lambda\nu}$, are

$$G^0{}_0 = -\frac{3}{c^2}\left[\left(\frac{\dot{a}}{a}\right)^2 + \frac{kc^2}{a^2 R_0^2}\right],$$ (2.134)

$$G^i{}_j = -\frac{1}{c^2}\left[2\frac{\ddot{a}}{a} + \left(\frac{\dot{a}}{a}\right)^2 + \frac{kc^2}{a^2 R_0^2}\right]\delta^i_j.$$ (2.135)

I leave it to you to verify that these components of the Einstein tensor follow from our results for the Ricci tensor.

2.3.5 Friedmann Equations

We have finally assembled all ingredients to evaluate the Einstein equation (2.84). Setting $G^0{}_0$ in (2.134) equal to $(8\pi G/c^4)\,T^0{}_0 = -(8\pi G/c^2)\,\rho$, we get the **Friedmann equation**

$$\left(\frac{\dot{a}}{a}\right)^2 = \frac{8\pi G}{3}\rho - \frac{kc^2}{a^2 R_0^2},$$ (2.136)

where ρ should be understood as the sum of all contributions to the energy density of the universe. We will write ρ_r for the contribution from radiation (with ρ_γ for photons and ρ_ν for neutrinos), ρ_m for the contribution by matter (with ρ_c for cold dark matter and ρ_b for baryons) and ρ_Λ for the vacuum energy contribution. Equation (2.136) is the fundamental equation describing the evolution of the scale factor. To write it as a closed form equation for $a(t)$, we have to specify the evolution of the density, $\rho(a)$, as in (2.107).

The spatial part of the Einstein equation, $G^i{}_j = (8\pi G/c^4)\,T^i{}_j$, leads to the second Friedmann equation (also known as the **Raychaudhuri equation**)

$$\frac{\ddot{a}}{a} = -\frac{4\pi G}{3}\left(\rho + \frac{3P}{c^2}\right).$$ (2.137)

It is easy to show that this equation also follows from taking a time derivative of the first Friedmann equation (2.136) and using the continuity equation (2.106) for $\dot{\rho}$.

Newtonian derivation Some intuition for the form of the Friedmann equations can be developed from a non-relativistic Newtonian analysis. Consider an expanding sphere of matter of uniform mass density $\rho(t)$ and radius $R(t) = a(t)R_0$. Let us study the dynamics of a test particle on the surface of the sphere. The acceleration of the particle is

$$\ddot{R} = -\frac{GM(R)}{R^2}, \quad M(R) = \frac{4\pi}{3}R^3\rho,$$ (2.138)

where $M(R)$ is the mass enclosed in the ball of radius R. Multiplying this by $\dot{R}$ and integrating, we get

$$\frac{1}{2}\dot{R}^2 - \frac{GM(R)}{R} = E\,, \tag{2.139}$$

where the integration constant E is the energy per unit mass of the particle. In terms of the scale factor and the density, this becomes

$$\left(\frac{\dot{a}}{a}\right)^2 = \frac{8\pi G}{3}\rho + \frac{2E}{a^2 R_0^2}\,. \tag{2.140}$$

Identifying $2E$ with $-kc^2$, this is precisely of the form of the Friedmann equation for a pressureless fluid. We see that the spatial curvature is related to the total energy, with flatness arising if the kinetic and potential energies precisely add up to zero. This interpretation will be relevant when we describe the flatness problem in Chapter 4.

The first Friedmann equation is often written in terms of the Hubble parameter, $H \equiv \dot{a}/a$,

$$\boxed{H^2 = \frac{8\pi G}{3}\,\rho - \frac{kc^2}{a^2 R_0^2}}\,. \tag{2.141}$$

We will use the subscript '0' to denote quantities evaluated today, at $t = t_0$. A flat universe ($k = 0$) corresponds to the following **critical density** today

$$\rho_{\mathrm{crit},0} = \frac{3H_0^2}{8\pi G} = 1.9 \times 10^{-29}\,h^2\,\mathrm{g\,cm^{-3}}$$

$$= 2.8 \times 10^{11}\,h^2\,M_\odot\,\mathrm{Mpc^{-3}}$$

$$= 1.1 \times 10^{-5}\,h^2\,\mathrm{protons\,cm^{-3}}\,. \tag{2.142}$$

It is convenient to measure all densities relative to the critical density and work with the following dimensionless density parameters

$$\Omega_{i,0} \equiv \frac{\rho_{i,0}}{\rho_{\mathrm{crit},0}}\,, \quad i = r, m, \Lambda, \ldots \tag{2.143}$$

In the literature, the subscript '0' on the density parameters $\Omega_{i,0}$ is often dropped, so that Ω_i denotes the density *today* in terms of the critical density *today*. From now on, I will follow this convention. The Friedmann equation (2.141) can then be written as

$$\boxed{\frac{H^2}{H_0^2} = \Omega_r\,a^{-4} + \Omega_m\,a^{-3} + \Omega_k\,a^{-2} + \Omega_\Lambda}\,, \tag{2.144}$$

where we have introduced the curvature "density" parameter, $\Omega_k \equiv -kc^2/(R_0 H_0)^2$. Note that $\Omega_k < 0$ for $k > 0$. Evaluating both sides of the Friedmann equation at the present time, with $a(t_0) \equiv 1$, leads to the constraint

$$1 = \underbrace{\Omega_r + \Omega_m + \Omega_\Lambda}_{\equiv \Omega_0} + \Omega_k \,, \tag{2.145}$$

and the curvature parameter can be written as $\Omega_k = 1 - \Omega_0$. A central task in cosmology is to measure the parameters occurring in the Friedmann equation (2.144) and hence determine the composition of the universe. I will describe these measurements in Section 2.4.

2.3.6 Exact Solutions

In general, the Friedmann equation is a complicated nonlinear differential equation for the scale factor $a(t)$. In this section, we will study special cases for which exact solutions are possible. As we will see, many of these solutions are good approximations to the real universe for certain periods of time.

Single-component universes

Consider first a flat universe ($k = 0$) with just a single fluid component. In fact, the different scalings of the densities of radiation (a^{-4}), matter (a^{-3}) and vacuum energy (a^0) imply that for most of its history the universe was dominated by a single component (first radiation, then matter, then vacuum energy; see Fig. 2.10). Parameterizing this component by its equation of state w_i captures all cases of interest. The Friedmann equation (2.144) then reduces to

$$\frac{\mathrm{d}\ln a}{\mathrm{d}t} \approx H_0 \sqrt{\Omega_i}\, a^{-\frac{3}{2}(1+w_i)} \,. \tag{2.146}$$

Integrating this equation, we obtain the time dependence of the scale factor

$$a(t) \propto \begin{cases} t^{2/3(1+w_i)} \quad w_i \neq -1 & \begin{array}{ll} t^{2/3} & \mathrm{MD} \\ t^{1/2} & \mathrm{RD} \end{array} \\ e^{H_0 \sqrt{\Omega_\Lambda} t} \quad w_i = -1 & \Lambda\mathrm{D} \end{cases} \tag{2.147}$$

where MD, RD and ΛD stand for matter dominated, radiation dominated and dark energy dominated, respectively.

In conformal time, the Friedmann equation becomes

$$\frac{\mathrm{d}\ln a}{\mathrm{d}\eta} \approx H_0 \sqrt{\Omega_i}\, a^{-\frac{1}{2}(1+3w_i)} \,, \tag{2.148}$$

and we get

$$a(\eta) \propto \eta^{2/(1+3w_i)} \,, \tag{2.149}$$

which reads η^2, η and $(-\eta)^{-1}$ for a matter-dominated, radiation-dominated and dark energy-dominated universe, respectively. The special case $w_i = -1/3$ corresponds to a universe dominated by spatial curvature.

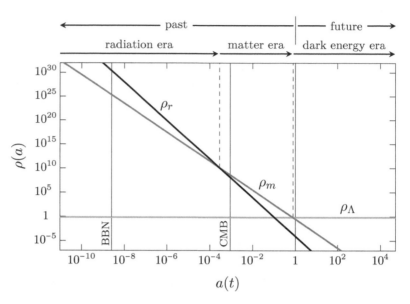

Fig. 2.10 Evolution of the energy densities in the universe. We see that there is often one dominant component: first radiation, then matter and finally dark energy. Sometimes two components are relevant during the transitions between the different eras.

Let me make a few comments:

- The solution for a pure matter universe is known as the **Einstein–de Sitter universe**:

$$a(t) = \left(\frac{t}{t_0}\right)^{2/3}. \tag{2.150}$$

This solution was first derived by Einstein and de Sitter in an influential two-page paper [11]. Apparently, neither Einstein nor de Sitter were very enthusiastic about the paper. In a conversation with Eddington, Einstein remarked "I did not think the paper very important myself, but de Sitter was keen on it." To which de Sitter responded "You will have seen the paper by Einstein and myself. I do not myself consider the result of much importance, but Einstein seemed to think it was." Nevertheless, for a long time, this Einstein–de Sitter universe was the standard cosmological model and it is still a good approximation to the long matter-dominated period in our universe.

Equation (2.150) implies the following relation between the Hubble constant and the age of the universe:

$$H_0 = \frac{2}{3}\frac{1}{t_0}. \tag{2.151}$$

Taking the observed value of the Hubble constant, $H_0 \approx 70\,\mathrm{km\,s^{-1}Mpc^{-1}}$, then leads to

$$t_0 = \frac{2}{3}H_0^{-1} \approx 9\,\text{billion yrs}. \tag{2.152}$$

This is the famous "age problem": the age of a pure matter universe is shorter than that of the oldest stars.

- The above results also apply to a universe without any matter and only a curvature term. The a^{-2} scaling of the curvature contribution implies that we must set $w_k = -1/3$. In the absence of additional fluid contributions, the Friedmann equation then only admits a solution for $k = -1$. In this case, we get

$$a(t) = \frac{t}{t_0}, \tag{2.153}$$

which is called the **Milne universe**.

- The solution for a cosmological constant, $w_\Lambda = -1$, is known as **de Sitter space**. This de Sitter solution turns out to be a good approximation to the evolution of our universe in both the far past and the far future. I will have much more to say about this below and in Chapter 4. For now, let me just remark that the time dependence in the de Sitter solution in (2.147) is somewhat of a fake. All physical quantities—like the density ρ—are independent of time in an exact de Sitter space. To have any cosmological evolution requires that this time-translation invariance of the de Sitter background is broken by a time-dependent form of stress-energy.

A singularity theorem

With the exception of the de Sitter solution, the solutions above all have a singularity at $t = 0$. In particular, the scale factor goes to zero and the density diverges as $\rho \propto t^{-2}$. As famously proven by Hawking and Penrose [12], this singularity is actually a generic feature of all Big Bang cosmologies. We can get a glimpse of this singularity theorem for the special case of our FRW spacetimes.

Consider the second Friedmann equation

$$\frac{\ddot{a}}{a} = -\frac{4\pi G}{3}\left(\rho + \frac{3P}{c^2}\right). \tag{2.154}$$

We will now show that this implies a singularity in the past, as long as the matter obeys the **strong energy condition** (SEC), $\rho c^2 + 3P \geq 0$, so that $\ddot{a} < 0$ at all times. This makes sense: gravity is attractive for all familiar matter sources, so it slows down the expansion. We want to show that $a(t)$ must have started from zero at some point in the past. It is easiest to prove this graphically (see Fig. 2.11).

Let us pick some fiducial time $t = t_0$ (which for our universe may be the present time) and specify two conditions on the scale factor at that time

$$a(t_0) = 1, \tag{2.155}$$
$$\dot{a}(t_0) = H_0.$$

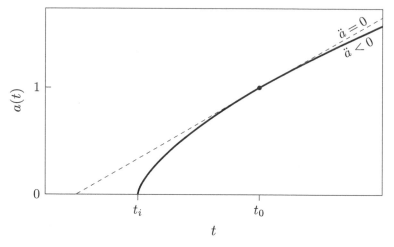

Fig. 2.11 Graphical illustration that a universe with $\ddot{a} < 0$ must have a Big Bang singularity. The dashed line corresponds to the special case $\ddot{a} = 0$. The solid line is a generic universe with $\ddot{a} < 0$ at all times, which has the same a and $\dot{a}$ as the reference universe at $t = t_0$.

We assume that the universe is expanding at t_0, so that $H_0 > 0$, and then use (2.154) to evolve the universe backwards in time. Consider first the special case $\ddot{a} = 0$. The solution is $a(t) = 1 + H_0(t - t_0)$, which has a singularity at $t_0 - H_0^{-1}$ (see the dashed line in Fig. 2.11). Now, let us compare this to solutions with $\ddot{a} < 0$. Any such solution has the same tangent at t_0, but *smaller* $a(t)$ at $t < t_0$. The solid line in Fig. 2.11 shows an example. Any solution with $\ddot{a} < 0$, for all $t < t_0$, therefore also must have a singularity at some $t_i > t_0 - H_0^{-1}$.

This singularity theorem assumed the FRW solution. The singularity theorem by Hawking and Penrose is more general, but also much more complicated to prove.

Two-component universes

It will also be convenient to have solutions for the case of two fluid components. This is relevant in the early universe, which contained a mixture of matter and radiation, and in the late universe, where both matter and dark energy are important.

To obtain a solution for the scale factor, it will be helpful to work in conformal time η. We will use primes to denote derivatives with respect to η. Using $\dot{a} = a'/a$ and $\ddot{a} = a''/a^2 - (a')^2/a^3$, the Friedmann equations (2.136) and (2.137) then are

$$(a')^2 + \frac{kc^2}{R_0^2}a^2 = \frac{8\pi G}{3}\rho a^4 , \tag{2.156}$$

$$a'' + \frac{kc^2}{R_0^2}a = \frac{4\pi G}{3}\left(\rho - \frac{3P}{c^2}\right)a^3 . \tag{2.157}$$

In the following, we will apply this to a few examples.

- **Matter and radiation**

Consider a flat universe filled with a mixture of matter and radiation. The total density can be written as

$$\rho \equiv \rho_m + \rho_r = \frac{\rho_{eq}}{2} \left[\left(\frac{a_{eq}}{a} \right)^3 + \left(\frac{a_{eq}}{a} \right)^4 \right], \tag{2.158}$$

where the subscript 'eq' denotes quantities evaluated at matter–radiation equality. Notice that radiation doesn't contribute as a source term in (2.157), since $\rho_r c^2 - 3P_r = 0$. Moreover, since $\rho_m a^3 = \text{const} = \frac{1}{2}\rho_{eq}a_{eq}^3$ and $k = 0$, we can write (2.157) as

$$a'' = \frac{2\pi G}{3} \rho_{eq} a_{eq}^3 . \tag{2.159}$$

This equation has the following solution

$$a(\eta) = \frac{\pi G}{3} \rho_{eq} a_{eq}^3 \eta^2 + C\eta + D , \tag{2.160}$$

where C and D are integration constants. Using $a(\eta = 0) \equiv 0$, we have $D = 0$. Substituting (2.160) and (2.158) into (2.156) then gives

$$C = \left(\frac{4\pi G}{3} \rho_{eq} a_{eq}^4 \right)^{1/2} , \tag{2.161}$$

and the solution (2.160) can be written as

$$\boxed{a(\eta) = a_{eq} \left[\left(\frac{\eta}{\eta_*} \right)^2 + 2 \left(\frac{\eta}{\eta_*} \right) \right]}, \tag{2.162}$$

where

$$\eta_* \equiv \left(\frac{\pi G}{3} \rho_{eq} a_{eq}^2 \right)^{-1/2} = \frac{\eta_{eq}}{\sqrt{2} - 1} . \tag{2.163}$$

For $\eta \ll \eta_{eq}$, we recover the radiation-dominated limit, $a \propto \eta$, while for $\eta \gg \eta_{eq}$, we agree with the matter-dominated limit, $a \propto \eta^2$.

- **Matter and curvature**

A historically important solution is a universe containing a mixture of matter and spatial curvature. Before the discovery of dark energy, it was believed that pressureless matter would dominate at late times. In that case, the fate of the universe would be determined by its spatial curvature. A positively curved universe would recollapse and end in a Big Crunch, while a negatively curved universe would expand forever. Let us derive these facts.

Defining the conformal Hubble parameter as $\mathcal{H} \equiv a'/a$, the two Friedmann equations (2.156) and (2.157) can be written as

$$\mathcal{H}^2 + \frac{kc^2}{R_0^2} = \frac{8\pi G}{3}\rho_m a^2 \,, \tag{2.164}$$

$$\frac{\mathrm{d}\mathcal{H}}{\mathrm{d}\eta} = -\frac{4\pi G}{3}\rho_m a^2 \,. \tag{2.165}$$

Combining (2.165) and (2.164) allows us to eliminate the dependence on ρ_m, and we find

$$2\frac{\mathrm{d}\mathcal{H}}{\mathrm{d}\eta} + \mathcal{H}^2 + \frac{kc^2}{R_0^2} = 0\,. \tag{2.166}$$

Introducing the dimensionless time coordinate $\theta \equiv c\eta/R_0$ and the rescaled Hubble parameter $\widetilde{\mathcal{H}} \equiv R_0\mathcal{H}/c$, this becomes

$$2\frac{\mathrm{d}\widetilde{\mathcal{H}}}{\mathrm{d}\theta} + \widetilde{\mathcal{H}}^2 + k = 0\,. \tag{2.167}$$

The solutions of this equation are

$$\widetilde{\mathcal{H}}(\theta) = \frac{1}{a}\frac{\mathrm{d}a}{\mathrm{d}\theta} = \begin{cases} \cot(\theta/2) & k = +1\,, \\ 2/\theta & k = 0\,, \\ \coth(\theta/2) & k = -1\,, \end{cases} \tag{2.168}$$

which is easily integrated to give the evolution of the scale factor

$$a(\theta) = A\begin{cases} \sin^2(\theta/2) & k = +1\,, \\ \theta^2 & k = 0\,, \\ \sinh^2(\theta/2) & k = -1\,, \end{cases} \tag{2.169}$$

where A is an integration constant. At early times, all solutions scale as $\theta^2 \propto \eta^2$, as required for a matter-dominated universe. Using the solution for the scale factor, we obtain the following relation between the conformal and physical time coordinates

$$t = \frac{R_0}{c}\int a(\theta)\,\mathrm{d}\theta = \frac{R_0 A}{2c}\begin{cases} \theta - \sin(\theta) & k = +1\,, \\ \theta^3 & k = 0\,, \\ \sinh(\theta) - \theta & k = -1\,. \end{cases} \tag{2.170}$$

Equations (2.169) and (2.170) then determine $a(t)$ in parametric form. A plot of the solutions is shown in Fig. 2.12. As advertised, a universe with positive curvature recollapses in the future.

The maximal expansion of the positive-curvature universe can be found from the Friedmann equation

$$\frac{H^2}{H_0^2} = \frac{\Omega_0}{a^3} + \frac{1 - \Omega_0}{a^2}\,. \tag{2.171}$$

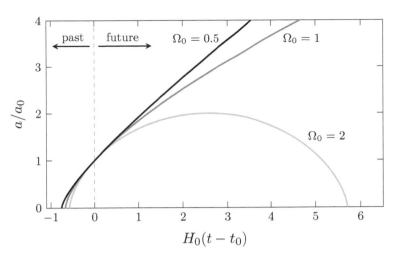

Fig. 2.12 Evolution of the scale factor for a model universe containing a mixture of matter and spatial curvature. We see that the fate of the universe depends on the total amount of matter. If the present matter density is greater than the critical density, $\Omega_0 > 1$, then the universe recollapses in a Big Crunch, while if it is less than the critical density, $\Omega_0 < 1$, it expands forever.

At the turn-around point, the expansion rate must vanish, $H(t) = 0$, which occurs when

$$a_{\max} = \frac{\Omega_0}{\Omega_0 - 1}.$$ (2.172)

This also fixes the integration constant in (2.169) to be $A = \Omega_0/|\Omega_0 - 1|$. Explaining why our universe lives so close to the boundary separating recollapse and eternal expansion is called the **flatness problem** (see Chapter 4).

- **Cosmological constant and curvature**
  Next, we consider a universe with a cosmological constant and spatial curvature. The Friedmann equation is

$$H^2 = \frac{\Lambda c^2}{3} - \frac{kc^2}{a^2 R_0^2}.$$ (2.173)

For $\Lambda > 0$, this has the following solutions

$$a(t) = A \begin{cases} \cosh(\alpha t) & k = +1, \\ \exp(\alpha t) & k = 0, \\ \sinh(\alpha t) & k = -1, \end{cases}$$ (2.174)

where $\alpha \equiv \sqrt{\Lambda c^2/3}$. The normalization A equals $\sqrt{3/\Lambda}/R_0$ for the $k = \pm 1$ solutions, and is arbitrary for the $k = 0$ solution. These solutions are shown in Fig. 2.13. We see that the $k = +1$ solution doesn't have a singularity, while

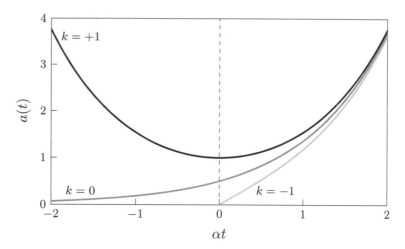

Fig. 2.13   Evolution of the scale factor for the three different slicings of de Sitter space.

the scale factor seems to vanish at $t = -\infty$ and $t = 0$, for the $k = 0$ and $k = -1$ solutions, respectively. The latter singularities, however, are coordinate artifacts. In fact, the three solutions represent different slicings of the same de Sitter space.[7]

For $\Lambda < 0$, solutions only exist for $k = -1$. In that case, the solution of the Friedmann equation is

$$a(t) = A \sin(\alpha t),\tag{2.175}$$

where $\alpha \equiv \sqrt{-\Lambda c^2 / 3}$. This so-called **anti-de Sitter space** plays an important role in toy models of quantum gravity [13].

- **Matter and cosmological constant**
  Finally, we look at a flat universe with matter and a positive cosmological constant, which is a good approximation to our universe today. The Friedmann equation is

$$\frac{H^2}{H_0^2} = \frac{\Omega_m}{a^3} + \Omega_\Lambda,\tag{2.176}$$

where $\Omega_m + \Omega_\Lambda = 1$. As you are invited to show in Problem 2.9 at the end of the chapter, this has the following solution

$$a(t) = \left(\frac{\Omega_m}{\Omega_\Lambda}\right)^{1/3} \sinh^{2/3}\left(\frac{3}{2}\alpha t\right),\tag{2.177}$$

where $\alpha \equiv \sqrt{\Omega_\Lambda} H_0$. At early times, this solution reduces to $a \sim t^{2/3}$, as expected for the early matter-dominated period. Similarly, at late times, we recover the

---

[7] While in a matter-dominated universe the spatial curvature is determined by the energy density, for a Λ-dominated universe all three types of solutions exist for any given value of ρ_Λ. The different solutions therefore describe the same physical spacetime in different coordinates.

scaling of the de Sitter solution in the flat-slicing, $a \sim \exp(\alpha t)$. Using $a(t_0) \equiv 1$, we can solve (2.177) for the age of the universe

$$t_0 = \frac{2}{3\sqrt{\Omega_\Lambda}H_0} \sinh^{-1}\left(\sqrt{\frac{\Omega_\Lambda}{\Omega_m}}\right). \tag{2.178}$$

Substituting the observed values of the cosmological parameters ($\Omega_m = 0.32$ and $\Omega_\Lambda = 0.68$), we get

$$t_0 = 0.96\, H_0^{-1} \approx 14\, \text{billion yrs}, \tag{2.179}$$

which solves the age problem of the Einstein–de Sitter universe; cf. (2.152).

2.4 Our Universe

Our universe is simple, but strange [14]. Simple, because it is characterized by just a handful of parameters. Strange, because the physical interpretation of many of these parameters remains puzzling. In the following, I will give a short description of the key parameters that shape our universe and their measured values (see Table 2.1).[8] In the rest of this book, you will learn where this knowledge comes from.

One of the most precisely measured quantities is the photon density. The COBE satellite found the temperature of the CMB blackbody spectrum to be [15]

$$T_0 = (2.7255 \pm 0.0006)\,\text{K}. \tag{2.180}$$

In Chapter 3, you will learn how this temperature relates to the number density and energy density of the relic photons:

$$n_{\gamma,0} = 0.24 \times \left(\frac{k_B T_0}{\hbar c}\right)^3 \approx 410\,\text{photons cm}^{-3}, \tag{2.181}$$

$$\rho_{\gamma,0} = 0.66 \times \frac{(k_B T_0)^4}{(\hbar c)^3} \approx 4.6 \times 10^{-34}\,\text{g cm}^{-3}, \tag{2.182}$$

where k_B is the Boltzmann constant and $\hbar$ is the reduced Planck constant. In terms of the critical density, the energy density of the photons is

$$\Omega_\gamma \approx 5.4 \times 10^{-5}. \tag{2.183}$$

You will also learn in the next chapter that the universe is filled with a background of relic neutrinos. As long as the neutrinos are relativistic, their energy density is

[8] The values and uncertainties for the cosmological parameters depend on the exact data sets, priors, and parameters used in the fit. Unless stated otherwise, the quoted results on cosmological parameters in this book are derived from a 6-parameter ΛCDM cosmology fit to the Planck 2018 temperature, polarization and lensing data [5]. I will quote 68% confidence regions on measured parameters and 95% on upper limits.

Table 2.1 Key parameters of our universe and their observed values [5]

Parameter	Meaning	Value
H_0	expansion rate	$67.74 \pm 0.46 \, \mathrm{km \, s^{-1} Mpc^{-1}}$
ρ_{crit}	critical density	$(8.62 \pm 0.12) \times 10^{-27} \, \mathrm{kg \, m^{-3}}$
Ω_r	radiation density	$(9.02 \pm 0.21) \times 10^{-5}$
Ω_γ	photon density	$(5.38 \pm 0.15) \times 10^{-5}$
Ω_ν	neutrino density	< 0.003
Ω_b	baryon density	0.0493 ± 0.0006
Ω_m	matter density	0.3153 ± 0.0073
Ω_Λ	dark energy density	0.6847 ± 0.0073
$|\Omega_k|$	spatial curvature	< 0.005
t_0	age of the universe	13.799 ± 0.021 billion yrs
t_{eq}	matter–radiation equality	$51\,100 \pm 800$ yrs
z_{eq}	redshift of equality	3402 ± 26

68% that of the relic photons. Extrapolating this to the present time gives $\Omega_\nu \approx 3.6 \times 10^{-5}$, so that the total radiation density is

$$\Omega_r \approx 9.0 \times 10^{-5} \,. \tag{2.184}$$

In reality, neutrinos have a small mass and become non-relativistic at late times, which increases their density relative to that of the photons. Current observational constraints are $0.0012 < \Omega_\nu < 0.003$, where the lower limit comes from neutrino oscillation experiments and the upper limit is from cosmology.

The COBE satellite also discovered that the CMB temperature varies with position on the sky. Since then these anisotropies have been measured more and more accurately by many CMB experiments. The fluctuations are small, $\Delta T/T \sim 10^{-5}$, but contain an enormous amount of information about the composition of our universe. Figure 2.14 shows the latest measurement by the Planck satellite of the two-point correlation function of the CMB temperature fluctuations. As we will discuss in detail in Chapter 7, the precise shape of this correlation function depends on the cosmological model. Fitting the theory to the data points therefore allows us to measure many of the key cosmological parameters. For example, the peak positions depend on the spatial curvature of the universe. This is because curvature changes the angular diameter distance to the surface of last-scattering, and hence affects the angle at which a fixed physical scale is observed. These measurements suggest an upper bound for any amount of spatial curvature

$$|\Omega_k| < 0.005 \,. \tag{2.185}$$

Recalling the definition $\Omega_k \equiv -kc^2/(R_0 H_0)^2$, this constraint can also be phrased as a lower bound on the curvature scale, $R_0 > 14 \, c H_0^{-1}$. We see that even today curvature makes up less than 1% of the cosmic energy budget. At earlier times, the

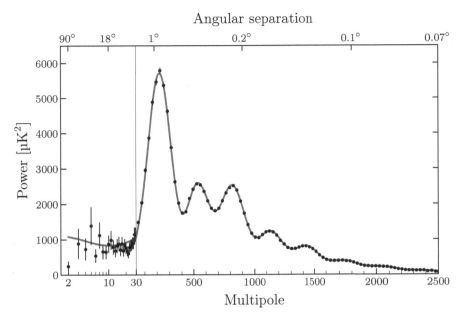

Angular separation

Fig. 2.14 The power spectrum of CMB temperature anisotropies as measured by the Planck satellite [5]. The position of the first peak depends on the spatial curvature of the universe. The height of the first peak is a measure of the matter density and the relative heights of the other peaks determine the baryon density.

effects of curvature are then completely negligible (recall that matter and radiation scale as a^{-3} and a^{-4}, respectively, while the curvature contribution only increases as a^{-2}). In Chapter 4, we will see that inflation indeed predicts that the universe should be very close to spatially flat.

Ordinary matter makes up only a small fraction of the mass and energy in the universe. The density of baryons is known because it affects the abundances of the light chemical elements produced in the hot Big Bang. It also determines detailed features of the CMB spectrum. Both Big Bang nucleosynthesis and the CMB observations show that baryons only make up 5% of the critical density

$$\Omega_b \approx 0.05 \,. \tag{2.186}$$

Taking the typical baryon mass to be the proton mass (most baryons—by number—are protons), the number density of baryons today is

$$n_{b,0} \approx \frac{\rho_{b,0}}{m_p c^2} = \frac{\Omega_b \rho_{\text{crit},0}}{m_p c^2} \approx 0.3 \times 10^{-6} \,\text{cm}^{-3} \,. \tag{2.187}$$

Comparing this to the photon density (2.181), we find the baryon-to-photon ratio to be

$$\eta \equiv \frac{n_b}{n_\gamma} \approx 6 \times 10^{-10} \,. \tag{2.188}$$

This dimensionless ratio is of fundamental importance to cosmology. Why it has this small value remains a mystery.

Most of the matter in the universe is in the form of dark matter. Its gravitational effects are observed in the dynamics of galaxies and clusters of galaxies, as well as in the formation of the large-scale structure of the universe. Moreover, the pattern of CMB fluctuations depends sensitively on the amount of dark matter. The inferred dark matter density today is

$$\Omega_c \approx 0.27 \,, \tag{2.189}$$

where the subscript (c) indicates that we are assuming a "cold" form of dark matter with equation of state $w_c \approx 0$. The sum of the densities of baryons and dark matter gives the total matter density,

$$\Omega_m \approx 0.32 \,. \tag{2.190}$$

Going back in time, the radiation density becomes more and more important relative to the matter density. The scale factor at matter–radiation equality is

$$a_{\mathrm{eq}} = \frac{\Omega_r}{\Omega_m} \approx 2.9 \times 10^{-4} \,, \tag{2.191}$$

where we have used the extrapolated radiation density defined in (2.184). The corresponding redshift at matter–radiation equality is $z_{\mathrm{eq}} \approx 3400$.

Most of the energy density of the universe today is in the form of dark energy. This energy causes the present expansion of the universe to accelerate, as inferred from the apparent brightnesses of distant supernovae. These supernovae appear fainter than expected in a pure matter universe (see Fig. 2.6). In Section 2.2.3, I explained how type Ia supernovae are used as standard candles to obtain measurements of their luminosity distances as a function of their redshifts. A compilation of such measurements is shown in Fig. 2.15.[9] Assuming a flat universe (as suggested by the CMB observations), this data can only be fit if the universe contains a significant amount of dark energy

$$\Omega_\Lambda \approx 0.68 \,. \tag{2.192}$$

The equation of state of the dark energy seems to be that of a cosmological constant, $w_\Lambda \approx -1$. The scale factor at matter–dark energy equality is

$$a_{m\Lambda} = \left(\frac{\Omega_m}{\Omega_\Lambda} \right)^{1/3} \approx 0.77 \,, \tag{2.193}$$

which corresponds to a redshift of $z_{m\Lambda} \approx 0.3$.

Integrating the Friedmann equation, we obtain a relation between time and scale factor,

$$H_0 t = \int_0^a \frac{\mathrm{d}a}{\sqrt{\Omega_r a^{-2} + \Omega_m a^{-1} + \Omega_\Lambda a^2 + \Omega_k}} \,. \tag{2.194}$$

---

[9] The figure plots the **distance modulus**, $\mu \equiv m - M$, where m is the apparent brightness of the object and M is its absolute brightness. The relation between the distance modulus and the luminosity distance is $\mu = 5 \log(d_L/\mathrm{Mpc}) + 25$.

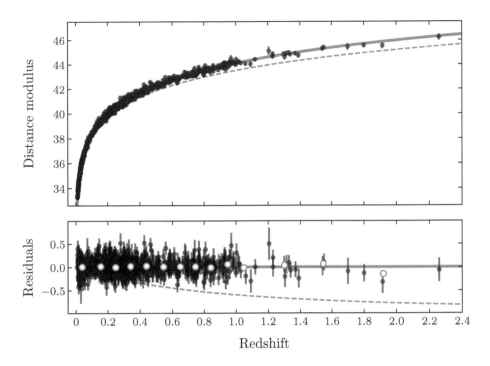

Distance–redshift relationship of 1048 supernovae from the Pantheon sample [16]. The solid line is the best-fit ΛCDM model (with $\Omega_m = 0.32$ and $\Omega_\Lambda = 0.68$), while the dashed line is for a flat matter-only universe (with $\Omega_m = 1.0$). The supernovae clearly appear fainter (or more distant) than predicted in a matter-only universe. The lower panel shows the residuals relative to the ΛCDM best-fit model, with binned data given by the white points.

Using the cosmological parameters cited above, we can then infer a few important timescales in the history of the universe. First of all, integrating (2.194) to $a = 1$ gives the age of the universe as

$$t_0 \approx 13.8 \,\text{Gyrs}\,. \tag{2.195}$$

Evaluating it at $a_{\text{eq}} = 2.9 \times 10^{-4}$ and $a_{m\Lambda} = 0.77$ puts the time of matter–radiation and matter–dark energy equality at

$$t_{\text{eq}} \approx 50\,000 \,\text{yrs}\,, \tag{2.196}$$

$$t_{m\Lambda} \approx 10.2 \,\text{Gyrs}\,. \tag{2.197}$$

Why dark energy came to dominate so close to the present time is called the **coincidence problem**. As we see from (2.193), it is related to the question why the amounts of matter and dark energy are almost equal today, while matter dominated in the past and dark energy will dominate in the future.

The above timescales show that most of the time is spent at low redshift in the matter-dominated era. However, as we will see in the next chapter, a lot of the action was in the hot radiation-dominated beginning of the universe. This is because the rate of interaction goes up with temperature, so that a lot of things can happen in a short amount of time.

Exercise 2.5 At early times, the universe is a mixture of matter and radiation. Using the measured values of the cosmological parameters, show that

$$t = 130\,000\,\text{yrs} \left[\frac{4}{3} + \frac{2}{3}(1+y)^{3/2} - 2(1+y)^{1/2} \right], \quad y \equiv \frac{1+z_{eq}}{1+z}. \quad (2.198)$$

Use this to determine the times of matter–radiation equality ($z_{eq} = 3400$) and recombination ($z_{rec} = 1100$).

At late times, the universe is a mixture of matter and dark energy. Show that

$$t = 11.5\,\text{Gyrs} \times \sinh^{-1}(y^{3/2}), \quad y \equiv \frac{1+z_{m\Lambda}}{1+z}. \quad (2.199)$$

Use this to determine the time of matter–dark energy equality ($z_{m\Lambda} = 0.3$) and the age of the universe ($z_0 = 0$).

2.5 Summary

In this chapter, we have studied the geometry and dynamics of the universe, as well as the propagation of particles within it. We showed that a homogeneous and isotropic spacetime is described by the Robertson–Walker metric

$$ds^2 = -c^2 dt^2 + a^2(t) \left[\frac{dr^2}{1 - kr^2/R_0^2} + r^2 \left(d\theta^2 + \sin^2 \theta \, d\phi^2 \right) \right], \quad (2.200)$$

where $k = 0, +1, -1$ for flat, spherical and hyperbolic spatial slices with curvature radius R_0. The light of distant galaxies is stretched by the expansion of the universe, with the fractional shift in the wavelength given by

$$z \equiv \frac{\lambda_{obs} - \lambda_{em}}{\lambda_{em}} = \frac{a(t_{obs})}{a(t_{em})} - 1. \quad (2.201)$$

The physical velocities of galaxies receive contributions both from the expansion and from their peculiar motions, $\mathbf{v}_{phys} = H\mathbf{r}_{phys} + \mathbf{v}_{pec}$, where $H \equiv \dot{a}/a$ is the Hubble parameter.

The evolution of the scale factor $a(t)$ is determined by the Friedmann equation

$$\left(\frac{\dot{a}}{a} \right)^2 = \frac{8\pi G}{3}\rho - \frac{kc^2}{a^2 R_0^2} + \frac{\Lambda c^2}{3}, \quad (2.202)$$

where ρ is the total energy density of the universe. The cosmological constant Λ is often interpreted as a vacuum energy and absorbed into the density parameter. Each component of the density satisfies the continuity equation

$$\frac{\mathrm{d}\rho}{\mathrm{d}\ln a} + 3(1+w)\rho = 0 \,, \tag{2.203}$$

with $w = 0$ for non-relativistic matter (baryons and dark matter), $w = 1/3$ for relativistic matter (radiation like photons and neutrinos), and $w = -1$ for dark energy. The densities therefore scale as $\rho \propto a^{-3(1+w)}$ and the Friedmann equation can be written as

$$H^2 = H_0^2 \left[\Omega_r a^{-4} + \Omega_m a^{-3} + \Omega_\Lambda + (1 - \Omega_0) a^{-2} \right], \tag{2.204}$$

where H_0 is the Hubble constant and Ω_i are the dimensionless density parameters today. Observations have shown that $\Omega_r \approx 9.0 \times 10^{-5}$, $\Omega_m \approx 0.32$, $\Omega_\Lambda \approx 0.68$ and $\Omega_0 \equiv \sum_i \Omega_i \approx 1$.

Further Reading

The material in this chapter is treated in every cosmology textbook. My derivation of the FRW metric and the energy-momentum tensor is based on a similar analysis in Weinberg's book [17]. My treatment of exact solutions to the Friedmann equations was inspired by Tong's lecture notes [18] and Ryden's book [19]. A nice review of the various distance measures used in cosmology is [20]. Common misconceptions about the expansion of the universe are treated carefully in [21].

The cosmological constant problem is subtle and sometimes not described very accurately. The classic review on the cosmological constant problem is by Weinberg [22]. Nice descriptions can also be found in the article by Polchinski [23], the review by Carroll [24] and the lecture notes by Bousso [25], Burgess [26] and Padilla [27].

The discovery of the expanding universe has an interesting history. It is fascinating to read the original papers through the lens of our modern understanding of cosmology. On the website for this book (www.cambridge.org/baumann), I describe some of the most important developments. This is not meant to be a rigorous history of cosmology (for this see e.g. [28, 29]), but only a pointer to some classic papers.

Students who haven't had a course in general relativity before may have found parts of this chapter challenging. Fortunately, there is an abundance of good resources to learn GR. Appendix A of this book contains a brief introduction to the main ideas. More details can be found in many excellent textbooks (e.g. [30–32]) and lecture notes (e.g. [33–35]).

Problems

2.1 Robertson–Walker metric

In this problem, you will be guided through an alternative derivation of the Robertson–Walker metric.

1. Explain why the most general metric for a homogeneous and isotropic universe is

$$ds^2 = -dt^2 + a^2(t)\gamma_{ij}(\mathbf{x})\,dx^i dx^j \,,$$

   where we have set $c \equiv 1$. In particular, explain why $g_{00} = -1$ and $g_{0i} = 0$.

2. Assume *isotropy* of the universe about a *fixed* point $r = 0$. Show that the most general spatial metric takes the form

$$d\ell^2 \equiv \gamma_{ij}dx^i dx^j = e^{2\alpha(r)}dr^2 + r^2 d\Omega^2 \,,$$

   where $d\Omega^2 \equiv d\theta^2 + \sin^2\theta\,d\phi^2$. Show that the scalar curvature associated with this metric is

$$R_{(3)} = \frac{2}{r^2}\left[1 - \frac{d}{dr}\left(re^{-2\alpha(r)}\right)\right].$$

   Warning: this part is a bit tedious.

3. If the universe is *homogeneous* then the scalar curvature must be a constant. Show that this implies

$$e^{2\alpha(r)} = \frac{1}{1 - Kr^2 - br^{-1}}\,,$$

   where K and b are constants. Show that requiring the geometry to be locally flat at $r = 0$ implies $b = 0$ and hence

$$ds^2 = -dt^2 + a^2(t)\left[\frac{dr^2}{1 - Kr^2} + r^2 d\Omega^2\right].$$

   Let $K \equiv k/R_0^2$, with $k = 0, +1, -1$. Interpret the three different values of k and explain the physical meaning of the scale R_0.

4. Consider the transformations $\rho = a(t)r$ and $T = t + \frac{1}{2}\dot{a}ar^2$. The new coordinates (T, ρ, θ, ϕ) are called *Fermi coordinates*. Show that for small ρ, the line element in the new coordinates takes the form

$$ds^2 \approx -\left(1 - \frac{\ddot{a}}{a}\rho^2\right)dT^2 + \left(1 + \frac{k}{a^2 R_0^2}\rho^2 + \frac{\dot{a}^2}{a^2}\rho^2\right)d\rho^2 + \rho^2 d\Omega^2 \,.$$

   What is the effective Newtonian potential?

2.2 Geodesics from a Lagrangian

In Appendix A, we derive the geodesic equation from the relativistic action of a point particle. In this problem, you will discover a simpler way to obtain the same result.

1. Consider the "Lagrangian"

$$\mathcal{L} \equiv -g_{\mu\nu}\dot{x}^\mu \dot{x}^\nu \,,$$

where $\dot{x}^\mu \equiv dx^\mu/d\lambda$, for a general parameter λ. Show that the Euler–Lagrange equation

$$\frac{d}{d\lambda}\left(\frac{\partial\mathcal{L}}{\partial\dot{x}^\mu}\right) = \frac{\partial\mathcal{L}}{\partial x^\mu}$$

leads to the geodesic equation.

2. If $\mathcal{L}$ has no explicit dependence on λ, then $\partial\mathcal{L}/\partial\lambda = 0$. Show that this implies that the "Hamiltonian" is a constant along the geodesics:

$$\mathcal{H} \equiv \mathcal{L} - \frac{\partial\mathcal{L}}{\partial\dot{x}^\mu}\dot{x}^\mu = g_{\mu\nu}\dot{x}^\mu \dot{x}^\nu \,.$$

For massive particles, we can set λ equal to the proper time τ, and the constraint becomes $g_{\mu\nu}\dot{x}^\mu \dot{x}^\nu = -1$. A nice feature of the Lagrangian method is that it also applies to massless particles, in which case we must have $g_{\mu\nu}\dot{x}^\mu \dot{x}^\nu = 0$.

2.3 Christoffel symbols from a Lagrangian

In this problem, you will learn a neat trick to compute Christoffel symbols using a Lagrangian method. Write down the Lagrangian $\mathcal{L} \equiv -g_{\mu\nu}\dot{x}^\mu \dot{x}^\nu$ for the flat FRW metric

$$ds^2 = -dt^2 + a^2(t)\delta_{ij}dx^i dx^j \,.$$

Applying the Euler–Lagrange equation, find an equation for $d^2t/d\lambda^2$. Comparing this to the geodesic equation, read off $\Gamma^0_{\alpha\beta}$. Similarly calculate all other Christoffel symbols.

2.4 Geodesics in de Sitter space

The line element of de Sitter space (in static patch coordinates) is

$$ds^2 = -\left(1 - \frac{r^2}{R^2}\right)dt^2 + \left(1 - \frac{r^2}{R^2}\right)^{-1}dr^2 + r^2(d\theta^2 + \sin^2\theta\, d\phi^2)\,,$$

where $R^2 \equiv 3/\Lambda$. Use the Lagrangian method introduced in Problem 2.2 to study the motion of a massive test particle in this spacetime.

1. Derive the conserved energy E and angular momentum L of the particle.

2. Show that the radial motion is described by the potential

$$V(r) = 1 - \frac{L^2}{R^2} + \frac{L^2}{r^2} - \frac{r^2}{R^2}\,.$$

Sketch this potential for $L = 0$ and $L = 0.5R$.

3. The particle is released with a small radial velocity near $r = 0$. Show that its trajectory is

$$r(\tau) = R\sqrt{E^2 - 1} \, \sinh(\tau/R),$$

where τ is the proper time along the geodesic. We see that the particle reaches the horizon at $r = R$ in a finite amount of proper time $\Delta\tau$. Show that the corresponding time Δt measured by an observer at $r = 0$ is infinite.

2.5 Distances

In this problem, you will study some unusual properties of the luminosity distance d_L and the angular diameter distance d_A.

1. Consider the relationship between d_L and the proper separation ℓ_0 in a flat matter-dominated universe. Show that $\ell_0 = 3t_0(1 - 1/\sqrt{1 + z})$ and hence

$$d_L = \ell_0 \left(1 - \frac{\ell_0}{3t_0}\right)^{-2}.$$

What happens as $\ell_0 \to 3t_0$? How do you interpret this result?

2. Now consider objects of a fixed physical size D in a flat matter-dominated universe. Show that the angular diameter of these objects at first decreases with distance, but then becomes larger beyond a critical distance. What is the redshift corresponding to this critical distance? Why does the observed angular size $\delta\theta$ diverge as $z \to 0$ and $z \to \infty$?

2.6 Flatland cosmology

Suppose you are a flatlander living in a universe with only two spatial and one time dimensions. A spatially homogeneous and isotropic $2+1$ dimensional spacetime can be described by the following metric:

$$ds^2 = -dt^2 + a^2(t)(dx^2 + dy^2),$$

where $a(t)$ is the scale factor.

1. Calculate the Christoffel symbols for the above metric.

2. Show that the nonzero components of the Ricci tensor are

$$R_{tt} = -2\ddot{a}/a, \quad R_{xx} = R_{yy} = \dot{a}^2 + a\ddot{a},$$

where the overdots denote derivatives with respect to the time t.

3. Assume that the universe is filled with a perfect fluid whose energy-momentum tensor is

$$T_{\mu\nu} = \begin{pmatrix} \rho & 0 & 0 \\ 0 & a^2 P & 0 \\ 0 & 0 & a^2 P \end{pmatrix},$$

where $\rho(t)$ and $P(t)$ are the density and pressure of the fluid. Show that the Einstein equations imply the following equations for the scale factor:

$$\left(\frac{\dot{a}}{a}\right)^2 = 8\pi G\rho, \quad \frac{\ddot{a}}{a} = -8\pi G P.$$

4. Consider a fluid with a constant equation of state w, such that $P = w\rho$. Using the conservation equation $\nabla_\mu T^\mu{}_\nu = 0$, show that ρ scales as

$$\rho \propto a^{-n},$$

and find n as a function of w. Give an interpretation of the scaling for a pressureless fluid.

5. Show that the scale factor evolves as

$$a(t) \propto t^q,$$

and find q as a function of w.

2.7 Friedmann universes

Consider a universe with a cosmological constant, spatial curvature and a perfect fluid with density ρ and constant equation of state $w \geq 0$. The curvature radius today is R_0. Use units where $8\pi G = c \equiv 1$.

1. Show that the Friedmann equation can be written as the equation of motion of a particle moving in one dimension with vanishing total energy and potential

$$V(a) = -\frac{\rho_0}{6}\frac{1}{a^{(1+3w)}} + \frac{K}{2} - \frac{\Lambda}{6}a^2,$$

where $K \equiv k/R_0^2$.

Sketch $V(a)$ for the following cases: (i) $k = 0$, $\Lambda < 0$, (ii) $k = \pm 1$, $\Lambda = 0$, and (iii) $k = 0$, $\Lambda > 0$. Assuming that the universe "starts" with $da/dt > 0$ near $a = 0$, describe the evolution in each case. Where applicable, determine the maximal value of the scale factor.

2. Now consider the case $\Lambda = 0$ and $k = +1$.
   Show that the scale factor obeys the differential equation

$$a'' + \frac{1}{R_0^2}a = \frac{\rho_0}{6}(1 - 3w)a^{-3w},$$

where the primes denote derivatives with respect to conformal time η. You may assume (or show) that this equation has the following solution

$$a(\eta) = A\left[\sin\left(\frac{1+3w}{2}\frac{\eta}{R_0} + B\right)\right]^{2/(1+3w)},$$

where A and B are integration constants. On physical grounds, determine the constant A in terms of ρ_0, R_0 and w.

Defining $a(\eta = 0) \equiv 0$, give the solution for (i) pressureless matter ($w = 0$) and (ii) radiation ($w = \frac{1}{3}$). In each case, determine the time of the "Big Crunch."

Consider a photon leaving the origin at $\eta = 0$. For the case of pressureless matter, how many times can the photon circle the universe before the universe ends? How far does the photon get in the case of radiation?

2.8 Einstein's biggest blunder

Einstein introduced the cosmological constant into his field equations to avoid the conclusion that the universe is expanding. In this problem, you will see that this was misguided.

1. Show that for a perfect fluid with positive density and pressure there is no static solution to the Einstein equations.

2. Consider now a universe with pressureless matter and a cosmological constant. Show that it is possible to obtain a static solution—called the *Einstein static universe*—if

$$\Lambda = 4\pi G \rho_m .$$

What is the spatial curvature of this solution?

3. Show that the Einstein static universe is unstable to small perturbations.

   Hint: Consider a small perturbation around the static solution

   $$\rho_m(t) = \rho_{m,0}[1 + \delta(t)] ,$$
   $$a(t) = 1 + \epsilon(t) ,$$

   with $|\delta| \ll 1$ and $|\epsilon| \ll 1$. Show that the perturbations δ and ϵ are related to each other and that they grow exponentially with time.

2.9 The accelerating universe

Consider a universe with spatial curvature, pressureless matter and a positive cosmological constant.

1. The *deceleration parameter* is defined as

$$q(t) \equiv -\frac{\ddot{a}a}{\dot{a}^2} .$$

Plot $q(t)$ as a function of $a(t)$ for our universe, with $\Omega_m = 0.3$ and $\Omega_\Lambda = 0.7$. Show that today

$$q_0 = \frac{\Omega_m}{2} - \Omega_\Lambda .$$

What is the value for our universe?

2. The *jerk* is defined as

$$J(t) \equiv \frac{\dddot{a}\, a^2}{\dot{a}^3} .$$

Show that

$$J(t) = 1 + \frac{1}{a^2 H^2} \frac{kc^2}{R_0^2} ,$$

so that the jerk is one for a flat universe.

Now specialize to the case of a flat universe.

3. Find the redshift at which the energy density of the universe becomes dominated by the cosmological constant. What is the value for our universe?

4. Find the redshift at which the universe begins accelerating.

5. Show that the solution for the scale factor is of the form

$$a(t) = A(\sinh \alpha t)^{2/3},$$

where A and α should be expressed in terms of Ω_m and H_0.

- Verify that your solution has the correct limits at early and late times.
- Compute the deceleration parameter and jerk of the solution.
- Use the solution to determine the age of the universe.

6. Compute the luminosity distance as a function of the redshift. Plot $d_L(z)$ for $\Omega_m = 1.0$ and $\Omega_m = 0.3$; compare your result to Fig. 2.6. Consider supernova measurements with a minimum redshift of 0.5. How accurate do the measurements of d_L have to be in order to distinguish the two models?

2.10 Phantom dark energy

Dark energy is most likely a cosmological constant with equation of state $w = -1$. Radical alternatives have nevertheless been considered. In this problem, you will study the dynamics of "phantom dark energy," with $w < -1$.

1. Consider a flat universe with matter (m) and phantom dark energy (X). Show that the energy density of the dark energy increases with time. If the scale factor today is $a_0 \equiv 1$, show that in the future

$$\Omega_X(a) \equiv \frac{\rho_X(a)}{\rho_{\rm crit}(a)} = \left(1 + \frac{\Omega_{m,0}}{\Omega_{X,0}} a^{3w_X}\right)^{-1}.$$

If $\Omega_{X,0} = 0.75$ and $w_X = -2$, at what scale factor is 99.9% of the energy density in dark energy?

2. If the dark energy density dominates the matter density at a time t_*, show that $a \to \infty$ in a finite time Δt. This divergence of the scale factor has been called a "Big Rip." Find Δt in terms of w_X and the Hubble parameter H_* at the time t_*.

3. What would happen to the observed wavelengths of CMB photons as the Big Rip is approached?

3   The Hot Big Bang

The early universe was hot and dense. At temperatures above 10^4 K, stable atoms didn't exist because the average energy of particles was larger than the binding energies of atoms. The universe was therefore a plasma of free electrons and nuclei, with high-energy photons scattering between them. At higher temperatures, above 10^9 K, even the nuclei had dissolved into their constituent protons and neutrons. The rate of interactions in the primordial plasma was very high and the universe was in a state of thermal equilibrium. This equilibrium state provides the initial condition for the hot Big Bang. It is a very simple state in which the abundances of all particle species are determined by the temperature of the universe.

Fortunately, this simplicity of the primordial universe did not persist. As the temperature decreased, the rate of some particle interactions eventually became smaller than the expansion rate. These particles dropped out of thermal equilibrium and decoupled from the thermal bath. These departures from thermal equilibrium are what make life interesting. In particular, non-equilibrium dynamics is required in order for massive particles to maintain significant abundances. Deviations from equilibrium are also crucial for understanding the origin of the CMB and the formation of the light chemical elements.

Some key events in the thermal history of the universe are listed in Table 3.1. Our story in this chapter will begin one second after the Big Bang. Almost all of the energy density of the universe was in the form of free electrons (e^-), positrons (e^+), photons (γ) and neutrinos (ν). A small, but important, amount were protons (p^+) and neutrons (n).[1] The photons were trapped by the large density of free electrons.[2] Neutrinos interacted with the rest of the primordial plasma through the weak nuclear interaction. The rate of these interactions soon dropped below the expansion rate of the universe and the neutrinos decoupled. This created the **cosmic neutrino background** ($C\nu B$), which still fills the universe today, but has such low energy that it is hard to detect directly. Electrons and positrons annihilated into photons shortly after neutrino decoupling. The energy of the electrons and positrons got transferred to the photons, but not to the neutrinos. This resulted in a slight increase in the photon temperature relative to the neutrino temperature. Around three minutes later, **Big Bang nucleosynthesis** (BBN) took place, which

---

[1] As we will see below, when the universe was one second old, the density of baryons was about the density of air! Of course, the energy density of the universe at that time was much higher, because it was dominated by radiation (mostly photons and neutrinos), and not matter.

[2] The mean free path of a photon was about the size of an atom.

Event	Time	Redshift	Temperature
Inflation	?	–	–
Baryogenesis	?	?	?
Dark matter freeze-out	?	?	?
EW phase transition	20 ps	10^{15}	100 GeV
QCD phase transition	20 μs	10^{12}	150 MeV
Neutrino decoupling	1 s	6×10^9	1 MeV
Electron–positron annihilation	6 s	2×10^9	500 keV
Big Bang nucleosynthesis	3 min	4×10^8	100 keV
Matter–radiation equality	50 kyr	3400	0.80 eV
Recombination	290–370 kyr	1090–1270	0.25–0.29 eV
Photon decoupling	370 kyr	1090	0.25 eV

Table 3.1 Key events in the history of the early universe

fused protons and neutrons into the light elements—mostly deuterium, helium and lithium. Heavier elements were formed much later, in the interior of stars. The next significant event, called **recombination**, occurred 370 000 years after the Big Bang. At this moment, the temperature became low enough for hydrogen atoms to form through the reaction $e^- + p^+ \rightarrow H + \gamma$. Before recombination, the strongest coupling between the photons and the rest of the plasma was through Thomson scattering, $e^- + \gamma \rightarrow e^- + \gamma$. The sharp drop in the free electron density after recombination meant that this process became very inefficient and the photons decoupled. They have since streamed freely through the universe and are today observed as the **cosmic microwave background** (CMB).

It is remarkable that this story of the hot Big Bang can now be told as a scientific fact. In this chapter, I will explain in more detail where this knowledge comes from and develop the precise theoretical framework for the thermal history of the universe.

$$* * *$$

Starting in this chapter, we will switch to so-called **natural units**, where the speed of light and the reduced Planck constant are set to unity

$$c = \hbar \equiv 1 \,. \tag{3.1}$$

This will reduce the clutter in many of our expressions. In these units, length and time have the same units, and are inverses of mass and energy. We will also introduce the reduced Planck mass

$$M_{\rm Pl} \equiv \sqrt{\frac{\hbar c}{8\pi G}} = 2.4 \times 10^{18} \text{ GeV} \,, \tag{3.2}$$

so that the Friedmann equation for a flat universe reads $H^2 = \rho/3M_{\rm Pl}^2$. Note that appearance of the Planck mass in the Friedmann equation is just for notational convenience and does not imply any relation to quantum gravity. Finally, we will often set Boltzmann's constant equal to unity, $k_{\rm B} \equiv 1$, so that temperature has units of energy. Useful conversions are

$$
\begin{aligned}
m_p \approx 1\,{\rm GeV} &= 1.60 \times 10^{-10}\,{\rm J} \\
&= 1.16 \times 10^{13}\,{\rm K} \\
&= 1.78 \times 10^{-27}\,{\rm kg} \\
&= (1.97 \times 10^{-16}\,{\rm m})^{-1} \\
&= (6.65 \times 10^{-25}\,{\rm s})^{-1},
\end{aligned}
\tag{3.3}
$$

where m_p is the proton mass. More about the concept of natural units can be found at the beginning of Appendix C.

3.1 Thermal Equilibrium

The blackbody spectrum of the CMB is strong observational evidence that the early universe was in a state of thermal equilibrium.[3] Moreover, on theoretical grounds, we expect the interactions of the Standard Model to have established thermal equilibrium at temperatures above 100 GeV. In this section, I will describe this initial state of the hot Big Bang and its subsequent evolution using the methods of thermodynamics and statistical mechanics, suitably generalized to apply to an expanding universe.

3.1.1 Some Statistical Mechanics

The early universe was a hot gas of weakly interacting particles. It is impractical to describe this gas by the positions and velocities of each particle. Instead, we will use a coarse-grained description of the gas using the principles of statistical mechanics. In other words, rather than following the evolution of each individual particle, we will characterize the properties of the gas statistically. In this section, I will give a lightning introduction to the relevant concepts of statistical mechanics and equilibrium thermodynamics. Further details can be found in your favorite textbooks on these subjects.

Distribution functions

A key concept in statistical mechanics is the probability that a particle chosen at random has a momentum $\mathbf{p}$. In general, this (probability) distribution function,

---

[3] Strictly speaking, the universe can never truly be in equilibrium since the FRW spacetime doesn't possess a time-like Killing vector. But this is physics and not mathematics: if the expansion is slow enough, particles have enough time to reach a state of approximate local equilibrium.

$f(\mathbf{p}, t)$, can be very complicated.[4] However, if we wait long enough (relative to the typical interaction timescale), then the system will reach *equilibrium* and is characterized by a time-independent distribution function. At this point, the gas has reached a state of maximum entropy in which the distribution function is given by either the **Fermi–Dirac distribution** (for fermions) or the **Bose–Einstein distribution** (for bosons)

$$\boxed{f(p, T) = \frac{1}{e^{(E(p)-\mu)/T} \pm 1}} \,, \tag{3.4}$$

where the $+$ sign is for fermions and the $-$ sign for bosons. A derivation of these equilibrium distribution functions can be found in any textbook on statistical mechanics. The function in (3.4) has two parameters: the temperature, T, and the chemical potential, μ. The latter describes the response of a system to a change in the particle number and can be positive or negative (see Section 3.1.5). The chemical potential may be temperature dependent, and since the temperature changes in an expanding universe, even the equilibrium distribution functions depend implicitly on time.

Density of states

To relate this microscopic description of the gas to its macroscopic properties, we must sum over all possible momentum states of the particles weighted by their probabilities. For example, the number density of particles in the gas is

$$n = \sum_{\mathbf{p}} f(p, T) \,. \tag{3.5}$$

To define this sum over states as an integral over the continuous variable $\mathbf{p}$, requires the **density of states**. It is easiest to derive this density of states by considering the gas as a quantum system. Recall that, in quantum mechanics, the momentum eigenstates of a particle in a box of side length L have a discrete spectrum. Solving the Schrödinger equation with periodic boundary conditions gives

$$\mathbf{p} = \frac{h}{L}(r_1 \hat{\mathbf{x}} + r_2 \hat{\mathbf{y}} + r_3 \hat{\mathbf{z}}) \,, \tag{3.6}$$

where $r_i = 0, \pm 1, \pm 2, \ldots$ and $h = 4.14 \times 10^{-15}$ eV s is Planck's constant. In momentum space, the states of the particle are therefore represented by a discrete set of points. The density of states in momentum space $\{\mathbf{p}\}$ then is $L^3/h^3 = V/h^3$, and the state density in phase space $\{\mathbf{x}, \mathbf{p}\}$ is $1/h^3$. If the particle has g internal degrees of freedom (for example, due to the intrinsic spin of elementary particles), then the density of states becomes

$$\frac{g}{h^3} = \frac{g}{(2\pi)^3} \,, \tag{3.7}$$

where in the second equality we have used natural units with $\hbar = h/(2\pi) \equiv 1$.

---

[4] In principle, the distribution function can also be a function of the position $\mathbf{x}$. However, in a homogeneous universe such a dependence is forbidden by translation invariance. Moreover, isotropy requires that the momentum dependence is only in terms of the magnitude of the momentum $p \equiv |\mathbf{p}|$ (or the energy E).

Densities and pressure

Weighting each state by its probability distribution, and integrating over momentum, we obtain the number density of particles

$$n(T) = \frac{g}{(2\pi)^3} \int d^3p\, f(p, T)\,. \tag{3.8}$$

Moreover, the energy density and pressure of the gas are then given by the following integrals

$$\rho(T) = \frac{g}{(2\pi)^3} \int d^3p\, f(p, T)\, E(p)\,, \tag{3.9}$$

$$P(T) = \frac{g}{(2\pi)^3} \int d^3p\, f(p, T)\, \frac{p^2}{3E(p)}\,, \tag{3.10}$$

where $E(p) = \sqrt{m^2 + p^2}$, if we can ignore the interaction energies between the particles.[5] The expression for the energy density should be fairly intuitive. Each momentum eigenstate is simply weighted by its energy. The origin of the factor of $p^2/3E$ in the pressure, on the other hand, requires a bit more explanation. Recall that pressure is defined as force per unit area, or momentum change per unit time per unit area. Let the momentum in the x-direction be p_x per particle. This results in a change in momentum of $2|p_x|$ if a particle hits a perpendicular area element dA. The volume swept out in unit time is $|v_x|dA = |p_x|dA/E$, so we want the average of $2p_x^2/E$ over the distribution for particles moving in the right direction ($p_x > 0$), which for an isotropic distribution is $\frac{1}{2} \times 2\langle p_x^2 \rangle/E = p^2/3E$.

Each particle species i (with possibly distinct m_i, μ_i, T_i) has its own distribution function f_i and hence its own densities and pressure, n_i, ρ_i, P_i. Species that are in thermal equilibrium share a common temperature, $T_i = T$. Their densities and pressures can then only differ because of differences in their masses and chemical potentials.

At early times, the chemical potentials of all particles are much smaller than the temperature, $|\mu_i| \ll T$, and can hence be neglected. For electrons and protons this is a provable fact (see Problem 3.1), for photons it holds by definition, and for neutrinos it is likely true (see Problem 3.3), but not proven. We will drop the chemical potential from our discussion for now, but return to it in Section 3.1.5 and thereafter.

3.1.2 The Primordial Plasma

Let us now use the results from the previous section to describe the state of the early universe in thermal equilibrium. Concretely, we will relate the densities and pressures of the different species in the primordial plasma to the overall temperature of the universe.

[5] Of course, these interactions are important to establish equilibrium, so they cannot literally be zero.

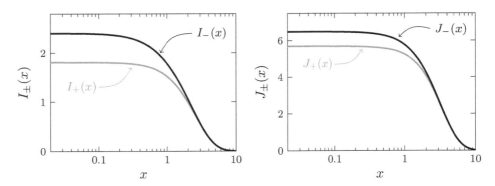

Fig. 3.1   Numerical evaluation of the functions $I_\pm(x)$ and $J_\pm(x)$ defined in (3.13) and (3.14).

Setting the chemical potential to zero, we get

$$n = \frac{g}{2\pi^2} \int_0^\infty dp \, \frac{p^2}{\exp\left[\sqrt{p^2 + m^2}/T\right] \pm 1} \,, \tag{3.11}$$

$$\rho = \frac{g}{2\pi^2} \int_0^\infty dp \, \frac{p^2 \sqrt{p^2 + m^2}}{\exp\left[\sqrt{p^2 + m^2}/T\right] \pm 1} \,. \tag{3.12}$$

Defining the dimensionless variables $x \equiv m/T$ and $\xi \equiv p/T$, this can be written as

$$n = \frac{g}{2\pi^2} T^3 \, I_\pm(x)\,, \qquad I_\pm(x) \equiv \int_0^\infty d\xi \, \frac{\xi^2}{\exp\left[\sqrt{\xi^2 + x^2}\right] \pm 1} \,, \tag{3.13}$$

$$\rho = \frac{g}{2\pi^2} T^4 \, J_\pm(x)\,, \qquad J_\pm(x) \equiv \int_0^\infty d\xi \, \frac{\xi^2 \sqrt{\xi^2 + x^2}}{\exp\left[\sqrt{\xi^2 + x^2}\right] \pm 1} \,. \tag{3.14}$$

In general, the functions $I_\pm(x)$ and $J_\pm(x)$ have to be evaluated numerically (see Fig. 3.1), but in the relativistic and non-relativistic limits, we can determine them analytically.

Relativistic limit

At temperatures much larger than the particle mass, we can take the limit $x \to 0$ and the integral in (3.13) reduces to

$$I_\pm(0) = \int_0^\infty d\xi \, \frac{\xi^2}{e^\xi \pm 1} \,. \tag{3.15}$$

The denominator can be written as a geometric series

$$\frac{1}{e^\xi \pm 1} = \frac{e^{-\xi}}{1 \pm e^{-\xi}} = \sum_{j=1}^\infty (\mp 1)^{j-1} e^{-j\xi} \,, \tag{3.16}$$

so that the integral in (3.15) becomes

$$I_\pm(0) = \sum_{j=1}^{\infty} (\mp 1)^{j-1} \int_0^\infty d\xi\, \xi^2 e^{-j\xi} = \sum_{j=1}^{\infty} (\mp 1)^{j-1} \frac{2}{j^3}. \tag{3.17}$$

For bosons, we get

$$I_-(0) = 2\left(1 + \frac{1}{2^3} + \frac{1}{3^3} + \frac{1}{4^3} + \cdots\right) \equiv 2\zeta(3), \tag{3.18}$$

where the Riemann zeta function is $\zeta(3) \approx 1.20205 \cdots$. For fermions, we instead have

$$\begin{aligned}
I_+(0) &= 2\left(1 - \frac{1}{2^3} + \frac{1}{3^3} - \frac{1}{4^3} + \cdots\right) \\
&= 2\left(1 + \frac{1}{2^3} + \frac{1}{3^3} + \frac{1}{4^3} + \cdots\right) - 4\left(\frac{1}{2^3} + \frac{1}{4^3} + \cdots\right) \\
&= 2\left(1 + \frac{1}{2^3} + \frac{1}{3^3} + \frac{1}{4^3} + \cdots\right) - \frac{4}{2^3}\left(1 + \frac{1}{2^3} + \frac{1}{3^4} + \cdots\right) \\
&= \left(1 - \frac{1}{4}\right) 2\zeta(3) \\
&= \frac{3}{4} I_-(0). \tag{3.19}
\end{aligned}$$

Alternatively, the proportionality between $I_+(0)$ and $I_-(0)$ can be found by noting that

$$\frac{1}{e^\xi + 1} = \frac{1}{e^\xi - 1} - \frac{2}{e^{2\xi} - 1}, \tag{3.20}$$

so that

$$I_+(0) = I_-(0) - 2 \times \left(\frac{1}{2}\right)^3 I_-(0) = \frac{3}{4} I_-(0). \tag{3.21}$$

Substituting (3.18) and (3.21) into (3.13), we get

$$n = \frac{\zeta(3)}{\pi^2} g T^3 \begin{cases} 1 & \text{bosons} \\ \frac{3}{4} & \text{fermions} \end{cases}. \tag{3.22}$$

A similar computation for the energy density gives

$$\rho = \frac{\pi^2}{30} g T^4 \begin{cases} 1 & \text{bosons} \\ \frac{7}{8} & \text{fermions} \end{cases}, \tag{3.23}$$

where we have used that $\zeta(4) = \pi^4/90$.

Exercise 3.1 Show that $J_-(0) = 6\zeta(4)$ and $J_+(0) = \frac{7}{8}J_-(0)$.

Using the observed temperature of the CMB, $T_0 \approx 2.73\,\mathrm{K}$, we find that the number density and energy density of relic photons today are

$$n_{\gamma,0} = \frac{2\zeta(3)}{\pi^2} T_0^3 \approx 410 \text{ photons cm}^{-3},\qquad(3.24)$$

$$\rho_{\gamma,0} = \frac{\pi^2}{15} T_0^4 \approx 4.6 \times 10^{-34}\,\mathrm{g\,cm}^{-3}.\qquad(3.25)$$

In terms of the critical density, the photon energy density is then found to be $\Omega_\gamma h^2 \approx 2.5 \times 10^{-5}$.

Finally, taking $p = E$ in (3.10), we get

$$P = \frac{1}{3}\rho,\qquad(3.26)$$

as expected for a gas of relativistic particles ("radiation").

Non-relativistic limit

At temperatures below the particle mass, we take the limit $x \gg 1$ and the integral in (3.13) is the same for bosons and fermions

$$I_\pm(x) \approx \int_0^\infty \mathrm{d}\xi \, \frac{\xi^2}{e^{\sqrt{\xi^2+x^2}}}.\qquad(3.27)$$

Most of the contribution to the integral comes from $\xi \ll x$ and we can Taylor expand the square root in the exponential to lowest order in ξ,

$$I_\pm(x) \approx \int_0^\infty \mathrm{d}\xi \, \frac{\xi^2}{e^{x+\xi^2/(2x)}} = e^{-x} \int_0^\infty \mathrm{d}\xi \, \xi^2 e^{-\xi^2/(2x)}$$

$$= (2x)^{3/2} e^{-x} \int_0^\infty \mathrm{d}\xi \, \xi^2 e^{-\xi^2}.\qquad(3.28)$$

Performing the Gaussian integral, we then get

$$I_\pm(x) = \sqrt{\frac{\pi}{2}}\, x^{3/2} e^{-x},\qquad(3.29)$$

and, using (3.18), we find

$$\frac{I_\pm(x)}{I_-(0)} \approx 0.5\, x^{3/2} e^{-x} \ll 1.\qquad(3.30)$$

As expected, massive particles are exponentially rare at low temperatures.

Substituting (3.29) into (3.13), we can write the density of the non-relativistic gas as a function of its temperature

$$n = g \left(\frac{mT}{2\pi} \right)^{3/2} e^{-m/T} . \tag{3.31}$$

To determine the energy density in the non-relativistic limit, we write $E(p) = \sqrt{m^2 + p^2} \approx m + p^2/2m$. We then find

$$\rho \approx mn + \frac{3}{2} nT , \tag{3.32}$$

where the leading term is simply equal to the mass density (recall that $c \equiv 1$).

Finally, from (3.10), it is easy to show that the pressure of a non-relativistic gas of particles is

$$P = nT , \tag{3.33}$$

which is nothing but the ideal gas law, $PV = Nk_{\mathrm{B}}T$ (for $k_{\mathrm{B}} \equiv 1$). Since $T \ll m$, we have $P \ll \rho$, so that the gas acts like a pressureless fluid ("matter").

Exercise 3.2 Derive (3.32) and (3.33).

By comparing the relativistic limit ($T \gg m$) and the non-relativistic limit ($T \ll m$), we see that the number density, energy density, and pressure of a particle species fall exponentially (are "Boltzmann suppressed") as the temperature drops below the mass of the particles. This can be interpreted as the annihilation of particles and antiparticles. At higher energies these annihilations also occur, but they are balanced by particle–antiparticle pair production. At low temperatures, the thermal energies of the particles aren't sufficient for pair production.

Relativistic species

The early universe was a collection of different species and the total energy density ρ is the sum over all contributions

$$\rho = \sum_i \frac{g_i}{2\pi^2} T_i^4 J_\pm(x_i) , \tag{3.34}$$

where we have allowed for the possibility that the different species have different temperatures T_i. For the Standard Model, this complication is only relevant for neutrinos after electron–positron annihilation (see Section 3.1.4). It is common to write the density in terms of the "temperature of the universe" T (typically chosen to be the photon temperature T_γ),

$$\rho = \frac{\pi^2}{30} g_*(T) T^4 , \tag{3.35}$$

where we have defined the "effective number of relativistic degrees of freedom" at the temperature T as

$$g_*(T) \equiv \sum_i g_i \left(\frac{T_i}{T}\right)^4 \frac{J_\pm(x_i)}{J_-(0)}.$$ (3.36)

Since the energy density of relativistic species is much greater than that of non-relativistic species, it often suffices to include only the relativistic species in (3.36). Moreover, for $T_i \gg m_i$, we have $J_\pm(x_i \ll 1) \approx$ const and (3.36) reduces to

$$g_*(T) \equiv \sum_{i=b} g_i \left(\frac{T_i}{T}\right)^4 + \frac{7}{8} \sum_{i=f} g_i \left(\frac{T_i}{T}\right)^4.$$ (3.37)

When all particles are in equilibrium at a common temperature T, determining $g_*(T)$ is then simply a counting exercise.

Learning to count

At early times, $T \gtrsim 100\,\text{GeV}$, all particles of the Standard Model were relativistic (see Table 3.2). To determine the corresponding value of g_*, we need to sum over the internal degrees of freedom of each particle species.

Let us start with the gauge bosons, the force carriers of the Standard Model. Photons—the mediators of the electromagnetic force—have $g_\gamma = 2$ corresponding to two polarizations transverse to the direction of propagation. This is a general feature, any massless particle with spin has exactly two polarization states. A massive particle can have additional longitudinal polarizations, i.e. polarizations along the direction of propagation. In total, a massive particle of spin s has $g = 2s+1$ polarization states. For the massive spin-1 gauge bosons associated with the weak nuclear force, we therefore have $g_{W^\pm, Z} = 3$ and hence a total of $3 \times 3 = 9$ internal degrees of freedom. Gluons—the mediators of the strong nuclear force—are massless and therefore contribute $g_g = 2$ internal degrees of freedom, like the photons. There are 8 of them, corresponding to the 8 generators of the group $SU(3)$, so we get $8 \times 2 = 16$.

Next, we consider the fermions—the matter particles of the Standard Model. The charged leptons $(e^\pm, \mu^\pm, \tau^\pm)$ are massive spin-$\frac{1}{2}$ particles and therefore contribute two spin states each. Including a factor of 2 for the antiparticles, we have $3 \times 2 \times 2 = 12$. Similarly, each quark contributes two spin states. There are 6 flavors of quark (t, b, c, s, d, u) and each comes in 3 different colors. Including a factor of 2 for the antiparticles, we then have $6 \times 2 \times 3 \times 2 = 72$. Lastly, we must talk about neutrinos. Although neutrinos are massive spin-$\frac{1}{2}$ particles, they only contribute 1 internal degree of freedom. The explanation is somewhat involved and will be given in the box below.

Type		Mass	Spin	g
gauge bosons	γ	0		2
	$W^\pm$	80 GeV	1	3
	Z	91 GeV		
gluons	g_i	0	1	$8 \times 2 = 16$
Higgs boson	H	125 GeV	0	1
quarks	$t, \bar{t}$	173 GeV	$\frac{1}{2}$	$2 \times 3 \times 2 = 12$
	$b, \bar{b}$	4 GeV		
	$c, \bar{c}$	1 GeV		
	$s, \bar{s}$	100 MeV		
	$d, \bar{s}$	5 MeV		
	$u, \bar{u}$	2 MeV		
leptons	$\tau^\pm$	1777 MeV	$\frac{1}{2}$	$2 \times 2 = 4$
	$\mu^\pm$	106 MeV		
	$e^\pm$	511 keV		
	$\nu_\tau, \bar{\nu}_\tau$	< 0.6 eV	$\frac{1}{2}$	$2 \times 1 = 2$
	$\nu_\mu, \bar{\nu}_\mu$	< 0.6 eV		
	$\nu_e, \bar{\nu}_e$	< 0.6 eV		

Table 3.2 Particle content of the Standard Model

An aside on neutrinos For a long time, it was thought that neutrinos are massless and, in fact, the gauge symmetries of the Standard Model require them to be massless. Massless particles travel at the speed of light and their spin can be either aligned or anti-aligned with the direction of travel. These two options correspond to the particle having positive or negative *helicity*. Alternatively, we say that the particle is left-handed or right-handed. It is a striking fact that only left-handed neutrinos have been observed in nature and it was long believed that right-handed neutrinos simply do not exist. This would then explain why each neutrino only contributes 1 internal degree of freedom and not 2. However, things are a bit more complicated because we now know that neutrinos do, in fact, have a small mass. Theoretically, there are two different kinds of masses that neutrinos could have—a *Majorana mass* and a *Dirac mass*. Unfortunately, we don't know which option is realized in nature. In the case of a Majorana mass, the neutrino is its own antiparticle—like the photon is its own antiparticle. Instead of $1 + 1 = 2$ from each left-handed neutrino and right-handed antineutrino, we then get 2 spin states for each neutrino and no contribution from antineutrinos. The end result, of course, is the same; only the words justifying it are different. The situation

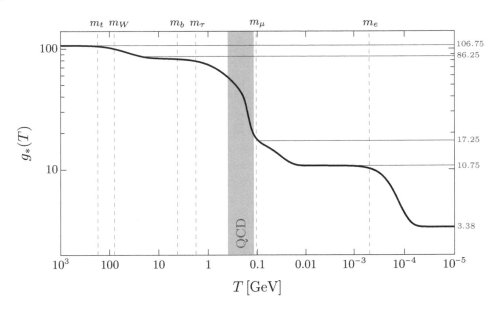

Fig. 3.2 Evolution of the effective number of relativistic degrees of freedom assuming the Standard Model particle content. The gray band indicates the QCD phase transition.

would be more complicated if the neutrino had a Dirac mass. In that case, we get 2 spin states for each neutrino and 2 for each antineutrino for a total of $2 + 2 = 4$, which is inconsistent with measurements from BBN. This means that either neutrinos are Majorana particles or half of the degrees of freedom of the Dirac neutrinos somehow decoupled in the very early universe and their energy density diluted, so that they made a negligible contribution by the time of BBN. Models of such Dirac neutrinos exist.

Adding up the internal degrees of freedom, we get

$$g_b = 28 \qquad \text{photons}\,(2),\ W^\pm \text{ and } Z\,(3 \times 3),\ \text{gluons}\,(8 \times 2),\ \text{and Higgs}\,(1)$$
$$g_f = 90 \qquad \text{quarks}\,(6 \times 12),\ \text{charged leptons}\,(3 \times 4),\ \text{and neutrinos}\,(3 \times 2)$$

and hence

$$g_* = g_b + \frac{7}{8} g_f = 106.75\,. \tag{3.38}$$

As the temperature drops, various particle species become non-relativistic and annihilate. This leads to the evolution of $g_*(T)$ shown in Fig. 3.2. To estimate g_* at a temperature T, we simply add up the contributions from all relativistic degrees of freedom (with $m \ll T$) and discard the rest.

Being the heaviest particles of the Standard Model, the top quarks annihilate first. At $T \sim \frac{1}{6} m_t \sim 30$ GeV,[6] the effective number of relativistic species is then reduced to $g_* = 106.75 - \frac{7}{8} \times 12 = 96.25$. The Higgs boson and the gauge bosons

---

[6] The transition from relativistic to non-relativistic behavior isn't instantaneous. About 80% of the particle–antiparticle annihilations take place in the interval $T = m \rightarrow \frac{1}{6}m$.

$W^{\pm}$, Z^0 disappear next. This happens roughly at the same time. At $T \sim 10$ GeV, we have $g_* = 96.25 - (1 + 3 \times 3) = 86.25$. Next, the bottom quarks annihilate ($g_* = 86.25 - \frac{7}{8} \times 12 = 75.75$), followed by the charm quarks and the tau leptons ($g_* = 75.75 - \frac{7}{8} \times (12 + 4) = 61.75$). Before the strange quarks have time to annihilate, something else happens: matter undergoes the QCD phase transition. At $T \sim 150$ MeV, the quarks combine into baryons (protons, neutrons, ...) and mesons (pions, ...). Although there are many different species of baryons and mesons, all except the pions ($\pi^{\pm}, \pi^0$) are non-relativistic below the temperature of the QCD phase transition and are therefore Boltzmann suppressed. Thus, the only particle species left in large numbers are pions, electrons, muons, neutrinos and photons. The three types of pions are spin-0 bosons, which carry a total of $g = 3$ internal degrees of freedom. We therefore get $g_* = 2 + 3 + \frac{7}{8} \times (4 + 4 + 6) = 17.25$. Soon after the QCD phase transition, the muons and pions annihilate, leading to $g_* = 10.75$. Next, electrons and positrons will annihilate. However, to understand this process we first need to talk about entropy.

3.1.3 Entropy and Expansion History

To describe the evolution of the universe it is useful to track a conserved quantity. As we will see, in cosmology, **entropy** is more informative than energy, because it is conserved in equilibrium.

Conservation of entropy

In statistical mechanics, a precise definition of entropy can be given in terms of the microstates of the system. Here, we will instead determine the entropy of the primordial plasma from the first law of thermodynamics.

The first law states that the change in the entropy (S) of a system is related to changes in its internal energy (U) and volume (V) as

$$T \mathrm{d}S = \mathrm{d}U + P \mathrm{d}V \,, \tag{3.39}$$

where we have assumed that any chemical potentials are small. Defining the **entropy density** as $s \equiv S/V$, we can write

$$\begin{aligned} T \, \mathrm{d}(sV) &= \mathrm{d}(\rho V) + P \, \mathrm{d}V \\ Ts \, \mathrm{d}V + TV \, \mathrm{d}s &= \rho \, \mathrm{d}V + V \, \mathrm{d}\rho + P \, \mathrm{d}V \,. \end{aligned} \tag{3.40}$$

Since s and ρ depend only on the temperature T, and not on the volume V, this implies

$$(Ts - \rho - P) \, \mathrm{d}V + V \left(T \frac{\mathrm{d}s}{\mathrm{d}T} - \frac{\mathrm{d}\rho}{\mathrm{d}T} \right) \mathrm{d}T = 0 \,. \tag{3.41}$$

In order for this to be satisfied for arbitrary variations dV and dT, the two brackets have to vanish separately: The vanishing of the first bracket implies that the entropy density can be written as

$$s = \frac{\rho + P}{T}, \tag{3.42}$$

while the vanishing of the second bracket enforces that

$$\frac{ds}{dT} = \frac{1}{T}\frac{d\rho}{dT}. \tag{3.43}$$

Using the continuity equation, $d\rho/dt = -3H(\rho + P) = -3H\,Ts$, the last equation can also be written in the following instructive form

$$\frac{d(sa^3)}{dt} = 0. \tag{3.44}$$

This means that the total entropy is conserved in equilibrium and that the entropy density evolves as $s \propto a^{-3}$. This conservation law will be very useful for describing the expansion history of the universe.

Exercise 3.3 Including a nonzero chemical potential, the first law of thermodynamics becomes

$$T dS = dU + P\,dV - \mu\,dN. \tag{3.45}$$

Show that the entropy density is

$$s = \frac{\rho + P - \mu n}{T}, \tag{3.46}$$

and evolves as

$$\frac{d(sa^3)}{dt} = -\frac{\mu}{T}\frac{d(na^3)}{dt}. \tag{3.47}$$

Entropy now is conserved either if the chemical potential is small, $\mu \ll T$, or if no particles are created or destroyed.

Relativistic species

Integrating (3.43), we get

$$s(T) = \int_0^T \frac{d\tilde{T}}{\tilde{T}}\frac{d\rho}{d\tilde{T}} = \frac{\rho(T)}{T} + \int_0^T \frac{\rho(\tilde{T})}{\tilde{T}^2}\,d\tilde{T}, \tag{3.48}$$

where we have integrated by parts and used that $\rho/T \to 0$ as $T \to 0$. Comparing this to (3.42), we see that the second term is P/T. The equation of state of the plasma can then be written as

$$w(T) \equiv \frac{P(T)}{\rho(T)} = \int_0^1 \frac{g_*(yT)}{g_*(T)}\,y^2 dy. \tag{3.49}$$

If all particles are relativistic, then $g_*(T) = $ const and we recover the equation of state of radiation, $w = 1/3$.

For a collection of different species, the total entropy density is

$$s = \sum_i \frac{\rho_i + P_i}{T_i} \equiv \frac{2\pi^2}{45} g_{*S}(T) \, T^3 \,, \qquad (3.50)$$

where we have defined $g_{*S}(T)$ as the "effective number of degrees of freedom in entropy." Away from mass thresholds, $T_i \gg m_i$, we have

$$g_{*S}(T) \approx \sum_{i=b} g_i \left(\frac{T_i}{T}\right)^3 + \frac{7}{8} \sum_{i=f} g_i \left(\frac{T_i}{T}\right)^3. \qquad (3.51)$$

When all species are in equilibrium at the same temperature, $T_i = T$, then g_{*S} is simply equal to g_*. In our universe, this is the case until $t \approx 1\,\mathrm{s}$. Since s is proportional to the number density of relativistic particles, it is sometimes useful to write $s \approx 1.8 \, g_{*S}(T) \, n_\gamma$, where n_γ is the number density of photons. In general, $g_{*S}(T)$ depends on temperature, so that s and n_γ cannot be used interchangeably. However, after electron–positron annihilation (see below), we have $g_{*S} = 3.94$ and hence $s \approx 7n_\gamma$.

Since $s \propto a^{-3}$, the number of particles in a comoving volume is proportional to the number density n_i divided by the entropy density

$$N_i \equiv \frac{n_i}{s} \,. \qquad (3.52)$$

If particles are neither produced nor destroyed, then $n_i \propto a^{-3}$ and N_i is a constant. An important example of a conserved species is the total baryon number after baryogenesis, $n_B/s \equiv (n_b - n_{\bar{b}})/s$. A related quantity is the baryon-to-photon ratio

$$\eta \equiv \frac{n_B}{n_\gamma} = 1.8 g_{*S} \frac{n_B}{s} \,. \qquad (3.53)$$

After electron–positron annihilation, $\eta \approx 7 \, n_B/s$ becomes a conserved quantity and is therefore a useful measure of the baryon content of the universe.

Another important consequence of entropy conservation is that

$$g_{*S}(T) \, T^3 \, a^3 = \text{const} \qquad \text{or} \qquad T \propto g_{*S}^{-1/3} a^{-1} \,. \qquad (3.54)$$

Away from particle mass thresholds, g_{*S} is approximately constant and the temperature has the expected scaling, $T \propto a^{-1}$. The factor of $g_{*S}^{-1/3}$ accounts for the fact that whenever a particle species becomes non-relativistic and disappears, its entropy is transferred to the other relativistic species still present in the thermal plasma, causing T to decrease slightly more slowly than a^{-1}. We will see an example of this phenomenon in the next section.

Expansion history

As we have seen in Chapter 2, the Friedmann equation relates the Hubble expansion rate to the energy density of the universe. At early times, the universe is dominated by relativistic species and curvature is negligible. Hence, the Friedmann equation reads

$$H^2 = \left(\frac{1}{a}\frac{da}{dt}\right)^2 = \frac{\rho}{3M_{\mathrm{Pl}}^2} \approx \frac{\pi^2}{90} g_* \frac{T^4}{M_{\mathrm{Pl}}^2}. \tag{3.55}$$

This is a single equation relating the expansion history of the universe to its temperature. We need one more equation to close the system for $a(t)$ and $T(t)$. Previously, we used the approximate equation of state for radiation, $w \approx 1/3$, which through the continuity equation determines $\rho(a)$. More precisely, we can substitute $T \propto g_{*S}^{-1/3}a^{-1}$ into (3.55). Away from mass thresholds, this reproduces the result for a radiation-dominated universe, $a \propto t^{1/2}$, but we see that there is a slight change in this scaling every time $g_{*S}(T)$ changes.

When $a \propto t^{1/2}$, we have $H = 1/(2t)$ and the Friedmann equation leads to

$$\frac{T}{1\,\mathrm{MeV}} \approx 1.5\, g_*^{-1/4} \left(\frac{1\,\mathrm{s}}{t}\right)^{1/2}. \tag{3.56}$$

It is a useful rule of thumb that the temperature of the universe 1 second after the Big Bang was about $1\,\mathrm{MeV}$ (or $10^4\,\mathrm{K}$), and evolved as $t^{-1/2}$ before that. As we will show next, one second after the Big Bang was, in fact, an interesting moment in the history of the universe.

3.1.4 Cosmic Neutrino Background

The most weakly interacting particles of the Standard Model are neutrinos. We therefore expect them to decouple first from the thermal plasma. In the following, I will show how this produces the cosmic neutrino background (CνB).

Neutrino decoupling

Neutrinos were coupled to the thermal bath through weak interaction processes like

$$\begin{aligned} \nu_e + \bar{\nu}_e &\leftrightarrow e^+ + e^-, \\ e^- + \bar{\nu}_e &\leftrightarrow e^- + \bar{\nu}_e. \end{aligned} \tag{3.57}$$

The interaction rate (per particle) is $\Gamma \equiv n\sigma|v|$, where n is the number density of the target particles, σ is the cross section, and v is the relative velocity (which in the relativistic limit can be approximated by the speed of light). By dimensional analysis, we infer that the cross section for weak scale interactions is $\sigma \approx G_F^2 T^2$, where $G_F \approx 1.2 \times 10^{-5}\,\mathrm{GeV}^{-2}$ is Fermi's constant. Taking the number density to be $n \approx T^3$, the interaction rate becomes

$$\Gamma = n\sigma|v| \approx G_F^2 T^5. \tag{3.58}$$

As the temperature decreases, the interaction rate drops much more rapidly than the Hubble rate $H \approx T^2/M_{\mathrm{Pl}}$:

$$\frac{\Gamma}{H} \approx \left(\frac{T}{1\,\mathrm{MeV}}\right)^3. \tag{3.59}$$

We conclude that neutrinos decouple around 1 MeV. (A more accurate computation gives a decoupling temperature of 0.8 MeV.) After decoupling, the neutrinos move freely along geodesics and preserve the *relativistic* Fermi–Dirac distribution (even after they become non-relativistic at later times). In Section 2.2.1, we showed that the physical momentum of free-streaming particles scales as $p \propto a^{-1}$. It is therefore convenient to define the time-independent combination $q \equiv ap$, so that the neutrino number density is

$$n_\nu \propto a^{-3} \int \mathrm{d}^3 q \, \frac{1}{\exp(q/aT_\nu) + 1}. \tag{3.60}$$

After decoupling, particle number conservation requires $n_\nu \propto a^{-3}$, which is only consistent with (3.60) if the neutrino temperature evolves as $T_\nu \propto a^{-1}$. As long as the photon temperature[7] T_γ scales in the same way, we still have $T_\nu = T_\gamma$. However, particle annihilations will cause a deviation from $T_\gamma \propto a^{-1}$ in the photon temperature.

Electron–positron annihilation

Shortly after the neutrinos decouple, the temperature drops below the electron mass, so that electrons and positrons can annihilate into photons:

$$e^- + e^+ \rightarrow \gamma + \gamma.$$

The energy density and entropy of the electrons and positrons are transferred to the photons, but not to the decoupled neutrinos. The photons are thus "heated" (the photon temperature decreases more slowly) relative to the neutrinos (see Fig. 3.3). To quantify this effect, we consider the change in the effective number of degrees of freedom in entropy. If we neglect neutrinos and other decoupled species,[8] then we have

$$g_{*S} = \begin{cases} 2 + \frac{7}{8} \times 4 = \frac{11}{2} & T \gtrsim m_e, \\ 2 & T < m_e. \end{cases} \tag{3.61}$$

The annihilation of electrons and positrons occurs on a timescale of $\alpha^2/m_e \sim 10^{-18}\,\mathrm{s}$ (where α is the fine-structure constant), which is much less than the age of the universe ($\sim 1\,\mathrm{s}$) at the time. This means that the $e^\pm$–γ plasma evolves quasi-adiabatically into the γ-only plasma. Entropy is therefore conserved during the process. Taking $g_{*S}(aT_\gamma)^3$ to remain constant, we find that aT_γ increases after

---

[7] For the moment, I will restore the subscript on the photon temperature to highlight the difference with the neutrino temperature.

[8] Obviously, entropy is separately conserved for the thermal bath and the decoupled species.

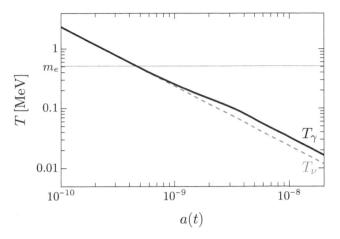

Fig. 3.3 Evolution of the photon and neutrino temperatures through electron–positron annihilation. Neutrinos are decoupled and their temperature redshifts simply as $T_\nu \propto a^{-1}$. The energy density of the electron–positron pairs is transferred to the photon gas whose temperature therefore redshifts more slowly, $T_\gamma \propto g_{*S}^{-1/3} a^{-1}$.

electron–positron annihilation by a factor $(11/4)^{1/3}$, while aT_ν remains the same. This means that, after $e^+ e^-$ annihilation, the neutrino temperature is slightly lower than the photon temperature,

$$
\boxed{T_\nu = \left(\frac{4}{11}\right)^{1/3} T_\gamma} \; . \tag{3.62}
$$

For $T \ll m_e$, the effective number of relativistic species (in energy density and entropy) therefore is

$$
g_* = 2 + \frac{7}{8} \times 2N_{\text{eff}} \left(\frac{4}{11}\right)^{4/3} = 3.36 \, , \tag{3.63}
$$

$$
g_{*S} = 2 + \frac{7}{8} \times 2N_{\text{eff}} \left(\frac{4}{11}\right) = 3.94 \, , \tag{3.64}
$$

where we have introduced the parameter N_{eff} as the *effective* number of neutrino species in the universe. If neutrino decoupling was instantaneous then we would simply have $N_{\text{eff}} = 3$. However, neutrino decoupling was not quite complete when $e^+ e^-$ annihilation began, so some of the energy and entropy did leak to the neutrinos. Taking this into account[9] raises the effective number

---

[9] To get the precise value of N_{eff}, one also has to consider the fact that the neutrino spectrum after decoupling deviates slightly from the Fermi–Dirac distribution. This spectral distortion arises because the energy dependence of the weak interaction causes neutrinos in the high-energy tail to interact more strongly.

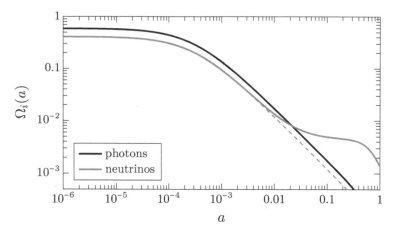

$\Omega_i(a)$

photons
neutrinos

a

Fig. 3.4 Evolution of the fractional energy densities of photons and neutrinos. Massless neutrinos (dashed line) are always a fixed fraction of the photon density, while massive neutrinos (solid gray line) can dominate over photons at late times.

of neutrinos to $N_{\rm eff} = 3.046$.[10] Using this value in (3.63) explains the final value of $g_*(T)$ in Fig. 3.2.

Neutrino density

The relation (3.62) holds until the present. The cosmic neutrino background therefore has a slightly lower temperature, $T_{\nu,0} = 1.95\,{\rm K}$, than the cosmic microwave background, $T_0 = 2.73\,{\rm K}$. The number density of neutrinos (per flavor) is

$$n_\nu \approx \frac{3}{4} \times \frac{4}{11} n_\gamma \,. \tag{3.65}$$

Using (3.24), we see that this corresponds to 112 neutrinos cm^{-3} per flavor. The present energy density of neutrinos depends on whether the neutrinos are relativistic or non-relativistic today. It used to be believed that neutrinos were massless, in which case we would have

$$\rho_\nu = \frac{7}{8} N_{\rm eff} \left(\frac{4}{11} \right)^{4/3} \rho_\gamma \quad \Rightarrow \quad \Omega_\nu h^2 \approx 1.7 \times 10^{-5} \quad (m_\nu = 0)\,. \tag{3.66}$$

Neutrino oscillation experiments have since shown that neutrinos do have a mass. The minimum sum of the neutrino masses is $\sum m_{\nu,i} > 0.06\,{\rm eV}$ [2]. Massive neutrinos behave as radiation-like particles in the early universe (for $m_\nu < 0.2\,{\rm eV}$, neutrinos are relativistic at recombination) and as matter-like particles in the late

---

[10] The Planck constraint on $N_{\rm eff}$ is 2.99 ± 0.17 [1]. This still leaves room for discovering that $N_{\rm eff} \neq 3.046$, which is one of the avenues in which cosmology could discover new physics beyond the Standard Model.

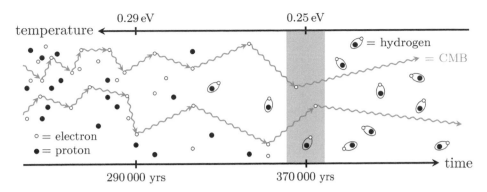

Cartoon of the recombination of protons and electrons into neutral hydrogen atoms and the corresponding decoupling of photons.

universe (see Fig. 3.4). In Problem 3.4, you will show that energy density of massive neutrinos, $\rho_\nu = \sum m_{\nu,i} n_{\nu,i}$, corresponds to

$$\Omega_\nu h^2 \approx \frac{\sum m_{\nu,i}}{94\,\text{eV}} \,. \tag{3.67}$$

By demanding that neutrinos don't overclose the universe, $\Omega_\nu < 1$, a cosmological upper bound can be placed on the sum of the neutrino masses, $\sum m_{\nu,i} < 15\,\text{eV}$ (using $h \approx 0.7$). Massive neutrinos also affect the late-time expansion rate, which is constrained by CMB and BAO measurements. The current Planck constraint on the sum of the neutrino masses is $\sum m_{\nu,i} < 0.13\,\text{eV}$, which implies $\Omega_\nu < 0.003$ [1]. Future observations promise to be sensitive enough to measure the neutrino masses [3].

3.1.5 Cosmic Microwave Background

An important event in the history of the early universe is the formation of the first atoms and the associated decoupling of photons (see Fig. 3.5). At temperatures above about $1\,\text{eV}$, the universe still consisted of a plasma of free electrons and nuclei. Photons were tightly coupled to the electrons via Thomson scattering, which in turn strongly interacted with protons via Coulomb scattering. There was very little neutral hydrogen. When the temperature dropped below $0.3\,\text{eV}$, the electrons and nuclei combined to form neutral atoms and the density of free electrons decreased sharply. The photon mean free path grew rapidly and became longer than the Hubble length, H^{-1}. Around $0.25\,\text{eV}$, the photons decoupled from the matter and the universe became transparent. Today, these photons are observed as the cosmic microwave background.

Table 3.3 Key events in the formation of the cosmic microwave background				
Event	Redshift	Temp (eV)	Temp (K)	Time (yrs)
Matter–radiation equality	3400	0.80	9270	50 000
Recombination	1270	0.29	3460	290 000
Photon decoupling	1090	0.25	2970	370 000
Last-scattering	1090	0.25	2970	370 000

Key events in the formation of the CMB are summarized in Table 3.3. In the following, we will derive these facts.

Chemical equilibrium

During recombination, the number of each particle species wasn't fixed, because hydrogen atoms were formed, while the number of free electrons and protons decreased. In thermodynamics, we describe such a situation with the help of the **chemical potential**.

Consider the generic reaction

$$1 + 2 \leftrightarrow 3 + 4 .$$

Each particle species has a chemical potential μ_i. The second law of thermodynamics implies that particles flow to the side of the reaction where the total chemical potential is lower. Chemical equilibrium is reached when the sum of the chemical potentials on each side is equal, in which case the rates of the forward and reverse reactions are equal. For the above example, this implies that

$$\mu_1 + \mu_2 = \mu_3 + \mu_4 . \tag{3.68}$$

It is useful to establish a few facts:

- There is no chemical potential for photons, because photon number is not conserved (e.g. double Compton scattering $e^- + \gamma \leftrightarrow e^- + \gamma + \gamma$ happens in equilibrium at high temperatures). Sometimes, this is expressed as

$$\mu_\gamma = 0 , \tag{3.69}$$

but, more accurately, the concept of a chemical potential doesn't exist for photons.
- If the chemical potential of a particle X is μ_X, then the chemical potential of the corresponding antiparticle $\bar{X}$ is

$$\mu_{\bar{X}} = -\mu_X . \tag{3.70}$$

To see this, just consider particle–antiparticle annihilation, $X + \bar{X} \leftrightarrow \gamma + \gamma$, and use that $\mu_\gamma = 0$.

We will now apply the condition of chemical equilibrium to recombination. The equilibrium assumption will be sufficient to describe the onset of recombination, but will not capture the correct dynamics shortly thereafter (such as the freeze-out of electrons). We will revisit these non-equilibrium aspects of the physics of recombination in Section 3.2.5.

Hydrogen recombination

Recombination proceeds in two stages. The formation of helium atoms is followed by that of hydrogen atoms. For simplicity, we will focus on hydrogen recombination. Specifically, we will assume that the universe was filled only with free electrons, protons and photons. Over 90% (by number) of the nuclei are protons, so this isn't a terrible approximation to reality. Moreover, helium recombination is completed before hydrogen recombination, so that the two events can be treated separately.

The formation of hydrogen atoms occurs via the reaction

$$e^- + p^+ \leftrightarrow \mathrm{H} + \gamma \,.$$

Initially, this reaction keeps the particles in equilibrium, and since $T < m_i$, $i = \{e, p, \mathrm{H}\}$, we have the following equilibrium abundances

$$n_i^{\mathrm{eq}} = g_i \left(\frac{m_i T}{2\pi} \right)^{3/2} \exp\left(\frac{\mu_i - m_i}{T} \right), \tag{3.71}$$

where $\mu_p + \mu_e = \mu_\mathrm{H}$ (recall that $\mu_\gamma = 0$). To remove the dependence on the chemical potentials, we consider the following ratio

$$\left(\frac{n_\mathrm{H}}{n_e n_p} \right)_{\mathrm{eq}} = \frac{g_\mathrm{H}}{g_e g_p} \left(\frac{m_\mathrm{H}}{m_e m_p} \frac{2\pi}{T} \right)^{3/2} e^{(m_p + m_e - m_\mathrm{H})/T} \,. \tag{3.72}$$

In the prefactor, we can use $m_\mathrm{H} \approx m_p$, but in the exponential the small difference between m_H and $m_p + m_e$ is crucial: it is the ionization energy of hydrogen

$$E_I \equiv m_p + m_e - m_\mathrm{H} = 13.6\,\mathrm{eV} \,. \tag{3.73}$$

The numbers of internal degrees of freedom are $g_p = g_e = 2$ and $g_\mathrm{H} = 4$.[11] As far as we know, the universe isn't electrically charged, so that we have $n_e = n_p$. Equation (3.72) then becomes

$$\left(\frac{n_\mathrm{H}}{n_e^2} \right)_{\mathrm{eq}} = \left(\frac{2\pi}{m_e T} \right)^{3/2} e^{E_I/T} \,. \tag{3.74}$$

It is convenient to describe the process of recombination in terms of the **free-electron fraction**:

$$X_e \equiv \frac{n_e}{n_p + n_\mathrm{H}} = \frac{n_e}{n_e + n_\mathrm{H}} \,. \tag{3.75}$$

---

[11] The spins of the electron and proton in a hydrogen atom can be aligned or anti-aligned, giving one singlet state and one triplet state, so $g_\mathrm{H} = 1 + 3 = 4$.

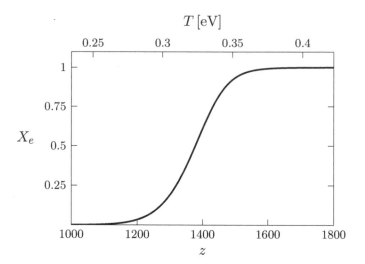

Fig. 3.6 Free-electron fraction as a function of redshift according to the Saha equation (3.78).

A fully ionized universe then corresponds to $X_e = 1$, while a universe of only neutral atoms has $X_e = 0$. Our goal is to understand how X_e evolves.

If we neglect the small number of helium atoms, then the denominator in (3.75) can be approximated by the baryon density

$$n_b = \eta\, n_\gamma = \eta \times \frac{2\zeta(3)}{\pi^2}\, T^3\,, \tag{3.76}$$

where η is the baryon-to-photon ratio. We can then write

$$\frac{1 - X_e}{X_e^2} = \frac{n_\mathrm{H}}{n_e^2}\, n_b\,, \tag{3.77}$$

and substituting (3.74), we arrive at the so-called **Saha equation**

$$\boxed{\left(\frac{1 - X_e}{X_e^2}\right)_{\mathrm{eq}} = \frac{2\zeta(3)}{\pi^2}\, \eta \left(\frac{2\pi T}{m_e}\right)^{3/2} e^{E_I/T}}\,. \tag{3.78}$$

The solution to this equation is

$$X_e = \frac{-1 + \sqrt{1 + 4f}}{2f}\,, \quad \text{with} \quad f(T, \eta) = \frac{2\zeta(3)}{\pi^2}\, \eta \left(\frac{2\pi T}{m_e}\right)^{3/2} e^{E_I/T}\,, \tag{3.79}$$

which is shown in Fig. 3.6 as a function of temperature (or equivalently redshift).

Let us define the recombination temperature T_{rec} as the temperature at which $X_e = 0.5$ in (3.78).[12] For $\eta \approx 6 \times 10^{-10}$, we get

$$T_{\mathrm{rec}} \approx 0.32\,\mathrm{eV} \approx 3760\,\mathrm{K}\,. \tag{3.80}$$

---

[12] There is nothing deep about the choice $X_e(T_{\mathrm{rec}}) = 0.5$. It is as arbitrary as it looks. However, since X_e is exponentially sensitive to T, we don't change T_{rec} much by changing this criterion (see Fig. 3.6).

The reason why the recombination temperature is significantly below the binding energy of hydrogen, $T_{\rm rec} \ll E_I = 13.6\,{\rm eV}$, is that there are many photons for each hydrogen atom. Even when $T < E_I$, the high-energy tail of the photon distribution contains photons with energy $E_\gamma > E_I$, which can ionize the hydrogen atoms. Concretely, although the mean photon energy is $\langle E_\gamma \rangle \approx 2.7\,T$, one in 500 photons has $E_\gamma > 10\,T$, one in 3×10^6 has $E_\gamma > 20\,T$, and one in 3×10^{10} has $E_\gamma > 30\,T$. Since there are over 10^9 photons per baryon, rare high-energy photons are still present in sufficient numbers, unless the temperature drops far below the binding energy.

Using $T_{\rm rec} = T_0(1 + z_{\rm rec})$, with $T_0 = 2.73\,{\rm K}$, gives $z_{\rm rec} \approx 1380$ for the redshift of recombination.[13] As we will see in Section 3.2.5, in reality the details of recombination are slightly more complex and the moment of recombination is delayed relative to the Saha prediction [45], with $X_e = 0.5$ only being reached at

$$z_{\rm rec} \approx 1270\,,$$
$$t_{\rm rec} \approx 290\,000\,{\rm yrs}\,. \tag{3.81}$$

Since matter–radiation equality is at $z_{\rm eq} \approx 3400$, we conclude that recombination occurred in the matter-dominated era. Of course, as can be seen from Fig. 3.6, recombination was not an instantaneous process. It took about $\Delta t \approx 70\,000\,{\rm yrs}$ (or $\Delta z \approx 80$) for the ionization fraction to drop from $X_e = 0.9$ to $X_e = 0.1$.

Photon decoupling

At early times, photons are strongly coupled to the primordial plasma through their interactions with the free electrons

$$e^- + \gamma \leftrightarrow e^- + \gamma\,,$$

with the interaction rate given by $\Gamma_\gamma \approx n_e \sigma_T$, where $\sigma_T \approx 2 \times 10^{-3}\,{\rm MeV}^{-2}$ is the Thomson cross section. It is interesting to put this into more familiar units. At $a = 10^{-5}$ (prior to matter–radiation equality), the rate of photon scattering is $\Gamma_\gamma \approx 5.0 \times 10^{-6}\,{\rm s}^{-1}$, or three times per week. This doesn't seem like much, but this interaction rate was much larger than the expansion rate at the time (which was $H \approx 2 \times 10^{-10}\,{\rm s}^{-1}$), so that electrons and photons were in equilibrium.

---

[13] It is useful to compare this to the redshift of helium recombination. This proceeds in two stages: First, He^{2+} captures one electron to create He^+. This process occurs in equilibrium and takes place around $z \approx 6000$. Then, He^+ captures a second electron to become a neutral helium atom. This part of recombination is slower than predicted by Saha equilibrium and occurs around $z \approx 2000$. We see that helium recombination is indeed completed before hydrogen recombination begins. This means that we were justified to treat hydrogen recombination as a separate process. It also means that the details of helium recombination don't have a big effect on the predictions of the CMB, since the universe was still optically thick after helium recombination was completed and before hydrogen recombination started.

Since $\Gamma_\gamma \propto n_e$, the interaction rate decreases as the density of free electrons drops during recombination. At some point, this rate becomes smaller than the expansion rate and the photons decouple. We define the approximate moment of **photon decoupling** as $\Gamma_\gamma(T_{\rm dec}) \approx H(T_{\rm dec})$. Writing

$$\Gamma_\gamma(T_{\rm dec}) = n_b X_e(T_{\rm dec})\,\sigma_T = \frac{2\zeta(3)}{\pi^2}\,\eta\,\sigma_T\,X_e(T_{\rm dec})T_{\rm dec}^3\,, \tag{3.82}$$

$$H(T_{\rm dec}) = H_0\sqrt{\Omega_m}\left(\frac{T_{\rm dec}}{T_0}\right)^{3/2}\,, \tag{3.83}$$

we get

$$X_e(T_{\rm dec})T_{\rm dec}^{3/2} \approx \frac{\pi^2}{2\zeta(3)}\frac{H_0\sqrt{\Omega_m}}{\eta\,\sigma_T\,T_0^{3/2}}\,. \tag{3.84}$$

Using the Saha equation for $X_e(T_{\rm dec})$, and substituting the standard values for the cosmological parameters on the right-hand side, we find $T_{\rm dec} \approx 0.27\,{\rm eV}$. In the more precise treatment in Section 3.2.5, we find that decoupling occurs at a slightly lower temperature,

$$T_{\rm dec} \approx 0.25\,{\rm eV} \approx 2970\,{\rm K}\,, \tag{3.85}$$

with the corresponding redshift and time of decoupling being

$$\begin{aligned} z_{\rm dec} &\approx 1090\,, \\ t_{\rm dec} &\approx 370\,000\,{\rm yrs}\,. \end{aligned} \tag{3.86}$$

After decoupling, the photons stream freely through the universe.

Notice that although $T_{\rm dec}$ isn't far from $T_{\rm rec}$, the ionization fraction decreases significantly between recombination and decoupling, $X_e(T_{\rm rec}) \approx 0.5 \rightarrow X_e(T_{\rm dec}) \approx 0.001$. This shows that a large degree of neutrality is necessary before the universe becomes transparent to photons.

Exercise 3.4 Imagine that recombination did not occur, so that $X_e = 1$. At what redshift would the CMB photons now decouple?

Last-scattering

The scattering of photons off electrons essentially stops at photon decoupling. To define the precise moment of **last-scattering**, we have to consider the probability of photon scattering. Let dt be a small time interval around the time t. The probability that a photon will scatter during this time is $\Gamma_\gamma(t)\,dt$, and the integrated probability between the times t and $t_0 > t$ is

$$\tau(t) = \int_t^{t_0} \Gamma_\gamma(\tilde{t})\,d\tilde{t}\,. \tag{3.87}$$

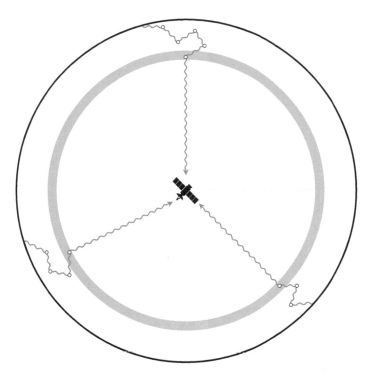

Fig. 3.7   Cartoon of the last-scattering of CMB photons after recombination. Today we observe the CMB photons from the spherical surface of last-scattering.

This probability is also called the **optical depth**. Taking t_0 to be the present time, the moment of last-scattering is defined by $\tau(t_*) \equiv 1$. To a good approximation, last-scattering coincides with photon decoupling, $t_* \approx t_{\mathrm{dec}}$. However, the optical depth is sensitive to the evolution of the free electron density at the end of recombination, which isn't captured well by the equilibrium treatment of this section. A precise evaluation of t_* must therefore await our more involved non-equilibrium analysis of recombination in Section 3.2.5.

When we observe the CMB, we are detecting photons from the **surface of last-scattering** (see Fig. 3.7). Given the age of the universe, and taking into account the expansion of the universe, the distance between us and the spherical last-scattering surface is 42 billion light-years. Of course, last-scattering is a probabilistic concept—not all photons experienced their last scattering event at the same time—so there is some thickness to the last-scattering surface.

Blackbody spectrum

The CMB is often presented as key evidence that the early universe began in a state of thermal equilibrium. I will now briefly explain why we can draw this conclusion from the observed frequency spectrum of the cosmic background radiation.

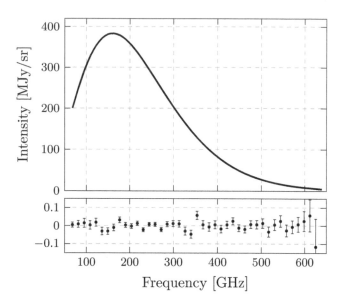

Fig. 3.8   Measurement of the CMB blackbody spectrum by the COBE satellite [4]. The error bars are too small to be seen in the top panel. In the bottom panel, we therefore also show the residuals relative to a blackbody spectrum with temperature $T_0 = 2.725$ K. Notice the change in scale between the two panels.

Before decoupling, the number density of photons with frequencies in the range f and $f + \mathrm{d}f$ is

$$n(f, T)\,\mathrm{d}f = \frac{2}{c^3} \frac{4\pi f^2}{e^{hf/k_\mathrm{B}T} - 1}\,\mathrm{d}f, \tag{3.88}$$

where I have restored factors of c and k_B for clarity. This frequency distribution is called the **blackbody spectrum** and is characteristic of objects in thermal equilibrium. After decoupling, the photons propagate freely, with their frequencies redshifting as $f(t) \propto a(t)^{-1}$ and the number density decreasing as $a(t)^{-3}$. The spectrum therefore maintains its blackbody form as long as we take the temperature to scale as $T \propto a(t)^{-1}$. It is in this sense that the relic radiation encodes the early equilibrium phase of the hot Big Bang.

CMB experiments observe the so-called **spectral radiation intensity**, I_f, which is the flux of energy per unit area per unit frequency. Let us see how this is related to the spectrum in (3.88). We first pick a specific direction and consider photons traveling in a solid angle $\delta\Omega$ around this direction. In a given time interval δt, these photons move through a volume $\delta V = (c\delta t)^3\,\delta\Omega$ and cross a cap of area $\delta A = (c\delta t)^2\,\delta\Omega$. The number of photons in this volume is

$$\delta N = \frac{n(f)\,\mathrm{d}f}{4\pi}\,\delta V = \frac{2}{c^3} \frac{f^2\,\mathrm{d}f}{e^{hf/k_\mathrm{B}T} - 1}\,(c\delta t)^3\,\delta\Omega, \tag{3.89}$$

and the number of photons crossing the surface per unit area and per unit time is

$$\frac{\delta N}{\delta A \, \delta t} = \frac{2}{c^2} \frac{f^2 \, \mathrm{d}f}{e^{hf/k_\mathrm{B}T} - 1} \,. \tag{3.90}$$

Since each photon has energy hf, the flux of energy across the surface (per unit frequency) is

$$I_f = \frac{2h}{c^2} \frac{f^3}{e^{hf/k_\mathrm{B}T} - 1} \,. \tag{3.91}$$

Figure 3.8 shows a measurement of the CMB frequency spectrum by the FIRAS instrument on the COBE satellite [4]. What you are seeing here is the most perfect blackbody ever observed in nature, proving that the early universe indeed started in a state of thermal equilibrium.

3.2 Beyond Equilibrium

To understand the world around us, it is crucial to understand deviations from equilibrium. For example, we saw that a massive particle species in thermal equilibrium becomes exponentially rare when the temperature drops below the mass of the particles, $N_i \equiv n_i/s \sim (m/T)^{3/2} \exp(-m/T)$. In order for these particles to survive until the present time, they must drop out of thermal equilibrium before m/T becomes much larger than unity. This decoupling, and the associated freeze-out of massive particles, occurs when the interaction rate of the particles becomes smaller the expansion rate (see Fig. 3.9).

The main tool to describe the evolution beyond equilibrium is the **Boltzmann equation**. In this section, I will first introduce the Boltzmann equation and then apply it to three important examples: (i) the production of dark matter; (ii) the formation of the light elements during Big Bang nucleosynthesis; and (iii) the freeze-out of electrons during recombination. I will also make some comments about the role of non-equilibrium processes in baryogenesis.

3.2.1 The Boltzmann Equation

In Chapter 2, we saw that, in the absence of interactions, the number density of a particle species i evolves as

$$\frac{\mathrm{d}n_i}{\mathrm{d}t} + 3\frac{\dot{a}}{a}n_i = 0 \,, \tag{3.92}$$

which simply means that the number of particles in a fixed physical volume ($V \propto a^3$) is conserved, so that the density dilutes as $n_i \propto a^{-3}$. To include the effects of interactions, we add a collision term to the right-hand side of (3.92),

$$\frac{1}{a^3} \frac{\mathrm{d}(n_i a^3)}{\mathrm{d}t} = C_i[\{n_j\}] \,. \tag{3.93}$$

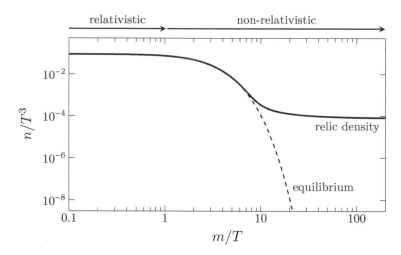

Fig. 3.9 Schematic illustration of particle freeze-out. Initially, the particle abundance tracks its equilibrium value and hence decreases when the particles become non-relativistic. Once the interaction rate drops below the expansion rate, however, the particles freeze out and maintain a constant relic density that is much larger than the Boltzmann-suppressed equilibrium abundance.

This is the **Boltzmann equation**.[14] The form of the collision term depends on the specific interactions under consideration. Interactions between three or more particles are very unlikely, so we can limit ourselves to single-particle decays and two-particle scatterings (or annihilations). For concreteness, let us consider the following process

$$1 + 2 \leftrightarrow 3 + 4 \,,$$

i.e. particle 1 interacts with particle 2 to produce particles 3 and 4, or the inverse process can produce 1 and 2. This reaction will capture all processes studied in this chapter. Suppose we are interested in tracking the number density n_1 of species 1. Obviously, the rate of change in the abundance of species 1 is given by the difference between the rates for producing and eliminating the species. The Boltzmann equation simply formalizes this statement,

$$\frac{1}{a^3} \frac{\mathrm{d}(n_1 a^3)}{\mathrm{d}t} = -\alpha \, n_1 n_2 + \beta \, n_3 n_4 \,. \tag{3.94}$$

We understand the right-hand side as follows: The first term, $-\alpha n_1 n_2$, describes the destruction of particles 1, while that second term, $+\beta \, n_3 n_4$, accounts for their production. Notice that the first term is proportional to n_1 and n_2, while the second term is proportional to n_3 and n_4. The parameter $\alpha = \langle \sigma v \rangle$ is the

---

[14] Strictly speaking, this is the "integrated Boltzmann equation." A more fundamental Boltzmann equation for the evolution of the distribution functions f_i will be introduced in Appendix B. Integrating that Boltzmann equation over phase space gives (3.93).

thermally-averaged cross section.[15] The second parameter β can be related to α by noting that the collision term has to vanish in (chemical) equilibrium

$$\beta = \left(\frac{n_1 n_2}{n_3 n_4}\right)_{\text{eq}} \alpha \,, \tag{3.95}$$

where n_i^{eq} are the equilibrium number densities that we calculated above. The relation in (3.95) is sometimes called **detailed balance**. We therefore find

$$\boxed{\frac{1}{a^3}\frac{\mathrm{d}(n_1 a^3)}{\mathrm{d}t} = -\langle\sigma v\rangle\left[n_1 n_2 - \left(\frac{n_1 n_2}{n_3 n_4}\right)_{\text{eq}} n_3 n_4\right].} \tag{3.96}$$

It is instructive to write this in terms of the number of particles in a comoving volume, as defined in (3.52), $N_i \equiv n_i/s \propto n_i a^3$. This gives

$$\frac{\mathrm{d}\ln N_1}{\mathrm{d}\ln a} = -\frac{\Gamma_1}{H}\left[1 - \left(\frac{N_1 N_2}{N_3 N_4}\right)_{\text{eq}}\frac{N_3 N_4}{N_1 N_2}\right], \tag{3.97}$$

where $\Gamma_1 \equiv n_2\langle\sigma v\rangle$ is the interaction rate of species 1. The right-hand side of (3.97) contains a factor describing the *interaction efficiency*, Γ_1/H, and a factor characterizing the *deviation from equilibrium*, $[1 - \cdots]$.

When the interaction rate is large, $\Gamma_1 \gg H$, the natural state of the system is to be in equilibrium. Imagine that we start with $N_1 \gg N_1^{\text{eq}}$ (while $N_i \sim N_i^{\text{eq}}$, $i = 2, 3, 4$). The right-hand side of (3.97) then is negative, particles of type 1 are destroyed and N_1 is reduced towards the equilibrium value N_1^{eq}. Similarly, if $N_1 \ll N_1^{\text{eq}}$, the right-hand side of (3.97) is positive and N_1 is driven towards N_1^{eq}. The same conclusion applies if several species deviate from their equilibrium values. As long as the interaction rates are large, the system therefore quickly relaxes to a steady state where the source in (3.97) vanishes and the particles assume their equilibrium abundances. At some point, however, the reaction rate drops below the Hubble scale, $\Gamma_1 < H$. The right-hand side of (3.97) then gets suppressed and the comoving density of particles approaches a constant relic density, $N_1 \to \text{const}$.

3.2.2 Dark Matter Freeze-Out

We will begin with a simple application of the Boltzmann equation (3.96): the freeze-out of massive particles. This may be how dark matter was produced in the early universe.

---

[15] Cross sections are the fundamental observables in particle physics. They describe the probability of a certain scattering process. In this book, we will simply use dimensional analysis to estimate the few cross sections that we will need. The cross section may depend on the *relative velocity* v of particles 1 and 2. The angle brackets in $\langle\sigma v\rangle$ denote an average over v.

Decoupling and freeze-out

Consider a heavy fermion X that can annihilate with its antiparticle $\bar{X}$ to produce two light (essentially massless) particles,

$$X + \bar{X} \;\leftrightarrow\; \ell + \bar{\ell}.$$

The particles X might be the dark matter, while ℓ could be particles of the Standard Model.

To illustrate the essential physics, we will make a few simplifying assumptions: First, we will take the light particles ℓ to be tightly coupled to the rest of the plasma, as would be the case, for instance, if the particles were electrically charged. The light particles will then maintain their equilibrium densities throughout, $n_\ell = n_\ell^{\text{eq}}$. Second, we will neglect any initial asymmetry between X and its antiparticle $\bar{X}$, so that $n_X = n_{\bar{X}}$. Finally, we will assume that there are no other particle annihilations during the freeze-out of the species X. This will allow us to take $T \propto a^{-1}$ for the times relevant to the freeze-out.

With these assumptions, the Boltzmann equation (3.96) for the evolution of the particles X becomes

$$\frac{1}{a^3}\frac{\mathrm{d}(n_X a^3)}{\mathrm{d}t} = -\langle\sigma v\rangle\left[n_X^2 - (n_X^{\text{eq}})^2\right]. \tag{3.98}$$

It is convenient to introduce the quantity $Y_X \equiv n_X/T^3$, which is proportional to the number of particles in a comoving volume, $N_X \equiv n_X/s \propto Y_X$. In fact, as long as $T \propto a^{-1}$, we can treat Y_X and N_X interchangeably. In terms of Y_X, the left-hand side of (3.98) becomes $T^3 \mathrm{d}Y_X/\mathrm{d}t$. Since most of the interesting dynamics will take place when the temperature is of order the particle mass, $T \sim M_X$, it will be convenient to define a new measure of time,

$$x \equiv \frac{M_X}{T}, \quad \text{where} \quad \frac{\mathrm{d}x}{\mathrm{d}t} = Hx. \tag{3.99}$$

For weakly interacting particles, the decoupling occurs at very early times, during the radiation-dominated era, where the Hubble parameter can be written as $H = H(M_X)/x^2$. Equation (3.98) then becomes the so-called **Riccati equation**,

$$\frac{\mathrm{d}Y_X}{\mathrm{d}x} = -\frac{\lambda}{x^2}\left[Y_X^2 - (Y_X^{\text{eq}})^2\right], \tag{3.100}$$

where we have defined the dimensionless parameter

$$\lambda \equiv \frac{\Gamma(M_X)}{H(M_X)} = \frac{M_X^3 \langle\sigma v\rangle}{H(M_X)}. \tag{3.101}$$

For simplicity, we will take λ to be a constant. Unfortunately, even then, there are no analytic solutions to (3.100). Nevertheless, it is straightforward to solve the Riccati equation numerically, using the equilibrium abundance as an initial condition.

Figure 3.10 shows numerical solutions of (3.100) for different values of λ. As expected, at high temperatures, $x \ll 1$, we have $Y_X \approx Y_X^{\text{eq}} = 3\zeta(3)/(4\pi^2) \approx 0.1$,

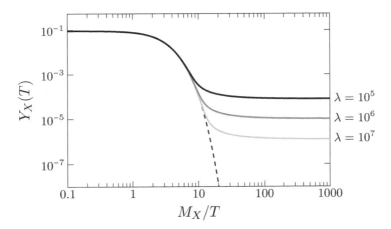

Fig. 3.10   Solutions of the Riccati equation (3.100) for different values of the coupling λ. The relic abundance scales inversely with the coupling, as predicted in (3.104).

while, at low temperatures, $x \gg 1$, the equilibrium abundance becomes exponentially suppressed, $Y_X^{\text{eq}} \sim (x/2\pi)^{3/2} e^{-x}$. Eventually, the massive particles will become so rare that they will not be able to find each other fast enough to maintain the equilibrium abundance. We see that this freeze-out happens at about $x_f \sim 10$.

The final relic abundance, $Y_X^\infty \equiv Y_X(x = \infty)$, determines the freeze-out density of the massive particles. Let us estimate its magnitude as a function of the interaction strength λ. Well after freeze-out, Y_X will be much larger than Y_X^{eq}. Thus, at late times, we can drop Y_X^{eq} from the Boltzmann equation, so that

$$\frac{\mathrm{d}Y_X}{\mathrm{d}x} \approx -\frac{\lambda Y_X^2}{x^2} \,. \tag{3.102}$$

Integrating this from $x = x_f$ to $x = \infty$, we find

$$\frac{1}{Y_X^\infty} - \frac{1}{Y_X^f} = \frac{\lambda}{x_f} \,, \tag{3.103}$$

where $Y_X^f \equiv Y_X(x_f)$. Typically, $Y_X^f \gg Y_X^\infty$ (see Fig. 3.10), so a simple analytic approximation for the relic abundance is

$$\boxed{Y_X^\infty \approx \frac{x_f}{\lambda}} \,. \tag{3.104}$$

Of course, this answer still depends on the unknown freeze-out time x_f. We have seen that a good order-of-magnitude estimate is $x_f \sim 10$. Using this value, the estimate in (3.104) is in excellent agreement with the numerical results shown in Fig. 3.10. Moreover, the value of x_f isn't terribly sensitive to the precise value of λ, namely $x_f(\lambda) \propto |\ln \lambda|$, where the logarithmic scaling can be understood from the decoupling condition $\Gamma(x_f) \approx H(x_f)$.

Equation (3.104) predicts that the freeze-out abundance Y_X^∞ decreases as the interaction rate λ increases. This makes intuitive sense: larger interactions maintain equilibrium for a longer time, deeper into the Boltzmann-suppressed regime. Since the estimate in (3.104) works very well, we will use it in the following.

Dark matter density

We would like to relate the freeze-out abundance of the dark matter relics to their density today. To do this, we note that the number density of the particles after freeze-out decreases as $n_X \propto a^{-3}$. We can then write the number density today as

$$n_{X,0} = n_{X,1} \left(\frac{a_1}{a_0}\right)^3 = Y_X^\infty T_0^3 \left(\frac{a_1 T_1}{a_0 T_0}\right)^3 , \tag{3.105}$$

where a_1 is an arbitrary time that is late enough for the species to have reached their relic abundance, but before any additional particle annihilations have become relevant. As we have seen, such particle annihilations lead to deviations from the scaling $T \propto a^{-1}$, so that the final factor in (3.105) isn't simply unity. Using the conservation of entropy, $g_{*S}(aT)^3 = \text{const}$, we instead get

$$n_{X,0} = Y_X^\infty T_0^3 \frac{g_{*S}(T_0)}{g_{*S}(M_X)} . \tag{3.106}$$

The normalized energy density of the particles then is

$$\Omega_X \equiv \frac{\rho_{X,0}}{\rho_{\text{crit},0}} = \frac{M_X n_{X,0}}{3 M_{\text{Pl}}^2 H_0^2} = \frac{M_X T_0^3}{3 M_{\text{Pl}}^2 H_0^2} \frac{x_f}{\lambda} \frac{g_{*S}(T_0)}{g_{*S}(M_X)} ,$$

$$= \frac{H(M_X)}{M_X^2} \frac{T_0^3}{3 M_{\text{Pl}}^2 H_0^2} \frac{x_f}{\langle \sigma v \rangle} \frac{g_{*S}(T_0)}{g_{*S}(M_X)} , \tag{3.107}$$

where we have used (3.104) for Y_X^∞ and substituted (3.101) for the coupling λ. Using (3.55) for $H(M_X)$, this becomes

$$\Omega_X = \frac{\pi}{\sqrt{90}} \frac{T_0^3}{3 M_{\text{Pl}}^3 H_0^2} \frac{x_f}{\langle \sigma v \rangle} \sqrt{g_*(M_X)} \frac{g_{*S}(T_0)}{g_{*S}(M_X)} . \tag{3.108}$$

Finally, we insert the measured values of T_0 and H_0, and use $g_{*S}(T_0) = 3.91$ as well as $g_{*S}(M_X) = g_*(M_X)$, to get

$$\boxed{\Omega_X \sim 0.1 \frac{x_f}{\sqrt{g_*(M_X)}} \frac{10^{-8}\,\text{GeV}^{-2}}{\langle \sigma v \rangle}} . \tag{3.109}$$

This result is insensitive to the mass of the new particles and is mostly determined by their cross section. The observed dark matter density is reproduced if

$$\sqrt{\langle \sigma v \rangle} \sim 10^{-4}\,\text{GeV}^{-1} \sim 0.1 \sqrt{G_F} .$$

The fact that a thermal relic with a cross section characteristic of the weak interaction gives the right order of magnitude for the dark matter abundance today is called the **WIMP miracle**.[16]

---

[16] Here, WIMP stands for **weakly interacting massive particle**. WIMPs used to be a very popular dark matter candidate, not the least because of the result in (3.109). However, recent direct

3.2.3 Baryogenesis: A Sketch*

The analysis of the previous section can also be applied to the annihilation of baryons, $b + \bar{b} \leftrightarrow \gamma + \gamma$. Since the effective interaction between nucleons is mediated by pions, we can estimate the annihilation cross section as $\langle \sigma v \rangle \sim 1/m_\pi^2$, with $m_\pi \approx 140\,\text{MeV}$. Substituting this into (3.109), we then find

$$\Omega_b \approx 10^{-11}\,, \qquad\qquad (3.110)$$

which is much smaller than the observed value. This illustrates the need for **baryogenesis** to produce the abundance of baryons in the early universe. A lazy option would be to simply impose the baryon asymmetry as an initial condition. Much more satisfying would be to start with a baryon-symmetric universe and then see the asymmetry between matter and antimatter arise dynamically.

Sakharov conditions

In 1967, Sakharov identified three necessary conditions that any successful theory of baryogenesis must satisfy [5]. These **Sakharov conditions** are:

1. *Violation of baryon number.* The interactions that generate the baryon asymmetry must violate baryon symmetry. It should be rather obvious that a universe with vanishing baryon number, $B = 0$, can only evolve into a universe with $B \neq 0$ if the interactions don't conserve the baryon number.

   In the Standard Model (SM), baryon number is an "accidental symmetry," meaning that all renormalizable interactions have this symmetry because of the gauge symmetries of the SM. Despite baryon number being a symmetry of the SM Lagrangian, baryon number violation occurs at the quantum level through the so-called "triangle anomaly" [6]. Nonperturbatively, this can lead "sphaleron transitions" which convert baryons to antileptons and thus change the baryon number.[17] Theories beyond the SM—such as Grand Unified Theories (GUTs)— may violate baryon number even in the Lagrangian (i.e. at tree level).

2. *Violation of C and CP.* Slightly less obvious is the fact that the interactions must violate C (charge conjugation) and CP (charge conjugation combined with

---

detection experiments have put pressure on the WIMP scenario, because the couplings required to explain the dark matter abundance via (3.109) are in tension with the experimental limits. Alternative dark matter candidates, such as non-thermal **axions**, have therefore increased in popularity.

[17] **Sphalerons** are nonperturbative solutions of the electroweak field equations [7, 8], corresponding to the collective excitation of W, Z and Higgs fields. They are rather spectacular objects. The typical size of a sphaleron is 10^{-17} m, so that its volume is a million times smaller than that of a proton. Its mass, however, is around $10\,\text{TeV}$, or 10 thousand times the mass of a proton. Sphalerons are therefore 10 billion times denser than protons and a teaspoon of sphalerons would weigh twice as much as the Moon [9].

parity). If C was an exact symmetry, then the probability of the process $i \to f$ would be equal to that involving the corresponding antiparticles, $\bar{i} \to \bar{f}$, and no net baryon number would be generated. To see why CP must be violated, we first note that CPT (with T being time reversal) is a symmetry in any relativistic quantum field theory [10]. CP invariance is therefore equivalent to invariance under time reversal. Consider then the process $i(\mathbf{p}_a, \mathbf{s}_a) \to f(\mathbf{p}_a, \mathbf{s}_a)$, where $\mathbf{p}_a$ are the momenta of the particles and $\mathbf{s}_a$ are their spins. Under time reversal, this becomes $f(-\mathbf{p}_a, -\mathbf{s}_a) \to i(-\mathbf{p}_a, -\mathbf{s}_a)$. If time reversal is unbroken, then integrating over all momenta and summing over all spins would lead to a vanishing net baryon number.

Both C and CP are violated by the weak interaction. The Standard Model therefore, in principle, contains the required symmetry breaking, but whether the breaking is strong enough to explain the size of the baryon asymmetry in our universe remains to be seen.

3. *Departure from equilibrium.* Finally, the processes that generate the baryon asymmetry must occur out of thermal equilibrium. For non-relativistic particles and antiparticles in equilibrium, we have

$$B \propto n_X - n_{\bar{X}} = 2g_X (M_X T)^{3/2} e^{-M_X/T} \sinh(\mu_X/T), \qquad (3.111)$$

where μ_X is the chemical potential. By the first Sakharov condition, X and $\bar{X}$ undergo B-violating reactions, such as

$$XX \to \bar{X}\bar{X}.$$

If these reactions occur in equilibrium, then $\mu_X = 0$ and $B = 0$ in (3.111). The necessary out-of-equilibrium conditions can be provided by the cosmological evolution, making baryogenesis an interesting interplay between cosmology and particle physics.

Out-of-equilibrium decay

In the following, I will sketch a simple toy model for baryogenesis: the out-of-equilibrium decay of a massive particle [11]. The decay rate of the particle is Γ_X, and most of the decay occurs when the age of the universe is of the order of the particle's lifetime, $t \sim H^{-1} \sim \Gamma_X^{-1}$. If the decay violates baryon number, then a baryon asymmetry will be generated.

For concreteness, we assume that the X particle has two decay channels a and b, with baryon numbers B_a and B_b. Similarly, the antiparticle $\bar{X}$ has the decay channels $\bar{a}$ and $\bar{b}$, with baryon numbers $-B_a$ and $-B_b$. The branching ratios for the decays are

$$r \equiv \frac{\Gamma(X \to a)}{\Gamma_X}, \qquad \bar{r} = \frac{\Gamma(\bar{X} \to \bar{a})}{\Gamma_X},$$
$$1 - r \equiv \frac{\Gamma(X \to b)}{\Gamma_X}, \qquad 1 - \bar{r} = \frac{\Gamma(\bar{X} \to \bar{b})}{\Gamma_X}, \qquad (3.112)$$

where we have used that the total decay rates of X and $\bar{X}$ are equal (because of CPT symmetry and unitarity). The net baryon number produced in the decays of X and $\bar{X}$ then is

$$
\begin{aligned}
\Delta B &\equiv B_X + B_{\bar{X}} \\
&= \left(rB_a + (1-r)B_b\right) + \left(-\bar{r}B_a - (1-\bar{r})B_b\right) \\
&= (r - \bar{r})(B_a - B_b) \, .
\end{aligned}
\tag{3.113}
$$

If the baryon number isn't violated in the individual decays, then $B_a = B_b = 0$ and hence $\Delta B = 0$. Similarly, if C and CP are preserved, then $r = \bar{r}$ and $\Delta B = 0$. This illustrates the necessity of the Sakharov conditions for baryogenesis.

The net baryon number density produced by the decays is $n_B = \Delta B \, n_X$. Using that $n_X \sim n_\gamma$ at the time of the decay, we get

$$
\eta = \frac{n_B}{n_\gamma} \sim \Delta B \, ,
\tag{3.114}
$$

which relates the cosmologically observed baryon-to-photon ratio to the amount of baryon number violation. Here, we have assumed that the entropy release in the decays is negligible, which is not always a good approximation. Finally, a more accurate computation would also involve explicit integration of the Boltzmann equation.

Models of baryogenesis

The treatment above was an extremely rough sketch of a particular mechanism for baryogenesis, but there are many other scenarios. In the following, I will comment very briefly on some popular models of baryogenesis. For more details, please consult the suggested references at the end of the chapter.

- **GUT baryogenesis**   Grand Unified Theories—theories that unify the strong and electroweak interactions—generically contain baryon number-violating interactions. In the effective theory of the Standard Model, these interactions appear as higher-dimensional operators suppressed by a large scale and are often constrained by the absence of proton decay. The baryon asymmetry can then be created by the out-of-equilibrium decay of a heavy particle, as sketched above. The details of GUT baryogenesis were developed in many papers; see e.g. [12–18]. The non-observation of proton decay puts a lower bound on the mass of the decaying particle, which can be in tension with the reheating temperature after inflation.

- **Electroweak baryogenesis**   It is tempting to ask whether baryogenesis can be realized purely within the Standard Model. At first sight, it looks like there is a good chance that all of the Sakharov conditions could be satisfied: (i) sphalerons provide baryon number violation, (ii) weak interactions violate C and CP, and (iii) the electroweak phase transition provides out-of-equilibrium conditions. On closer inspection, however, it turns out that life is not so simple.

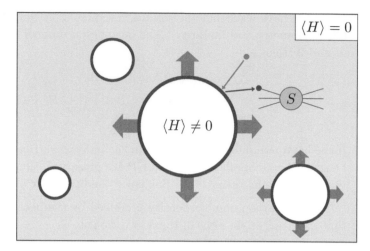

Fig. 3.11   Sketch of bubble nucleation during the electroweak phase transition. Scattering with the bubble walls can produce CP asymmetries in the particle densities which affect the B-violating sphaleron transitions in the symmetric phase. The excess baryons then diffuse into the broken phase where sphaleron transitions aren't effective and the baryon symmetry is therefore preserved.

The EW phase transition is *not* strongly first order [19] (as required because sphaleron transitions would wash out the baryon asymmetry in a second-order phase transition) and the size of the CP violation is too small [20–22]. New physics is therefore required to change the nature of the EW phase transition and provide new sources of CP violation. Nevertheless, since this new physics can be tested by current and future colliders, electroweak baryogenesis [23–25] remains an attractive option.

The details of electroweak baryogenesis are somewhat technical and I will provide only a very qualitative description of the basic physics. More details can be found, for example, in the reviews [11, 26–28].

During a first-order phase transition, bubbles of the broken phase (with nonzero Higgs vacuum expectation value, $\langle H \rangle \neq 0$) nucleate within the surrounding plasma in the unbroken phase (with $\langle H \rangle = 0$); see Fig. 3.11. These bubbles expand and coalesce until only the broken phase remains. If the underlying theory contains CP violation, then the scattering of particles with the bubble walls can generate CP asymmetries in the particle number densities. These asymmetries diffuse into the symmetric phase, where they lead to a biasing of electroweak sphaleron transitions which produce a net baryon asymmetry. Some of the excess baryons are then transferred to the broken phase where the rate of sphaleron transitions is strongly suppressed, so that the baryon asymmetry is preserved.

● **Leptogenesis**   Another interesting possibility is to generate baryogenesis via leptogenesis, as was first suggested by Fukugita and Yanagida [29]. Imagine a theory beyond the Standard Model with an extra heavy (but sterile) neutrino (like in the "seesaw mechanism" for neutrino masses). If the decay of this heavy neutrino violates CP, then it could produce more antileptons than leptons. These antileptons are then transformed into baryons by sphaleron transitions. The end result is an excess of baryons over antibaryons.[18]

As even this short discussion has shown, there is no shortage of candidate scenarios for baryogenesis. The challenge is to decide observationally which (if any) of these is realized in our universe.

3.2.4 Big Bang Nucleosynthesis

As our next example, we will use the Boltzmann equation to follow the evolution of baryonic matter in the early universe. Since baryon number is conserved, the total number of nucleons stays constant. However, weak nuclear reactions convert neutrons and protons into each other and strong nuclear reactions will build heavier nuclei from them. In this section, I will show you how the light elements—hydrogen, helium and lithium—were synthesized in the Big Bang. I won't give a complete account of all of the complicated details of BBN.[19] Instead, our goal will be more modest: I want to give you a theoretical understanding of a single number, the ratio of the (mass) density of helium to hydrogen,

$$Y_P \equiv \frac{4n_{\text{He}}}{n_{\text{H}}} \sim \frac{1}{4}. \tag{3.115}$$

This number is one of the key predictions of the Big Bang theory.

Let me briefly sketch the main sequence of events during BBN. In the rest of this section, I will then put some mathematical flesh on these bones. At early times, the baryonic matter in the universe was mostly in the form of protons and neutrons, which were coupled to each other by β-decay and inverse β-decay:

$$\begin{aligned} n + \nu_e &\leftrightarrow p^+ + e^-\,, \\ n + e^+ &\leftrightarrow p^+ + \bar{\nu}_e\,, \end{aligned} \tag{3.116}$$

where ν_e and $\bar{\nu}_e$ are electron neutrinos and antineutrinos. Initially, the relative abundances of the protons and neutrons were determined by equilibrium thermodynamics. Around $1\,\text{MeV}$ (close to the time of neutrino decoupling), the reactions in (3.116) became inefficient and the neutrons decoupled. We will determine the freeze-out abundance by solving the relevant Boltzmann equation. Free neutrons

---

[18] Interest in leptogenesis was recently revived by hints of CP violation for the ordinary neutrinos by the T2K experiment [30]. If confirmed, this would make it plausible that CP violation also plays an important role for sterile neutrinos (if they exist).

[19] The detailed predictions of BBN can be computed numerically using the code PRIMAT [31]. Thanks to Cyril Pitrou for making the code publicly available.

decay, which further reduced their abundance, until neutrons and protons combined into deuterium[20]

$$n + p^+ \leftrightarrow D + \gamma. \qquad (3.117)$$

This occurred at a temperature of around 0.1 MeV. Finally, the deuterium nuclei fused into helium

$$\begin{aligned} D + p^+ &\leftrightarrow {}^3He + \gamma, \\ D + {}^3He &\leftrightarrow {}^4He + p^+. \end{aligned} \qquad (3.118)$$

In this way, essentially all of the primordial neutrons got converted into helium. The formation of heavier elements was very inefficient, which is why most protons survived and became the hydrogen in the late universe.

Equilibrium

Above $T \approx 0.1$ MeV, only free protons and neutrons existed, while other light nuclei hadn't been formed yet. We can therefore first solve for the neutron-to-proton ratio and then use this abundance as an input for the synthesis of deuterium, helium, etc.

Consider the reactions in (3.116). Let us assume that the chemical potentials of the electrons and neutrinos are negligibly small, so that $\mu_n = \mu_p$. Using (3.71) for the equilibrium number densities, n_i^{eq}, we then have

$$\left(\frac{n_n}{n_p}\right)_{eq} = \left(\frac{m_n}{m_p}\right)^{3/2} e^{-(m_n - m_p)/T}. \qquad (3.119)$$

The small difference between the proton and neutron masses can be ignored in the prefactor, but has to be kept in the exponential. Hence, we find

$$\boxed{\left(\frac{n_n}{n_p}\right)_{eq} = e^{-Q/T}}, \qquad (3.120)$$

where $Q \equiv m_n - m_p = 1.30$ MeV. For $T \gg 1$ MeV, there are therefore as many neutrons as protons, while, for $T < 1$ MeV, the neutron fraction gets smaller. If the weak interactions would operate efficiently enough to maintain equilibrium indefinitely, then the neutron abundance would drop to zero. Luckily, in the real world the weak interactions are not so efficient.

Neutron freeze-out

The primordial ratio of neutrons to protons is of particular importance to the outcome of BBN, since essentially all the neutrons eventually become incorporated into ^4He. It is convenient to define the neutron fraction as

---

[20] This reaction is very similar to the reaction $e^- + p^+ \to H + \gamma$ studied in Section 3.1.5. Our analysis of the equilibrium abundance of deuterium will therefore be virtually identical to the Saha analysis of recombination.

$$X_n \equiv \frac{n_n}{n_n + n_p}, \tag{3.121}$$

and, using (3.120), we get

$$X_n^{\text{eq}}(T) = \frac{e^{-Q/T}}{1 + e^{-Q/T}}. \tag{3.122}$$

Neutrons follow this equilibrium abundance until weak interaction processes such as (3.116) effectively shut off at a temperature of around 1 MeV (roughly at the same time as neutrino decoupling).[21] To follow this evolution in detail, we must solve the Boltzmann equation (3.96), with 1 = neutron, 3 = proton, and 2, 4 = leptons (with $n_\ell = n_\ell^{\text{eq}}$):

$$\frac{1}{a^3} \frac{d(n_n a^3)}{dt} = -\Gamma_n \left[n_n - \left(\frac{n_n}{n_p} \right)_{\text{eq}} n_p \right], \tag{3.123}$$

where $\Gamma_n = n_\ell \langle \sigma v \rangle$ is the rate for neutron/proton conversion. Substituting (3.121) and (3.122) into (3.123), we find

$$\frac{dX_n}{dt} = -\Gamma_n \left[X_n - (1 - X_n) e^{-Q/T} \right]. \tag{3.124}$$

Our task is to integrate this equation to determine $X_n(t)$, using (3.122) as the initial condition.

The key input from particle physics is the rate Γ_n. For the reactions in (3.116), we have

$$\Gamma_n = n_\nu \langle \sigma(n\nu_e \to p^+ e^-) v \rangle + n_e \langle \sigma(n e^+ \to p^+ \bar{\nu}_e) v \rangle, \tag{3.125}$$

where the neutrino and positron densities are

$$n_e = 2n_\nu = \frac{3\zeta(3)}{2\pi^2} T^3. \tag{3.126}$$

Moreover, the relevant thermally-averaged cross sections are [32]

$$\begin{aligned}
\langle \sigma(n\nu_e \to p^+ e^-) v \rangle &= \frac{510\pi^2}{3\zeta(3)\,\tau_n} \frac{(12T^2 + 6QT + Q^2)}{Q^5}, \\
\langle \sigma(n e^+ \to p^+ \bar{\nu}_e) v \rangle &= \frac{255\pi^2}{3\zeta(3)\,\tau_n} \frac{(12T^2 + 6QT + Q^2)}{Q^5},
\end{aligned} \tag{3.127}$$

where $\tau_n = 886.7 \pm 0.8$ s is the neutron lifetime. The total neutron/proton conversion rate then is

$$\Gamma_n(x) = \frac{255}{\tau_n} \cdot \frac{12 + 6x + x^2}{x^5}, \tag{3.128}$$

---

[21] It is fortunate that $T_f \sim Q$. This seems to be a coincidence: Q is determined by the strong and electromagnetic interactions, while the value of T_f is fixed by the weak interaction. Imagine a world in which $T_f \ll Q$.

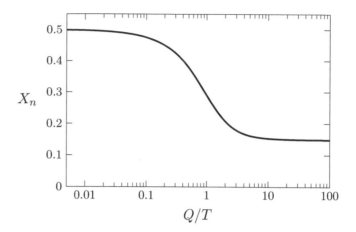

Fig. 3.12 Numerical solution of (3.131) giving the evolution of the fractional neutron abundance.

where $x \equiv Q/T$. During BBN, the universe is dominated by relativistic species, so the Hubble rate is

$$H(x) = \sqrt{\frac{\rho}{3M_{\mathrm{Pl}}^2}} = \frac{\pi}{3}\sqrt{\frac{g_*}{10}}\frac{Q^2}{M_{\mathrm{Pl}}}\frac{1}{x^2} \equiv \frac{H_1}{x^2}, \qquad (3.129)$$

where $g_* = 10.75$ and $H_1 \approx 1.13\,\mathrm{s}^{-1}$. At $T = 1\,\mathrm{MeV}$, we have $\Gamma_n = 1.66\,\mathrm{s}^{-1}$ and $H = 0.67\,\mathrm{s}^{-1}$. This shows that the conversion time $\Gamma_n^{-1} = 0.60\,\mathrm{s}$ is comparable to the age of the universe $t = (2H)^{-1} = 0.74\,\mathrm{s}$. Since $\Gamma_n^{-1} \propto T^{-3} \propto t^{3/2}$, the interaction time eventually becomes longer than the age of the universe and the neutron-to-proton ratio approaches a constant.

To determine the freeze-out abundance, we have to solve the Boltzmann equation (3.124). Unfortunately, this can only be done numerically. As in the case of the dark matter freeze-out, it is useful to choose an evolution variable that focuses on the dynamics at $T \sim Q$. Using $x = Q/T$, we can write the left-hand side of (3.124) as

$$\frac{\mathrm{d}X_n}{\mathrm{d}t} = \frac{\mathrm{d}x}{\mathrm{d}t}\frac{\mathrm{d}X_n}{\mathrm{d}x} = -\frac{x}{T}\frac{\mathrm{d}T}{\mathrm{d}t}\frac{\mathrm{d}X_n}{\mathrm{d}x} = xH\frac{\mathrm{d}X_n}{\mathrm{d}x}, \qquad (3.130)$$

where we have used $T \propto a^{-1}$ in the last equality. Substituting the Hubble rate (3.129), we get

$$\frac{\mathrm{d}X_n}{\mathrm{d}x} = \frac{\Gamma_n(x)}{H_1}\,x\left[e^{-x} - X_n(1 + e^{-x})\right]. \qquad (3.131)$$

The numerical solution of this equation is shown in Fig. 3.12. We see that the freeze-out abundance of neutrons is

$$\boxed{X_n^\infty \equiv X_n(x = \infty) = 0.15}. \qquad (3.132)$$

Table 3.4	Binding energies per nucleon
Element	B/A
D	1.1 MeV
^{3}H	2.8 MeV
^{3}He	2.6 MeV
^{4}He	7.1 MeV
^{6}Li	5.3 MeV
^{7}Li	5.7 MeV
^{7}Be	5.4 MeV
^{9}Be	6.5 MeV

A rough estimate of this freeze-out abundance can also be obtained by evaluating the equilibrium abundance (3.122) at the time of neutrino decoupling, $T_f \sim 0.8\,\mathrm{MeV}$, which gives $X_n^\infty \sim 1/6$.

Neutron decay

At temperatures below 0.2 MeV (or $t \gtrsim 100\,\mathrm{s}$) the finite lifetime of the neutron becomes important. To include this neutron decay in our computation, we simply multiply the freeze-out abundance (3.132) by an exponential decay factor

$$\boxed{X_n(t) = X_n^\infty e^{-t/\tau_n} \approx 0.15\, e^{-t/887\,\mathrm{s}}}\,. \tag{3.133}$$

It is now a race against time. If the onset of nucleosynthesis takes significantly longer than the neutron lifetime, then the number of neutrons becomes exponentially small and not much fusion will occur. Fortunately, this is not the case in our universe.

Deuterium

At this point, the universe is mostly protons and neutrons. Helium cannot form directly because the density is too low and the time available is too short for reactions involving three or more incoming nuclei to occur at any appreciable rate. The heavier nuclei therefore have to be built sequentially from lighter nuclei in two-particle reactions. The first nucleus to form via the reaction in (3.117) is deuterium.[22] Only when deuterium is available can helium be produced. Since deuterium is formed directly from neutrons and protons it can follow its equilibrium abundance as long as enough free neutrons are available. However, since the deuterium binding energy is rather small (see Table 3.4), it takes a while for the deuterium abundance to become large. Although heavier nuclei have larger

---

[22] The fusion of two protons is inefficient because the nuclei have to overcome the Coulomb repulsion before the nuclear force can take over. The fusion of two neutrons produces a very unstable nucleus that quickly breaks apart again.

binding energies and hence would have larger equilibrium abundances, they cannot be formed until sufficient deuterium has become available. This is the **deuterium bottleneck**.[23] To get a rough estimate for the time of nucleosynthesis, we determine the temperature T_{nuc} when the deuterium fraction in equilibrium would be of order one, $(n_D/n_p)_{\text{eq}} \sim 1$.

Deuterium is produced by the reaction in (3.117). Since $\mu_\gamma = 0$, we have $\mu_n + \mu_p = \mu_D$. To remove the dependence on the chemical potentials, we consider

$$\left(\frac{n_D}{n_n n_p}\right)_{\text{eq}} = \frac{3}{4}\left(\frac{m_D}{m_n m_p}\frac{2\pi}{T}\right)^{3/2} e^{-(m_D - m_n - m_p)/T}, \qquad (3.134)$$

where, as before, we have used (3.71) for n_i^{eq} (with $g_D = 3$ and $g_p = g_n = 2$). In the prefactor, m_D can be set equal to $2m_n \approx 2m_p \approx 1.9$ GeV, but in the exponential the small difference between $m_n + m_p$ and m_D is crucial: it is the binding energy of deuterium

$$B_D \equiv m_n + m_p - m_D = 2.22 \text{ MeV}. \qquad (3.135)$$

As long as equilibrium holds, the deuterium-to-proton ratio therefore is

$$\left(\frac{n_D}{n_p}\right)_{\text{eq}} = \frac{3}{4} n_n^{\text{eq}} \left(\frac{4\pi}{m_p T}\right)^{3/2} e^{B_D/T}. \qquad (3.136)$$

To get an order-of-magnitude estimate, we approximate the neutron density by the baryon density and write this in terms of the photon temperature and the baryon-to-photon ratio,

$$n_n \sim n_b = \eta\, n_\gamma = \eta \times \frac{2\zeta(3)}{\pi^2} T^3. \qquad (3.137)$$

Equation (3.136) then becomes

$$\left(\frac{n_D}{n_p}\right)_{\text{eq}} \approx \eta \left(\frac{T}{m_p}\right)^{3/2} e^{B_D/T}. \qquad (3.138)$$

As in the case of recombination, the smallness of the baryon-to-photon ratio η inhibits the production of deuterium until the temperature drops well below the binding energy B_D. The temperature has to drop enough so that the exponential factor $e^{B_D/T}$ can compete with $\eta \approx 6 \times 10^{-10}$. We find that $(n_D/n_p)_{\text{eq}} \sim 1$ at $T_{\text{nuc}} \sim 0.06$ MeV, which via (3.56), with $g_* = 3.38$, translates into

$$\boxed{t_{\text{nuc}} = 120\,\text{s} \left(\frac{0.1\,\text{MeV}}{T_{\text{nuc}}}\right)^2 \sim 330\,\text{s}}. \qquad (3.139)$$

We see from Fig. 3.13 that a better estimate would be $n_D^{\text{eq}}(T_{\text{nuc}}) \approx 10^{-3} n_p^{\text{eq}}(T_{\text{nuc}})$, which gives $T_{\text{nuc}} \approx 0.07$ MeV and $t_{\text{nuc}} \approx 250$ s. Notice that $t_{\text{nuc}} \ll \tau_n$, so the result

---

[23] Even in equilibrium the production of helium would occur rather late, at $T \sim 0.3$ MeV. The deuterium bottleneck leads to a further delay of this equilibrium expectation.

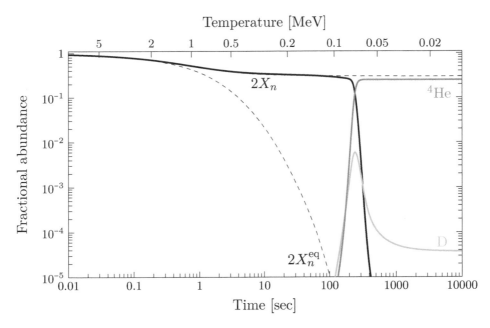

Fig. 3.13 Numerical result for the production of helium and deuterium in the early universe [31]. At early times, neutrons and protons are in equilibrium and their abundances are roughly equal. Around 1 MeV, the weak interactions become inefficient and the neutrons freeze out. Due to its low binding energy, it takes a while until neutrons and protons combine into deuterium. Once the fusion of deuterium has started, the subsequent production of helium is very rapid.

for $X_n(t_{\rm nuc})$ from (3.133) won't be very sensitive to the estimate for $t_{\rm nuc}$. Once the formation of deuterium starts, the fusion of the heavier nuclei can also begin.

Helium

Since the binding energy of helium is larger than that of deuterium, the Boltzmann factor $e^{B/T}$ favors helium over deuterium. Indeed, Fig. 3.13 shows that helium is produced almost immediately after deuterium. The reaction proceeds in two steps: First, helium-3, ^{3}He, and tritium, ^{3}H, are formed via

$$\mathrm{D} + p^+ \;\leftrightarrow\; {}^3\mathrm{He} + \gamma\,,$$
$$\mathrm{D} + \mathrm{D} \;\leftrightarrow\; {}^3\mathrm{H} + p^+\,,$$
$$\mathrm{D} + \mathrm{D} \;\leftrightarrow\; {}^3\mathrm{He} + n\,.$$

The rate of D–D fusion is much faster, but there are more protons than deuterium. Next, helium-4, ^{4}He, is produced through the following chain of reactions

$$^3\mathrm{H} + p^+ \;\leftrightarrow\; {}^3\mathrm{He} + n\,,$$
$$^3\mathrm{H} + \mathrm{D} \;\leftrightarrow\; {}^4\mathrm{He} + n\,,$$
$$^3\mathrm{He} + \mathrm{D} \;\leftrightarrow\; {}^4\mathrm{He} + p^+\,.$$

Virtually all of the neutrons at $t \sim t_{\mathrm{nuc}}$ are processed into $^4\mathrm{He}$ by these reactions. Substituting $t_{\mathrm{nuc}} \approx 250\,\mathrm{s}$ into (3.133), we find

$$X_n(t_{\mathrm{nuc}}) \approx 0.11 \,. \tag{3.140}$$

Since two neutrons go into one nucleus of $^4\mathrm{He}$, the final helium abundance is equal to half of the neutron abundance at t_{nuc}, so that $n_{\mathrm{He}} = \frac{1}{2}n_n(t_{\mathrm{nuc}})$. Hence, the mass fraction of helium is

$$\boxed{Y_P = \frac{4n_{\mathrm{He}}}{n_{\mathrm{H}}} = \frac{4n_{\mathrm{He}}}{n_p} \approx \frac{2X_n(t_{\mathrm{nuc}})}{1 - X_n(t_{\mathrm{nuc}})} \sim 0.25} \,, \tag{3.141}$$

as we wished to show. A more exact calculation—directly solving the coupled Boltzmann equations for all reactions—gives [31]

$$Y_P = 0.2262 + 0.0135 \ln(\eta/10^{-10}) \,, \tag{3.142}$$

where the logarithmic dependence on η can be traced back to the effect of the baryon-to-photon ratio on the time of nucleosynthesis in (3.139). Figure 3.14 shows the predicted helium mass fraction as a function of η. The abundance of $^4\mathrm{He}$ increases with increasing η because nucleosythesis starts earlier for larger baryon density.

In making the prediction (3.141), we have taken the effective number of relativistic species, g_*, to be fixed at the value predicted by the Standard Model. It is also interesting to let the value of g_* float and use measurements on the nuclear abundances to put constraints on any new physics contributing to g_* at the time of BBN. The dependence on g_* is through $H \propto \sqrt{g_*}\, T^2$, so that increasing the value of g_* leads to a larger expansion rate (at the same T) which implies an earlier neutron freeze-out, $T_f \propto g_*^{1/6}$. This increases the final abundance of $^4\mathrm{He}$.

Other light elements

To determine the abundances of other light elements, the *coupled* Boltzmann equations have to be solved numerically. I won't go into details, but will just provide a qualitative discussion of the most important reactions.

We see from Fig. 3.13 that the deuterium abundance drops after nucleosynthesis begins. Most of the deuterium is processed into $^3\mathrm{He}$ and $^4\mathrm{He}$. (Tritium is unstable and decays into $^3\mathrm{He}$.) A small residual amount of deuterium remains when the fusion reactions become inefficient. The final abundances of D and $^3\mathrm{He}$, as a function of η, are shown in Fig. 3.14. Since both deuterium and helium-3 are burnt by fusion, their abundances decrease as η increases.

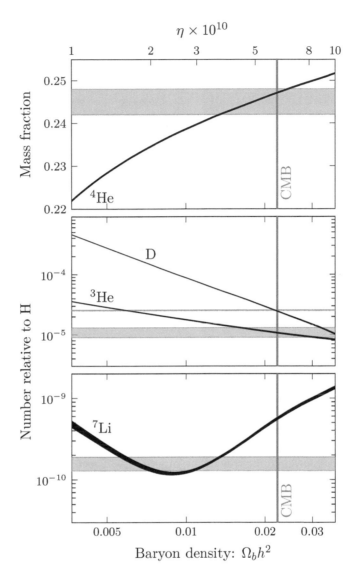

Fig. 3.14 Theoretical predictions for the light element abundances and observational constraints (gray bands) [31]. The line widths on the theory curves account for uncertainties in the neutron lifetime and the nuclear cross sections.

Nucleosynthesis is a very inefficient process, so very few nuclei beyond helium are formed. Small amounts of beryllium and lithium are created by the reactions

$$^3\text{He} + {}^4\text{He} \ \leftrightarrow \ {}^7\text{Be} + \gamma \,,$$
$$^3\text{H} + {}^4\text{He} \ \leftrightarrow \ {}^7\text{Li} + \gamma \,,$$

and can be converted into each other via

$$^7\text{Be} + n \ \leftrightarrow \ {}^7\text{Li} + p^+ \,.$$

Lithium can be destroyed by capturing a proton

$$^{7}\text{Li} + p^{+} \ \leftrightarrow \ ^{4}\text{He} + {}^{4}\text{He} \,.$$

This last reaction reduces the amount of lithium relative to beryllium. However, once the temperature of the universe gets low enough, beryllium can decay into lithium by capturing an electron

$$^{7}\text{Be} + e^{-} \ \rightarrow \ ^{7}\text{Li} + \nu_{e} \,.$$

The predicted amount of lithium is shown in Fig. 3.14. At low η, ^{7}Li is destroyed by protons with an efficiency that increases with η. On the other hand, its precursor ^{7}Be is produced more efficiently as η increases. This explains the valley in the curve for ^{7}Li.

Almost no nuclei with mass numbers $A > 7$ are produced in the hot Big Bang. The basic reason for this is that there are no stable nuclei with $A = 8$ that can be formed fast enough to sustain the chain reactions. The half-life of ^{8}Be is $10^{-12}\,\text{s}$. The merger of three helium-4 nuclei into carbon-12 is too slow. Similarly, the capture of neutrons and protons into ^{8}Li and ^{9}Be is also very inefficient.

Observations

To test the predictions of Big Bang nucleosynthesis, the element abundances must be measured in regions where very little post-processing of the primordial gas has taken place. In particular, nuclear fusion inside of stars changes the abundances and therefore complicates the measurements of the primordial abundances. The measurements cited below are those suggested by the Particle Data Group [33]. The fact that we find good quantitative agreement with observations remains one of the great triumphs of the Big Bang model.

- **Helium-4** can be measured from the light of ionized gas clouds, because the strength of some emission lines depends on the amount of helium. We have to correct for the fact that ^{4}He is also produced in stars. One way to do this is to correlate the measured helium abundance with the abundances of heavier elements, such as nitrogen and oxygen. The larger the amount of oxygen, the more helium has been created. Extrapolating the measurements to zero oxygen gives an improved estimate of the primordial helium abundance. Using this approach, the measured abundance of primordial helium-4 is found to be [34–36]

$$Y_P = 0.245 \pm 0.003 \,. \tag{3.143}$$

We see from Fig. 3.14 that this is consistent with the prediction from BBN, given the CMB measurement of the baryon density.

- **Deuterium** is very weakly bound and therefore easily destroyed in the late universe. BBN is the only source of significant deuterium in the universe. The best way to determine the primordial (unprocessed) value of the deuterium abundance is to measure the spectra of high-redshift quasars. These spectra contain absorption lines due to gas clouds along the line-of-sight, and, in particular, the Lyman-α line is a sensitive probe of the amount of deuterium (see e.g. [37]). A weighted mean of several such measurements is $D/H = (25.47\pm0.25)\times 10^{-6}$ [33]. As we see in Fig. 3.14, this agrees precisely with the prediction from BBN.

- **Helium-3** can be created and destroyed in stars. Its abundance is therefore hard to measure and interpret. Given these uncertainties, helium-3 is usually not used as a cosmological probe.
- **Lithium** is mostly destroyed by stellar nucleosynthesis. The best estimate of its primordial abundance follows from the measurement of metal-poor stars in the Galactic halo. Averaging the measurements of [38–41] gives $Li/H = (1.6\pm0.3)\times 10^{-10}$. As we see from Fig. 3.14, the measured lithium abundance is significantly smaller than the predicted value. It is unclear whether this *lithium problem* is due to systematic errors in the interpretation of the measurements or signals the need for new physics during BBN.

3.2.5 Recombination Revisited*

Finally, we will use the Boltzmann equation to provide a more accurate treatment of recombination. It is important to determine the evolution of the free-electron density accurately because it affects the predictions for the CMB anisotropies (see Chapter 7). Unfortunately, our treatment of recombination in Section 3.1.5 was a bit too simplistic and does not provide results that are accurate enough for the era of precision cosmology. In particular, we assumed that thermal equilibrium holds throughout the process. In reality, this assumption breaks down rather quickly and non-equilibrium effects become important. Moreover, we did not yet discuss the rather intricate dynamics of recombination. As we will see, recombination directly to the ground state of the hydrogen atom is very inefficient and a more complicated path needs to be considered.

Non-equilibrium recombination

As the free-electron fraction drops during recombination, the interaction rate also decreases and can fall below the expansion rate of the universe. When this happens, our equilibrium treatment becomes suspect and we must use the Boltzmann equation to follow the evolution. Applying (3.96) to the reaction $e^- + p^+ \leftrightarrow H + \gamma$, we get

$$\frac{1}{a^3}\frac{d(n_e a^3)}{dt} = -\langle\sigma v\rangle\left[n_e^2 - \left(\frac{n_e^2}{n_H}\right)_{eq} n_H\right], \qquad (3.144)$$

where we have used that $n_e = n_p$ and $n_\gamma = n_\gamma^{eq}$. For the factor of $(n_e^2/n_H)_{eq}$ we can use (3.74). Moreover, the electron and hydrogen densities can be written in terms of the free-electron fraction and the baryon density, $n_e = X_e n_b$ and $n_H = (1 - X_e)n_b$.

Using that $n_b a^3 = \text{const}$, we then get the following equation for the evolution of the free-electron fraction

$$\frac{dX_e}{dt} = \left[\beta(T)(1 - X_e) - \alpha(T) n_b X_e^2 \right], \tag{3.145}$$

where we have introduced

$$\alpha(T) \equiv \langle \sigma v \rangle, \tag{3.146}$$

$$\beta(T) \equiv \langle \sigma v \rangle \left(\frac{m_e T}{2\pi} \right)^{3/2} e^{-E_I/T}. \tag{3.147}$$

The parameter α characterizes the recombination rate, while β is associated with the ionization rate. When α is large, the right-hand side of (3.145) must vanish and the evolution of X_e is given by Saha equilibrium. The function $\alpha(T)$ depends on the precise way that the electrons get captured into the ground state of a hydrogen atom. When recombination occurs directly into the ground state, it releases photons with energy larger than 13.6 eV. These photons will then quickly ionize other atoms, leading to no net recombination.

The effective three-level atom

To avoid the instantaneous reionization of the hydrogen atoms, recombination must first occur to an excited state, which then decays to the ground state. The photons created during this multi-step recombination have lower energy and are therefore less likely to ionize the hydrogen atoms. The details were worked out by Peebles in 1968 [42] (see also Zel'dovich, Kurt and Sunyaev [43]). Here, I will provide a rough sketch of the rather involved analysis.

Peebles first argued that the hydrogen atom can be treated as an effective **three-level atom**. The three relevant states are the ground state ($1s$), the excited states (mostly $2s$ and $2p$), and the continuum states of ionized hydrogen (see Fig. 3.15). The excited states are in thermal equilibrium with each other since radiative excitations and decays are very fast. Peebles therefore considered all excited states as a single entity. In other words, the number of atoms in the first excited state, with $n = 2$, also determines the (smaller) fraction of atoms in the higher excited states. Since the direct recombination to the ground state is very inefficient, the rate of recombination will be determined by the rate of decay of the first excited state. A standard computation in quantum field theory gives [32]

$$\alpha(T) \approx 9.8 \frac{\alpha^2}{m_e^2} \left(\frac{E_I}{T} \right)^{1/2} \ln \left(\frac{E_I}{T} \right), \tag{3.148}$$

where $\alpha \approx 1/137$ is the fine-structure constant.

Unfortunately, this is still not the answer. When an atom in the first excited state decays to the ground state it produces a Lyman-α photon. This photon has a large probability to be absorbed by a nearby atom, which can then be ionized. To achieve

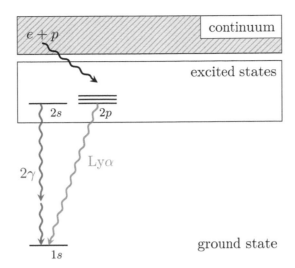

Fig. 3.15 Illustration of the Peebles model of recombination. Recombination directly to the ground state $1s$ is very inefficient. Instead, recombination proceeds first by capture of the electron to one of the excited states, which quickly cascade to the first excited states $2s$ and $2p$. These excited states then decay to the ground state at a much slower rate.

a significant level of recombination we must avoid these resonant excitations. There are two ways out:

- First, there is a small probability that the $2s$ state will decay to the $1s$ state through the emission of two photons. These photons then don't have enough energy to excite the atoms in the ground state back to the first exited state. The rate for this **two-photon decay** is

$$\Lambda_{2\gamma} = 8.227\,\mathrm{s}^{-1}\,. \tag{3.149}$$

About 57% of all neutral hydrogen is formed via this route.

- Second, as the universe expands, the energy of the Lyman-α photons that are created by the $2p \to 1s$ transition is redshifted. This moves these photons off resonance. If a photon avoids being reabsorbed for a sufficiently long time, then this effect allows them to escape. There is a small probability that this will happen. The rate of recombination via this **resonance escape** is [44][24]

$$\Lambda_\alpha = \frac{8\pi}{\lambda_\alpha^3 n_{1s}}\,H\,, \tag{3.150}$$

---

[24] This rate is a product of the rate of the decay $2p \to 1s + \gamma$ and the probability that the Lyman-α photon escapes to infinity without exciting another hydrogen atom. I hate pulling results like this one out of the hat. Unfortunately, deriving this from first principles would take us too much time and effort. You can find a more detailed discussion in Peebles' original paper [42] or in Weinberg's book [45].

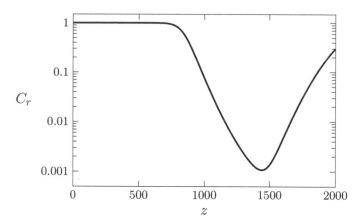

Fig. 3.16 Numerical solution of the Peebles factor defined in (3.152). Recombination is suppressed for $C_r \ll 1$.

where $\lambda_\alpha \equiv 2\pi E_\alpha^{-1} = 8\pi/(3E_I)$ is the wavelength of a Lyman-α photon and $n_{1s} \approx (1 - X_e)n_b$ is the abundance of hydrogen atoms in the $1s$ state. Substituting $n_b = \eta n_\gamma$, the recombination rate becomes

$$\Lambda_\alpha = \frac{27}{128\,\zeta(3)} \frac{H(T)}{(1 - X_e)\,\eta\,(T/E_I)^3}\,. \tag{3.151}$$

About 43% of all hydrogen atoms are formed in this way.

Although the details of the analysis are rather complex, the final answer is easy to state and interpret. The main new parameter is the effective branching ratio

$$C_r(T) \equiv \frac{\Lambda_{2\gamma} + \Lambda_\alpha}{\Lambda_{2\gamma} + \Lambda_\alpha + \beta_\alpha}\,, \tag{3.152}$$

where $\beta_\alpha \equiv \beta\,e^{3E_I/4T}$ is the ionization rate of the $n = 2$ state. I will call C_r the **Peebles factor**. It describes the probability that an atom in the first excited state reaches the ground state through either of the two ways described above before being ionized. When C_r is much smaller than unity, recombination is suppressed. The evolution of $C_r(z)$ is shown in Fig. 3.16. We see that $C_r \ll 1$ for $z > 900$, which implies that recombination will be delayed.

The branching ratio C_r appears as a multiplicative factor in the evolution equation (3.145), leading to the so-called **Peebles equation**

$$\boxed{\frac{dX_e}{dt} = C_r\left[\beta(1 - X_e) - \alpha n_b X_e^2\right]}\,. \tag{3.153}$$

This equation can be integrated numerically to obtain the evolution of the free-electron fraction.

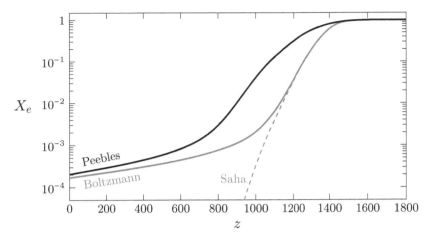

Fig. 3.17 Free-electron fraction as a function of redshift. The dashed curve labeled "Saha" is the prediction (3.79), which assumes that equilibrium holds throughout. The curves labeled "Boltzmann" and "Peebles" are numerical solutions to the Boltzmann equations (3.145) and (3.153), respectively. We see that recombination is delayed relative to the Saha prediction, and that the electron abundance freezes out at a constant value.

Electron freeze-out

It will be convenient to use redshift instead of time as the independent variable describing the evolution. Using

$$\frac{\mathrm{d}X_e}{\mathrm{d}t} = \frac{\mathrm{d}X_e}{\mathrm{d}z}\frac{\mathrm{d}z}{\mathrm{d}t} = -\frac{\mathrm{d}X_e}{\mathrm{d}z}H(1+z)\,, \tag{3.154}$$

the Peebles equation (3.153) can then be written as

$$\frac{\mathrm{d}X_e}{\mathrm{d}z} = -\frac{C_r(z)}{H(z)}\frac{\alpha(z)}{(1+z)}\left[(1-X_e)\left(\frac{m_eT}{2\pi}\right)^{3/2}e^{-E_I/T} - X_e^2\,\eta\,\frac{2\zeta(3)}{\pi^2}T^3\right]\,, \tag{3.155}$$

where $T = 0.235\,\mathrm{meV}(1+z)$. The evolution of the Hubble parameter is

$$H(z) = \sqrt{\Omega_m}H_0(1+z)^{3/2}\left(1+\frac{1+z}{1+z_{\mathrm{eq}}}\right)^{1/2}\,, \tag{3.156}$$

with $H_0 \approx 1.5 \times 10^{-33}\,\mathrm{eV}$. Figure 3.17 shows the evolution of the free-electron fraction as a function of redshift. We see that recombination is indeed delayed relative to the Saha expectation. In particular, we find $X_e = 0.5$ at $z = 1270$ (compared to $z = 1380$ for Saha). Moreover, unlike the Saha prediction, the electron density freezes out at a nonzero value of about 2×10^{-4}.

Decoupling and last-scattering

In Section 3.1.5, we defined last-scattering as the moment when the optical depth is equal to unity. This is very sensitive to the evolution of the free electron density, so it is worth revisiting our previous estimate.

The probability that a photon did not scatter off an electron in the redshift interval $[z_0, z_1]$ is

$$P(z_0, z_1) = e^{-\tau(z_0, z_1)}, \tag{3.157}$$

where τ is the optical depth

$$\tau(z_0, z_1) \equiv \int_{t_1}^{t_0} \mathrm{d}t\, \sigma_T n_e(t) = \int_{z_0}^{z_1} \mathrm{d}z \frac{\sigma_T\, n_b X_e(z)}{H(1+z)}. \tag{3.158}$$

The probability that a photon scattered for the last time in the interval $[z_1, z_1 + \mathrm{d}z_1]$ then is

$$P(z_0, z_1) - P(z_0, z_1 + \mathrm{d}z_1) \equiv g(z_0, z_1)\, \mathrm{d}z_1, \tag{3.159}$$

where $g(z_0, z_1)$ is the **visibility function**. We will be interested in $g(z) \equiv g(0, z)$, which is the probability that a photon observed today scattered last at redshift z. Using (3.157) and (3.159), we have[25]

$$g(z) = -\frac{\mathrm{d}}{\mathrm{d}z} e^{-\tau(0,z)} = \frac{\mathrm{d}\tau}{\mathrm{d}z} e^{-\tau}. \tag{3.160}$$

Note that $\tau(0,0) = 0$ and $\tau(0, \infty) = \infty$, so that the visibility function satisfies

$$\int_0^\infty \mathrm{d}z\, g(z) = -e^{-\tau(0,z)} \Big|_0^\infty = 1, \tag{3.161}$$

as required for a probability density. At early times, τ is large and the visibility function is exponentially suppressed. After recombination, $\mathrm{d}\tau/\mathrm{d}z$ is small because the density of free electrons is small. This means that $g(z)$ will be peaked at the moment of decoupling. Indeed, this is what we find in Fig. 3.18. The maximum of the visibility function is at $z_* = 1080$, which we take as our definition of the moment of last-scattering. The visibility function has a width of $\Delta z_* = 80$, which characterizes the finite width of the last-scattering surface.

Precision cosmology

It is remarkable how much of the complex physics of recombination was understood by Peebles and others in the 1960s. For the analysis of modern CMB experiments, however, the above analysis is still not precise enough. What used to be considered small corrections have become large effects when compared to the high precision of the CMB data. This has motivated a refined analysis of recombination. The details are beyond the scope of this book, but I will sketch the most important developments, since they made the modern era of precision cosmology possible.

---

[25] In Chapter 7, we will define the visibility function in terms of conformal time, $g(\eta) \equiv (e^{-\tau})' = -\tau' e^{-\tau}$, where $' \equiv \partial_\eta$. In that case, the peak of the visibility function is at $z_* = 1090$.

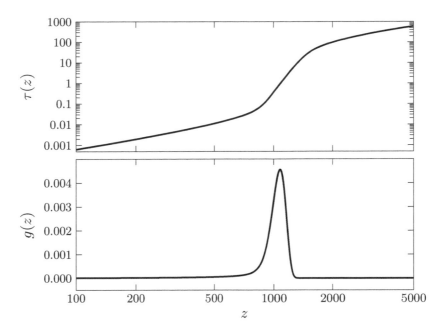

Fig. 3.18 Redshift evolution of the optical depth (3.158) and the visibility function (3.160). The peak of the visibility function at $z_* = 1080$ defines the moment of last-scattering. The width of the function, $\Delta z_* = 80$, measures the thickness of the last-scattering surface.

The effective three-level atom assumes that all excited states are in equilibrium with the $n = 2$ states. This assumption breaks down during the later stages of recombination, leading to a 10% change in the recombination history. The more precise **multi-level atom** was first studied by Seager, Sasselov and Scott in 1999 [46]. It was found that accounting for the additional excited states leads to an increase in the rate of recombination at late times (see Fig. 3.19). To follow the evolution of each state, a large number of coupled Boltzmann equations has to be solved, which numerically is very demanding. Fortunately, it was found that the dynamics can be mimicked by solving an effective three-level atom with an artificially enhanced recombination coefficient, $\alpha \rightarrow 1.14\,\alpha$. This approach became the basis of the recombination code RECFAST.

While RECFAST was precise enough for the analysis of the WMAP data, an unbiased analysis of the Planck data required knowing the recombination history to better than 0.1%. Many new effects were identified that change the predictions at the 1% level and therefore needed to be included in the analysis:

- Seager et al. [46] assumed that states with same energy, but different angular momentum quantum numbers ℓ, were in equilibrium. This assumption breaks down at the percent level [48]. As in the original analysis of Seager et al., solving for the abundances of all the excited states at every time-step is numerically very

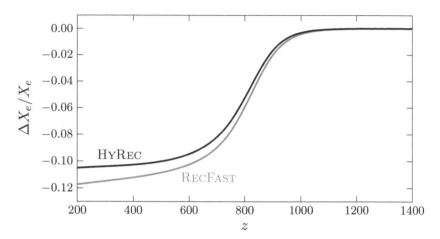

Fig. 3.19 Comparison of the recombination histories predicted by RECFAST and HYREC relative to the Peebles model. Figure adapted from [47].

expensive. An elegant solution to this problem was found by Ali-Haïmoud and Hirata [49]. They realized that only the populations of a few excited states had to be tracked, provided one uses precomputed *effective* transition rates, which capture all the information about the highly excited states of hydrogen. This **effective multi-level atom** is the basis of the recombination codes HYREC [47] and COSMOREC [50].

- Above we have shown that recombination proceeds mainly through the slow escape of redshifted Lyman-α photons and two-photon decays. At the precision of the Planck experiment, the rates for these processes have to be modeled more accurately than we have done. For example, stimulated two-photon emission cannot be ignored and changes the effective two-photon decay rate $\Lambda_{2\gamma}$ at the percent level [51]. Similarly, the absorption of non-thermal photons (produced in previous transitions) must be accounted for [52]. More generally, the time-dependent radiative transfer problem must be solved to high accuracy; see [53–57] for further relevant corrections. Finally, it was realized that two-photon decays from higher excited states, ns and nd, become relevant at this level of accuracy [58–60].

3.3 Summary

In this chapter, we have studied the thermal history of the universe. At early times, the rates of particle interactions, Γ, were much larger than the expansion rate, H, so that all particles were in equilibrium at a common temperature, T. The energy density was dominated by relativistic species

$$\rho = \frac{\pi^2}{30} g_* T^4 \,,$$

(3.162)

where g_* is the effective number of relativistic degrees of freedom (which is the sum of the internal degrees of freedom for each particle weighted by a factor of $7/8$ for fermions). The pressure of the primordial plasma was $P = \rho/3$, so that it behaved like radiation.

An important quantity is the entropy density

$$s = \frac{\rho + P}{T} = \frac{2\pi^2}{45} g_{*S} T^3 \,, \tag{3.163}$$

where g_{*S} is the effective number of relativistic degrees of freedom in entropy, which for most of the universe's history was equal to g_*. The conservation of entropy then implies $g_{*S} T^3 \propto a^{-3}$. We used this to relate the temperature of the cosmic neutrino background to that of the cosmic microwave background, $T_\nu = (4/11)^{1/3} T_\gamma$.

To study non-equilibrium effects, we introduced the Boltzmann equation. For processes of the form $1 + 2 \leftrightarrow 3 + 4$, the Boltzmann equation for the species 1 can be written as

$$\frac{\mathrm{d} \ln N_1}{\mathrm{d} \ln a} = -\frac{\Gamma_1}{H} \left[1 - \left(\frac{N_1 N_2}{N_3 N_4} \right)_{\mathrm{eq}} \frac{N_3 N_4}{N_1 N_2} \right] \,, \tag{3.164}$$

where $N_i = n_i/s$ is proportional to the number of particles in a comoving volume. As long as the interaction rate is larger than the expansion rate, $\Gamma_1 \gg H$, the particle abundance tracks its equilibrium value. Once the interaction rate drops below the expansion rate, however, the particles drop out of thermal equilibrium and decouple from the thermal bath. By integrating the Boltzmann equation, we can follow the non-equilibrium evolution of the particle abundances. We presented three examples: (i) dark matter freeze-out; (ii) Big Bang nucleosynthesis; and (iii) recombination.

As a simple example of dark matter production, we considered the reaction $X + \bar{X} \leftrightarrow \ell + \bar{\ell}$. We used the Boltzmann equation to follow the evolution of the density of dark matter particles X. The resulting dark matter density today depends inversely on the annihilation cross section.

We then studied nucleosynthesis. Initially, the most abundant nuclei were neutrons and protons, kept in equilibrium through reactions of the form

$$n + \nu_e \leftrightarrow p^+ + e^- \,.$$

Around 1 MeV, these reactions became inefficient and the neutrons decoupled. We derived the freeze-out abundance by solving the Boltzmann equation. At 0.1 MeV, neutrons and protons fused into deuterium, which then combined with protons into helium-3 and helium-4:

$$\begin{aligned}
n + p^+ &\to \mathrm{D} + \gamma \,, \\
\mathrm{D} + p^+ &\to {}^3\mathrm{He} + \gamma \,, \\
\mathrm{D} + {}^3\mathrm{He} &\to {}^4\mathrm{He} + p^+ \,.
\end{aligned}$$

The resulting mass fraction of helium-4 is $Y_P \equiv 4n_{\text{He}}/n_{\text{H}} \approx 0.25$, with a weak dependence on the baryon density of the universe.

Finally, we investigated the recombination reaction

$$e^- + p^+ \to \text{H} + \gamma \,.$$

We used the Boltzmann equation to follow the evolution of the free-electron fraction, $X_e \equiv n_e/n_b$. We took into account that recombination directly to the ground state is very inefficient and instead proceeds via two-photon decay and resonance escape from the first excited state. Defining recombination as the time when $X_e = 0.5$, we found

$$z_{\text{rec}} \approx 1270 \,,$$
$$t_{\text{rec}} \approx 290\,000\,\text{yrs} \,. \tag{3.165}$$

As the number of free electrons dropped, the reaction $e^- + \gamma \leftrightarrow e^- + \gamma$ became inefficient and the photons decoupled. This happened at

$$z_{\text{dec}} \approx 1090 \,,$$
$$t_{\text{dec}} \approx 370\,000\,\text{yrs} \,. \tag{3.166}$$

The photons from this era are observed today as the CMB.

Further Reading

The thermal history of the universe is an old topic that has been treated very well in many references.[26] In particular, despite its advanced age, the book by Kolb and Turner [61] still remains a valuable resource for this part of cosmology. My treatment of the thermal history beyond equilibrium was inspired by the presentation in Dodelson's book [32]. My discussion of nucleosynthesis and recombination furthermore benefited from insightful lecture notes by Hirata [62]. A nice overview of observational constraints on BBN can be found in the review by the Particle Data Group [63]. I have used the publicly available code PRIMAT to compute the predictions of BBN [31]. This chapter didn't contain enough about the important topic of baryogenesis. For this, I recommend the reviews by Cline [64] and Riotto [11]. A very clear description of baryogenesis at the popular level is given in [9]. I also didn't say much about phase transitions in the early universe. More on this subject can be found in [65, 66].

The chapter required some basic knowledge of thermodynamics and statistical mechanics. Although I reviewed the fundamentals, students who would like to get a deeper understanding of this material should consult one of the many good textbooks (e.g. [67–69]) or the lecture notes by Tong [70].

---

[26] A wonderful popular book describing this part of cosmology is Weinberg's *The First Three Minutes*.

Problems

3.1 Chemical potential of electrons

In this problem, you will show that the chemical potential of electrons was negligible in the early universe.

1. Consider fermions with $\mu \neq 0$ and $m \ll T$. Show that the difference between the number densities of particles and antiparticles is

$$n - \bar{n} = \frac{gT^3}{6\pi^2} \left[\pi^2 \left(\frac{\mu}{T} \right) + \left(\frac{\mu}{T} \right)^3 \right].$$

Note that this result is exact and not a truncated series.

Hint: You may use that

$$\int_0^\infty \mathrm{d}y \, \frac{y}{e^y + 1} = \frac{\pi^2}{12}.$$

2. Use the above result to show that the chemical potential of electrons satisfies $\mu_e/T \sim 10^{-9}$ and hence can be set to zero in the early universe.

3.2 Conservation of entropy

1. Starting from the distribution function f, and assuming vanishing chemical potential, show that the following holds in equilibrium

$$\frac{\partial P}{\partial T} = \frac{\rho + P}{T}.$$

Use this result to show that entropy is conserved in equilibrium.

Hint: First show that

$$\frac{\partial f}{\partial T} = -\frac{E^2}{Tp} \frac{\partial f}{\partial p}.$$

2. Assuming only Standard Model particles and interactions, we showed that the neutrino and photon temperatures today are related by $T_{\nu,0} = (4/11)^{1/3} T_{\gamma,0}$. Now imagine that after the neutrinos have decoupled from the SM particles, they remain in thermal equilibrium with a new massive spin-1 boson X (and its antiparticle $\bar{X}$). Eventually, this particle self-annihilates and transfers its energy to the neutrinos. How would this affect the neutrino temperature today?

3.3 Degenerate neutrinos

The distribution function for massless neutrinos is

$$f_\nu(p) = \left[\exp\left(\frac{p - \mu_\nu}{T_\nu} \right) + 1 \right]^{-1},$$

where T_ν is the neutrino temperature and μ_ν is the chemical potential. A neutrino species with $|\mu_\nu| \gg T_\nu$ is called *degenerate*.

1. Show that the energy density of degenerate neutrinos is

$$\rho_\nu \approx \frac{|\mu_\nu|^4}{8\pi^2},$$

   where $\mu_\nu \propto T_\nu$ in thermal equilibrium. What is the contribution of the corresponding antineutrinos?

2. Derive a bound on $|\mu_\nu|/T_\nu$ from the requirement that the present energy density of degenerate neutrinos doesn't exceed the critical density, $\rho_{\rm crit} = 6 \times 10^3\, T_0^4$, where T_0 is the CMB temperature.

3. Discuss qualitatively how degenerate neutrinos would affect the production of primordial helium.

3.4 Massive neutrinos

At least two of the three neutrino species in the Standard Model must have small masses. In this problem, you will explore the cosmological consequences of this neutrino mass.

1. Let us assume that the neutrino mass is small enough, so that the neutrinos are relativistic at decoupling. Show that the energy density after decoupling is

$$\rho_\nu = \frac{T_\nu^4}{\pi^2} \int_0^\infty \mathrm{d}\xi\, \frac{\xi^2\, \sqrt{\xi^2 + m_\nu^2/T_\nu^2}}{e^\xi + 1},$$

   where T_ν is the neutrino temperature.

2. By considering a series expansion for small m_ν/T_ν show that

$$\rho_\nu \approx \rho_{\nu 0} \left(1 + \frac{5}{7\pi^2} \frac{m_\nu^2}{T_\nu^2} \right),$$

   where $\rho_{\nu 0}$ is the energy density of massless neutrinos.

3. If ρ_ν is significantly larger than $\rho_{\nu 0}$ at recombination, then the mass of the neutrinos affects the CMB anisotropies. What is the smallest neutrino mass that is observable in the CMB?

If the neutrino mass is larger than the present photon temperature $T_{\gamma,0} \approx 0.235\,\mathrm{meV}$, then these neutrinos will be non-relativistic today.

4. Estimate the redshift at which the neutrinos become non-relativistic.

5. Compute the number density of these neutrinos today.

6. Show that their contribution to the energy density in the universe today is

$$\Omega_\nu h^2 \approx \frac{m_\nu}{94\,\mathrm{eV}}.$$

   Use the lower bound on the sum of the neutrino masses from oscillation experiments, $\sum m_\nu > 0.06\,\mathrm{eV}$, to derive a lower bound on the total neutrino density. How does this compare to the cosmological bound $\Omega_\nu h^2 < 0.001$?

7. Discuss qualitatively why a much larger mass would still be compatible with the standard cosmology.

3.5 Extra relativistic species

Theories beyond the Standard Model may contain extra relativistic species. These particles would add an extra radiation density ρ_X to the early universe. It is conventional to measure this density relative to the density of a neutrino species:

$$\Delta N_{\mathrm{eff}} \equiv \frac{\rho_X}{\rho_\nu}\,,$$

and define $N_{\mathrm{eff}} = 3.046 + \Delta N_{\mathrm{eff}}$ as the *effective number of neutrinos*, although ρ_X may have nothing to do with neutrinos.

1. Derive ΔN_{eff} as a function of the decoupling temperature $T_{\mathrm{dec},X}$ and the spin of the particles. What is the size of ΔN_{eff} if the decoupling occurs before or after the QCD phase transition (but before neutrino decoupling)?

2. Show that ΔN_{eff} increases the effective number of relativistic degrees of freedom in the primordial plasma to

$$g_* = 3.38 + 0.45 \Delta N_{\mathrm{eff}}\,.$$

Discuss qualitatively how an increase in g_* would affect the predictions of Big Bang nucleosynthesis.

3.6 Gravitinos as dark matter

In supergravity theories, the superpartner of the graviton, the *gravitino*, is a spin-3/2 dark matter candidate. Assuming a negligible energy density of gravitinos after inflation, these particles would be produced most efficiently just after reheating. The rate at which Standard Model particles in the primordial plasma would convert into gravitinos is $\Gamma_g \sim T^3/M_{\mathrm{Pl}}^2$. In this problem, you will estimate the relic density of gravitinos.

1. First, argue that the normalized number of gravitinos, $N_g \equiv n_g/s$, is given by

$$N_g \sim \int_0^{T_R} \frac{\mathrm{d}T}{T}\frac{\Gamma_g}{H}\frac{n_g}{s}\,,$$

where T_R is the reheating temperature, s is the total entropy density and n_g is the particle density. Using that $n_g/s \sim O(1)$ and $H \sim T^2/M_{\mathrm{Pl}}$, derive a simple expression for N_g in terms of T_R and M_{Pl}.

2. Assuming one massless neutrino species, show that the total entropy density of the universe today is $s_0 \approx 2000\,\mathrm{cm}^{-3}$.

3. Show that the energy density of gravitinos today would be

$$\Omega_g h^2 \sim 0.1 \left(\frac{T_R}{10^9 \, \text{GeV}} \right) \left(\frac{m_g}{1 \, \text{GeV}} \right).$$

For a reheating temperature of $T_R \sim 10^9 \, \text{GeV}$, gravitinos with masses around $m_g \sim 1 \, \text{GeV}$ could hence constitute all of the dark matter. Note, however, that this scenario is severely constrained by its effects on BBN through the decay of heavier supersymmetric particles.

3.7 Baryon asymmetry

Consider particles and antiparticles with mass m and number densities $n(t)$ and $\bar{n}(t)$. The Boltzmann equation for $n(t)$ is

$$\frac{\mathrm{d}n}{\mathrm{d}t} = -3\frac{\dot{a}}{a}n - n\bar{n}\langle \sigma v \rangle + P(t).$$

Describe the physical significance of each term appearing in this equation.

1. Show that $(n - \bar{n})a^3$ is a constant.

2. Assuming initial particle–antiparticle symmetry, show that the Boltzmann equation can be written as

$$\frac{1}{a^3} \frac{\mathrm{d}(na^3)}{\mathrm{d}t} = -\langle \sigma v \rangle \left[n^2 - n_{\text{eq}}^2 \right],$$

where n_{eq} denotes the equilibrium number density.

3. Taking $\langle \sigma v \rangle$ to be a constant, derive the late-time value of $Y = n/T^3$ in terms of the freeze-out temperature T_f.

4. If there was a speed-up in the expansion rate of the universe caused by the addition of extra relativistic species, what would happen to the abundance of the surviving massive particles and why?

Now apply this to proton–antiproton annihilation. You may use that $\langle \sigma v \rangle \approx 100 \, \text{GeV}^{-2}$ and $m_p \approx 1 \, \text{GeV}$.

5. Estimate the proton-to-photon ratio, n_p/n_γ. How does your result compare to observations?

3.8 Big Bang nucleosynthesis

BBN is sensitive to the physical conditions in the first three minutes of the hot Big Bang. In this problem, you will explore how this can be used to probe various types of physics beyond the Standard Model. You will be asked to assess how changes to the physics during BBN would affect the amount of helium being produced.

1. Let us approximate the freeze-out temperature of the neutrons, T_f, by the decoupling temperature of the neutrinos. Write down an expression for T_f in terms of the number of relativistic species, g_*, Newton's constant, G,

and Fermi's constant, G_F. Explain how a change in T_f affects the neutron-to-proton ratio and the final helium abundance.

2. Describe how a change in the baryon-to-photon ratio η would affect the amount of helium and deuterium produced by BBN.

3. Discuss the effect of the following suppositions on the production of helium:
   - g_*: There were more relativistic species during BBN than we expected.
   - G_F: The weak interaction strength was smaller than we thought.
   - G: Newton's constant was larger during BBN.
   - Q: The neutron–proton mass difference was larger than supposed.
   - τ_n: The neutron lifetime was shorter than assumed.
   - μ_ν: There were many more neutrinos than antineutrinos.

4 Cosmological Inflation

The standard Big Bang theory described in the previous two chapters is incomplete. Key features of the observed universe, such as its large-scale homogeneity and flatness, seem to require that the universe started with very special, finely tuned initial conditions. These initial conditions must be imposed by hand. In this chapter, I will show how inflation—an early period of accelerated expansion—drives the primordial universe towards this special state, even if it started from more generic initial conditions.

To specify the initial conditions of the hot Big Bang, we define the positions and velocities of all particles on an initial time slice, or in the fluid approximation, we specify the density and pressure as a function of position. The laws of gravity and fluid dynamics are then used to evolve the system forward in time. In the previous chapters, we assumed the distribution of matter to be homogeneous and isotropic. But, why is this a good assumption? How do we explain the extraordinary smoothness of the early universe? This is particularly surprising since, in the conventional Big Bang picture, most of the universe appears not to have been in causal contact and there is no dynamical reason why these causally-disconnected regions have such similar physical properties. This homogeneity problem is often called the **horizon problem**.

In addition to specifying the initial density distribution, a complete characterization of the initial conditions also requires the fluid velocities at every point in space. As we will see, in order for the universe to remain homogeneous at late times the initial fluid velocities must take very precise values. If the initial velocities are just slightly too small, the universe recollapses within a fraction of a second. If they are just slightly too big, the universe expands too rapidly and quickly becomes nearly empty. The fine-tuning of the initial velocities is made even more dramatic by considering it in combination with the horizon problem, since the fluid velocities need to be fine-tuned across causally-disconnected regions of space. In Section 2.3.5, I showed that local curvature of a region in space is determined by the sum of the potential and kinetic energies in that region. The fine-tuning of the initial velocities is therefore often called the **flatness problem** and can be phrased as the question why the spatial curvature of the universe is so small.

Maybe it could be argued on the basis of symmetry that perfect homogeneity and exact flatness are a preferred set of initial conditions and appeal to an unknown theory of quantum gravity to pick out this special state of the early universe. Even taking this attitude, however, one then faces the problem that the universe is not perfectly homogeneous, but has small fluctuations in the distribution of matter. Moreover, these fluctuations are not random, but display correlations over very large distances. For example, the correlations observed in the CMB extend over distances that are larger than the distance that light could have traveled since the beginning of the hot Big Bang. What created these **superhorizon correlations**?

One of the amazing features of inflation is that it contains a mechanism to produce the primordial density fluctuations and its superhorizon correlations. Small quantum fluctuations get stretched by the inflationary expansion and become the seeds for the formation of the large-scale structure of the universe. Chapter 8 is devoted to a detailed account of this. In this chapter, I will present the basic idea of cosmic inflation and explain how it solves the problems of the standard hot Big Bang.

4.1 Problems of the Hot Big Bang

Above I have sketched three problems of the hot Big Bang model: the horizon problem, the flatness problem and the problem of superhorizon correlations. In this section, I will describe these problems in more detail and make them quantitative.

4.1.1 The Horizon Problem

The size of a causally-connected patch of space is determined by the maximal distance from which light can be received. This is best studied in comoving coordinates, where null geodesics are straight lines and the distance between two points equals the corresponding difference in conformal time, $\Delta\eta$. Hence, if the Big Bang "started" with the singularity at $t_i \equiv 0$,[1] then the greatest comoving distance from which an observer at time t will be able to receive signals traveling at the speed of light is the (comoving) **particle horizon**:

$$d_h(\eta) = \eta - \eta_i = \int_{t_i}^{t} \frac{\mathrm{d}t}{a(t)} \,. \tag{4.1}$$

I use the term "particle horizon" to make a distinction with the concept of an "event horizon" which is the maximal distance that we can influence future events, rather than the maximal distance from which we can be influenced by past events. In the conventional Big Bang cosmology, we can set $\eta_i \equiv 0$ and the comoving particle horizon is simply equal to the conformal time. The size of the horizon at time η

---

[1] Notice that the Big Bang singularity is a *moment in time*, **not** a *point in space*. Indeed, in Figures 4.1 and 4.2 we depict the singularity by an extended (possibly infinite) spacelike hypersurface.

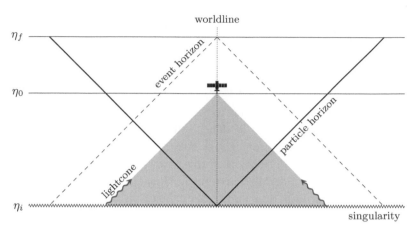

Fig. 4.1   Spacetime diagram illustrating the concept of cosmological horizons. The particle horizon describes the maximal distance from which an observer can receive signals. These signals must lie in the past lightcone of the observer. The event horizon is the maximal distance to which the observer can send signals in the future. Note that η_f can be finite even if t_f is infinite.

may be visualized by the intersection of the past lightcone of an observer with the spacelike surface at η_i (see Fig. 4.1). Causal influences have to come from within this region. Signals from outside this region would have to travel faster than the speed of light to reach the observer.

Equation (4.1) can be written in the following illuminating way

$$d_h(\eta) = \int_{t_i}^{t} \frac{dt}{a(t)} = \int_{a_i}^{a} \frac{da}{a\dot{a}} = \int_{\ln a_i}^{\ln a} (aH)^{-1}\, d\ln a\,, \tag{4.2}$$

where $a_i \equiv 0$ corresponds to the Big Bang singularity. The causal structure of the spacetime is hence related to the evolution of the **comoving Hubble radius**, $(aH)^{-1}$. For ordinary matter sources, the comoving Hubble radius is a monotonically increasing function of time (or scale factor), and the integral in (4.2) is dominated by the contributions from late times. This implies that in the standard cosmology $d_h \sim (aH)^{-1}$, which has led to the confusing practice of referring to both the particle horizon and the Hubble radius as the "horizon."

We conclude that the amount of conformal time between the initial singularity and the emission of the CMB[2] (or, equivalently, the comoving horizon at the time of recombination) was much smaller than the conformal age of the universe today (or, equivalently, the comoving distance to the last-scattering surface), $\eta_{\rm rec} \ll \eta_0$. A moment's thought will convince you that this implies a serious problem: it means that most parts of the CMB have non-overlapping past lightcones and hence never were in causal contact. This is illustrated by the spacetime diagram in Fig. 4.2.

---

[2] In this chapter, we won't be careful to distinguish between recombination and photon decoupling, and simply refer to both as "recombination."

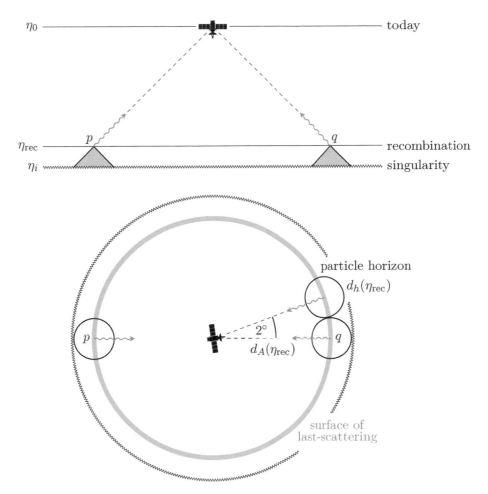

Fig. 4.2 Illustration of the horizon problem in the conventional Big Bang model. All events that we currently observe are on our past lightcone. The intersection of our past lightcone with the spacelike slice at the time of recombination is the surface of last-scattering. Points that are separated by more than 2° on the sky seem never to have been in causal contact, since their past lightcones don't overlap.

As an extreme example, consider two opposite directions on the sky. The CMB photons that we receive from these directions were emitted at the points labeled p and q in Fig. 4.2. We see that the photons were released sufficiently close to the Big Bang singularity that the past lightcones of p and q don't overlap. Since no point lies inside the horizons of both p and q, we have the following puzzle: how do the photons coming from p and q "know" that they should be at almost exactly the same temperature? There was simply not enough time for differences in the initial temperatures to be erased by heat transfer. As I will show below, the same question applies to any two points in the CMB that are separated by more than 2 degrees on the sky (about four times the angular size of the Moon, seen from Earth).

The homogeneity of the CMB spans scales that are much larger than the particle horizon at the time when the CMB was emitted. In fact, in the standard cosmology the CMB consists of over $40\,000$ causally-disconnected patches of space. If there wasn't enough time for these regions to communicate, why do they look so similar? This is the **horizon problem**.

To make this more explicit, consider a flat universe filled with only matter and radiation. The Friedmann equation gives us the evolution of the Hubble rate

$$H^2 = H_0^2 \left[\Omega_{m,0} a^{-3} + \Omega_{r,0} a^{-4} \right], \quad \text{with} \quad \Omega_{r,0} = a_{\rm eq} \Omega_{m,0}. \tag{4.3}$$

The comoving Hubble radius can then be written as

$$(aH)^{-1} = \frac{1}{\sqrt{\Omega_{m,0}}} H_0^{-1} \frac{a}{\sqrt{a + a_{\rm eq}}}. \tag{4.4}$$

We see that the Hubble radius is indeed growing, as we claimed earlier. The conformal time as a function of the scale factor then is

$$\eta = \int_0^a \frac{{\rm d}\ln a}{aH} = \frac{2}{\sqrt{\Omega_{m,0}}} H_0^{-1} \left(\sqrt{a + a_{\rm eq}} - \sqrt{a_{\rm eq}} \right). \tag{4.5}$$

Notice that this result has the correct limits: at early times ($a \ll a_{\rm eq} \approx 3400^{-1}$), we get $\eta \propto a$ (as expected for the radiation-dominated era), while at late times ($a \gg a_{\rm eq}$), we have $\eta \propto a^{1/2}$ (as expected for the matter-dominated era). The conformal times today ($a_0 = 1$) and at recombination ($a_{\rm rec} \approx a_{\rm dec} \approx 1100^{-1}$) are

$$\eta_0 \approx \frac{2}{\sqrt{\Omega_{m,0}}} H_0^{-1}, \tag{4.6}$$

$$\eta_{\rm rec} = \frac{2}{\sqrt{\Omega_{m,0}}} H_0^{-1} \left[\sqrt{1100^{-1} + 3400^{-1}} - \sqrt{3400^{-1}} \right] \approx 0.0175\, \eta_0. \tag{4.7}$$

Using the observed value of the Hubble constant, the comoving horizon at recombination is $d_h(\eta_{\rm rec}) = \eta_{\rm rec} \approx 265\,{\rm Mpc}$. This should be compared to the comoving distance to the surface of last-scattering, which in a flat universe is[3] $d_A(\eta_{\rm rec}) = \eta_0 - \eta_{\rm rec} \approx \eta_0 \approx 15.1\,{\rm Gpc}$. The angle subtended by the horizon at recombination then is

$$\theta_h = \frac{2 d_h(\eta_{\rm rec})}{d_A(\eta_{\rm rec})} = \frac{2\eta_{\rm rec}}{\eta_0 - \eta_{\rm rec}} = \frac{2 \times 0.0175}{1 - 0.0175} = 0.036\,{\rm rad} \approx 2.0°. \tag{4.8}$$

Including dark energy changes the distance to the surface of last-scattering by a factor of 0.9, so that $\theta_h \approx 2.3°$.[4]

Let me digress briefly to make an important qualifier: when we inferred that the total conformal time between the singularity and recombination is finite and small,

---

[3] Note that we used the same notation, d_A, for the *physical* angular diameter distance in Section 2.2.3.

[4] As we will see in Chapters 6 and 7, fluctuations in the primordial plasma actually propagate at the *speed of sound*, which is slightly smaller than the speed of light, $c_s = c/\sqrt{3}$. The angular size of the corresponding sound horizon is $\theta_s = \theta_h/\sqrt{3} \approx 1.3°$, which explains why the pattern of CMB fluctuations has special features on the scale of 1 degree.

we included times in the integral in (4.2) that were arbitrarily close to the initial singularity:

$$\Delta \eta = \int_0^{\delta t} \frac{dt}{a(t)} + \int_{\delta t}^t \frac{dt}{a(t)} . \tag{4.9}$$

However, in the first integral in (4.9) we have *no* reason to trust the classical geometry. By stating the horizon problem as we did, we implicitly assumed that the breakdown of general relativity in the regime close to the singularity does *not* lead to a large contribution to the conformal time: i.e. we assumed $\delta \eta \ll \Delta \eta$. This assumption may be incorrect and there may, in fact, be no horizon problem in a complete theory of quantum gravity. In the absence of an alternative solution to the horizon problem this is a completely reasonable attitude to take. However, we will soon see that inflation provides a simple and computable solution to the horizon problem. Effectively, this is achieved by modifying the scale factor evolution in the second integral in (4.9), i.e. in the *classical* regime. I then leave it to you to decide if this solution or a version of "quantum gravity magic" is preferable.

4.1.2 The Flatness Problem

Next, I will discuss the flatness problem. As we will see, it is closely related to the horizon problem, so that it seems likely that any solution to the horizon problem will also address the flatness problem.

Let us define the time-dependent critical density of the universe as $\rho_{\rm crit}(t) = 3M_{\rm Pl}^2 H^2$. The *time-dependent* curvature parameter then is

$$\Omega_k(t) = \frac{\rho_{\rm crit} - \rho}{\rho_{\rm crit}} = \frac{(a_0 H_0)^2}{(aH)^2} \Omega_{k,0} , \tag{4.10}$$

where we used that $\rho_{\rm crit} \propto H^2$ and $\rho_{\rm crit} - \rho \propto a^{-2}$. Recall that observations provide an upper bound on the size of the curvature parameter today, $|\Omega_{k,0}| < 0.005$. Because the comoving Hubble radius $(aH)^{-1}$ is growing during the conventional hot Big Bang, we expect that $|\Omega_k(t)|$ was even smaller in the past. Let me quantify this. Ignoring the short period of dark energy domination, the comoving Hubble radius is given by (4.4) and we can write (4.10) as

$$\Omega_k(t) = \frac{\Omega_{k,0}}{\Omega_{m,0}} \frac{a^2}{a + a_{\rm eq}} , \tag{4.11}$$

where we have set $a_0 \equiv 1$. At matter–radiation equality, this implies

$$|\Omega_k(t_{\rm eq})| = \frac{|\Omega_{k,0}|}{\Omega_{m,0}} \frac{a_{\rm eq}}{2} < 10^{-6} . \tag{4.12}$$

At earlier times, the universe was dominated by radiation and we can use

$$H^2 = H_{\rm eq}^2 \Omega_{r,{\rm eq}} \left(\frac{a_{\rm eq}}{a} \right)^4 , \tag{4.13}$$

with $\Omega_{r,\text{eq}} = 0.5$. The curvature parameter then becomes

$$\Omega_k(t) = \frac{(a_{\text{eq}}H_{\text{eq}})^2}{(aH)^2}\,\Omega_k(t_{\text{eq}}) = 2\Omega_k(t_{\text{eq}})\left(\frac{a}{a_{\text{eq}}}\right)^2. \tag{4.14}$$

Evaluating this at the time of Big Bang nucleosynthesis, $z_{\text{BBN}} \approx 4 \times 10^8$, or the electroweak phase transition, $z_{\text{EW}} \approx 10^{15}$, we find

$$|\Omega_k(t_{\text{BBN}})| < 10^{-16}, \tag{4.15}$$

$$|\Omega_k(t_{\text{EW}})| < 10^{-29}. \tag{4.16}$$

At even earlier times, the curvature parameter is constrained to be even smaller.

A useful way of rephrasing the problem is in terms of the curvature scale $R(t)$, which is related to $\Omega_k(t)$ as follows

$$R(t) = \frac{1}{\sqrt{|\Omega_k(t)|}}\,H^{-1}(t). \tag{4.17}$$

We have seen that observations constrain the curvature scale today to be $R(t_0) > 14H_0^{-1}$. The above constraints on $\Omega_k(t)$ then imply that the curvature scale in the early universe was many orders of magnitude larger than the Hubble radius at that time. Since, in the standard Big Bang cosmology, the Hubble radius is of the same order as the particle horizon, this suggest a fine-tuning of many causally-disconnected patches of space. Finally, in Section 2.3.5, I related the spatial curvature of a region of space to the sum of the kinetic and potential energies in that region. The flatness problem can therefore also be viewed as the problem of adjusting the initial velocities of all particles over distances that naively have never been in causal contact.

4.1.3 Superhorizon Correlations

It is possible to dismiss the horizon and flatness problems because perfect homogeneity and flatness are the simplest and most symmetrical options, to be picked out by an unknown theory of the initial conditions. It is what you might have guessed given the options. What cannot be dismissed, however, is the fact that the universe contains density fluctuations that are correlated over apparently acausal distances. This modern version of the horizon problem begs for a dynamical explanation. As David Tong has put it, "These detailed correlations make it more difficult to appeal to a creator without sounding like a young Earth creationist, arguing that the fossil record was planted to deceive us" [1].

To understand the severity of the problem, let us consider the fluctuations that we observe in the CMB and in the large-scale structure of the universe. Figure 4.3 shows the evolution of a representative fluctuation of wavelength λ relative to the Hubble radius $(aH)^{-1}$. In comoving coordinates, the wavelength λ is fixed and the Hubble radius increases. This means that any fluctuation that is inside the Hubble radius today was outside the Hubble radius at sufficiently early times. For the

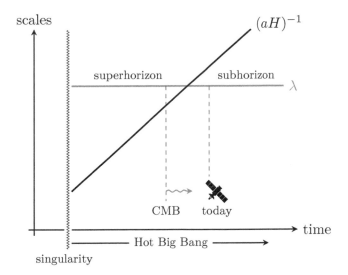

Fig. 4.3   Evolution of a representative fluctuation of fixed (comoving) wavelength λ relative to the Hubble radius $(aH)^{-1}$.

standard hot Big Bang, the Hubble radius is approximately equal to the particle horizon, so we call these regimes "subhorizon" and "superhorizon." Above I showed that the particle horizon at recombination was about 265 Mpc. Scales larger than this would not have been inside the horizon before the CMB was emitted. Yet, we find the CMB fluctuations to be correlated on scales that are larger than this apparent horizon. This is the modern version of the horizon problem. Not only is the CMB homogeneous on apparently acausal scales, it also has correlated fluctuations on these scales.

4.2 Before the Hot Big Bang

The causality issues described above suggest that there was a phase *before the hot Big Bang* during which the homogeneity of the universe and its correlated fluctuations were generated. In the following, I will describe what characteristics this early period needs to have in order to solve the puzzles of the conventional Big Bang cosmology.

4.2.1 A Shrinking Hubble Sphere

Our description of the horizon and flatness problems has highlighted the fundamental role played by the growing comoving Hubble radius of the standard Big

Bang cosmology. A simple solution to these problems therefore suggests itself: let us conjecture a phase of *decreasing comoving Hubble radius* in the early universe,

$$\boxed{\frac{\mathrm{d}}{\mathrm{d}t}(aH)^{-1} < 0} \, . \tag{4.18}$$

I like the shrinking Hubble sphere as the fundamental definition of inflation since it relates most directly to the horizon problem and is also a key feature of the inflationary mechanism for generating fluctuations (see Chapter 8). Using $aH = \dot{a}$, it is easy to see that this is equivalent to $\ddot{a} > 0$, leading to the familiar notion of inflation as a period of *accelerated expansion*. If inflation lasts long enough, then the horizon and flatness problems can be avoided.

It will now be important to distinguish clearly between the particle horizon and the Hubble radius. Both are related to causality: The Hubble radius is the distance that particles can travel in the course of one expansion time, H^{-1}, which is roughly the time in which the scale factor doubles. The Hubble radius is therefore another way of measuring whether particles are causally connected with each other: if they are separated by distances larger than the Hubble radius, then they cannot communicate with each other *now*. In contrast, particles that are separated by distances greater than the particle horizon could *never* have communicated with one another. Inflation is a mechanism to make the particle horizon much larger than the Hubble radius. This means that particles can't communicate now, but were in causal contact early on.

4.2.2 Horizon Problem Revisited

Let us show explicitly how the shrinking Hubble sphere solves the horizon problem. Consider again the expression (4.2) for the comoving particle horizon

$$d_h(\eta) = \int_{\ln a_i}^{\ln a} (aH)^{-1} \, \mathrm{d} \ln a \, . \tag{4.19}$$

In the standard Big Bang cosmology, the integral is dominated by late times and the particle horizon is of the same order as the Hubble radius, $d_h(\eta) \sim (aH)^{-1}$. However, if the comoving Hubble radius is instead a decreasing function of time, then the integral is dominated by early times and the particle horizon can be much larger than the Hubble radius, $d_h(\eta) \sim (a_i H_i)^{-1} \gg (aH)^{-1}$. As we will see, this is precisely what the doctor ordered.

To illustrate this further, consider a flat universe dominated by a fluid with a constant equation of state $w \equiv P/\rho$. From the Friedmann equation, we then infer that the evolution of the comoving Hubble radius is

$$(aH)^{-1} = H_0^{-1} a^{\frac{1}{2}(1+3w)} \, , \tag{4.20}$$

and the comoving particle horizon becomes

$$d_h(a) = \frac{2H_0^{-1}}{(1+3w)} \left[a^{\frac{1}{2}(1+3w)} - a_i^{\frac{1}{2}(1+3w)} \right] \equiv \eta - \eta_i \, . \tag{4.21}$$

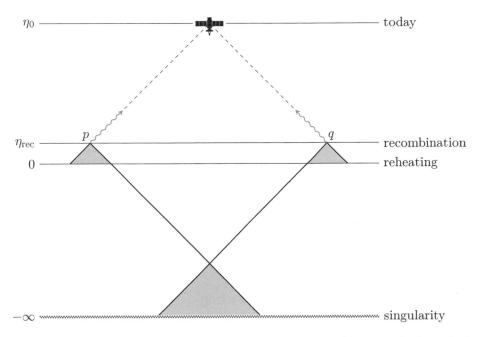

Fig. 4.4 Inflationary solution to the horizon problem. The spacelike singularity of the standard Big Bang is replaced by the reheating surface. Rather than marking the beginning of time, $\eta = 0$ now corresponds to the transition from inflation to the standard Big Bang evolution. All points in the CMB have overlapping past lightcones and therefore originated from a causally-connected region of space.

Ordinary matter satisfies the **strong energy condition** (SEC), $1 + 3w > 0$, so that the Hubble radius grows and the particle horizon is dominated by late times, $a \gg a_i$. Specifically, we can set $\eta_i = 0$ for $a_i = 0$ and get $d_h = \eta$. Now, take $1 + 3w < 0$ instead. The Hubble radius then shrinks and the particle horizon receives most of its contribution from early times. In fact, we now have

$$\eta_i \equiv \frac{2H_0^{-1}}{(1 + 3w)} \, a_i^{\frac{1}{2}(1+3w)} \xrightarrow{\;a_i \to 0\,,\;w < -\frac{1}{3}\;} -\infty\,. \qquad (4.22)$$

The Big Bang singularity has been pushed to *negative conformal time*, so that there was actually much more conformal time between the singularity and recombination than we had thought.

Figure 4.4 shows the new spacetime diagram. Let us denote the beginning and the end of inflation by η_i and η_e, respectively. If $|\eta_e - \eta_i| \gg \eta_0 - \eta_{\mathrm{rec}}$, then the past lightcones of widely separated points in the CMB had enough time to intersect. The uniformity of the CMB is not a mystery anymore. In inflationary cosmology, $\eta = 0$ is *not* the initial singularity, but instead is only a transition point between inflation and the hot Big Bang. There is time both before and after $\eta = 0$.[5]

---

[5] Notice that the Big Bang singularity is still at $t = 0$. In physical coordinates, the time between the singularity and recombination is fixed, but the lightcones get stretched drastically by the inflationary expansion allowing them to overlap before the end of inflation.

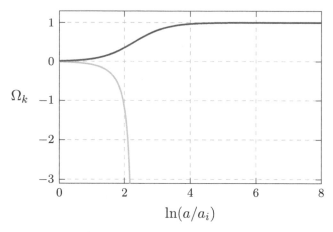

$$\ln(a/a_i)$$

Fig. 4.5 Evolution of the curvature parameter Ω_k in a radiation-dominated universe. Shown are cases with initial curvatures $\Omega_{k,i} = +0.01$ (*black*) and $\Omega_{k,i} = -0.01$ (*gray*). For $\Omega_{k,i} = +0.01$, the universe rapidly evolves towards an empty universe with $\Omega_k = 1$, while, for $\Omega_{k,i} = -0.01$, the curvature parameter diverges at $N_* \approx 2.3$, as predicted in (4.26).

4.2.3 Flatness Problem Revisited

Next, we show how the shrinking Hubble sphere also solves the flatness problem. We have seen that the evolution of the curvature parameter is related to the evolution of the comoving Hubble radius as

$$\Omega_k(t) = \frac{(a_i H_i)^2}{(aH)^2} \, \Omega_k(t_i) \, . \tag{4.23}$$

Any initial curvature will therefore decrease if the comoving Hubble radius decreases. If inflation lasts long enough it therefore solves the flatness problem.

To see this more explicitly, let us again consider the case of a perfect fluid with a constant equation of state w. The Friedmann equation then implies

$$(aH)^2 = (a_i H_i)^2 \left[(1 - \Omega_{k,i})(a/a_i)^{-(1+3w)} + \Omega_{k,i} \right], \tag{4.24}$$

and substituting this into (4.23), we get

$$\Omega_k(N) = \frac{\Omega_{k,i} \, e^{(1+3w)N}}{(1 - \Omega_{k,i}) + \Omega_{k,i} \, e^{(1+3w)N}} \, , \tag{4.25}$$

where $N \equiv \ln(a/a_i)$ is the number of "e-folds" of expansion. This solution has two fixed points at $\Omega_{k,i} = 0$ (a flat universe) and $\Omega_{k,i} = 1$ (an empty Milne universe). Figure 4.5 shows the evolution of Ω_k for a radiation-dominated universe, but the qualitative behavior is the same for any fluid with $1 + 3w > 0$. We see that any deviation from $\Omega_k = 0$ is unstable and grows with time. For a negatively curved

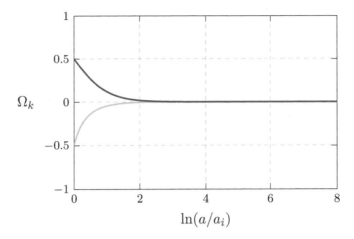

Fig. 4.6 Evolution of the curvature parameters Ω_k in a universe dominated by vacuum energy. Shown are two examples with initial curvatures $\Omega_{k,i} = \pm 0.5$. We see that a flat universe, with $\Omega_k = 0$, is an attractor.

universe (with $\Omega_k > 0$), the growth then slows as Ω_k approaches its second fixed point at $\Omega_k = 1$. This final state is an empty universe with negative curvature. For a positively curved universe (with $\Omega_k < 0$), the growth accelerates and the curvature parameter diverges at

$$N_* = \frac{1}{1 + 3w} \ln \left| \frac{\Omega_{k,i} - 1}{\Omega_{k,i}} \right|. \tag{4.26}$$

This divergence occurs when the Hubble parameter vanishes, cf. (4.24), and therefore corresponds to the turn-around point of the scale factor (see Fig. 2.12).

Exercise 4.1 Show that the curvature parameter in (4.23) satisfies the following evolution equation

$$\frac{\mathrm{d}\Omega_k}{\mathrm{d}N} = (1 + 3w)\, \Omega_k (1 - \Omega_k). \tag{4.27}$$

This makes it transparent that $\Omega_k = 0$ and $\Omega_k = 1$ are fixed points, and that the behavior away from these fixed points depends on the equation of state.

The evolution of the curvature parameter is drastically different if the fluid violates the SEC, $1 + 3w < 0$. The solution in (4.25) still applies and shows that $|\Omega_k|$ decreases as $e^{-|1+3w|N}$, so that the fixed point $\Omega_k = 0$ is now an **attractor** (see Fig. 4.6). The flatness problem is solved if the inflationary expansion lasts long enough to reduce any initial curvature to such a low value at the beginning of the hot Big Bang that the subsequent expansion doesn't increase it above the observational bound of $|\Omega_{k,0}| < 0.005$. In practice, this means between 40 and 60 e-folds of inflationary expansion (see Section 4.2.5).

4.2.4 Superhorizon Correlations

The shrinking Hubble sphere also explains how the fluctuations observed in the CMB can be correlated on apparently superhorizon scales. Figure 4.7 shows the evolution of a fixed comoving scale relative to the Hubble radius. If inflation lasted long enough, this scale started its life inside the Hubble radius, where causal processes were able to create nontrivial correlations. The fluctuations then exited the Hubble radius[6] and are measured today as apparently superhorizon correlations in the CMB. Notice the symmetry of the evolution. Scales just entering the Hubble radius today, 60 e-folds *after* the end of inflation, left the Hubble radius 60 e-folds *before* the end of inflation.

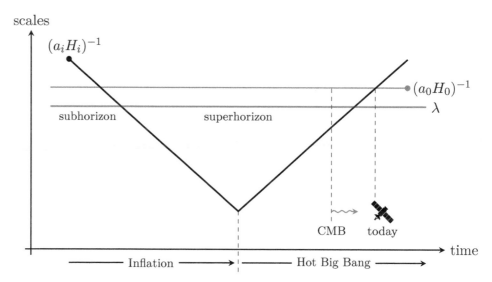

Fig. 4.7 Illustration of the way that inflation solves the problem of superhorizon correlations. Shown is a representative mode that appears to be outside of the horizon at the time of recombination, but was inside the horizon during inflation.

4.2.5 Duration of Inflation

How much inflation is needed to solve the problems of the hot Big Bang? At the very least, we require that all observed fluctuations were inside the particle horizon at early times. Since the Hubble radius is easier to calculate than the particle horizon

---

[6] There is a common, but confusing, practice of referring to the Hubble radius simply as the "horizon" and the exiting of the Hubble radius as the "horizon crossing." We will later adopt this terminology, but it is important to keep in mind that the Hubble radius and the horizon are vastly different during inflation.

it is common to use the Hubble radius as a means of judging the causality problems of the hot Big Bang. In particular, the problems are solved if the entire observable universe was smaller than the comoving Hubble radius at the beginning of inflation (see Fig. 4.7),

$$(a_0 H_0)^{-1} < (a_i H_i)^{-1}.$$ (4.28)

Notice that this is more conservative than using the particle horizon, since $d_h(t_i)$ is bigger than $(a_i H_i)^{-1}$. Using $(a_i H_i)^{-1}$ to assess the horizon problem also has the advantage that we don't have to assume anything about earlier times $t < t_i$, while the value of $d_h(t_i)$ depends on the entire history of the universe before inflation.

Below we will see that the physical Hubble rate is nearly constant during inflation, so that $H_i \approx H_e$. The amount by which the comoving Hubble radius decreases during inflation is therefore equal to amount by which the scale factor increases. The latter is typically expressed in terms of the number of e-foldings

$$N_{\text{tot}} \equiv \ln(a_e/a_i).$$ (4.29)

We would like to determine the minimal number of e-foldings that is required to solve the problems of the hot Big Bang.

As illustrated in Fig. 4.7, the decrease of the comoving Hubble radius during inflation must compensate for its increase during the hot Big Bang. The amount by which the Hubble radius has grown during the hot Big Bang depends on the maximal temperature of the thermal plasma at the beginning of the hot Big Bang. We will denote this so-called **reheating temperature** by T_R. For simplicity, we take the universe to be radiation dominated throughout and ignore the relatively recent periods of matter and dark energy domination. Remembering that $H \propto a^{-2}$ during the radiation-dominated era, we have

$$\frac{a_0 H_0}{a_R H_R} = \frac{a_0}{a_R} \left(\frac{a_R}{a_0}\right)^2 = \frac{a_R}{a_0} \frac{T_0}{T_R} \sim 10^{-28} \left(\frac{10^{15}\,\text{GeV}}{T_R}\right),$$ (4.30)

where we have introduced a reference value of 10^{15} GeV for the reheating temperature. We will furthermore assume that the energy density at the end of inflation was converted relatively quickly into the particles of the thermal plasma, so that the Hubble radius didn't experience significant growth between the end of inflation and the beginning of the hot Big Bang, $(a_e H_e)^{-1} \sim (a_R H_R)^{-1}$. The condition in (4.28) can then be written as

$$(a_i H_i)^{-1} > (a_0 H_0)^{-1} \sim 10^{28} \left(\frac{T_R}{10^{15}\,\text{GeV}}\right)(a_e H_e)^{-1},$$ (4.31)

and, using $H_i \approx H_e$, we get

$$N_{\text{tot}} \equiv \ln(a_e/a_i) > 64 + \ln\left(T_R/10^{15}\,\text{GeV}\right).$$ (4.32)

This is the famous statement that the solution of the horizon problem requires about 60 e-folds of inflation. Notice that fewer e-folds are needed if the reheating temperature is lower.

4.3 The Physics of Inflation

In the previous section, we presented the basic conditions under which the horizon and flatness problems can be avoided. To make this a theory of the early universe, we must introduce a physical mechanism that gives rise to this evolution.

A key characteristic of inflation is that all physical quantities are *slowly varying*, despite the fact that the space is expanding rapidly. To see this, let us write the time derivative of the comoving Hubble radius as

$$\frac{d}{dt}(aH)^{-1} = -\frac{\dot{a}H + a\dot{H}}{(aH)^2} = -\frac{1}{a}(1 - \varepsilon),$$
(4.33)

where we have introduced the **slow-roll parameter**

$$\boxed{\varepsilon \equiv -\frac{\dot{H}}{H^2} = -\frac{d\ln H}{dN}},$$
(4.34)

with $dN \equiv d\ln a = H dt$. This shows that a shrinking Hubble radius, $\partial_t(aH)^{-1} < 0$, is associated with $\varepsilon < 1$. In other words, inflation occurs if the fractional change of the Hubble parameter, $\Delta \ln H = \Delta H/H$, per e-folding of expansion, ΔN, is small. Moreover, as we will see in Chapter 8, the near scale-invariance of the observed fluctuations in fact requires that $\varepsilon \ll 1$. In the limit $\varepsilon \to 0$, the dynamics becomes time-translation invariant, $H = \text{const}$, and the spacetime is **de Sitter space**

$$ds^2 = -dt^2 + e^{2Ht}d\mathbf{x}^2.$$
(4.35)

Of course, inflation has to end, so the time-translation symmetry must be broken and the spacetime must deviate from a perfect de Sitter space. However, for small, but finite $\varepsilon \neq 0$, the line element (4.35) is still a good approximation to the inflationary background. This is why inflation is also often referred to as a *quasi-de Sitter period*.

Finally, we want inflation to last for a sufficiently long time (usually at least 40 to 60 e-folds), which requires that ε remains small for a sufficiently large number of Hubble times. This condition is measured by a second slow-roll parameter[7]

$$\boxed{\kappa \equiv \frac{d\ln\varepsilon}{dN} = \frac{\dot{\varepsilon}}{H\varepsilon}}.$$
(4.36)

For $|\kappa| < 1$, the fractional change of ε per e-fold is small and inflation persists. In the next section, we will discuss what microscopic physics can lead to the conditions $\varepsilon < 1$ and $|\kappa| < 1$.

---

[7] This parameter is often denoted η, but I have chosen to use η for conformal time. Moreover, the parameters ε and κ are often called **Hubble slow-roll parameters** to distinguish them from the **potential slow-roll parameters** defined in Section 4.3.1.

4.3.1 Scalar Field Dynamics

The simplest models of inflation implement the time-dependent dynamics during inflation in terms of the evolution of a scalar field, $\phi(t,\mathbf{x})$, called the **inflaton**. As indicated by the notation, the value of the field can depend on the time t and the position $\mathbf{x}$. Associated with each field value is a potential energy density $V(\phi)$ (see Fig. 4.8). When the field is dynamical (i.e. changes with time) then it also carries a kinetic energy density $\frac{1}{2}\dot{\phi}^2$. If the energy associated with the scalar field dominates the universe, then it sources the evolution of the FRW background. We want to determine under which conditions this can drive an inflationary expansion.

Let us begin with a scalar field in Minkowski space. Its action is

$$S = \int \mathrm{d}t\,\mathrm{d}^3x\left[\frac{1}{2}\dot{\phi}^2 - \frac{1}{2}(\nabla\phi)^2 - V(\phi)\right],\tag{4.39}$$

where we have also included the gradient energy, $\frac{1}{2}(\nabla\phi)^2$, associated with a spatially varying field. To determine the equation of motion of the scalar field, we consider the variation $\phi \to \phi + \delta\phi$. Under this variation, the action changes as

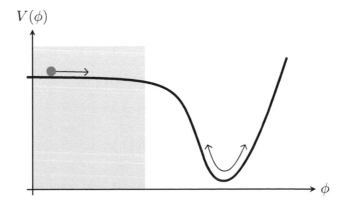

Fig. 4.8 Example of a slow-roll potential. Inflation occurs in the shaded part of the potential.

$$\delta S = \int \mathrm{d}t\, \mathrm{d}^3x \left[\dot\phi \delta\dot\phi - \nabla\phi \cdot \nabla\delta\phi - \frac{\mathrm{d}V}{\mathrm{d}\phi}\, \delta\phi \right] \tag{4.40}$$

$$= \int \mathrm{d}t\, \mathrm{d}^3x \left[-\ddot\phi + \nabla^2\phi - \frac{\mathrm{d}V}{\mathrm{d}\phi} \right] \delta\phi\,, \tag{4.41}$$

where we have integrated by parts and dropped a boundary term. If the variation is around the classical field configuration, then the principle of least action states that $\delta S = 0$. For this to be valid for an arbitrary field variation $\delta\phi$, the expression in the square brackets in (4.41) must vanish:

$$\ddot\phi - \nabla^2\phi = -\frac{\mathrm{d}V}{\mathrm{d}\phi}\,. \tag{4.42}$$

This is the **Klein–Gordon equation**.

It is straightforward to extend this analysis to the evolution in an expanding FRW spacetime. For simplicity, we will ignore spatial curvature. Remember that physical coordinates are related to comoving coordinates by the scale factor, $a(t)\mathbf{x}$. Taking this into account, it is easy to guess that the generalization of the action (4.39) to an FRW background is

$$S = \int \mathrm{d}t\, \mathrm{d}^3x\, a^3(t) \left[\frac{1}{2}\dot\phi^2 - \frac{1}{2a^2(t)}(\nabla\phi)^2 - V(\phi) \right]. \tag{4.43}$$

We are interested in the evolution of a homogeneous field configuration, $\phi = \phi(t)$, in which case the action reduces to

$$S = \int \mathrm{d}t\, \mathrm{d}^3x\, a^3(t) \left[\frac{1}{2}\dot\phi^2 - V(\phi) \right]. \tag{4.44}$$

Performing the same variation of the action as before, we find

$$\delta S = \int \mathrm{d}t\, \mathrm{d}^3x\, a^3(t) \left[\dot\phi \delta\dot\phi - \frac{\mathrm{d}V}{\mathrm{d}\phi}\, \delta\phi \right] \tag{4.45}$$

$$= \int \mathrm{d}t\, \mathrm{d}^3x \left[-\frac{\mathrm{d}}{\mathrm{d}t}\left(a^3\dot\phi \right) - a^3 \frac{\mathrm{d}V}{\mathrm{d}\phi} \right] \delta\phi\,, \tag{4.46}$$

and the principle of least action leads to the following Klein–Gordon equation

$$\boxed{\ddot\phi + 3H\dot\phi = -\frac{\mathrm{d}V}{\mathrm{d}\phi}\,.} \tag{4.47}$$

The expansion has introduced one new feature, the so-called **Hubble friction** associated with the term $3H\dot\phi$. This friction will play a crucial role in the inflationary dynamics.

Let us now assume that this scalar field dominates the universe and determine its effect on the expansion. To use the Friedmann equations, we need to know the energy density and pressure associated with the field. Given the form of the action in (4.44), it is natural to guess that the energy density is the sum of kinetic and potential energy densities

$$\boxed{\rho_\phi = \frac{1}{2}\dot{\phi}^2 + V(\phi)}\,.$$

(4.48)

Taking the time derivative of the energy density, we find

$$\dot{\rho}_\phi = \left(\ddot{\phi} + \frac{\mathrm{d}V}{\mathrm{d}\phi}\right)\dot{\phi} = -3H\dot{\phi}^2\,,$$

(4.49)

where the Klein–Gordon equation (4.47) was used in the second equality. Comparing this to the continuity equation, $\dot{\rho}_\phi = -3H(\rho_\phi + P_\phi)$, we infer that the pressure induced by the field is

$$\boxed{P_\phi = \frac{1}{2}\dot{\phi}^2 - V(\phi)}\,.$$

(4.50)

This pressure determines the acceleration of the expansion, $\ddot{a} \propto -(\rho_\phi + 3P_\phi)$. We see that the density and pressure are in general not related by a constant equation of state as for a perfect fluid. Notice, however, that if the kinetic energy of the inflaton is much smaller than its potential energy, then $P_\phi \approx -\rho_\phi$. The inflationary potential then acts like a temporary cosmological constant, sourcing a period of exponential expansion.

Exercise 4.3 The action of a scalar field in a general curved spacetime is

$$S = \int \mathrm{d}^4x\sqrt{-g}\left[-\frac{1}{2}g^{\mu\nu}\partial_\mu\phi\partial_\nu\phi - V(\phi)\right],$$

(4.51)

where $g \equiv \det g_{\mu\nu}$. It is easy to check that this reduces to (4.43) for a flat FRW metric. Under a variation of the (inverse) metric, $g^{\mu\nu} \to g^{\mu\nu} + \delta g^{\mu\nu}$, the action changes as

$$\delta S = -\frac{1}{2}\int \mathrm{d}^4x\sqrt{-g}\,T_{\mu\nu}\delta g^{\mu\nu}\,,$$

(4.52)

where $T_{\mu\nu}$ is the energy-momentum tensor.

Show that the energy-momentum tensor for a scalar field is

$$T_{\mu\nu} = \partial_\mu\phi\partial_\nu\phi - g_{\mu\nu}\left(\frac{1}{2}g^{\alpha\beta}\partial_\alpha\phi\partial_\beta\phi + V(\phi)\right),$$

(4.53)

and confirm that this leads to the expressions for ρ_ϕ and P_ϕ found above.

Hint: You have to use that $\delta\sqrt{-g} = -\frac{1}{2}\sqrt{-g}\,(g_{\mu\nu}\delta g^{\mu\nu})$.

Show that the conservation of the energy-momentum tensor, $\nabla_\mu T^{\mu\nu} = 0$, implies the Klein–Gordon equation (4.47).

4.3.2 Slow-Roll Inflation

The dynamics during inflation is then determined by a combination of the Friedmann and Klein–Gordon equations

$$H^2 = \frac{1}{3M_{\rm Pl}^2} \left[\frac{1}{2}\dot\phi^2 + V \right] , \tag{4.54}$$

$$\ddot\phi + 3H\dot\phi = -\frac{dV}{d\phi} . \tag{4.55}$$

These equations are coupled. The energy stored in the field determines the Hubble rate, which in turn induces friction and hence affects the evolution of the field. We would like to determine the precise conditions under which this feedback leads to inflation.

A slowly rolling field

First, we note that (4.54) and (4.55) can be combined into an expression for the evolution of the Hubble parameter

$$\dot H = -\frac{1}{2} \frac{\dot\phi^2}{M_{\rm Pl}^2} . \tag{4.56}$$

Taking the ratio of (4.56) and (4.54), we then find

$$\varepsilon \equiv -\frac{\dot H}{H^2} = \frac{\frac{1}{2}\dot\phi^2}{M_{\rm Pl}^2 H^2} = \frac{\frac{3}{2}\dot\phi^2}{\frac{1}{2}\dot\phi^2 + V} . \tag{4.57}$$

Inflation ($\varepsilon \ll 1$) therefore occurs if the kinetic energy density, $\frac{1}{2}\dot\phi^2$, only makes a small contribution to the total energy density, $\rho_\phi = \frac{1}{2}\dot\phi^2 + V$. For obvious reasons, this situation is called **slow-roll inflation**.

In order for the slow-roll behavior to persist, the acceleration of the scalar field also has to be small. To assess this, it is useful to define the dimensionless acceleration per Hubble time

$$\delta \equiv -\frac{\ddot\phi}{H\dot\phi} . \tag{4.58}$$

When δ is small, the friction term in (4.55) dominates and the inflaton speed is determined by the slope of the potential. Moreover, as long as δ is small, the inflaton kinetic energy stays subdominant and the inflationary expansion continues. To see this more explicitly, we take the time derivative of (4.57),

$$\dot\varepsilon = \frac{\dot\phi\ddot\phi}{M_{\rm Pl}^2 H^2} - \frac{\dot\phi^2 \dot H}{M_{\rm Pl}^2 H^3} , \tag{4.59}$$

and substitute it into (4.36):

$$\kappa = \frac{\dot\varepsilon}{H\varepsilon} = 2\frac{\ddot\phi}{H\dot\phi} - 2\frac{\dot H}{H^2} = 2(\varepsilon - \delta) . \tag{4.60}$$

This shows that $\{\varepsilon, |\delta|\} \ll 1$ implies $\{\varepsilon, |\kappa|\} \ll 1$. If both the speed and the acceleration of the inflaton field are small, then the inflationary expansion will last for a long time.

Slow-roll approximation

So far, no approximations have been made. We simply noted that in a regime where $\{\varepsilon, |\delta|\} \ll 1$, inflation occurs and persists. Now, we will use these conditions to simplify the equations of motion. This is called the **slow-roll approximation**.

First, we note that the condition $\varepsilon \ll 1$ implies $\dot{\phi}^2 \ll V$, which leads to the following simplification of the Friedmann equation (4.54),

$$H^2 \approx \frac{V}{3M_{\rm Pl}^2}. \tag{4.61}$$

In the slow-roll approximation, the Hubble expansion rate is therefore determined completely by the potential. Next, we see that the condition $|\delta| \ll 1$ simplifies the Klein–Gordon equation (4.55) to

$$3H\dot{\phi} \approx -V_{,\phi}, \tag{4.62}$$

where $V_{,\phi} \equiv dV/d\phi$. This provides a simple relationship between the slope of the potential and the speed of the inflaton. Finally, substituting (4.61) and (4.62) into (4.57) gives

$$\varepsilon = \frac{\frac{1}{2}\dot{\phi}^2}{M_{\rm Pl}^2 H^2} \approx \frac{M_{\rm Pl}^2}{2}\left(\frac{V_{,\phi}}{V}\right)^2, \tag{4.63}$$

which expresses the parameter ε purely in terms of the potential.

To evaluate the parameter δ, defined in (4.58), in the slow-roll approximation, we take the time derivative of (4.62), $3\dot{H}\dot{\phi} + 3H\ddot{\phi} = -V_{,\phi\phi}\dot{\phi}$. This leads to

$$\delta + \varepsilon = -\frac{\ddot{\phi}}{H\dot{\phi}} - \frac{\dot{H}}{H^2} \approx M_{\rm Pl}^2 \frac{V_{,\phi\phi}}{V}. \tag{4.64}$$

Hence, a convenient way to judge whether a given potential $V(\phi)$ can lead to slow-roll inflation is to compute the **potential slow-roll parameters**

$$\boxed{\varepsilon_V \equiv \frac{M_{\rm Pl}^2}{2}\left(\frac{V_{,\phi}}{V}\right)^2, \quad \eta_V \equiv M_{\rm Pl}^2 \frac{V_{,\phi\phi}}{V}.} \tag{4.65}$$

Successful inflation occurs when these parameters are much smaller than unity.

Exercise 4.4 Show that in the slow-roll regime, the potential slow-roll parameters and the Hubble slow-roll parameters are related as follows:

$$\varepsilon_V \approx \varepsilon \quad \text{and} \quad \eta_V \approx 2\varepsilon - \frac{1}{2}\kappa. \tag{4.66}$$

The total number of e-foldings of accelerated expansion is

$$N_{\text{tot}} \equiv \int_{a_i}^{a_e} \mathrm{d}\ln a = \int_{t_i}^{t_e} H(t)\,\mathrm{d}t = \int_{\phi_i}^{\phi_e} \frac{H}{\dot{\phi}}\,\mathrm{d}\phi\,, \tag{4.67}$$

where t_i and t_e are defined as the times when $\varepsilon(t_i) = \varepsilon(t_e) \equiv 1$. In the slow-roll regime, we can use (4.63) to write the integral over the field space as

$$N_{\text{tot}} \approx \int_{\phi_i}^{\phi_e} \frac{1}{\sqrt{2\varepsilon_V}} \frac{|\mathrm{d}\phi|}{M_{\text{Pl}}}\,, \tag{4.68}$$

where ϕ_i and ϕ_e are the field values at the boundaries of the interval where $\varepsilon_V < 1$. The absolute value around the integration measure in (4.68) indicates that we pick the overall sign of the integral in such a way that $N_{\text{tot}} > 0$. As we have seen above, a solution to the horizon problem requires $N_{\text{tot}} \gtrsim 60$, which provides an important constraint on successful inflationary models.

Case study: quadratic inflation

As an example, let us give the slow-roll analysis of arguably the simplest model of inflation: single-field inflation driven by a mass term

$$V(\phi) = \frac{1}{2} m^2 \phi^2\,. \tag{4.69}$$

As we will see in Chapter 8, this model is ruled out by CMB observations, but it still provides a useful example to illustrate the mechanism of slow-roll inflation. The slow-roll parameters for this potential are

$$\varepsilon_V(\phi) = \eta_V(\phi) = 2 \left(\frac{M_{\text{Pl}}}{\phi}\right)^2\,. \tag{4.70}$$

To satisfy the slow-roll conditions $\{\varepsilon_V, |\eta_V|\} < 1$, we therefore need to consider super-Planckian values for the inflaton

$$\phi > \sqrt{2} M_{\text{Pl}} \equiv \phi_e\,. \tag{4.71}$$

Let the initial field value be ϕ_i. As the field moves from ϕ_i to ϕ_e, the e-foldings of inflationary expansion are

$$N_{\text{tot}} = \int_{\phi_e}^{\phi_i} \frac{\mathrm{d}\phi}{M_{\text{Pl}}} \frac{1}{\sqrt{2\varepsilon_V}} = \left.\frac{\phi^2}{4M_{\text{Pl}}^2}\right|_{\phi_e}^{\phi_i} = \frac{\phi_i^2}{4M_{\text{Pl}}^2} - \frac{1}{2}\,. \tag{4.72}$$

To obtain $N_{\text{tot}} > 60$, the initial field value must satisfy

$$\phi_i > 2\sqrt{60}\, M_{\text{Pl}} \approx 15 M_{\text{Pl}}\,. \tag{4.73}$$

We note that the total field excursion is super-Planckian, $\Delta\phi = \phi_i - \phi_e \gg M_{\text{Pl}}$. In Section 4.4.1, we will discuss whether this large field variation should be a cause for concern.

4.3.3 Creating the Hot Universe

Most of the energy density during inflation is in the form of the inflaton potential $V(\phi)$. Inflation ends when the potential steepens and the field picks up kinetic energy. The energy in the inflaton sector then has to be transferred to the particles of the Standard Model. This process is called **reheating** and starts the hot Big Bang. I will only have time for a very brief and mostly qualitative description of the absolute basics of the reheating phenomenon. At the end of the chapter, I suggest a few review articles on reheating where you can learn more.

Once the inflaton field reaches the bottom of the potential, it begins to oscillate. Near the minimum, the potential can be approximated as $V(\phi) \approx \frac{1}{2}m^2\phi^2$ and the equation of motion of the inflaton is

$$\ddot{\phi} + 3H\dot{\phi} = -m^2\phi\,. \tag{4.74}$$

The energy density evolves according to the continuity equation

$$\dot{\rho}_\phi + 3H\rho_\phi = -3HP_\phi = \frac{3}{2}H(m^2\phi^2 - \dot{\phi}^2)\,, \tag{4.75}$$

where the right-hand side averages to zero over one oscillation period. This averaging has ignored the Hubble friction in (4.74), which is justified on timescales that are short compared to the expansion time. We see that the oscillating field behaves like pressureless matter, with $\rho_\phi \propto a^{-3}$ (see Problem 4.1 for a more rigorous derivation of this fact). As the energy density drops, the amplitude of the oscillations decreases.

To avoid that the universe ends up completely empty, the inflaton has to couple to Standard Model fields. The energy stored in the inflaton field will then be transferred to ordinary particles. If the decay is slow, then the inflaton's energy density follows the equation

$$\dot{\rho}_\phi + 3H\rho_\phi = -\Gamma_\phi\rho_\phi\,, \tag{4.76}$$

where Γ_ϕ is the inflaton decay rate. A slow decay of the inflaton typically occurs if the coupling is only to fermions. If the inflaton can also decay into bosons, on the other hand, then the decay rate may be enhanced by Bose condensation and parametric resonance effects. This kind of rapid decay is called **preheating**, since the bosons are created far from thermal equilibrium.

The new particles will interact with each other and eventually reach the thermal state that characterizes the hot Big Bang. The energy density at the end of the reheating epoch is $\rho_R < \rho_{\phi,e}$, where $\rho_{\phi,e}$ is the energy density at the end of inflation, and the reheating temperature T_R is determined by

$$\rho_R = \frac{\pi^2}{30}g_*(T_R)T_R^4\,. \tag{4.77}$$

If reheating takes a long time, then $\rho_R \ll \rho_{\phi,e}$ and the reheating temperature gets smaller. At a minimum, the reheating temperature has to be larger than $1\,\text{MeV}$ to allow for successful BBN, and most likely it is much larger than this to also allow for baryogenesis after inflation.

This completes my highly oversimplified sketch of the reheating phenomenon. In reality, the dynamics during reheating can be very rich, often involving nonperturbative effects that must be captured by numerical simulations (for reviews, see [2–5]). Describing this phase in the history of the universe is essential for understanding how the hot Big Bang began and the study of reheating remains a very active area of research. With accelerated expansion having ended, there is no "cosmic amplifier" to make the microscopic physics of reheating easily accessible on cosmological length scales. Moreover, the high energy scale associated with the end of inflation, as well as the subsequent thermalization, can further hide details of this era from our low-energy probes. Nevertheless, in some cases, the dynamics during this period can generate relics such as isocurvature perturbations, stochastic gravitational waves, non-Gaussianities, dark matter/radiation, primordial black holes, topological and non-topological solitons, matter/antimatter asymmetry, and primordial magnetic fields. Detecting any of these relics would give us an interesting window into the physics that created the hot Big Bang.

4.4 Open Problems*

Inflation is a successful model that solves the causality problems of the standard Big Bang scenario. In addition, it contains an elegant mechanism to generate the observed density fluctuations in the universe (see Chapter 8). Despite these phenomenological successes, however, inflation is not yet a complete theory. To achieve inflation we had to postulate new physics at energies far above those probed by particle colliders, and I will now show that the success of inflation is sensitive to assumptions about the physics at even higher energies.

4.4.1 Ultraviolet Sensitivity

In the absence of a complete microscopic theory, the most conservative way to describe the inflationary era is in terms of an **effective field theory** (EFT). In the EFT approach, we admit that we don't know the details of the high-energy theory and instead parameterize our ignorance. We begin by defining the degrees of freedom and symmetries of the inflationary theory. This theory is valid below a cutoff scale Λ, and we should ask how the unknown physics above the scale Λ can affect the low-energy dynamics during inflation. This means writing down all possible corrections to the low-energy theory that are consistent with the assumed symmetries. Schematically, the effective Lagrangian of the theory can then be written as

$$\mathcal{L}_{\text{eff}}[\phi] = \mathcal{L}_0[\phi] + \sum_n c_n \frac{O_n[\phi]}{\Lambda^{\delta_n - 4}}, \tag{4.78}$$

where $\mathcal{L}_0$ is the Lagrangian of the inflationary model and O_n are a set of "operators" that parameterize the corrections coming from the couplings to additional high-energy degrees of freedom. Unless there is a good reason to assume otherwise, the dimensionless parameters c_n are taken to be of order one.[8] If the cutoff scale Λ is much larger than the typical energies E at which the low-energy theory is being probed, then the corrections are small, suppressed by powers of E/Λ. This is why quantum gravity doesn't affect our everyday lives, or even those of our friends working at particle colliders.[9] A remarkable feature of inflation, however, is that it is extremely sensitive even to effects suppressed by the Planck scale.

We have seen that slow-roll inflation requires a flat potential (in units of the Planck scale), and I will now show that this is hard to control and sensitive to very small corrections. In the absence of any special symmetries, the EFT of slow-roll inflation takes the following form

$$\mathcal{L}_{\text{eff}}(\phi) = -\frac{1}{2}(\partial_\mu\phi)^2 - V(\phi) - \sum_n c_n V(\phi)\frac{\phi^{2n}}{\Lambda^{2n}} - \sum_n d_n \frac{(\partial\phi)^{2n}}{\Lambda^{4n}} + \cdots , \quad (4.79)$$

where I have written a representative set of higher-dimension operators. If the inflaton field value is smaller than the cutoff scale, then we can truncate the EFT expansion in (4.79) and the leading effect comes from the dimension-six operator[10]

$$\Delta V = c_1 V(\phi)\frac{\phi^2}{\Lambda^2} . \quad (4.80)$$

Since $\phi \ll \Lambda$, this is a small correction to the inflaton potential, $\Delta V \ll V$. Nevertheless, it affects the delicate flatness of the potential and hence is relevant for the dynamics during inflation. In particular, the slow-roll parameter η_V receives the correction

$$\Delta\eta_V = \frac{M_{\text{Pl}}^2}{V}(\Delta V)_{,\phi\phi} \approx 2c_1\left(\frac{M_{\text{Pl}}}{\Lambda}\right)^2 > 1 , \quad (4.81)$$

where the final inequality comes from $c_1 = O(1)$ and $\Lambda < M_{\text{Pl}}$. Notice that this problem is independent of the energy scale of inflation. All inflationary models have to address this so-called **eta problem**.

One of the most promising ways to solve the eta problem is to postulate a shift symmetry for the inflaton, $\phi \to \phi + c$, where c is a constant. If this symmetry is respected by the couplings to massive fields, then having $c_n \ll 1$ is *technically natural* [6]. Another way to ameliorate the eta problem (but not solve it) is supersymmetry. In that case, a minor miracle occurs. Above the scale H, the theory

---

[8] We know that some new degrees of freedom must appear below the Planck scale, $\Lambda \lesssim M_{\text{Pl}}$, because gravity is not renormalizable and needs a high-energy (or "ultraviolet") completion. If ϕ has order-one couplings to any heavy fields, with masses of order Λ, then integrating out these fields yields the effective action (4.78), with order-one couplings c_n.

[9] Note the largest energies probed at the Large Hadron Collider (LHC) are still tiny relative to the Planck scale, $E/M_{\text{Pl}} < 10^{-16}$.

[10] In principle, there can also be a dimension-five operator, which can induce a large correction to the slow-roll parameter ε_V. This operator, however, can be forbidden by the $\mathbb{Z}_2$ symmetry $\phi \to -\phi$ and therefore is usually less of a concern.

is supersymmetric and the contributions from bosons and fermions precisely cancel. However, supersymmetry is spontaneously broken during inflation, leading to an inflaton mass of order the Hubble scale, $m_\phi \sim H$, and an eta parameter of order one, $\Delta\eta_V \sim 1$. Although the eta problem is still there, it is much less severe than in models without supersymmetry, requiring just a percent-level fine-tuning of the inflaton mass parameter.

We have seen that in quadratic inflation the inflaton moves over a super-Planckian distance in field space, $\Delta\phi > M_{\rm Pl}$. In fact, as we will see in Chapter 8, this is a general feature of all inflationary models with observable gravitational wave signals. The super-Planckian excursion of the field should make us nervous. When the field value isn't smaller than the cutoff, we cannot trust the EFT expansion in (4.79). There is then no sense in which the problem can be reduced to a finite number of corrections. Fine-tuning is then not an option. This is to be contrasted with the eta problem, which, in principle, can be addressed by fine-tuning the inflaton mass parameter. Models of **large-field inflation** therefore must invoke symmetries—such as an approximate shift symmetry $\phi \to \phi + c$—to suppress the couplings of the inflaton to massive degrees of freedom and reduce the size of the corrections in (4.79). An important question is whether the assumed symmetries are respected by the UV-completion. Concrete examples of large-field inflation in string theory can be found in [7, 8].

The UV sensitivity of inflation is both a challenge and an opportunity. It is a challenge, because it means that we either need to work in a UV-complete theory of quantum gravity or make assumptions about the form that such a UV-completion might take. It is also an opportunity, because it suggests the exciting possibility of using cosmological observations to learn about fundamental aspects of quantum gravity. This feature of inflation is discussed at length in my book with Liam McAllister [9] and progress on inflation in string theory is reviewed in [10–14].

4.4.2 Initial Conditions

The main motivation for inflation was the fine-tuned initial conditions of the hot Big Bang. Explaining these initial conditions, however, cannot be viewed as a success if inflation itself requires fine-tuned initial conditions to get started.

Our analysis of slow-roll inflation made some implicit assumptions about initial conditions. First, we assumed that the inflaton field began its evolution at the top of a suitable potential. Who put it there? This question does not have a very satisfactory answer. The following argument is often made: Imagine that the inflaton field initially takes different values in different regions of space, some at the top of the potential, some at the bottom. The regions of space where the vacuum energy is large will inflate and therefore grow. Weighted by volume, these regions will dominate, explaining why most of space experiences inflation. Alternatively, it is sometimes assumed that slow-roll inflation was preceded by an earlier phase of false vacuum domination (see Fig. 4.9). In quantum mechanics, there is a small probability that the field will tunnel through the potential energy barrier.

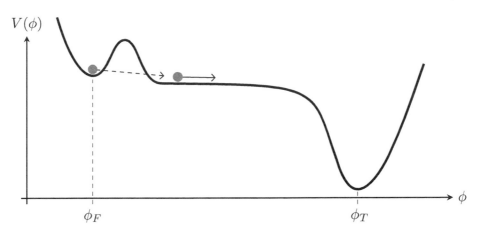

Fig. 4.9 Sketch of the inflaton potential with a metastable high-energy vacuum. Tunneling from this false vacuum state may set the initial conditions for slow-roll inflation. If the tunneling rate is smaller than the expansion rate, then this also provides an example of eternal inflation.

The question why inflation began at the top of the potential, then becomes the question why the quantum mechanical tunneling is more likely to put the inflaton field at the top of the potential than at its minimum. Of course, in this scenario, you should still ask why the universe started in the high-energy false vacuum. Finally, and most ambitiously, the no-boundary proposal by Hartle and Hawking gives a prescription for evaluating the probability that a universe is spontaneously created from nothing [15]. The hope is that this would explain why the universe started in the high-energy vacuum.

Second, we assumed that the initial velocity of the inflaton field was small and that the slow-roll solution was an attractor. If the initial inflaton velocity is non-negligible, then it is possible that the field will *overshoot* the region of the potential where inflation is supposed to occur, without actually sourcing accelerated expansion. This problem is stronger for small-field models where Hubble friction is often not efficient enough to slow down the field before it reaches the region of interest. In large-field models, on the other hand, Hubble friction is usually very efficient and the slow-roll solution becomes an attractor.

Finally, we also assumed that the initial inhomogeneities in the inflaton field were small. Since inflation is supposed to explain the homogeneous initial conditions of the hot Big Bang, we must ask what happens when we allow for large inhomogeneities in the inflaton field. These inhomogeneities carry gradient energy, which might hinder the accelerated expansion. The sensitivity of an inflationary model to initial perturbations of the inflaton field is best addressed by numerical simulations. The earliest such simulations were performed in 1989 by Goldwirth and Piran [16, 17]. Limited by the computational power at the time, the analysis was restricted to spherically symmetric perturbations using general relativistic simulations in $1 + 1$ dimensions. This work showed that inflation only starts if the universe is smooth on a scale of order the Hubble radius and that large-field models are more robust to initial perturbations than small-field models. Recently, the problem was revisited using simulations in $3 + 1$ dimensions [18, 19], allowing a more general class of initial perturbations to be studied. It was again found that large-field models are more robust than small-field models.

How severe the fine-tuning of initial conditions really is for inflation cannot be discussed outside of the incompletely understood topic of eternal inflation and the associated measure problem.

4.4.3 Eternal Inflation

A rather dramatic consequence of the quantum dynamics of inflation may be that globally it never ends, but is *eternal*. There are two types of **eternal inflation**, both of which are illustrated in Fig. 4.9. First, inflation can occur while the field is stuck in the false vacuum. Like for a radioactive substance, any region of space has a finite probability to decay and reach the true vacuum via quantum tunneling. If the decay rate is Γ, then the probability that the region will survive in the inflationary state is $e^{-\Gamma t}$. The special feature of inflation, and what makes it different from an ordinary radioactive substance, is that space is exponentially expanding during inflation. The volume of the inflating region therefore increases as e^{3Ht}. If the rate of exponential expansion is larger than the rate of decay, then inflation does not end globally. Of course, locally, there are many pockets of space where inflation does end—one of these pockets is what we call our "universe"—but they are embedded in a much larger "multiverse." The space between the pocket universes is rapidly expanding, creating new space, where new pocket universes can form. This never ends.

Eternal inflation may also occur as a limit of slow-roll inflation. So far, we have only discussed the classical evolution of the inflaton field as it rolls down the potential. In Chapter 8, we will study the effects of quantum mechanics on the dynamics of the inflaton. We will see that the field experiences random fluctuations up and down the potential. Eternal inflation occurs if these quantum fluctuations dominate over the classical rolling. To be a bit more quantitative, consider the evolution of the field in one Hubble volume H^{-3} during one Hubble time $\Delta t = H^{-1}$. Classically, the value of the field will change by $\Delta\phi_{\mathrm{cl}} \sim \dot\phi H^{-1}$. At the same time, quantum mechanics induces random jitters $\Delta\phi_{\mathrm{Q}}$, so that the total change of the field is

$$\Delta\phi = \Delta\phi_{\mathrm{cl}} + \Delta\phi_{\mathrm{Q}} . \tag{4.82}$$

There is a finite probability that this sum of the classical and quantum contributions is negative, keeping the field at the top of the potential. How big does this probability have to be in order for inflation to be eternal? To estimate this, let us first note that, during a Hubble time, the volume increases by a factor of $e^3 \approx 20$, meaning that the inflationary region breaks up into 20 Hubble-sized regions. In Chapter 8, we will see that, averaged over one of these 20 regions, the inflationary quantum fluctuations have a Gaussian probability distribution with width $H/2\pi$. If the probability to have $\Delta\phi < 0$ is greater than 1 in 20, then the volume of space that is inflating will increase. For a Gaussian probability distribution, this will be the case when

$$\Delta\phi_\mathrm{Q} = \frac{H}{2\pi} > 0.6\,\Delta\phi_\mathrm{cl} = 0.6\,|\dot\phi|H^{-1}\,. \qquad (4.83)$$

If this condition is satisfied, then inflation will continue forever. This argument for slow-roll eternal inflation was admittedly somewhat heuristic and it is still being debated how reliable its conclusions are. Concretely, once the quantum fluctuations become as important as the classical rolling, the backreaction on the inflationary dynamics may become large and we lose computational control.

Exercise 4.5 Show that, in quadratic inflation, slow-roll eternal inflation occurs for

$$\frac{\phi^2}{M_\mathrm{Pl}^2} \gtrsim 13\,\frac{M_\mathrm{Pl}}{m}\,. \qquad (4.84)$$

In Chapter 8, we will find that in order for the predicted CMB fluctuations to have the right amplitude, we must have $m \sim 10^{-6} M_\mathrm{Pl}$. This then implies that $\phi > 3600\,M_\mathrm{Pl}$. It is of course conceivable that quantum gravity censures such large field values, in which case the regime of eternal inflation would not exist.

Although the concept of eternal inflation is almost as old as inflation itself, it was recently revived in the context of string theory. We have learned that string theory doesn't have a unique vacuum, but seems to give rise to a large number of metastable solutions. The space of vacua is called the **string landscape**. The physics may be different in each of these vacua. For example, the effective cosmological constant can take on different values, providing a framework in which Weinberg's anthropic solution to the cosmological constant problem can be realized (see Section 2.3.3). Eternal inflation provides the mechanism by which this landscape is populated. The fact that eternal inflation combined with the string landscape allows for a solution to the cosmological constant problem is the single most important reason to take either of these ideas seriously.

Let me finally address the elephant in the room: the **measure problem** of the inflationary multiverse. We have seen that eternal inflation produces not just one universe, but an infinite number of universes. To make predictions in this multiverse requires being able to compute the relative probabilities of the different pocket universes. However, these probabilities are hard to define because there is an infinite number of everything. The fraction of universes with a certain observable property is equal to infinity divided by infinity. To make sense of this, we must regulate the infinities of the inflationary multiverse. Unfortunately, the answers tend to depend sensitively on the type of regulator that is being used. See [20–25] for further discussion of this subtle topic.

The problems discussed in this section illustrate that there is still room for improving our theoretical understanding of inflation and the cosmological initial conditions. Although we have remarkable observational evidence that something like inflation occurred in the early universe (see Chapter 8), inflation cannot yet be considered a part of the standard model of cosmology, with the same level of confidence as, for example, BBN is a fact about the early universe. Inflation is still a work in progress, leaving some things to be desired and making the search for alternatives a worthwhile enterprise [26, 27].

4.5 Summary

In this chapter, we have studied inflation. We started with a discussion of the causality problems of the hot Big Bang, highlighting the crucial role played by the evolution of the comoving Hubble radius, $(aH)^{-1}$. We showed that this evolution determines the particle horizon—the maximal distance that light can propagate between an initial time t_i and a later time t:

$$d_h = \int_{t_i}^{t} \frac{dt}{a(t)} = \int_{N_i}^{N} (aH)^{-1} dN \,, \tag{4.85}$$

where $N = \ln a$. Since the comoving Hubble radius is monotonically increasing in the conventional Big Bang theory, the particle horizon is dominated by the contributions from late times. Most parts of the cosmic microwave background would then never have been in causal contact, yet they somehow have nearly the same temperature. This is the horizon problem.

Inflation is a period of accelerated expansion during which the comoving Hubble radius is shrinking. The particle horizon then receives most of its contribution from early times and can be much larger than naively assumed. Causal contact is established in the time before the hot Big Bang. The shrinking of the comoving Hubble radius also solves the closely related flatness problem and allows a causal mechanism to generate superhorizon correlations.

Inflation occurs if all physical quantities vary slowly with time. For example, the fractional change of the Hubble rate during inflation must be small,

$$\varepsilon \equiv -\dot{H}/H^2 \ll 1 \,. \tag{4.86}$$

To solve the problems of the hot Big Bang, this condition must be maintained for about 60 e-foldings of expansion. A popular way to achieve this is through a slowly rolling scalar field ϕ (the inflaton). The evolution of the inflaton is governed by the Klein–Gordon equation

$$\ddot{\phi} + 3H\dot{\phi} = -V_{,\phi} \,. \tag{4.87}$$

The field rolls slowly if the dynamics is dominated by the friction term, $|\ddot{\phi}| \ll |3H\dot{\phi}| \approx |V_{,\phi}|$. The size of the friction is determined by the energy density associated with the inflaton itself, which by the Friedmann equation determines the expansion rate

$$H^2 = \frac{1}{3M_{\mathrm{Pl}}^2}\left[\frac{1}{2}\dot{\phi}^2 + V\right].$$

(4.88)

Inflation requires that the kinetic energy is much smaller than the potential energy, $\frac{1}{2}\dot{\phi}^2 \ll V$. The conditions for successful slow-roll inflation can be characterized by the slow-roll parameters

$$\varepsilon_V \equiv \frac{M_{\mathrm{Pl}}^2}{2}\left(\frac{V_{,\phi}}{V}\right)^2, \quad \eta_V \equiv M_{\mathrm{Pl}}^2 \frac{V_{,\phi\phi}}{V}.$$

(4.89)

Inflation will occur in regions of the potential where $\{\varepsilon_V, |\eta_V|\} \ll 1$. Achieving such flat potentials in a theory of quantum gravity is challenging.

Further Reading

There are many good reviews of inflation [28–31]. My own contributions have been in the form of two TASI lectures [32, 33] and a book [9]. Nice reviews of effective field theory are [34, 35] and its application to inflation is discussed in [9, 36, 37]. Inflation in string theory is reviewed in [9–14]. Guth has written a nice essay on eternal inflation [38]. Details on reheating can be found in [2–5], and the seminal paper of Kofman, Linde and Starobinsky is also well worth your time [39]. Finally, you should read Guth's original paper [40] both for historical reasons and because it contains an exceptionally clear description of the problems of the standard Big Bang model.

Problems

4.1 Oscillating scalar field

The action for a scalar field in a curved spacetime is

$$S = \int \mathrm{d}^4x \sqrt{-g}\left[-\frac{1}{2}g^{\mu\nu}\partial_\mu\phi\partial_\nu\phi - V(\phi)\right],$$

where $g \equiv \det g_{\mu\nu}$.

1. Evaluate the action for a homogeneous field $\phi = \phi(t)$ in a flat FRW spacetime and determine the equation of motion for the field.

2. Near the minimum of the potential, we have $V(\phi) \approx \frac{1}{2}m^2\phi^2$. Making the ansatz $\phi(t) = a^{-3/2}\chi(t)$, show that the equation of motion becomes

$$\ddot{\chi} + \left(m^2 - \frac{3}{2}\dot{H} - \frac{9}{4}H^2\right)\chi = 0\,.$$

Assuming that $m^2 \gg H^2 \sim \dot{H}$, find $\phi(t)$ and write your answer in terms of the maximum amplitude of the oscillations. What does this result imply for the evolution of the energy density during the oscillating phase after inflation?

4.2   Quadratic inflation

The equation of motion of the inflaton is

$$\ddot{\phi} + 3H\dot{\phi} + V_{,\phi} = 0\,, \quad \text{with} \quad 3M_{\text{Pl}}^2 H^2 = \frac{1}{2}\dot{\phi}^2 + V\,.$$

1. For the potential $V(\phi) = \frac{1}{2}m^2\phi^2$, use the slow-roll approximation to obtain the inflationary solutions

$$\phi(t) = \phi_i - \sqrt{\frac{2}{3}}mM_{\text{Pl}}\,t\,,$$

$$a(t) = a_i \exp\left[\frac{\phi_i^2 - \phi^2(t)}{4M_{\text{Pl}}^2}\right],$$

where $\phi_i > 0$ is the field value at the start of inflation ($t_i \equiv 0$).

2. What is the value of ϕ where inflation ends? Find an expression for the number of e-folds. If $V(\phi_i) \sim M_{\text{Pl}}^4$, estimate the total number of e-folds of inflation.

4.3   Power law inflation

Consider the potential [41]

$$V(\phi) = V_0 \exp\left(-c\frac{\phi}{M_{\text{Pl}}}\right),$$

where c and V_0 are constants. Find the solution for the field and the scale factor *without* making the slow-roll approximation. For what values of c does this solution correspond to inflation? What is the problem with this model of inflation?

Hint: Try an ansatz of the form $\phi(t) = \alpha \ln(\beta t)$ and $a(t) = (t/t_0)^{\gamma}$.

4.4   Natural inflation

An influential idea in inflationary model-building is that the inflaton could be a pseudoscalar axion. At the perturbative level, axions enjoy a continuous shift symmetry, but nonperturbatively this is broken to a discrete symmetry, leading to a potential of the form

$$V(\phi) = \frac{V_0}{2}\left[1 - \cos\left(\frac{\phi}{f}\right)\right],$$

where f is the axion decay constant. At what value ϕ_i close to $\phi = \pi f$ does the field have to start in order for the evolution to give more than 50 e-folds of inflation? The model is called *natural inflation* [42].

4.5 Kinetic inflation

Consider single-field inflation with a more general kinetic term [43]

$$S = \int d^4x \sqrt{-g}\, P(X, \phi),$$

where $P(X, \phi)$ is an arbitrary function of $X \equiv -\frac{1}{2}g^{\mu\nu}\partial_\mu\phi\partial_\nu\phi$ and ϕ.

1. By varying the action with respect to the metric, show that this corresponds to a perfect fluid with pressure P and energy density

$$\rho = 2XP_{,X} - P,$$

where $P_{,X} \equiv dP/dX$. You may easily check that this gives the expected result for the case of slow-roll inflation, $P = X - V(\phi)$, namely $\rho = X + V$.

2. By varying the action with respect to ϕ, show that the equation of motion for the inflaton is

$$-\frac{d}{dt}\left(a^3 P_{,X}\dot\phi\right) + a^3 P_{,\phi} = 0.$$

For slow-roll inflation, this gives the Klein–Gordon equation.

3. Show that the slow-roll parameter is

$$\varepsilon = -\frac{\dot H}{H^2} = \frac{3XP_{,X}}{2XP_{,X} - P}.$$

For suitable $P(X)$, this may lead to inflation even without a flat potential.

4.6 Starobinsky inflation

One of the earliest models of inflation was written down by Starobinsky [44]. He considered the following action for the Ricci scalar

$$S = \frac{M_{\rm Pl}^2}{2}\int d^4x \sqrt{-g}\, f(R), \quad \text{with} \quad f(R) \equiv R + \frac{R^2}{6M^2}. \tag{1}$$

The linear term in $f(R)$ gives the standard Einstein–Hilbert action for gravity, while the quadratic term is a correction that becomes relevant for large spacetime curvatures. In this problem, you will determine when this action leads to inflation.

1. Consider first the alternative action

$$S = \frac{M_{\rm Pl}^2}{2}\int d^4x \sqrt{-g}\left(f(\chi) + (R - \chi)f_{,\chi}\right), \tag{2}$$

where χ is an auxiliary scalar field. Under what condition is this action equivalent to the action in (1)?

2. Defining $\varphi \equiv f_{,\chi}$, the action in (2) can be written as

$$S = \int d^4x \sqrt{-g}\left(\frac{M_{\rm Pl}^2}{2}\varphi R - U(\varphi)\right). \tag{3}$$

Determine the potential $U(\varphi)$ for the specific function f defined in (1).

3. To analyze the slow-roll dynamics of the scalar field, we must first bring the action for the metric into the canonical Einstein–Hilbert form. Only then does the Einstein equation have the familiar form.

Show that under a conformal transformation, $\tilde{g}_{\mu\nu} = \Omega^2(x) g_{\mu\nu}$, with appropriately chosen $\Omega(x)$, the action (3) can be put into the following form

$$S = \int \mathrm{d}^4 x \sqrt{-\tilde{g}} \left(\frac{M_{\mathrm{Pl}}^2}{2} \tilde{R} - \frac{1}{2} \tilde{g}^{\mu\nu} \partial_\mu \phi \partial_\nu \phi - V(\phi) \right), \qquad (4)$$

where the field ϕ and the potential $V(\phi)$ are to be determined.

Hint: The Ricci scalar can be written as

$$R = \Omega^2(x) \left(\tilde{R} + 6\tilde{\Box}\omega - 6\tilde{g}^{\mu\nu} \partial_\mu \omega \partial_\nu \omega \right),$$

where $\omega \equiv \ln \Omega$ and the tilde denotes quantities constructed from $\tilde{g}_{\mu\nu}$. Make a choice for $\Omega(x)$, so that $\sqrt{-g}\,\varphi R$ becomes $\sqrt{-\tilde{g}}\,\tilde{R}$.

4. Compute the slow-roll parameter $\varepsilon_V(\phi)$ for the model. When does this theory lead to inflation?

4.7 Higgs inflation

Consider a scalar field φ with a φ^4 potential and a non-minimal coupling to gravity $\xi\varphi^2 R$ [45, 46]. The complete action is

$$S = \int \mathrm{d}^4 x \sqrt{-g} \left[\frac{M_{\mathrm{Pl}}^2}{2} \left(1 + \xi \frac{\varphi^2}{M_{\mathrm{Pl}}^2} \right) R - \frac{1}{2} g^{\mu\nu} \partial_\mu \varphi \partial_\nu \varphi - \frac{\lambda}{4!} \varphi^4 \right].$$

This action has played an important role in attempts to build inflationary models using the Higgs field [47] (see Problem 8.6).

1. Show that under a conformal transformation, $\tilde{g}_{\mu\nu} = \Omega^2(x) g_{\mu\nu}$, this action becomes

$$S = \int \mathrm{d}^4 x \sqrt{-\tilde{g}} \left[\frac{M_{\mathrm{Pl}}^2}{2} \tilde{R} - \frac{1}{2} k(\varphi)(\partial_\mu \varphi)^2 - V(\varphi) \right],$$

where

$$k(\varphi) \equiv \frac{1 + (6\xi + 1)\psi^2}{(1 + \psi^2)^2}, \quad \psi^2 \equiv \xi \frac{\varphi^2}{M_{\mathrm{Pl}}^2},$$

$$V(\varphi) \equiv \frac{\lambda}{4!} \frac{M_{\mathrm{Pl}}^4}{\xi^2} \frac{\psi^4}{(1 + \psi^2)^2}.$$

2. Show that, for $\xi \gg 1$, the potential for the canonically normalized field ϕ is the same as in the Starobinsky model.

PART II

THE INHOMOGENEOUS UNIVERSE

5 Structure Formation

So far, we have treated the universe as perfectly homogeneous. However, to understand the large-scale structure of the universe, we have to introduce inhomogeneities and follow their evolution. As long as these perturbations remain relatively small, we can treat them in perturbation theory. In the next two chapters, we will develop the formalism of cosmological perturbation theory, first in Newtonian gravity and then in full GR.

For most of the history of the universe, the growth of structure was dominated by the evolution of dark matter. Although the microscopic nature of the dark matter is still unknown, its long-wavelength properties are captured by a small number of parameters, such as an equation of state and a sound speed. The insensitivity of the physics at long distances to effects on small scales is a general feature of most physical systems. In fact, physics has progressed precisely because we can study "effective theories" that are valid for a certain range of scales and coarse grain over the microscopic details. In the present context, this means that the fluctuations in the unknown dark matter obey the equations of fluid mechanics. In Chapter 2, we studied the evolution of the universe filled with a homogeneous fluid. Now, we would like to understand what happens when this fluid is perturbed.

In this chapter, we will describe the evolution of perturbations in a non-relativistic fluid in Newtonian gravity. This will provide intuition for the subhorizon dynamics of dark matter fluctuations. To describe the dynamics on larger scales and for relativistic fluids, we have to develop a fully relativistic formalism, which we will do in the next chapter.

In Section 5.1, we will derive the basic equations governing structure formation in the Newtonian approximation. In Section 5.2, we will use these equations to describe the dynamics of dark matter perturbations in the linear regime. We will find that density perturbations grow linearly with the scale factor during the matter era, but only logarithmically during the radiation era. The suppression of growth during the radiation era implies a nontrivial transfer function for the Fourier modes of the dark matter density. In Section 5.3, we will introduce the main statistical properties of the cosmological density fluctuations. Finally, in Section 5.4, we will describe the essential features of the nonlinear clustering of dark matter. This last section is outside of the main development of this book and can therefore be omitted without loss of continuity.

5.1 Newtonian Perturbation Theory

Newtonian gravity is a good approximation when all velocities are small compared to the speed of light and the departures from a flat spacetime geometry are small. The latter is satisfied on scales smaller than the Hubble radius, H^{-1}. The treatment of this section will therefore be restricted to sub-Hubble fluctuations in non-relativistic matter.

5.1.1 Fluid Dynamics

Consider a fluid with a mass density ρ, pressure $P \ll \rho$ and velocity $\mathbf{u}$. We will first study the evolution of perturbations in this fluid while ignoring gravity and the expansion of the universe. After we have developed some intuition for this simplified situation, we will add these complications.

Mass conservation implies the **continuity equation**

$$\frac{\partial \rho}{\partial t} + \boldsymbol{\nabla} \cdot (\rho \mathbf{u}) = 0 \,. \tag{5.1}$$

This equation simply reflects the fact that the mass density in a fixed volume can only change if there is a flux of particles leaving or entering the volume. Momentum conservation leads to the **Euler equation** (see Problem 5.1):

$$\rho \frac{\mathrm{d}\mathbf{u}}{\mathrm{d}t} = \rho \left(\frac{\partial}{\partial t} + \mathbf{u} \cdot \boldsymbol{\nabla} \right) \mathbf{u} = -\boldsymbol{\nabla} P \,, \tag{5.2}$$

which is simply "$F = ma$" for a fluid element. Notice that the acceleration is not given by $\partial_t \mathbf{u}$ (which measures how the velocity changes at a fixed position), but by the "convective time derivative" $(\partial_t + \mathbf{u} \cdot \boldsymbol{\nabla}) \mathbf{u}$ which follows the fluid element as it moves. To close the system of equations, we also have to specify an **equation of state**:

$$P = P(\rho, T) \,, \tag{5.3}$$

i.e. a relation between the pressure and density of the fluid. As indicated, the equation of state may also depend on the temperature of the system.

For a static fluid, with $\mathbf{u} = 0$, the above equations simply imply a constant density and pressure, $\rho = \bar{\rho}$ and $P = \bar{P}$. We wish to see how small perturbations around this solution evolve in time. We take the velocity of the perturbed fluid to be small, $|\mathbf{u}| \ll c$, and write

$$\rho(t, \mathbf{x}) = \bar{\rho} + \delta\rho(t, \mathbf{x}) \,, \tag{5.4}$$

$$P(t, \mathbf{x}) = \bar{P} + \delta P(t, \mathbf{x}) \,. \tag{5.5}$$

As long as all perturbations are small, we can linearize the fluid equations. The concept of "linearizing" the equations will be important, so let me be clear what it means: Consider a product of two quantities A and B. Expanding each factor into a background value and a perturbation, we get

$$AB = (\bar{A} + \delta A)(\bar{B} + \delta B)$$
$$= \bar{A}\bar{B} + \bar{A}\,\delta B + \delta A\,\bar{B} + \delta A\delta B. \tag{5.6}$$

The result has three sets of terms: $\bar{A}\bar{B}$ contains only background quantities and is therefore called *zeroth order* in perturbations; $\bar{A}\,\delta B + \delta A\,\bar{B}$ is *first order* in perturbations; $\delta A\,\delta B$ is a product of two perturbations and hence is called *second order*. Linearizing the equation means dropping all terms that are higher than first order in perturbations. In the example above, this means writing

$$AB \approx \bar{A}\bar{B} + \bar{A}\,\delta B + \delta A\,\bar{B}. \tag{5.7}$$

The linearized answer is a good approximation to the true answer because the second-order term is smaller than the first-order terms (which themselves are smaller than the zeroth-order term).

Applying this procedure to the continuity equation (5.1) and the Euler equation (5.2), we get

$$\partial_t \delta\rho = -\boldsymbol{\nabla}\cdot(\bar{\rho}\mathbf{u}), \tag{5.8}$$
$$\bar{\rho}\,\partial_t\mathbf{u} = -\boldsymbol{\nabla}\delta P. \tag{5.9}$$

Moreover, combining ∂_t (5.8) and $\boldsymbol{\nabla}\cdot$ (5.9) gives

$$\partial_t^2 \delta\rho - \nabla^2 \delta P = 0. \tag{5.10}$$

To write this purely in terms of the density fluctuation $\delta\rho$, we use the relation between P and ρ provided by the equation of state. For simplicity, we will consider a **barotropic fluid**, such that $P = P(\rho)$. In that case, we can write the pressure perturbation as

$$\delta P = \frac{\partial P}{\partial \rho}\delta\rho \equiv c_s^2\,\delta\rho, \tag{5.11}$$

where c_s is the **sound speed** of the fluid. Equation (5.10) then takes the form

$$\boxed{\left(\frac{\partial^2}{\partial t^2} - c_s^2\nabla^2\right)\delta\rho = 0}. \tag{5.12}$$

This is a wave equation and its solutions are **sound waves**

$$\delta\rho(\mathbf{x}, t) = A_\mathbf{k}\sin(\omega t - \mathbf{k}\cdot\mathbf{x}) + B_\mathbf{k}\cos(\omega t - \mathbf{k}\cdot\mathbf{x}), \tag{5.13}$$

where $\mathbf{k}$ is the wavevector and $\omega = c_s|\mathbf{k}|$ is the frequency. The wavevector points in the wave's direction of travel and its magnitude is inversely related to the wavelength $\lambda = 2\pi/|\mathbf{k}|$. Since (5.12) is a linear equation, we can add solutions with different wavevectors $\mathbf{k}$.

The more formal way of solving a partial differential equation (PDE) like (5.12) is to write $\delta\rho$ as a Fourier expansion

$$\delta\rho(\mathbf{x},t) = \int \frac{\mathrm{d}^3 k}{(2\pi)^3} \, e^{i\mathbf{k}\cdot\mathbf{x}} \, \delta\rho(\mathbf{k},t) \,. \tag{5.14}$$

Acting with a spatial derivative ∂_i pulls down a factor of $i\mathbf{k}$ from $e^{i\mathbf{k}\cdot\mathbf{x}}$ and the Laplacian in (5.12) becomes $-k^2$, with $k \equiv |\mathbf{k}|$, in Fourier space. This turns the wave equation into an ordinary differential equation (ODE) for each Fourier mode

$$\left(\frac{\partial^2}{\partial t^2} + c_s^2 k^2 \right) \delta\rho(\mathbf{k},t) = 0 \,. \tag{5.15}$$

The solution for each Fourier mode then is

$$\delta\rho(\mathbf{k},t) = C_{\mathbf{k}} \, e^{-i\omega_k t} + D_{\mathbf{k}} \, e^{i\omega_k t} \,, \tag{5.16}$$

with $\omega_k \equiv c_s k$. Substituting this back into (5.14) gives

$$
\begin{aligned}
\delta\rho(\mathbf{x},t) &= \int \frac{\mathrm{d}^3 k}{(2\pi)^3} \left(C_{\mathbf{k}} \, e^{-i(\omega_k t - \mathbf{k}\cdot\mathbf{x})} + D_{\mathbf{k}} \, e^{i(\omega_k t + \mathbf{k}\cdot\mathbf{x})} \right) \\
&= \int \frac{\mathrm{d}^3 k}{(2\pi)^3} \left(C_{\mathbf{k}} \, e^{-i(\omega_k t - \mathbf{k}\cdot\mathbf{x})} + D_{-\mathbf{k}} \, e^{i(\omega_k t - \mathbf{k}\cdot\mathbf{x})} \right) \\
&= \int \frac{\mathrm{d}^3 k}{(2\pi)^3} \left(C_{\mathbf{k}} \, e^{-i(\omega_k t - \mathbf{k}\cdot\mathbf{x})} + C_{\mathbf{k}}^* \, e^{i(\omega_k t - \mathbf{k}\cdot\mathbf{x})} \right) ,
\end{aligned}
\tag{5.17}
$$

where we have sent $\mathbf{k}$ to $-\mathbf{k}$ in the second term of the second line and used $D_{-\mathbf{k}} = C_{\mathbf{k}}^*$ in the last line. The latter is required in order for $\delta\rho(\mathbf{x},t)$ to be real. Defining $C_{\mathbf{k}} \equiv [B_{\mathbf{k}} + iA_{\mathbf{k}}]/2$, we can also write this as

$$\delta\rho(\mathbf{x},t) = \int \frac{\mathrm{d}^3 k}{(2\pi)^3} \left(A_{\mathbf{k}} \sin(\omega_k t - \mathbf{k}\cdot\mathbf{x}) + B_{\mathbf{k}} \cos(\omega_k t - \mathbf{k}\cdot\mathbf{x}) \right) , \tag{5.18}$$

which shows that the most general solution to the wave equation is indeed a sum (or integral) over the plane wave solutions in (5.13).

5.1.2 Adding Gravity

Let us now turn on gravity, but still keep the spacetime static. As we will see, this is not a completely consistent thing to do, since the spacetime is forced to be dynamical. We will ignore this problem for now and fix it later.

The continuity equation (5.1) remains unchanged, but the Euler equation (5.2) receives an extra force term

$$\rho \left(\frac{\partial}{\partial t} + \mathbf{u}\cdot\boldsymbol{\nabla} \right) \mathbf{u} = -\boldsymbol{\nabla}P - \rho\boldsymbol{\nabla}\Phi \,, \tag{5.19}$$

where Φ is the gravitational potential. This potential is determined by the local density through the **Poisson equation**

$$\nabla^2 \Phi = 4\pi G \rho \,. \tag{5.20}$$

For a static fluid, with $\mathbf{u} = 0$, equations (5.1) and (5.19) are again solved by a constant density and pressure, $\rho = \bar{\rho}$ and $P = \bar{P}$. However, this implies that $\nabla\bar{\Phi} = 0$, which is *not* consistent with the Poisson equation (5.20). This is a reflection of the fact that there are no infinite, static self-gravitating fluids.

Although we haven't found a consistent solution for the background, we will continue and study small perturbations. We will make up for this sin in the next subsection. The linearized continuity and Euler equations now are

$$\partial_t \delta\rho = -\boldsymbol{\nabla}\cdot(\bar{\rho}\mathbf{u})\,, \tag{5.21}$$

$$\bar{\rho}\,\partial_t \mathbf{u} = -\boldsymbol{\nabla}\delta P - \bar{\rho}\,\boldsymbol{\nabla}\delta\Phi\,, \tag{5.22}$$

where $\delta\Phi = \Phi - \bar{\Phi}$, which satisfies

$$\nabla^2 \delta\Phi = 4\pi G\,\delta\rho\,. \tag{5.23}$$

As before, these equations can be combined into a single equation for the density fluctuation:

$$\boxed{\left(\frac{\partial^2}{\partial t^2} - c_s^2 \nabla^2 - 4\pi G\bar{\rho}\right)\delta\rho = 0}\,, \tag{5.24}$$

so that each Fourier mode now obeys

$$\left(\frac{\partial^2}{\partial t^2} + c_s^2 k^2 - 4\pi G\bar{\rho}\right)\delta\rho(\mathbf{k}, t) = 0\,. \tag{5.25}$$

We see that there is a critical wavenumber, called the **Jeans scale**, for which the frequency of oscillations is zero:

$$k_{\mathrm{J}} \equiv \frac{\sqrt{4\pi G\bar{\rho}}}{c_s}\,. \tag{5.26}$$

The nature of the solution depends on the size of the wavenumber k relative to k_{J}. For large wavenumbers (i.e. small wavelengths), $k \gg k_{\mathrm{J}}$, the pressure dominates and we find the same oscillations as before. However, on large scales, $k < k_{\mathrm{J}}$, gravity dominates and (5.25) becomes

$$\left(\frac{\partial^2}{\partial t^2} - 4\pi G\bar{\rho}\right)\delta\rho = 0\,. \tag{5.27}$$

The solutions to this equation take the form $\delta\rho \propto \exp(\pm t/\tau)$, where $\tau \equiv 1/\sqrt{4\pi G\bar{\rho}}$. The exponential growth of the fluctuations is called the **Jeans instability**.

5.1.3 Adding Expansion

Finally, we will account for the expansion of the universe. We do this by interpreting the coordinates appearing in the fluid equations as physical coordinates $\mathbf{r}$, and writing them in terms of the comoving coordinates $\mathbf{x}$,

$$\mathbf{r}(t) = a(t)\,\mathbf{x}\,. \tag{5.28}$$

The velocity field is then given by

$$\mathbf{u}(t) = \dot{\mathbf{r}} = H\mathbf{r} + \mathbf{v}\,,\tag{5.29}$$

where $H\mathbf{r}$ is the Hubble flow and $\mathbf{v} = a\dot{\mathbf{x}}$ is the (physical) peculiar velocity. In a static spacetime, the derivatives ∂_t and $\boldsymbol{\nabla}$ are independent, but in an expanding spacetime they are not. Consider, for example, the time derivative of a function $f(t, \mathbf{r}) = f(t, a(t)\mathbf{x})$ at fixed $\mathbf{x}$,

$$\left(\frac{\partial f}{\partial t}\right)_{\mathbf{x}} = \left(\frac{\partial f(t, a(t)\mathbf{x})}{\partial t}\right)_{\mathbf{x}} = \left(\frac{\partial f}{\partial t}\right)_{\mathbf{r}} + \frac{\partial (a(t)\mathbf{x})}{\partial t} \cdot \left(\frac{\partial f}{\partial \mathbf{r}}\right)_t$$
$$= \left[\left(\frac{\partial}{\partial t}\right)_{\mathbf{r}} + H\mathbf{r} \cdot \boldsymbol{\nabla}_{\mathbf{r}}\right] f\,,\tag{5.30}$$

where $\boldsymbol{\nabla}_{\mathbf{r}}$ is the gradient with respect to $\mathbf{r}$ at fixed t. In terms of $\boldsymbol{\nabla}_{\mathbf{x}}$, the gradient with respect to $\mathbf{x}$, we have $\boldsymbol{\nabla}_{\mathbf{r}} = a^{-1}\boldsymbol{\nabla}_{\mathbf{x}}$. Equation (5.30) can also be written as

$$\left(\frac{\partial f}{\partial t}\right)_{\mathbf{r}} = \left[\left(\frac{\partial}{\partial t}\right)_{\mathbf{x}} - H\mathbf{x} \cdot \boldsymbol{\nabla}_{\mathbf{x}}\right] f\,.\tag{5.31}$$

With this in mind, let us look at the fluid equations in an expanding universe.

- Applying these substitutions to the *continuity equation* (5.1), we get

$$\left(\frac{\partial \rho}{\partial t}\right)_{\mathbf{r}} + \boldsymbol{\nabla}_{\mathbf{r}} \cdot (\rho\mathbf{u}) = 0\,,$$
$$\left(\left(\frac{\partial}{\partial t}\right)_{\mathbf{x}} - H\mathbf{x} \cdot \boldsymbol{\nabla}_{\mathbf{x}}\right)\rho + \frac{1}{a}\boldsymbol{\nabla}_{\mathbf{x}} \cdot (\rho\mathbf{u}) = 0\,,\tag{5.32}$$

where we have used (5.31) in the second line. Using (5.29) for $\mathbf{u}$ then gives

$$\boxed{\frac{\partial \rho}{\partial t} + 3H\rho + \frac{1}{a}\boldsymbol{\nabla} \cdot (\rho\mathbf{v}) = 0}\,,\tag{5.33}$$

where we have dropped the subscripts $\mathbf{x}$ and used $\boldsymbol{\nabla} \cdot \mathbf{x} = 3$. In the following, all time derivatives will be at fixed $\mathbf{x}$ and $\boldsymbol{\nabla}$ will always stand for $\boldsymbol{\nabla}_{\mathbf{x}}$.

- Similar manipulations of the *Euler equation* (5.19) lead to

$$\boxed{\left(\frac{\partial}{\partial t} + \frac{\mathbf{v}}{a} \cdot \boldsymbol{\nabla}\right)\mathbf{u} = -\frac{1}{a}\frac{\boldsymbol{\nabla} P}{\rho} - \frac{1}{a}\boldsymbol{\nabla}\Phi}\,,\tag{5.34}$$

where we have again used (5.31) and (5.29).

- The *Poisson equation* (5.20) simply becomes

$$\nabla^2 \Phi = 4\pi G\, a^2 \rho\,,\tag{5.35}$$

where the factor of a^2 accounts for the difference between physical and comoving distances.

Our task is to solve the coupled equations (5.33), (5.34) and (5.35). We again first find the background solution and then consider linear perturbations.

- Setting $\mathbf{v} = 0$ in (5.33) and (5.34), we get

$$\frac{\partial \bar{\rho}}{\partial t} = -3H\bar{\rho} \,, \tag{5.36}$$

$$\frac{\partial \bar{\mathbf{u}}}{\partial t} = -\frac{1}{a} \frac{\boldsymbol{\nabla} \bar{P}}{\bar{\rho}} - \frac{1}{a} \boldsymbol{\nabla} \bar{\Phi} \,. \tag{5.37}$$

Equation (5.36) is the familiar continuity equation for a homogeneous non-relativistic fluid. As expected, the background density therefore evolves as $\bar{\rho} \propto a^{-3}$. The Euler equation (5.37) has the following solution

$$\bar{\mathbf{u}} = Ha\mathbf{x} \,, \quad \bar{P} = \mathrm{const} \,, \quad \boldsymbol{\nabla} \bar{\Phi} = -\ddot{a}a\,\mathbf{x} \,. \tag{5.38}$$

This solution is now consistent with the Poisson equation, as long as the scale factor satisfies

$$\frac{\ddot{a}}{a} = -\frac{4\pi G}{3} \bar{\rho} \,, \tag{5.39}$$

which is the acceleration equation (2.137) found in Chapter 2.

- Next, we introduce perturbations around the background solution

$$\rho = \bar{\rho}[1 + \delta] \,, \quad \mathbf{u} = Ha\mathbf{x} + \mathbf{v} \,, \quad P = \bar{P} + \delta P \,, \quad \Phi = \bar{\Phi} + \delta\Phi \,, \tag{5.40}$$

where the **density contrast** is

$$\delta \equiv \frac{\delta\rho}{\bar{\rho}} \,. \tag{5.41}$$

As long as $\delta \ll 1$, we can linearize the equations.

Expanding the continuity equation (5.33) to linear order in the fluctuations, we find

$$\frac{\partial(\bar{\rho}\delta)}{\partial t} + 3H\bar{\rho}\delta + \frac{1}{a}\boldsymbol{\nabla}\cdot(\bar{\rho}\mathbf{v}) = 0 \,. \tag{5.42}$$

Using (5.36) for the evolution of the background density, we obtain the following equation for the density contrast

$$\boxed{\dot{\delta} = -\frac{1}{a}\boldsymbol{\nabla}\cdot\mathbf{v}} \,, \tag{5.43}$$

where the overdot denotes a time derivative at fixed $\mathbf{x}$. We see that the growth of the density perturbations is sourced by the divergence of the fluid velocity. At linear order, the vorticity of the fluid, $\boldsymbol{\nabla} \times \boldsymbol{\omega}$, therefore does not contribute.

Similarly, linearizing the Euler equation (5.34) gives

$$\boxed{\dot{\mathbf{v}} + H\mathbf{v} = -\frac{1}{a\bar{\rho}}\boldsymbol{\nabla}\delta P - \frac{1}{a}\boldsymbol{\nabla}\delta\Phi} \,, \tag{5.44}$$

where we have used that $(\mathbf{v}\cdot\boldsymbol{\nabla})\mathbf{x} = \mathbf{v}$. In the absence of pressure and gravitational perturbations, this equation simply says that $\mathbf{v} \propto a^{-1}$.

As before, it will be instructive to combine the two first-order equations (5.43) and (5.44) into a single second-order equation. First, we take the time derivative of (5.43) to get

$$\ddot{\delta} - \frac{H}{a}\boldsymbol{\nabla}\cdot\mathbf{v} + \frac{1}{a}\boldsymbol{\nabla}\cdot\dot{\mathbf{v}} = 0\,. \tag{5.45}$$

Substituting (5.44) for $\dot{\mathbf{v}}$ and (5.43) for $\boldsymbol{\nabla}\cdot\mathbf{v}$, this becomes

$$\ddot{\delta} + 2H\dot{\delta} - \frac{1}{a^2}\left(\frac{1}{\bar{\rho}}\nabla^2\delta P + \nabla^2\delta\Phi\right) = 0\,. \tag{5.46}$$

Finally, we use $\delta P = c_s^2\,\bar{\rho}\delta$ for the pressure perturbation and $\nabla^2\delta\Phi = 4\pi G a^2\bar{\rho}\delta$ for the gravitational potential. The final form of our evolution equation for the density contrast then is

$$\boxed{\ddot{\delta} + 2H\dot{\delta} - \left(\frac{c_s^2}{a^2}\nabla^2 + 4\pi G\bar{\rho}(t)\right)\delta = 0}\,. \tag{5.47}$$

This is the main equation describing the growth of density fluctuations in an expanding universe.

5.2 Growth of Matter Perturbations

The Newtonian treatment of the previous section applies to non-relativistic fluids. Two important examples for such fluids are cold dark matter and baryons after decoupling. In this section, we will study the evolution of fluctuations in these fluids.

5.2.1 Jeans Instability

We already encountered the Jeans instability for matter perturbations in a static spacetime. Now, we will show how the nature of this instability is modified by the expansion of the universe. As before, it is convenient to study the evolution in Fourier space, so that (5.47) becomes

$$\ddot{\delta}(\mathbf{k}, t) + 2H\dot{\delta}(\mathbf{k}, t) + c_s^2\left(\frac{k^2}{a^2} - k_{\rm J}^2\right)\delta(\mathbf{k}, t) = 0\,, \tag{5.48}$$

where we have introduced the *physical* Jeans wavenumber

$$k_{\rm J}(t) \equiv \frac{\sqrt{4\pi G\bar{\rho}(t)}}{c_s(t)}\,. \tag{5.49}$$

We see that the Jeans scale takes the same form as in (5.26), but now depends on time through the evolution of $\bar{\rho}(t)$ and possibly $c_s(t)$.

On small scales, $k/a \gg k_J$, equation (5.48) reduces to the equation of a damped harmonic oscillator with a friction term, $2H\dot{\delta}$, provided by the expansion of the universe. On these scales, we again get oscillatory solutions, but this time with a decreasing amplitude. On large scales, $k/a \ll k_J$, we can ignore the pressure term and (5.48) becomes

$$\ddot{\delta} + 2H\dot{\delta} - 4\pi G \bar{\rho}(t)\,\delta = 0 \,, \tag{5.50}$$

which should be compared to (5.27) in static space. A consequence of the extra friction term is that the fluctuations now don't grow exponentially.

For baryons, the sound speed and hence the Jeans scale are strong functions of time. Before recombination, the baryons are strongly coupled to the photons, forming a relativistic photon–baryon fluid with $c_s \approx 1/\sqrt{3}$. The Jeans scale is then of the order of the Hubble radius, so that there is no growth of baryonic fluctuations on subhorizon scales. After recombination, the sound speed of the baryon fluid drops drastically and its fluctuations start to grow.

For cold dark matter, on the other hand, the sound speed is negligible at all times, so that subhorizon modes can start to grow earlier. The details of the growth depend on the evolution of the background density. We explore this in the next section.

5.2.2 Linear Growth Function

We have seen that above the Jeans scale, the density contrast satisfies the second-order linear ODE (5.50) involving only time derivatives. Any solution of (5.50) can therefore be written as

$$\delta(\mathbf{k}, t) = \delta_+(\mathbf{k})D_+(t) + \delta_-(\mathbf{k})D_-(t) \,, \tag{5.51}$$

where $\delta_\pm(\mathbf{k})$ are the Fourier modes of the initial density field and the k-independent functions $D_\pm(t)$ describe the linear evolution. We refer to $D_+(t)$ and $D_-(t)$ as the growing and decaying modes, respectively. The growing mode is often called the **linear growth function** and is normalized so that $D_+(t_0) \equiv 1$. The functional form of $D_\pm(t)$ will depend on the evolution of the background density.

Let us start with the case of matter perturbations in a flat matter-dominated universe. This is relevant for the growth of structure in our universe after matter–radiation equality. Since $a \sim t^{2/3}$, we have

$$H(t) = \frac{2}{3t} \,,$$

$$4\pi G \bar{\rho}_m(t) = \frac{3}{2}H^2 = \frac{2}{3t^2} \,, \tag{5.52}$$

and hence (5.50) leads to

$$\ddot{D} + \frac{4}{3t}\dot{D} - \frac{2}{3t^2}D = 0 \,. \tag{5.53}$$

The two solutions of this equation are

$$D_-(t) \propto t^{-1} \propto a^{-3/2}\,,$$
$$D_+(t) \propto t^{2/3} \propto a\,,$$

(5.54)

so that the density contrast grows in proportion to the scale factor, $\delta \propto a$. Note that this means that the density fluctuation $\delta\rho$ is still decreasing, since $\delta\rho = \bar\rho\delta \propto a^{-2}$, but not as fast as the background density, $\bar\rho \propto a^{-3}$.

Exercise 5.1 Using the Poisson equation, show that the gravitational potential is a constant during the matter era.

Exercise 5.2 After decoupling, baryons behave as a non-relativistic fluid and can therefore also be studied in the Newtonian approximation. The coupled equations of baryons and cold dark matter are

$$\ddot\delta_b + \frac{4}{3t}\dot\delta_b = 4\pi G\left(\bar\rho_b\delta_b + \bar\rho_c\delta_c\right),$$
$$\ddot\delta_c + \frac{4}{3t}\dot\delta_c = 4\pi G\left(\bar\rho_b\delta_b + \bar\rho_c\delta_c\right).$$

(5.55)

Show that δ_b approaches δ_c at late times. We can then treat dark matter and baryons as a single pressureless fluid with total density contrast

$$\delta_m = \frac{\bar\rho_b\delta_b + \bar\rho_c\delta_c}{\bar\rho_b + \bar\rho_c}.$$

(5.56)

The growth function in (5.54) therefore applies to both baryons and dark matter after decoupling.

Next, we want to determine how the dark matter fluctuations evolve during the radiation era. In this case, we must modify the evolution equation (5.47) by including the perturbations in the radiation fluid in the Poisson equation for the gravitational potential. This leads to the following equation for the dark matter density contrast:

$$\ddot\delta_c + 2H\dot\delta_c - 4\pi G\sum_a \bar\rho_a\delta_a = 0\,.$$

(5.57)

During the radiation era, baryons are strongly coupled to the photons and are therefore included in the radiation fluid. The sum in (5.57) therefore has two contributions—from dark matter and radiation, $\bar\rho_c\delta_c + \bar\rho_r\delta_r$. Since the radiation fluid has a large sound speed, we expect its fluctuations to oscillate on scales smaller than the horizon. We will show this more rigorously in the relativistic treatment in the next chapter (see also [1]). These oscillations imply that the *time-averaged* density contrast of the radiation vanishes. Dark matter is therefore the only clustered component and we can write (5.57) as

$$\ddot{\delta}_c + 2H\dot{\delta}_c - 4\pi G\bar{\rho}_c\delta_c = 0 \,. \tag{5.58}$$

Of course, the homogeneous radiation density still plays an important role in determining the Hubble expansion rate, $H = 1/(2t)$ for $a \sim t^{1/2}$. Since the dark matter fluctuations evolve on the scale of a Hubble time, we have $\ddot{\delta}_c \sim H^2\delta_c \gg 4\pi G\bar{\rho}_c\delta_c$, where the final inequality follows because $\bar{\rho}_r \gg \bar{\rho}_c$ during the radiation era. We can therefore ignore the last term in (5.58), so that the growth function of the dark matter perturbations satisfies

$$\ddot{D} + \frac{1}{t}\dot{D} = 0 \,. \tag{5.59}$$

The two solutions of this equation are

$$\begin{aligned} D_-(t) &\propto \text{const} \,, \\ D_+(t) &\propto \ln t \propto \ln a \,. \end{aligned} \tag{5.60}$$

We see that the dark matter perturbations only grow logarithmically during the radiation era. This suppressed growth is called the **Mészáros effect**.

Exercise 5.3 Show that the gravitational potential decays as $\Phi \propto a^{-2}$ inside the horizon during the radiation era. This leaves an imprint in the CMB called the **early ISW effect** (see Chapter 7).

In Problem 5.2, you will derive a solution for the growth function that holds even in the transition from the radiation era to the matter era:

$$D \propto \begin{cases} 1 + \dfrac{3}{2}y \,, \\[2mm] \left(1 + \dfrac{3}{2}y\right)\ln\left(\dfrac{\sqrt{1+y}+1}{\sqrt{1+y}-1}\right) - 3\sqrt{1+y} \,, \end{cases} \tag{5.61}$$

where $y \equiv a/a_{\text{eq}}$. At early times ($y \ll 1$), the growing mode solution is $D_+ \propto \ln y$, while at late times ($y \gg 1$), it becomes $D_+ \propto y$. As expected, these limits reproduce the results in (5.60) and (5.54), respectively.

Finally, let us show that the growth of perturbations stops when dark energy starts to dominate the universe. Since dark energy doesn't cluster, it doesn't contribute as a source in the Poisson equation. The evolution equation for matter fluctuations (dark matter and baryons) therefore is

$$\ddot{\delta}_m + 2H\dot{\delta}_m - 4\pi G\bar{\rho}_m\delta_m = 0 \,, \tag{5.62}$$

where $H \approx \text{const}$ during dark energy domination. Since $4\pi G\bar{\rho}_m \ll H^2$, we can drop the last term and the equation for the growth function becomes

$$\ddot{D} + 2H\dot{D} = 0 \,. \tag{5.63}$$

The two solutions of this equation are

$$D_-(t) \propto a^{-2} \, ,$$
$$D_+(t) \propto \text{const} \, ,$$

(5.64)

which confirms that the growth of structure halts in the dark energy era.

Exercise 5.4 Show that the gravitational potential decays as $\Phi \propto a^{-1}$ inside the horizon during the dark energy era. This leaves an imprint in the CMB called the **late ISW effect** (see Chapter 7).

In Problem 5.2, you will derive a solution for the growth function that holds even when the universe contains a mixture of matter and dark energy:

$$D \propto \begin{cases} H \, , \\ H \displaystyle\int \frac{\mathrm{d}a}{(aH)^3} \, . \end{cases}$$

(5.65)

You will also be asked to confirm that this solution reduces to (5.54) at early times and to (5.64) at late times.

5.2.3 Transfer Function

The results of the previous section apply only to perturbations that are well within the Hubble radius. To describe the dynamics of perturbations outside the horizon requires relativistic perturbation theory, which will be the subject of the next chapter. The details are somewhat involved, but the result is easy to state. We will find that the superhorizon evolution of δ is

$$\delta(\mathbf{k}, t) \propto \frac{1}{(aH)^2} \propto \begin{cases} a^2 & \text{radiation era} \, , \\ a & \text{matter era} \, . \end{cases}$$

(5.66)

A heuristic way to obtain this result is to *assume* that the Poisson equation, $\nabla^2 \Phi = 4\pi G a^2 \bar\rho \, \delta$ continues to hold on superhorizon scales[1] and that the gravitational potential Φ is a constant on these scales. This then implies that $\delta \propto (a^2 \bar\rho)^{-1} \propto (aH)^{-2}$, as we asserted in (5.66).

We see that, in the matter era, the density contrast grows as $a(t)$ both inside and outside the horizon. For modes entering the horizon in the matter era, the growth is therefore independent of scale. In contrast, in the radiation era, the perturbations evolve very differently inside and outside the horizon. Subhorizon modes experience only logarithmic growth, while superhorizon modes grow as a^2. Since the moment

---

[1] As we will see in Chapter 6, on superhorizon scales, the form of the Poisson equation depends on the choice of coordinates. Assuming the ordinary Poisson equation to hold on superhorizon scales amounts to a specific coordinate choice—the so-called *comoving gauge*. Note that, in Chapter 6, we will denote the density contrast in comoving gauge by Δ *not* δ.

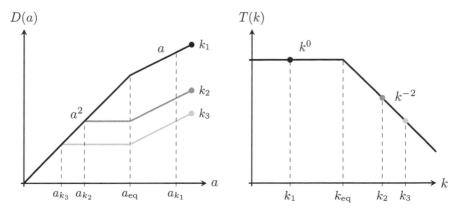

Fig. 5.1 Linear growth function (*left*) and transfer function (*right*) of the matter density perturbations for three different Fourier modes. Superhorizon modes grow as a^2 in the radiation era and as a in the matter era. After horizon crossing, the modes grow as $\ln a$ in the radiation era and as a in the matter era. The shape of the transfer function is a consequence of the suppressed growth of subhorizon modes during the radiation era.

of horizon entry depends on the wavenumber of the mode, $k = aH$, this leads to a k-dependent growth of the fluctuations. We describe this effect by the so-called **transfer function**:

$$T(k) \equiv \frac{D_+(t_i)}{D_+(t)} \frac{\delta(\mathbf{k}, t)}{\delta(\mathbf{k}, t_i)}, \qquad (5.67)$$

where D_+ is the *scale-independent* growth function. The function $T(k)$ captures all deviations in the evolution due to the scale-dependent growth in the radiation era. Because the evolution is isotropic, the transfer function depends only on the magnitude of the wavevector. The initial time t_i is taken to be some time after inflation, when all modes of interest were still outside the horizon. In the following, I will briefly sketch the functional form of the transfer function for dark matter. A more rigorous discussion will be given in Section 7.4.3.

Let us denote by $k_{\rm eq} \equiv (aH)_{\rm eq}$ the wavenumber of the mode that entered the horizon at the time of matter–radiation equality. This scale separates modes that entered the horizon in the radiation era ($k > k_{\rm eq}$) from those that entered in the matter era ($k < k_{\rm eq}$). We will discuss these two regimes in turn.

First, we consider long-wavelength modes with $k < k_{\rm eq}$, like the mode k_1 shown in Fig. 5.1. These modes were outside the horizon during the radiation era and hence grew as $\delta \sim a^2$. When the universe became dominated by matter, the growth slowed to $\delta \sim a$. After the mode enters the horizon in the matter era, the growth is still proportional to the scale factor. These considerations imply that

$$\delta(\mathbf{k}, t) = \left(\frac{a_{\rm eq}}{a_i}\right)^2 \frac{a}{a_{\rm eq}} \delta(\mathbf{k}, t_i) \equiv \frac{D_+(t)}{D_+(t_i)} \delta(\mathbf{k}, t_i), \qquad \text{for} \quad k < k_{\rm eq}, \qquad (5.68)$$

where we have identified the k-independent growth function. As expected, the transfer function for modes entering the horizon during the matter era is trivial, $T(k < k_{\rm eq}) = 1$.

Next, we follow the evolution of short-wavelength modes with $k > k_{\rm eq}$, like k_2 and k_3 in Fig. 5.1. These modes enter the horizon at a time t_k in the radiation era. Before horizon entry, the modes grow as $\delta \sim a^2$, while after horizon entry, the growth becomes logarithmic, $\delta \sim \ln a$. For simplicity, we will assume that the growth has stopped completely on subhorizon scales, so that $\delta \sim$ const. The modes start to grow again as $\delta \sim a$ only after the universe has become matter dominated. We therefore get

$$
\begin{aligned}
\delta(\mathbf{k}, t) &= \left(\frac{a_k}{a_i}\right)^2 \frac{a}{a_{\rm eq}} \delta(\mathbf{k}, t_i) \\
&= \left(\frac{a_k}{a_{\rm eq}}\right)^2 \times \left[\left(\frac{a_{\rm eq}}{a_i}\right)^2 \frac{a}{a_{\rm eq}}\right] \delta(\mathbf{k}, t_i), \qquad \text{for} \quad k > k_{\rm eq}, \quad (5.69)
\end{aligned}
$$

where the term in the square brackets is the same k-independent factor as in (5.68). We see that the amplitude is suppressed by an additional factor of $(a_k/a_{\rm eq})^2$. Since $H \propto a^{-2}$ in the radiation era, a mode with wavenumber k enters the horizon at $k = (aH)_k \propto 1/a_k$. The suppression factor can therefore also be written as $(a_k/a_{\rm eq})^2 = (k_{\rm eq}/k)^2$ and the transfer function is $T(k > k_{\rm eq}) = (k_{\rm eq}/k)^2$.

In summary, the results in (5.68) and (5.69) imply the following transfer function

$$
T(k) \approx \begin{cases} 1 & k < k_{\rm eq}, \\ (k_{\rm eq}/k)^2 & k > k_{\rm eq}. \end{cases} \quad (5.70)
$$

Including the logarithmic growth during the radiation era would give $T(k) \approx (k_{\rm eq}/k)^2 \ln(k/k_{\rm eq})$ for $k > k_{\rm eq}$. Our treatment has also ignored the effects of baryons on the transfer function. In Section 6.4.3, we will show that sound waves in the photon–baryon fluid before decoupling lead to an oscillatory feature in the dark matter transfer function.

5.3 Statistical Properties

It is a crucial fact the large-scale structure in the universe isn't distributed randomly, but has interesting correlations between spatially separated points. In this section, I will provide a brief introduction to these correlations. I will have much more to say about them in the rest of the book.

5.3.1 Correlation Functions

By definition, the mean value of the density perturbations is zero, $\langle \delta \rangle = 0$. The first nontrivial statistical measure of the density field (at a fixed time t) is therefore the two-point correlation function

$$\xi(|\mathbf{x} - \mathbf{x}'|, t) \equiv \langle \delta(\mathbf{x}, t)\delta(\mathbf{x}', t)\rangle = \int \mathcal{D}\delta\, \mathbb{P}[\delta]\, \delta(\mathbf{x}, t)\delta(\mathbf{x}', t)\,. \tag{5.71}$$

To avoid clutter, I will often suppress the time dependence. The integral is a functional integral over field configurations and $\mathbb{P}[\delta]$ is the probability of realizing some field configuration $\delta(\mathbf{x})$. The expectation value $\langle \ldots \rangle$ denotes an ensemble average of the stochastic process that created the random field δ (e.g. quantum fluctuations during inflation; see Chapter 8). The homogeneity of the universe implies that the two-point function should be invariant under a translation of the spatial coordinates, i.e. it can only depend on the separation between the two points, $\mathbf{x} - \mathbf{x}'$. Moreover, by isotropy the correlation function must be invariant under rotations and hence can only depend on the distance between the points, $r \equiv |\mathbf{x} - \mathbf{x}'|$. Equation (5.71) has a natural generalization to correlations between more than two points, but for now we will focus on the two-point function.

While the ensemble average is the natural object to compute from a theory of the initial conditions, it is *not* what we actually measure in observations. The observed large-scale structure of the universe is a single realization of a random process. To relate these observations to our theoretical predictions, we assume **ergodicity**, which states that ensemble averages are equal to spatial averages as the volume becomes infinitely large. For a Gaussian random field, ergodic behavior holds if $\xi(r) \to 0$ for $r \to \infty$. In that case, different parts of the universe can be viewed as different realizations of the underlying random process. The volume average therefore approximates the average over different members of the ensemble. Of course, in reality, we don't measure an infinite volume. For example, galaxy surveys measure objects only out to a maximal redshift. The finite volume of the survey will introduce statistical fluctuations called **sample variance**. In the case of the CMB, the volume is bounded by the size of the observable universe itself. The associated sample variance is then also called **cosmic variance** and arises because we only have one finite universe to observe.

Above we have seen that the Fourier modes of small fluctuations evolve independently. It is therefore very useful to define the two-point function in Fourier space

$$\begin{aligned}
\langle \delta(\mathbf{k})\delta^*(\mathbf{k}')\rangle &= \int \mathrm{d}^3x\, \mathrm{d}^3x'\, e^{-i\mathbf{k}\cdot\mathbf{x}} e^{i\mathbf{k}'\cdot\mathbf{x}'} \langle \delta(\mathbf{x})\delta(\mathbf{x}')\rangle \\
&= \int \mathrm{d}^3r\, \mathrm{d}^3x'\, e^{-i\mathbf{k}\cdot\mathbf{r}} e^{-i(\mathbf{k}-\mathbf{k}')\cdot\mathbf{x}'} \xi(r) \\
&= (2\pi)^3\delta_{\mathrm{D}}(\mathbf{k}-\mathbf{k}') \int \mathrm{d}^3r\, e^{-i\mathbf{k}\cdot\mathbf{r}} \xi(r) \\
&\equiv (2\pi)^3\delta_{\mathrm{D}}(\mathbf{k}-\mathbf{k}')\, \mathcal{P}(k)\,,
\end{aligned} \tag{5.72}$$

where, in the second line, we introduced $\mathbf{r} \equiv \mathbf{x} - \mathbf{x}'$ and then performed the integral over $\mathbf{x}'$. The fact that the result is proportional to the Dirac delta function $\delta_{\mathrm{D}}(\mathbf{k}-\mathbf{k}')$

is a consequence of translation invariance. This delta function implies that Fourier modes with different wavevectors are independent of each other. The function $\mathcal{P}(k)$ is called the **power spectrum** and is the three-dimensional Fourier transform of the correlation function $\xi(r)$. Because of rotational invariance, the power spectrum depends only on the magnitude of the wavevector, $k \equiv |\mathbf{k}|$. Working in spherical polar coordinates, we have $\mathbf{k}\cdot\mathbf{r} = kr\cos\theta$ and the power spectrum can be written as

$$
\begin{aligned}
\mathcal{P}(k) &= \int \mathrm{d}^3 r\, e^{-i\mathbf{k}\cdot\mathbf{r}}\, \xi(r) \\
&= \int_0^{2\pi} \mathrm{d}\phi \int_{-1}^{+1} \mathrm{d}(\cos\theta) \int_0^\infty \mathrm{d}r\, r^2 e^{-ikr\cos\theta} \xi(r) \\
&= 2\pi \int_0^\infty \mathrm{d}r\, \frac{r^2}{ikr}\big[e^{ikr} - e^{-ikr}\big]\xi(r) \\
&= \frac{4\pi}{k} \int_0^\infty \mathrm{d}r\, r \sin(kr)\, \xi(r)\,.
\end{aligned}
\tag{5.73}
$$

While the correlation function $\xi(r)$ is the natural object to consider in observations, the power spectrum $\mathcal{P}(k)$ is easier to predict theoretically. The two are related by (5.73) and are therefore completely equivalent descriptions of the statistics.

Exercise 5.5 Show that

$$
\xi(r) = \int \frac{\mathrm{d}k}{k} \frac{k^3}{2\pi^2} \mathcal{P}(k)\, j_0(kr)\,,
\tag{5.74}
$$

where $j_0(x) = \sin x / x$.

5.3.2 Gaussian Random Fields

For a Gaussian random field, the probability distribution $\mathbb{P}[\delta(\mathbf{x})]$ is a Gaussian functional of $\delta(\mathbf{x})$. The probability density function for N points in space, $\mathbf{x}_1, \ldots, \mathbf{x}_N$, is then a multi-variate Gaussian

$$
\mathbb{P}(\delta_1, \ldots, \delta_N) \propto \frac{1}{\sqrt{\det(\xi_{ij})}} \exp\left(-\frac{1}{2}\delta_i \xi_{ij}^{-1} \delta_j\right),
\tag{5.75}
$$

where $\delta_i \equiv \delta(\mathbf{x}_i)$ and $\xi_{ij} \equiv \xi(|\mathbf{x}_i - \mathbf{x}_j|)$. All N-point functions are then determined by functional integrals over $\mathbb{P}$ and are hence fixed completely in terms of the two-point correlation function $\xi(r)$. As we will see in Chapter 8, inflation predicts that the initial conditions of the universe were described to a good approximation by a Gaussian random field. Moreover, as long as the evolution is linear, the statistics remains Gaussian. The CMB probes the fluctuations mostly in the linear regime and they indeed look very Gaussian. It is for those reasons that most of this book will focus on the Gaussian statistics described by the two-point function (5.71), or equivalently the power spectrum (5.72).

At late times, however, structure formation is a highly nonlinear process and the probability distribution of the density field becomes non-Gaussian. A simple way to see this is to note that, by definition, the density contrast must satisfy $\delta > -1$, while there is no such constraint for positive values of δ. As δ becomes large, the probability distribution function must therefore become asymmetric and non-Gaussian.

5.3.3 Harrison–Zel'dovich Spectrum

In Sections 5.2.2 and 5.2.3, we described the linear evolution of a single Fourier mode of wavevector $\mathbf{k}$. Using these results, the power spectrum at the time t is

$$\mathcal{P}(k, t) = T^2(k) \frac{D_+^2(t)}{D_+(t_i)} \, \mathcal{P}(k, t_i), \tag{5.76}$$

where the transfer function satisfies $T(k = 0) = 1$. The primordial power spectrum is often written as a power law

$$\mathcal{P}(k, t_i) = A k^n, \tag{5.77}$$

where A and n are constants. The exponent n (often also denoted by $n_{\rm s}$) is called the **spectral index**. In 1970, well before inflation was introduced, it was argued by Harrison, Zel'dovich and Peebles that the initial perturbations of our universe are likely to have taken a power law form with spectral index $n \approx 1$ [2–4]. This is now called the **Harrison–Zel'dovich spectrum**.

To see why the value $n = 1$ is special, we note that the Poisson equation, $\nabla^2 \Phi = 4\pi G a^2 \bar{\rho} \, \delta$, implies the following relation between the power spectra of Φ and δ:

$$\mathcal{P}_\Phi(k, t_i) \propto k^{-4} \mathcal{P}(k, t_i) \propto k^{n-4}. \tag{5.78}$$

Using (5.74), we furthermore see that the variance of the gravitational potential is

$$\sigma_\Phi^2 \equiv \xi_\Phi(0) = \int \frac{dk}{k} \frac{k^3}{2\pi^2} \mathcal{P}_\Phi(k, t_i) \equiv \int \frac{dk}{k} \Delta_\Phi^2(k), \tag{5.79}$$

where we have introduced the dimensionless power spectrum

$$\Delta_\Phi^2(k) \equiv \frac{k^3}{2\pi^2} \mathcal{P}_\Phi(k, t_i) \propto k^{n-1}. \tag{5.80}$$

For $n = 1$, the dimensionless power spectrum of the gravitational potential therefore becomes k-independent, so that the variance receives equal contributions from every decade in k. This property of the field is called **scale invariance**.

Observations of the cosmic microwave background have found that the spectral index is [5]

$$n = 0.965 \pm 0.004. \tag{5.81}$$

The observed percent-level difference from the Harrison–Zel'dovich spectrum is precisely what is expected for fluctuations generated by inflation. As we will see in

Chapter 8, inflation predicts fluctuations that are naturally close to scale invariant because the inflationary dynamics is approximately time-translation invariant. However, we also know that inflation must end, so there has to be a small time dependence in the evolution. This translates into a small deviation from a perfectly scale-invariant spectrum.

5.3.4 Matter Power Spectrum

Combining the transfer function (5.70) with the primordial power spectrum (5.77), we predict that the late-time matter power spectrum should have the following asymptotic scalings:

$$\mathcal{P}(k,t) \propto \begin{cases} k^n & k < k_{\rm eq}, \\ k^{n-4} & k > k_{\rm eq}. \end{cases} \tag{5.82}$$

For a scale-invariant initial spectrum, we therefore expect the matter power spectrum to scale as k on large scales and as k^{-3} on small scales. Including the logarithmic growth of the fluctuations during the radiation era would give $\mathcal{P}(k,t) \propto k^{n-4}\ln(k/k_{\rm eq})^2$ for $k > k_{\rm eq}$.

Figure 5.2 shows a compilation of measurements of the linear matter power spectrum. We see that the observed spectrum indeed has the expected asymptotic scalings. On large scales, the measurements come from observations of the CMB anisotropies. As we will see in Chapter 7, the dark matter perturbations affect the evolution of the photon–baryon fluid before recombination, as well as the free streaming of the photons after decoupling. The latter effect arises because the photons lose energy as they climb out of the gravitational potential wells created by the dark matter. Taking these effects into account, the measurements of the CMB anisotropies can be related to the underlying matter density perturbations.[2]

Exercise 5.6 Show that

$$k_{\rm eq} = H_0\sqrt{\frac{2\Omega_m}{a_{\rm eq}}} \approx 0.015\,h\,{\rm Mpc}^{-1}, \tag{5.83}$$

where the second equality assumes standard values for the cosmological parameters. This explains the position of the turnover in Fig. 5.2.

---

[2] The matter perturbations also affect the observed CMB fluctuations through gravitational lensing. For clarity, these lensing constraints are not shown in Fig. 5.2. Also omitted from the figure are constraints from galaxy shear. The full set of constraints can be found in [6]. There it is also explained how the measurements of the nonlinear power spectrum are related back to the *linear* power spectrum at $z = 0$.

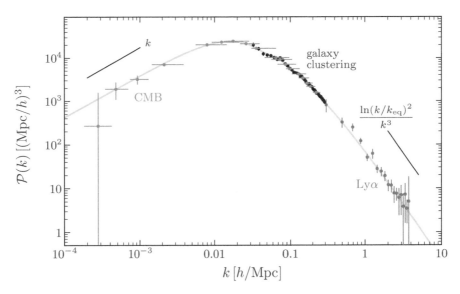

Fig. 5.2 Measurements of the linear matter power spectrum (figure adapted from [6]). The "galaxy clustering" constraints are from the luminous red galaxy sample of the Sloan Digital Sky Survey. The "CMB" constraints are derived from the Planck measurements of the CMB temperature anisotropies. "Lyα" refers to the Lyman-alpha forest. To avoid clutter, the plot doesn't include constraints from CMB polarization, CMB lensing and galaxy shear. These can be found in [6].

Other important tracers of the matter density fluctuations are galaxies. As we will see in Section 5.4.4, galaxies form inside dark matter **halos**, which themselves are created at peaks of the dark matter density field. We will show that the density fluctuations in the distribution of galaxies are related to fluctuations in the matter through $\delta_g = b\delta$, where the parameter b is called the **galaxy bias**. Taking this biasing into account, the galaxy power spectrum becomes a measure of the underlying matter power spectrum

$$\mathcal{P}_g(k) = b^2 \mathcal{P}(k) \,. \tag{5.84}$$

Note that the bias parameter can be different for different galaxy populations and that it becomes scale dependent when linear theory stops being a good approximation.

5.4 Nonlinear Clustering*

So far, our treatment has been restricted to the linear regime where $\delta \ll 1$. To understand the formation of small-scale structures (like galaxies and clusters of galaxies), with $\delta \gg 1$, we must go beyond the linearized analysis and tackle the

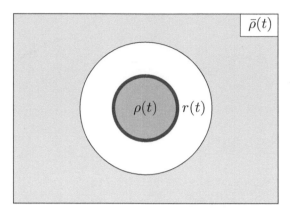

Fig. 5.3 Schematic illustration of a spherical overdensity of radius $r(t)$ and density $\rho(t)$ embedded in a homogeneous matter-dominated universe with density $\bar{\rho}(t)$. The evolution of the overdensity can be determined analytically.

nonlinear regime. This is challenging[3] and ultimately requires numerical simulations. In this section, we instead study a simple toy model with a high degree of symmetry that will allow us to follow the nonlinear evolution analytically [7]. As we will see, this toy model captures some key features of the nonlinear clustering.

5.4.1 Spherical Collapse

Consider a flat, matter-dominated universe. The average density is equal to the critical density and evolves as

$$\bar{\rho}(t) = \frac{1}{6\pi G}\frac{1}{t^2}\,.\qquad(5.85)$$

At some time t_i, we create a spherically symmetric overdensity by compressing a region of radius R_i to one with radius $r_i < R_i$; see Fig. 5.3. The initial density of the perturbation is

$$\rho_i = \frac{\bar{\rho}_i R_i^3}{r_i^3} \equiv \bar{\rho}_i(1 + \delta_i)\,,\qquad(5.86)$$

where we have used that the mass of the compressed region is conserved. In fact, mass conservation will be an important constraint. Even as the density and the size of the perturbation evolve, the mass will stay constant. As before, we have introduced the density contrast δ, but this time we will *not* assume that it stays small throughout the evolution.

What do we gain from the spherical symmetry of the situation? First, the evolution of the background will not be affected by the presence of the perturbation.

[3] A few of the complications are: (i) different Fourier modes start to couple to each other; (ii) the evolution cannot be described by a simple growth function; and (iii) the density field becomes non-Gaussian.

Second, the perturbation also evolves independently from the background. Third, we can think of the overdensity as consisting of many distinct mass shells. Our analysis will remain valid as long as these thin shells evolve independently, but breaks down when "shell crossing" occurs.

Let us study the evolution of a single mass shell of radius $r(t)$. In Newtonian gravity, the conservation of energy for the mass shell reads

$$\frac{1}{2}\dot{r}^2 - \frac{GM(r)}{r} = E, \tag{5.87}$$

where $M(r)$ is the mass enclosed by the shell. Both $M(r)$ and E are constant throughout the evolution. Inside the overdensity, we have $E < 0$, so that the overdensity acts like a closed universe with positive curvature. In fact, we have solved this problem in Section 2.3.6, where we have seen that a closed universe initially expands, before recollapsing under the gravitational attraction of the excess density (cf. Fig. 2.12). The parametric solution for the evolution of the radius of the closed universe was given in (2.169) and (2.170). The solution of (5.87) is of the same form

$$\begin{aligned} r(\theta) &= A(1 - \cos\theta), \\ t(\theta) &= B(\theta - \sin\theta), \end{aligned} \tag{5.88}$$

with

$$A \equiv \frac{GM}{2|E|}, \quad B \equiv \frac{GM}{(2|E|)^{3/2}}, \quad A^3 = GMB^2. \tag{5.89}$$

At early times, $\theta \ll 1$, we find that

$$\left.\begin{aligned} r(\theta) &\approx \frac{A}{2}\theta^2 \\ t(\theta) &\approx \frac{B}{6}\theta^3 \end{aligned}\right\} \quad \Rightarrow \quad r(t) \approx \frac{A}{2}\left(\frac{6}{B}\right)^{2/3} t^{2/3}. \tag{5.90}$$

We see that, initially, the overdense region evolves in the same way as the background. Eventually, however, the expansion will slow down and the overdense region will collapse just like a closed universe. The time at which the expansion halts is $\theta_{\text{turn}} = \pi$, while the time of collapse is $\theta_{\text{col}} = 2\pi$.

Using (5.88), we find that the density of the perturbation and the density of the background evolve as

$$\rho(\theta) = \frac{M}{(4\pi/3)r^3(\theta)} = \frac{3M}{4\pi A^3}\frac{1}{(1 - \cos\theta)^3}, \tag{5.91}$$

$$\bar{\rho}(\theta) = \frac{1}{6\pi G}\frac{1}{t^2(\theta)} = \frac{1}{6\pi GB^2}\frac{1}{(\theta - \sin\theta)^2}. \tag{5.92}$$

The ratio of (5.91) and (5.92) determines the density contrast

$$1 + \delta = \frac{\rho}{\bar{\rho}} = \frac{9}{2}\frac{(\theta - \sin\theta)^2}{(1 - \cos\theta)^3}, \tag{5.93}$$

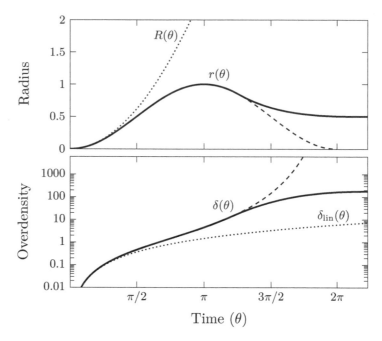

Fig. 5.4 Illustration of the evolution of a spherical overdensity embedded in a flat Einstein–de Sitter universe. The dashed lines show the would-be evolution of a perfectly spherical overdensity. In that idealized case, the overdensity collapses to a singularity at $\theta = 2\pi$. In reality, small deviations from spherical symmetry will grow during the collapse and the overdensity virializes. (Figure adapted from [8].)

where we have used that $A^3 = GMB^2$. It is useful to consider three distinct stages of the evolution (see Fig. 5.4 and Table 5.1):

- **Linear regime** At early times, while δ is still small, we expect this result to recover the solution found in linear perturbation theory. Expanding (5.93) for small θ, we get

$$\delta \approx \frac{3}{20}\theta^2 = \frac{3}{20}\left(\frac{6}{B}\right)^{2/3}t^{2/3} \equiv \delta_{\text{lin}}(t)\,, \tag{5.94}$$

which indeed is the expected scaling, $\delta_{\text{lin}} \propto a$. The solution in (5.93), however, holds even for large δ.

- **Turn-around** At the turn-around point, we find

$$\delta(\theta_{\text{turn}} = \pi) = \frac{9\pi^2}{16} - 1 \approx 4.55\,. \tag{5.95}$$

It is also useful to ask what the extrapolated linear solution would have been at the turn-around point. Using (5.94), we get

$$\delta_{\text{lin}}(t) = \frac{3}{20}(6\pi)^{2/3}\left(\frac{t}{t_{\text{turn}}}\right)^{2/3} \quad \Rightarrow \quad \delta_{\text{lin}}(t_{\text{turn}}) = \frac{3}{20}(6\pi)^{2/3} \approx 1.06\,, \tag{5.96}$$

Table 5.1 Stages in the evolution of the spherical collapse model

Stage	θ	δ_{lin}	δ
Turn-around	π	$\dfrac{3}{20}(6\pi)^{2/3} \approx 1.06$	$\dfrac{9\pi^2}{16} - 1 \approx 4.55$
Virialization	2π	$\dfrac{3}{20}(12\pi)^{2/3} \approx 1.69$	$\dfrac{9}{2}(2\pi)^2 - 1 \approx 177$

where we used that $t_{\text{turn}} = B\pi$. Of course, $\delta_{\text{lin}}(t_{\text{turn}})$ is a bit of an artificial concept, but it provides an easy way to judge by extrapolation of the linear solution whether a region of space should be viewed as decoupled from the Hubble flow.

- **Collapse** At the collapse time, the solution formally diverges,

$$\delta(\theta_{\text{col}} = 2\pi) = \infty \, , \tag{5.97}$$

signaling a point of infinite density. We will discuss the meaning of this divergence in the next section. Before we get to this, however, let us ask again what the linear solution would be if we extrapolated it to the collapse time. Using $t_{\text{col}} = 2t_{\text{turn}} = 2B\pi$, we find

$$\delta_{\text{lin}}(t_{\text{col}}) = \frac{3}{20}(12\pi)^{2/3} \approx 1.69 \, . \tag{5.98}$$

Once a naive extrapolation of linear perturbation theory reaches $\delta_{\text{lin}} = 1.69$, we should therefore think of the region as completely collapsed.

5.4.2 Virialization and Halos

The divergence that we found in the above analysis is an artifact of assuming perfect spherical symmetry for the initial perturbation. Such perfect symmetry does not exist in the real world. Instead, any small deviation from spherical symmetry will get amplified during the collapse. The individual shells will stop evolving independently and start crossing. Ultimately, the matter will **virialize**, meaning that it will settle into an equilibrium configuration with balanced kinetic and potential energies. The final object is an extended **dark matter halo**. These halos are the locations where baryons will eventually cluster into galaxies.

We can use the **virial theorem** to estimate the density of the dark matter halos. The theorem states that, for virialized objects, the average kinetic energy, T, is related to the average potential energy, V, by

$$T = -\frac{1}{2}V \, . \tag{5.99}$$

We will combine this with the conservation of energy, $E = T + V = \text{const}$. At the turn-around point, the kinetic energy vanishes and the total energy is $E = V_{\text{turn}}$. After virialization, we then have

$$T_{\text{vir}} + V_{\text{vir}} = \frac{1}{2}V_{\text{vir}} = V_{\text{turn}}\,. \tag{5.100}$$

This implies that the radius of the virialized halo is half the radius of the perturbation at turn-around, $r_{\text{vir}} = r_{\text{turn}}/2$, and the density of the halo is eight times the density at turn-around, $\rho_{\text{vir}} = 8\rho_{\text{turn}}$. Let us compare this to the background density at virialization. For simplicity, we take the time of virialization to be the collapse time, $t_{\text{vir}} \sim t_{\text{col}} = 2t_{\text{turn}}$. Since $\bar{\rho} \propto t^{-2}$, this then leads to $\bar{\rho}_{\text{vir}} = \bar{\rho}_{\text{turn}}/4$. Assembling all the pieces, the density contrast after virialization is

$$1 + \delta_{\text{vir}} = \frac{\rho_{\text{vir}}}{\bar{\rho}_{\text{vir}}} = \frac{8\rho_{\text{turn}}}{\bar{\rho}_{\text{turn}}/4} = 32(1 + \delta_{\text{turn}}) = 18\pi^2 \approx 178\,, \tag{5.101}$$

where we used (5.95) for δ_{turn}. The expected density of dark matter halos is therefore roughly 200 times greater than the background density. This agrees with the results of numerical simulations.

5.4.3 A Bound on Lambda

It is interesting to extend the spherical collapse model to the case of a universe with dark energy. Whether a spherical overdensity will recollapse will then depend on the size of the cosmological constant. In this section, we will derive an upper bound on the cosmological constant by demanding that the initial overdensity can form dark matter halos, a necessary requirement for the formation of galaxies and ultimately life itself. Interestingly, this upper bound will be close to the observed value of the cosmological constant [9].

Equation (5.87) now reads

$$\frac{1}{2}\dot{r}^2 - \frac{GM}{r} - \frac{1}{6}\Lambda r^2 = E\,, \tag{5.102}$$

where the functional form of the extra term can be understood by going back to our Newtonian derivation of the Friedmann equation in Section 2.3.5. In order for a spherical overdensity to turn around and collapse, there must be a time when the velocity of the mass shell vanishes, $\dot{r} = 0$. Such a moment only exists if there is a solution $r(t)$ to the following cubic equation

$$\frac{1}{6}\Lambda r^3 - |E|r + GM = 0\,. \tag{5.103}$$

However, if Λ is too large then this equation doesn't have a solution.

Exercise 5.7 Show that (5.103) only has a root with $r > 0$ if

$$\sqrt{\Lambda} < \frac{(2|E|)^{3/2}}{3GM} = \frac{1}{3B}, \tag{5.104}$$

where B was defined in (5.89).

Next, we want to write the bound in (5.104) in terms of the initial overdensity. At early times, the cosmological constant is negligible and our earlier analysis in a matter-dominated universe applies. We can therefore use (5.94) to write (5.104) as

$$\Lambda < 0.01 \frac{\delta^3}{t^2}. \tag{5.105}$$

To quantify this for our universe, we use that $\delta \approx 10^{-3}$ at the time of last-scattering, $t \approx 10^{13}\,\mathrm{s}$ (see Chapter 7). This then gives $\Lambda < 10^{-37}\,\mathrm{s}^{-2}$, which implies the following bound on the vacuum energy density

$$\rho_\Lambda = \frac{\Lambda}{8\pi G} < 10^{10}\,\mathrm{eV\,m^{-3}}. \tag{5.106}$$

Rather remarkably, this bound is only a factor of 10 larger than the observed size of the vacuum energy, which provides the basis for the anthropic solution to the cosmological constant problem [9].

5.4.4 Press–Schechter Theory

We have just learned that whenever a density perturbation in linear theory exceeds the threshold $\delta_c = 1.69$, a virialized halo with $\delta_{\mathrm{vir}} = 178$ will form. This simple fact was exploited by Press and Schechter to develop a theory for the statistics of dark matter halos [10]. The following is a rough sketch of the **Press–Schechter theory**. Since this is far outside of the main development of this book, my description will be brief and schematic. Further details can be found in the suggested readings at the end of this chapter.

Filtered density field

The first step is to introduce a smoothing (or "filtering") of the *linearly evolved* density field. This means that we remove contributions to the density field below a certain scale R. We denote the filtered field by $\delta_R(\mathbf{x}) \equiv \delta(\mathbf{x}; R)$. Roughly speaking, the smoothed density field is the average density in a volume of size R^3. At points where the filtered field δ_R exceeds the threshold δ_c, we declare a halo of size R, or mass $M \sim \bar{\rho} R^3$, to have formed. By thinking about this as a function of the smoothing scale R, we can derive the mass distribution of these halos.

Mathematically, the smoothed density field is defined as a convolution with a certain window function

$$\delta_R(t, \mathbf{x}) = \int \mathrm{d}^3 x'\, W(|\mathbf{x} - \mathbf{x}'|; R)\, \delta(t, \mathbf{x}'). \tag{5.107}$$

The choice of window function is not unique and a number of different possibilities are used in the literature. We will work with a simple top-hat function[4]

$$
W(r; R) = \begin{cases} 3/(4\pi R^3) & r < R, \\ 0 & r > R. \end{cases} \tag{5.108}
$$

The smoothed field is then the average density in spheres of radius R, around a point $\mathbf{x}$. The mass associated to each region is $M = (4\pi/3)R^3\bar{\rho}$, where $\bar{\rho}(t) \equiv \langle \rho(t, \mathbf{x}) \rangle$ is the average density. Often this mass scale is used to label the smoothed density field as δ_M. Being a convolution in real space, the smoothed density becomes a product in Fourier space, $\delta_R(t, \mathbf{k}) = W(k; R)\,\delta(t, \mathbf{k})$, where $W(k; R)$ is the Fourier transform of the window function.

Exercise 5.8 Show that the Fourier transform of the top-hat filter is

$$
W(k; R) = \frac{3}{(kR)^3}\Big[\sin(kR) - kR\cos(kR)\Big], \tag{5.109}
$$

and that the variance of the smoothed density can be written as

$$
\sigma^2(t; R) = \langle \delta_R^2(t, \mathbf{x}) \rangle = \int \mathrm{d}\ln k\, \Delta_{\mathrm{lin}}^2(t, k)\,|W(k; R)|^2, \tag{5.110}
$$

where

$$
\Delta_{\mathrm{lin}}^2(t, k) \equiv \frac{k^3}{2\pi^2}\mathcal{P}_{\mathrm{lin}}(t, k). \tag{5.111}
$$

Evaluating this variance for the specific scale $R = 8h^{-1}\mathrm{Mpc}$ gives the parameter σ_8 which is often used as a measure of the fluctuation amplitude. Its measured value today is $\sigma_8(t_0) = 0.811 \pm 0.006$.

Figure 5.5 shows the predictions of the standard ΛCDM cosmology for the dimensionless power spectrum of the linearly evolved density field and the variance of the smoothed density field. Over a small range of scales, the linear power spectrum can be written as a power law

$$
\mathcal{P}_{\mathrm{lin}}(t, k) = Ak^n T^2(k)\frac{D_+^2(t)}{D_+(t_i)} \equiv A(t)k^{n_{\mathrm{eff}}(k)}, \tag{5.112}
$$

where $T(k)$ is the dark matter transfer function and $n_{\mathrm{eff}}(k)$ is the effective spectral index. Substituting this into (5.110), we find

$$
\begin{aligned}
\sigma(R) &\propto D_+(t)R^{(n_{\mathrm{eff}}^* + 3)/2}, \\
\sigma(M) &\propto D_+(t)M^{-(n_{\mathrm{eff}}^* + 3)/6},
\end{aligned} \tag{5.113}
$$

---

[4] A drawback of this window function is that the sharp edge at $r = R$ creates power on all scales in Fourier space. Sometimes the window function is therefore defined as a sharp cutoff in Fourier space, with $W(k; R) = 1$ for $kR < 1$ and 0 otherwise. This has the disadvantage that there isn't a well-defined volume associated to the window function. A compromise is a Gaussian window function which has a smooth cutoff in both position and Fourier space.

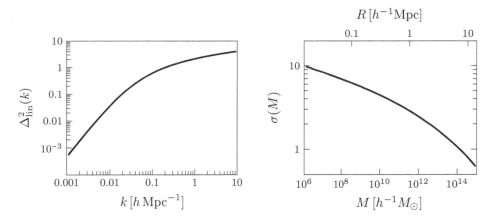

Fig. 5.5 *Left*: Dimensionless power spectrum of the linearly evolved density field. The shape of $\Delta^2_{\rm lin}(k)$ is determined by the dark matter transfer function. *Right*: Variance of the smoothed density field as a function of the smoothing scale R or mass M. Both plots were created for the standard ΛCDM cosmology.

where $n^*_{\rm eff} \equiv n_{\rm eff}(k_*)$ is the spectral index evaluated at $k_* = 2\pi/R$. As we see from Fig. 5.5, in a cold dark matter cosmology, we have $n^*_{\rm eff} + 3 > 0$ and $\sigma(M)$ is a monotonically decreasing function of mass. Hot dark matter leads to an extra suppression of power on small scales and can therefore give $n^*_{\rm eff} + 3 < 0$, for large k_*. This qualitative distinction between hot and cold dark matter will become important in a moment.

Mass function

The first statistical quantity that we will discuss is the number of halos in a given mass range. Let $n_h(t, \mathbf{x}, M)$ be the number of halos of mass M at a position $\mathbf{x}$ and time t, and $\bar{n}_h(t, M) \equiv \langle n_h(t, \mathbf{x}, M) \rangle$ be its mean value. To reduce clutter, I will often suppress the time dependence.

We will assume that the smoothed density is a Gaussian random field. The probability that a region of space has an overdensity δ_M is then given by

$$\mathbb{P}(\delta_M) = \frac{1}{\sqrt{2\pi\sigma^2(M)}} \exp\left[-\frac{1}{2}\frac{\delta_M^2}{\sigma^2(M)}\right], \tag{5.114}$$

where $\sigma^2(M)$ is the variance of the filtered field defined in (5.110). The probability for a region to exceed the density threshold δ_c is

$$\mathbb{P}(\delta_M > \delta_c) = \int_{\delta_c}^{\infty} \mathrm{d}\delta_M\, \mathbb{P}(\delta_M) = \int_{\nu}^{\infty} \mathrm{d}x\, e^{-x^2/2} = \frac{1}{2}\mathrm{erfc}\left(\frac{\nu}{\sqrt{2}}\right), \tag{5.115}$$

where $\nu(M) \equiv \delta_c/\sigma(M)$ is the so-called **peak height** and $\mathrm{erfc}(x)$ is the complementary error function. As we have seen in Fig. 5.5, in the standard ΛCDM cosmology, the variance $\sigma(M)$ is a decreasing function of mass M. This implies that small-scale fluctuations are the first to collapse. This type of structure formation is called "bottom up," because small-scale structures form first and then merge into larger objects. In the once popular hot dark matter models, structure would instead form in a "top down" fashion.

There is a problem with the result in (5.115). In the limit $R \to 0$, it should give the fraction of all mass in virialized objects. However, since $\sigma(R) \to \infty$ in this limit and $\mathrm{erfc}(0) = 1$, the answer in (5.115) implies that only half of the mass collapses into halos. This isn't consistent with the results of numerical simulations. The problem arises because only overdense regions, with $\mathbb{P}(\delta_M > 0) = \frac{1}{2}$, end up in collapsed objects. However, underdense regions can be enclosed in larger overdense regions and therefore become part of larger halos. Moreover, small overdense regions could be contained in bigger ones. These features aren't accounted for in (5.115). Press and Schechter "solved" this so-called **cloud-in-cloud problem** by multiplying the result in (5.115) by a factor of 2, $\tilde{\mathbb{P}} \equiv 2\mathbb{P}$. The justification for this fudge factor wasn't very satisfactory. Some time later, Bond, Cole, Efstathiou and Kaiser [12] introduced an extension of the naive Press–Schechter theory—using **excursion set theory**[5]—that explains the factor of 2. The final result for the mass function is the same as that of Press and Schechter, so we will proceed with their simpler (but less rigorous) analysis.

Given (5.115), the probability that a halo formed in the range $[M, M + \mathrm{d}M]$ is

$$P([M, M + \mathrm{d}M]) = \left| \tilde{\mathbb{P}}(\delta_{M+\mathrm{d}M} > \delta_c) - \tilde{\mathbb{P}}(\delta_M > \delta_c) \right| \approx -\frac{\mathrm{d}\tilde{\mathbb{P}}}{\mathrm{d}M}. \tag{5.116}$$

The abundance of halos of mass M—called the **mass function**—is then obtained by multiplying this by the maximum number density of such halos in a region of mean density $\bar{\rho}$. This is given by $\bar{\rho}/M$ and we find

$$\boxed{\frac{\mathrm{d}\bar{n}_h}{\mathrm{d}M} = -\frac{\bar{\rho}}{M}\frac{\mathrm{d}\tilde{\mathbb{P}}}{\mathrm{d}M} = -\sqrt{\frac{2}{\pi}}\,\nu \exp\left[-\frac{\nu^2}{2}\right]\frac{\bar{\rho}}{M^2}\frac{\mathrm{d}\ln\sigma}{\mathrm{d}\ln M}.} \tag{5.117}$$

For small masses (small ν), the mass function is a power law, while for large masses (large ν) it has an exponential fall-off. The function of the peak height ν appearing in (5.117) is called the **halo multiplicity**

$$\boxed{f_{\mathrm{PS}}(\nu) \equiv \sqrt{\frac{2}{\pi}}\,\nu \exp\left[-\frac{\nu^2}{2}\right].} \tag{5.118}$$

Although the Press–Schechter mass function captures essential qualitative features of halo formation, such as the exponential suppression at large masses, it disagrees with the results of N-body simulations at a more quantitative level. In particular, it

---

[5] Very roughly speaking, the idea is to consider the density field δ_R as a function of the smoothing scale R. At a fixed location, this produces a curve that looks like a random walk, starting at $\delta_R = 0$, for $R \to \infty$. The cloud-in-cloud problem is solved by identifying a halo with the *largest* smoothing scale at which the random walk first crosses the threshold δ_c for collapse, which removes the double-counting of smaller halos from the problem.

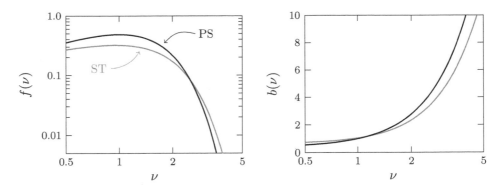

Fig. 5.6 Plots of the halo multiplicity (*left*) and the halo bias (*right*) as a function of the peak height $\nu(M) = \delta_c/\sigma(M)$. Shown are the Press–Schechter (*black*) and Sheth–Tormen (*gray*) predictions.

underpredicts the abundance of rare high-mass halos by a factor of 10 and overpredicts that of low-mass halos by a factor of 2. Given the crude assumptions that went into the Press–Schechter treatment, however, it is still remarkable that it captures the correct shape and overall normalization of the mass function. More importantly, the analysis of Press and Schechter was the starting point for many more precise theories that derive the more detailed properties of halo formation [11, 13].

Exercise 5.9 Using (5.113), show that the Press–Schechter mass function can be written as

$$\frac{\mathrm{d}\bar{n}_h}{\mathrm{d}M} \approx \gamma \frac{\beta(t)}{\sqrt{\pi}} \frac{\bar{\rho}}{M^2} M^{\gamma/2} \exp\left(-\beta^2(t) M^\gamma\right), \qquad (5.119)$$

where $\gamma(M) \equiv 1 + n_{\mathrm{eff}}(M)/3$.

Inspired by the form of the Press–Schechter mass function, Sheth and Tormen proposed a fitting function which could be tuned to allow for a better agreement with the data from N-body simulations [14]

$$f_{\mathrm{ST}}(\nu) \equiv A\sqrt{\frac{2}{\pi}} \left[1 + \left(a\nu^2\right)^{-p}\right] \sqrt{a\nu^2} \exp\left[-\frac{a\nu^2}{2}\right], \qquad (5.120)$$

where $A = 0.32$, $a = 0.75$ and $p = 0.3$ to match simulations. Figure 5.6 shows a comparison between the Press–Schechter and Sheth–Tormen predictions.

Biasing

The mass function tells us how many halos of a given mass M we should expect to find in a given volume. Next, we would like to determine how these halos are distributed in that volume. In particular, we want to compute the correlations in the halo positions. These correlations are inherited from the correlations in the primordial density field.

We define the density contrast of halos as

$$\delta_h(\mathbf{x}; M) \equiv \frac{n_h(\mathbf{x}; M) - \bar{n}_h(M)}{\bar{n}_h(M)}, \tag{5.121}$$

and the corresponding two-point correlation function is

$$\xi_{hh}(r; M) = \langle \delta_h(\mathbf{x}; M)\delta_h(\mathbf{x} + \mathbf{r}; M)\rangle. \tag{5.122}$$

Our task is to relate ξ_{hh} to the two-point function of the linear density field ξ_{lin}. The simplest way to do this uses the so-called **peak background split** [15–18].

Let us separate a density perturbation into a short-wavelength part ("peaks"), δ_h, and a long-wavelength part ("background"), δ_b, so that

$$\delta = \delta_h + \delta_b. \tag{5.123}$$

The peaks will eventually form halos, while the long-wavelength fluctuations modulate the local conditions for this halo formation. Since $\delta_b \ll 1$, the latter can be treated in linear theory. We can assume that δ_b is essentially constant over the scales that δ_h collapses into halos. The long-wavelength perturbation then affects the halo formation in two ways:

- First, it shifts the background density seen by the peaks to

$$\bar{\rho}' = \bar{\rho}(1 + \delta_b). \tag{5.124}$$

- Second, it perturbs the threshold for halo collapse. The linear part of δ_h now forms a halo when it reaches the effective threshold

$$\delta_c' = \delta_c - \delta_b, \tag{5.125}$$

corresponding to the original threshold δ_c for the total density perturbation δ. Since the effective threshold δ_c' depends on the value of the long-wavelength fluctuation δ_b, the local number density of halos becomes modulated by the field δ_b. In regions with positive δ_b, halos form earlier than in regions with negative δ_b. This effect is illustrated in Fig. 5.7.

Expanding the mass function to linear order in δ_b, we get

$$\begin{aligned}
\frac{dn_h}{dM}(\delta_b) &= \frac{d\bar{n}_h}{dM} + \left[\frac{\partial}{\partial\bar{\rho}'}\left(\frac{d\bar{n}_h}{dM}\right)\frac{d\bar{\rho}'}{d\delta_b} + \frac{\partial}{\partial\delta_c'}\left(\frac{d\bar{n}_h}{dM}\right)\frac{d\delta_c'}{d\delta_b}\right]\delta_b \\
&= \frac{d\bar{n}_h}{dM}\left[1 + \left(\bar{\rho}\frac{\partial}{\partial\bar{\rho}'}\ln\left(\frac{d\bar{n}_h}{dM}\right) - \frac{\partial}{\partial\delta_c'}\ln\left(\frac{d\bar{n}_h}{dM}\right)\right)\delta_b\right].
\end{aligned} \tag{5.126}$$

Using (5.117), we have

$$\bar{\rho}\frac{\partial}{\partial\bar{\rho}'}\ln\left(\frac{d\bar{n}_h}{dM}\right) = \frac{\bar{\rho}}{\bar{\rho}'} = \frac{1}{1 + \delta_b} \approx 1, \tag{5.127}$$

$$-\frac{\partial}{\partial\delta_c'}\ln\left(\frac{d\bar{n}_h}{dM}\right) = -\frac{1}{\sigma(M)}\frac{d\ln f(\nu)}{d\nu} = \frac{\nu^2 - 1}{\delta_c}, \tag{5.128}$$

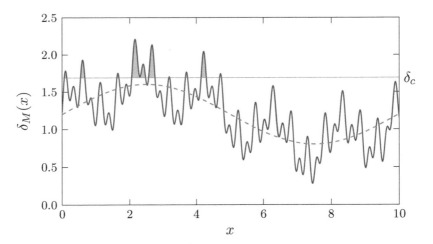

Fig. 5.7 One-dimensional illustration of the smoothed density field $\delta_M(x)$. The gray shaded peaks are the regions where halos form. The presence of a long-wavelength fluctuation modulates the large-scale halo density, leading to a biasing between the halo and matter perturbations.

where we used the Press–Schechter multiplicity function (5.118) in the final equality. Equation (5.126) then becomes

$$\frac{\mathrm{d}n_h}{\mathrm{d}M}(\delta_b) = \frac{\mathrm{d}\bar{n}_h}{\mathrm{d}M}\left[1 + \left(1 + \frac{\nu^2 - 1}{\delta_c}\right)\delta_b\right], \tag{5.129}$$

and substituting this into (5.121), we find

$$\delta_h(\mathbf{x}; M) = \left(1 + \frac{\nu^2 - 1}{\delta_c}\right)\delta_b(\mathbf{x}). \tag{5.130}$$

The factor relating δ_h and δ_b is called the **linear bias**:

$$\boxed{b_{\mathrm{PS}}(\nu) = 1 + \frac{\nu^2 - 1}{\delta_c}}. \tag{5.131}$$

This result applies to scales on which the long-wavelength mode can be treated in linear theory, so that the expansion in (5.129) is a good approximation. On smaller scales, nonlinear corrections are important and the biasing becomes more complicated [19].

Exercise 5.10 Repeating the above analysis for the Sheth–Tormen mass function (5.120), show that the bias parameter is

$$b_{\mathrm{ST}}(\nu) = 1 + \frac{a\nu^2 - 1}{\delta_c} + \frac{2p}{\delta_c[1 + (a\nu^2)^p]}. \tag{5.132}$$

A comparison between the bias parameters derived from the Press–Schechter and Sheth–Tormen mass functions is shown in Fig. 5.6.

Given (5.130), the halo two-point function can be written as

$$\boxed{\xi_{hh}(r; M) = b^2(M)\, \xi_{\rm lin}(r)}\,,\tag{5.133}$$

where $\xi_{\rm lin}(r)$ is the dark matter two-point function predicted in linear perturbation theory. We say the dark matter halos are "biased tracers" of the underlying density field. Since $b(M)$ is a monotonically increasing function, the biasing is stronger for more massive halos. Finally, because galaxies form inside of dark matter halos, the halo correlations are closely related to the observed galaxy correlations, and an ansatz like (5.133) is also used for the galaxy two-point function $\xi_{gg}(r)$.

5.5 Summary

In this chapter, we have studied the growth of structure in Newtonian theory. The Newtonian approximation is valid for non-relativistic fluids—such as dark matter or baryons after decoupling—and on subhorizon scales. For relativistic fluids and on scales larger than the horizon, the relativistic treatment of the next chapter is required.

We began with the non-relativistic fluid equations

$$\text{Continuity equation:}\quad \frac{\partial \rho}{\partial t} + \boldsymbol{\nabla}\cdot(\rho\mathbf{u}) = 0\,,\tag{5.134}$$

$$\text{Euler equation:}\quad \frac{\partial \mathbf{u}}{\partial t} + (\mathbf{u}\cdot\boldsymbol{\nabla})\mathbf{u} = -\frac{\boldsymbol{\nabla}P}{\rho} - \boldsymbol{\nabla}\Phi\,,\tag{5.135}$$

$$\text{Poisson equation:}\quad \nabla^2\Phi = 4\pi G\rho\,,\tag{5.136}$$

and incorporated the expansion of the universe by writing the physical coordinates appearing in these equations as $\mathbf{r} = a(t)\mathbf{x}$. We then expressed all quantities in terms of fluctuations around the homogeneous background solution

$$\rho = \bar{\rho}[1+\delta]\,,\quad \mathbf{u} = Ha\mathbf{x}+\mathbf{v}\,,\quad P = \bar{P}+\delta P\,,\quad \Phi = \bar{\Phi}+\delta\Phi\,,\tag{5.137}$$

and expanded the equations of motion to linear order in these fluctuations:

$$\text{Continuity equation:}\quad \dot{\delta} = -\frac{1}{a}\boldsymbol{\nabla}\cdot\mathbf{v}\,,\tag{5.138}$$

$$\text{Euler equation:}\quad \dot{\mathbf{v}} + H\mathbf{v} = -\frac{1}{a\bar{\rho}}\boldsymbol{\nabla}\delta P - \frac{1}{a}\boldsymbol{\nabla}\delta\Phi\,,\tag{5.139}$$

$$\text{Poisson equation:}\quad \nabla^2\delta\Phi = 4\pi Ga^2\bar{\rho}\delta\,,\tag{5.140}$$

where ∇ is now a derivative with respect to $\mathbf{x}$ and the overdot denotes a time derivative at fixed $\mathbf{x}$. The linearized equations can be combined into a single equation for the evolution of the density contrast

$$\ddot{\delta} + 2H\dot{\delta} - \left(\frac{c_s^2}{a^2}\nabla^2 + 4\pi G\bar{\rho}(t)\right)\delta = 0\,, \tag{5.141}$$

where we have used that $\delta P = c_s^2 \delta\rho$ for a barotropic fluid. On small scales, the pressure term dominates and we get oscillations with a frequency set by the speed of sound c_s. On larger scales, the pressure is negligible and gravity dominates. The resulting growth of the fluctuations is called the Jeans instability. We applied (5.141) to the evolution of dark matter perturbations. During the radiation era, the rapid expansion of the space counteracts the gravitational instability and the growth of the perturbations is only logarithmic, $\delta \sim \ln a$. Once the universe becomes dominated by matter, the perturbations start to grow as $\delta \sim a$.

Observations measure the spatial correlations in the large-scale structure of the universe. In linear theory, these correlations are easiest to predict in Fourier space, since the different Fourier modes evolve independently. The simplest correlation function in Fourier space is the power spectrum

$$\langle\delta(\mathbf{k},t)\delta^*(\mathbf{k}',t)\rangle = (2\pi)^3\delta_{\mathrm{D}}(\mathbf{k}-\mathbf{k}')\,\mathcal{P}(k,t)\,, \tag{5.142}$$

where the Dirac delta function is a consequence of the homogeneity of the background. For a Gaussian random field, the power spectrum contains all the statistical information of the density field. The late-time power spectrum can be written as

$$\mathcal{P}(k,t) = T^2(k)\frac{D_+^2(t)}{D_+^2(t_i)}\,\mathcal{P}(k,t_i)\,, \tag{5.143}$$

where $D_+(t)$ is the linear growth factor and $T(k)$ is the transfer function. A nontrivial transfer function arises because of the suppressed growth of subhorizon modes during the radiation era. The asymptotic scalings of the transfer function are

$$T(k) \approx \begin{cases} 1 & k < k_{\mathrm{eq}}\,, \\ (k_{\mathrm{eq}}/k)^2\ln(k/k_{\mathrm{eq}}) & k > k_{\mathrm{eq}}\,, \end{cases} \tag{5.144}$$

where k_{eq} is the wavenumber corresponding to the horizon at matter–radiation equality. The primordial power spectrum is

$$\mathcal{P}(k,t_i) = Ak^n\,, \tag{5.145}$$

with $n \approx 1$ for the Harrison–Zel'dovich spectrum. A key property of the Harrison–Zel'dovich spectrum is that the gravitational potential is scale invariant. Inflation predicts percent-level deviations from a perfectly scale-invariant spectrum. This prediction of inflation was verified by observations of the CMB, with the measured value of the spectral index being $n = 0.965 \pm 0.004$.

Finally, we studied a simple toy model for the nonlinear growth of density perturbations. We showed that the evolution of a spherically symmetric density perturbation in a flat Einstein–de Sitter universe is the same as that of a closed FRW universe and hence can be described analytically. Initially, the overdensity expands with the rest of the universe and the growth of the density contrast is consistent with the prediction of linear perturbation theory, $\delta \propto a(t)$. As for a universe with positive curvature, however, the expansion eventually stops and the matter recollapses. Small deviations from spherical symmetry get amplified during the collapse and the matter settles into a state of virial equilibrium. The predicted overdensity of the virialized halos is $\delta_{\mathrm{vir}} = 177$. Virialization occurs when the linearly evolved density contrast crosses the threshold $\delta_c = 1.69$. Press and Schechter used these insights from the spherical collapse model to derive the halo mass function, i.e. the abundance of halos as a function of their mass.

Further Reading

Structure formation is a big topic and I have only given a brief overview of its most basic features. Fortunately, there are many excellent resources with further details. A nice review of cosmological structure formation in both Newtonian gravity and general relativity is by Knobel [8]. A comprehensive treatment can be found in the book by Mo, van den Bosch and White [20] and a classic text is the book by Peebles [21]. My section on nonlinear clustering was inspired by a similar presentation in Tong's lecture notes [22], which was based on lecture notes by van den Bosch [23]. Nice reviews of Press–Schechter theory are [11, 13], and more on galaxy biasing can be found in [19]. The anthropic upper bound on the cosmological constant was derived by Weinberg in [9].

Problems

5.1 Fluid equations

In Chapter 2, we showed that the energy-momentum tensor of a perfect fluid in a curved spacetime is

$$T^{\mu\nu} = (\rho + P)\, U^\mu U^\nu + P g^{\mu\nu}\,,$$

where the four-velocity satisfies $g_{\mu\nu} U^\mu U^\nu = -1$.

1. Show that

$$\nabla_\alpha g_{\mu\nu} = 0\,,$$
$$U_\nu \nabla_\mu U^\nu = 0\,,$$

for arbitrary $g_{\mu\nu}$.

2. Show that the conservation of the energy-momentum tensor, $\nabla_\mu T^{\mu\nu} = 0$, implies

$$\text{Continuity:} \quad U^\mu \nabla_\mu \rho + (\rho + P)\nabla_\mu U^\mu = 0\,,$$
$$\text{Euler:} \qquad (\rho + P)U^\mu \nabla_\mu U^\nu = -(g^{\mu\nu} + U^\mu U^\nu)\nabla_\mu P\,.$$

Do the fluid elements move on geodesics? If not, under what condition would they move on geodesics?

3. Let $g_{\mu\nu} \approx \eta_{\mu\nu}$, $U^\mu \approx (1, \mathbf{u})$, with $|\mathbf{u}| \ll 1$, and $P \ll \rho$. Show that the above equations become

$$\text{Continuity:} \quad \dot{\rho} + \boldsymbol{\nabla} \cdot (\rho \mathbf{u}) = 0\,,$$
$$\text{Euler:} \qquad \rho(\dot{\mathbf{u}} + \mathbf{u} \cdot \boldsymbol{\nabla}\mathbf{u}) = -\boldsymbol{\nabla}P\,.$$

These non-relativistic versions of the fluid equations were the starting point for the Newtonian analysis in this chapter.

5.2 Growth of matter perturbatons

Above the Jeans scale, but below the Hubble radius, the evolution of matter perturbations is governed by

$$\ddot{\delta}_m + 2H\dot{\delta}_m = 4\pi G\bar{\rho}_m \delta_m\,,$$

where $\bar{\rho}_m = \bar{\rho}_{m,0}\, a^{-3}$. We have ignored the fluctuations in the radiation and the dark energy on the right-hand side, but still include their homogeneous densities as a source for the Hubble rate.

1. Show that the above evolution equation can be written as

$$\frac{\mathrm{d}}{\mathrm{d}a}\left(a^3 H \frac{\mathrm{d}\delta_m}{\mathrm{d}a}\right) = 4\pi G\bar{\rho}_{m,0} \frac{\delta_m}{Ha^2}\,. \tag{1}$$

2. Assuming that the universe contains a combination of matter, curvature and a cosmological constant, show that the solutions to (1) are

$$\delta_m \propto \begin{cases} H\,, \\ H \displaystyle\int \frac{\mathrm{d}a}{(aH)^3}\,. \end{cases}$$

What is the scaling of the growing mode solution when the expansion of the universe is dominated by matter, by curvature, or by a cosmological constant?

3. At early times, the universe was dominated by radiation and pressureless matter. You may ignore baryons. Let $y \equiv a/a_{\mathrm{eq}}$, where a_{eq} is the scale factor at matter–radiation equality. Show that the Hubble parameter can be written as

$$H(y) = \frac{A}{y^2}\sqrt{1+y}\,,$$

where the constant A should be determined in terms of the energy density at equality, ρ_{eq}. Show that (1) is equivalent to

$$\frac{\mathrm{d}^2\delta_m}{\mathrm{d}y^2} + \frac{2+3y}{2y(y+1)}\frac{\mathrm{d}\delta_m}{\mathrm{d}y} = \frac{3}{2}\frac{\delta_m}{y(y+1)}.$$

Verify that the solutions of this equation are

$$\delta_m \propto \begin{cases} 1 + \dfrac{3}{2}y, \\ \left(1 + \dfrac{3}{2}y\right)\ln\left(\dfrac{\sqrt{1+y}+1}{\sqrt{1+y}-1}\right) - 3\sqrt{1+y}. \end{cases}$$

What are the growing and decaying modes at early and late times?

5.3 Jeans mass

The baryonic Jeans mass is

$$M_{\mathrm{J}} \equiv \rho_b\left(\frac{4\pi}{3}\lambda_{\mathrm{J}}^3\right).$$

Perturbations with $M > M_{\mathrm{J}}$ will experience gravitational collapse.

1. Show that the baryonic Jeans mass just before decoupling is

$$M_{\mathrm{J}} \approx 10^{19}\,M_\odot,$$

and interpret the result.

2. After decoupling, the sound speed of the baryons is

$$c_s = \sqrt{\frac{k_{\mathrm{B}}T}{\mu}},$$

where μ is the mean mass of the atoms in the gas, which you may approximate by the mass of a hydrogen atom. Show that the baryonic Jeans mass just after decoupling is

$$M_{\mathrm{J}} \approx 10^5\,M_\odot,$$

and interpret the result.

5.4 Jeans instability

Consider a universe filled with a non-relativistic fluid with $P = \alpha\rho^{4/3}$, where $P \ll \rho$, so that the background evolution is well-approximated as matter dominated. Inside the Hubble radius, a Fourier mode of the density contrast $\delta \equiv \delta\rho/\rho$ evolves as

$$\delta'' + \mathcal{H}\delta' + \left(c_s^2 k^2 - \frac{3}{2}\mathcal{H}^2\right)\delta = 0,$$

where $c_s^2 \equiv \partial P/\partial\rho$, primes denote derivatives with respect to conformal time η and $\mathcal{H} \equiv a'/a$.

1. Show that $c_s^2 = c_0^2(\eta_0/\eta)^2$, where c_0 and η_0 are constants.

2. Describe qualitatively how you expect the solution for the density contrast to behave on small and large scales.

3. Find power law solutions of the form $\delta \propto \eta^\beta$, where β depends on the wavenumber k. At what value of k does the solution change from monotonic to oscillatory? For what k is the growing mode a constant?

6 Relativistic Perturbation Theory

The Newtonian analysis of the previous chapter is inadequate on scales larger than the Hubble radius and for relativistic fluids (like photons and neutrinos). The correct description requires a general-relativistic treatment which we will develop in this chapter. This will allow us to follow the evolution of all fluctuations in the primordial plasma, from their frozen initial conditions on very large scales to their complex evolution after horizon entry on small scales.

Figure 6.1 illustrates the interactions of all known species in the universe. At low energies—meaning after neutrino decoupling and nucleosynthesis—the only forces that play a significant role in the cosmological evolution are gravity and electromagnetism. By the equivalence principle, all species interact through a universal gravitational force. In addition, photons, electrons and protons are strongly coupled by their electromagnetic interactions and can be treated as a single

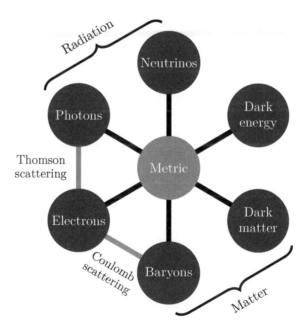

Fig. 6.1 Interactions between the different species in the universe. Before recombination, photons and electrons are tightly coupled by Thomson scattering, while electrons and baryons (mostly protons) interact through Coulomb scattering. All species interact gravitationally.

photon–baryon fluid. After recombination, the electromagnetic coupling becomes insignificant and the evolution is determined by gravity alone. There are two types of gravitational effects: the homogeneous density of the species determines the expansion rate of the universe, while their perturbations source the local gravitational potential. Both of these effects feed back into the evolution of the species. This feedback is nonlinear and can lead to rather complex dynamics.

Two important times in structure formation are matter–radiation equality, $z_{\mathrm{eq}} \approx 3400$, and photon decoupling, $z_{\mathrm{dec}} \approx 1100$. Before equality, the growth of structure is suppressed by the large radiation pressure of the photons. The clustering of dark matter perturbations therefore only starts to become significant when the universe becomes matter dominated. Before decoupling, baryons are tightly coupled to the photons and their density fluctuations oscillate. These oscillations arise because of two opposing forces: gravity and radiation pressure. As gravity makes the perturbations collapse, the radiation pressure increases and causes the perturbation to re-expand. The pressure then drops, so that gravity wins the battle again and the process repeats, leading to sustained oscillations or "sound waves" in the primordial plasma. The frequency of these oscillations depends on the wavelength of the fluctuations. At decoupling, modes with different wavelengths are captured at different phases in their evolution, giving rise to a characteristic pattern in the CMB. After decoupling, the baryons lose the pressure support of the photons and fall into the gravitational potential wells created by the dark matter. Eventually, stars and galaxies will form.

✲✲✲

In this chapter, I will tell this fascinating story of cosmic structure formation in mathematically precise terms. We will begin, in Section 6.1, by defining the perturbations of the metric and the energy-momentum tensor. We will show that the form of the perturbations depends on the choice of coordinates (or the "gauge choice"). We will derive the linearized evolution equations in Newtonian gauge, and show that, on small scales, they reduce to the Newtonian equations of the previous chapter.

In Section 6.2, we will discuss the initial conditions of perturbations on superhorizon scales. We will introduce the concept of "adiabatic perturbations" and show that the "comoving curvature perturbation" $\mathcal{R}$ is constant outside the horizon for such perturbations. Our task in the rest of the chapter will be to determine how these initial fluctuations give rise to the observed correlations in the late universe (see Fig. 6.2). In Section 6.3, we determine the evolution of matter fluctuations and derive the late-time power spectrum. In Section 6.4, we solve for the dynamics of the tightly-coupled photon–baryon fluid before decoupling, and show how it is imprinted in the CMB anisotropies and the matter power spectrum (see also Chapter 7). Finally, in Section 6.5, we review the evolution of tensor metric perturbations in an expanding universe, which will be useful for our discussion of gravitational waves in Chapters 7 and 8.

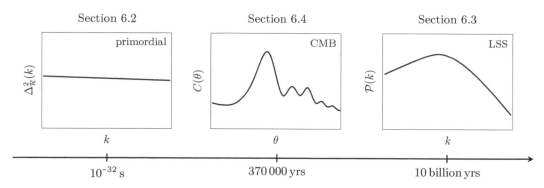

Fig. 6.2   In this chapter, we show how a scale-invariant spectrum of primordial curvature perturbations, $\Delta_{\mathcal{R}}^2(k)$, evolves into the angular power spectrum of the CMB anisotropies, $C(\theta)$, and the matter power spectrum, $\mathcal{P}(k)$.

All numerical computations in this chapter were performed with the Boltzmann code CLASS. As aspiring cosmologists you should download CLASS[1]—or the alternative CAMB[2]—and try to reproduce some of these results. Playing with CLASS will give you intuition for the dynamics of cosmological perturbations.

6.1 Linear Perturbations

Conceptually, perturbation theory in GR is the same as in Newtonian gravity. We write the metric and the energy-momentum tensor as[3]

$$g_{\mu\nu}(t,\mathbf{x}) = \bar{g}_{\mu\nu} + \delta g_{\mu\nu}(t,\mathbf{x}),$$
$$T_{\mu\nu}(t,\mathbf{x}) = \bar{T}_{\mu\nu} + \delta T_{\mu\nu}(t,\mathbf{x}),$$

(6.1)

and expand $\nabla^\mu T_{\mu\nu} = 0$ and $G_{\mu\nu} = 8\pi G\, T_{\mu\nu}$ to linear order in the perturbations. In practice, however, two features make it more complicated than the Newtonian analysis of the previous chapter: First, the algebra is much more involved. Computing the perturbed Einstein tensor is not for the faint of heart. This has earned cosmological perturbation theory the unfortunate reputation of being a hard subject, although conceptually there is nothing hard about it. To address this challenge, I will perform all algebraic manipulations in this chapter very explicitly and encourage you to work through it actively. After a while, this will actually become a fun (somewhat meditative) exercise. I will also provide a Mathematica notebook on the book's website that will allow you to perform all of these calculations without any effort. A second complication in relativistic

---

[1] http://class-code.net
[2] http://camb.info
[3] To reduce clutter, we will often drop the argument $(t,\mathbf{x})$ on the perturbations. We will also find it convenient to work in conformal time η. Derivatives with respect to η will be denoted by primes and the conformal Hubble rate will be written as $\mathcal{H} = a'/a$.

perturbation theory is the fact that, on large scales, there is an ambiguity in what we call matter perturbations and what are metric perturbations—by a change of coordinates we can transform one into the other. This subtlety did not arise in the Newtonian treatment.

6.1.1 Metric Perturbations

To avoid unnecessary technical distractions, we will take the background metric $\bar{g}_{\mu\nu}$ to be the flat FRW metric. The perturbed spacetime can then be written as

$$\mathrm{d}s^2 = a^2(\eta)\big[-(1+2A)\mathrm{d}\eta^2 + 2B_i\,\mathrm{d}x^i\mathrm{d}\eta + (\delta_{ij} + 2E_{ij})\mathrm{d}x^i\mathrm{d}x^j\big]\,, \qquad (6.2)$$

where A, B_i and E_{ij} are functions of space and (conformal) time. The factors of 2 are introduced for later convenience. We will adopt the useful convention that the indices on spatial vectors and tensors are raised and lowered with δ_{ij}.

It will be extremely useful to perform a scalar-vector-tensor (SVT) decomposition of the perturbations. For three-vectors, this should be familiar. It simply means that we can split any three-vector into the gradient of a scalar and a divergenceless vector

$$B_i = \underbrace{\partial_i B}_{\text{scalar}} + \underbrace{\hat{B}_i}_{\text{vector}}\,, \qquad (6.3)$$

with $\partial^i \hat{B}_i = 0$. In Fourier space, the decomposition in (6.3) becomes[4] $B_i = i\hat{k}_i B + \hat{B}_i$, where $\hat{k} \equiv \mathbf{k}/|\mathbf{k}|$. We see that the vector has been split into a piece that is parallel to the direction of the Fourier mode $\mathbf{k}$ and a piece orthogonal to it. Similarly, any rank-2 symmetric tensor can be written

$$E_{ij} = \underbrace{C\delta_{ij} + \partial_{\langle i}\partial_{j\rangle}E}_{\text{scalar}} + \underbrace{\partial_{\langle i}\hat{E}_{j\rangle}}_{\text{vector}} + \underbrace{\hat{E}_{ij}}_{\text{tensor}}\,, \qquad (6.4)$$

where

$$\partial_{\langle i}\partial_{j\rangle}E \equiv \left(\partial_i\partial_j - \frac{1}{3}\delta_{ij}\nabla^2\right)E\,, \qquad (6.5)$$

$$\partial_{\langle i}\hat{E}_{j\rangle} \equiv \frac{1}{2}\left(\partial_i\hat{E}_j + \partial_j\hat{E}_i\right)\,. \qquad (6.6)$$

The first term in (6.4) contains the trace of the spatial perturbation, $E^i{}_i = C\delta^i_i = 3C$, while the remaining terms are traceless. As before, the hatted quantities have vanishing divergence, $\partial^i\hat{E}_i = \partial^i\hat{E}_{ij} = 0$, and so correspond to transverse vector and tensor perturbations. The tensor perturbations describe gravitational waves (see Section 6.5) and following Georgi's dictum[5] I will also denote them by $h_{ij} \equiv 2\hat{E}_{ij}$. In Fourier space, we write (6.5) as $\partial_{\langle i}\partial_{j\rangle}E \to -\left(\hat{k}_i\hat{k}_j - \frac{1}{3}\delta_{ij}\right)E \equiv -\hat{k}_{\langle i}\hat{k}_{j\rangle}E$.

---

[4] The magnitude of the wavevector, $k \equiv |\mathbf{k}|$, has been absorbed into $B(\eta, \mathbf{k})$, so that B and B_i have the same dimensions.

[5] "Notation should be your slave, not the other way around."

The 10 degrees of freedom of the metric have thus been decomposed into $4+4+2$ SVT degrees of freedom:

- *scalars*: A, B, C, E
- *vectors*: $\hat{B}_i$, $\hat{E}_i$
- *tensors*: $\hat{E}_{ij}$

What makes the SVT decomposition so powerful is the fact that the Einstein equations for scalars, vectors and tensors don't mix at linear order and can therefore be treated separately. We will mostly be interested in scalar fluctuations and the associated density perturbations, and at linear order it is consistent to set vectors and tensors to zero. Vector perturbations aren't produced by inflation and even if they were, they would decay quickly with the expansion of the universe. Tensor perturbations are an important prediction of inflation and we will discuss them at various points in the text.

Gauge problem

The metric perturbations in (6.2) aren't uniquely defined, but depend on our choice of coordinates or the **gauge choice**. In particular, when we wrote down the perturbed metric, we implicitly introduced a specific time slicing of the spacetime and defined specific spatial coordinates on these time slices. Making a different choice of coordinates can change the values of the perturbation variables. It may even introduce "fictitious perturbations," which are fake perturbations that can arise from an inconvenient choice of coordinates even if the background is perfectly homogeneous.

Consider, for example, a homogeneous FRW spacetime and make the following change of the spatial coordinates, $x^i \mapsto \tilde{x}^i = x^i + \xi^i(\eta, \mathbf{x})$. We assume that ξ^i is small, so that it can also be treated as a perturbation. Using $\mathrm{d}x^i = \mathrm{d}\tilde{x}^i - \partial_\eta \xi^i \mathrm{d}\eta - \partial_k \xi^i \mathrm{d}\tilde{x}^k$, the line element becomes

$$
\begin{aligned}
\mathrm{d}s^2 &= a^2(\eta) \left[-\mathrm{d}\eta^2 + \delta_{ij} \mathrm{d}x^i \mathrm{d}x^j \right] \\
&= a^2(\eta) \left[-\mathrm{d}\eta^2 - 2\xi_i' \mathrm{d}\tilde{x}^i \mathrm{d}\eta + \left(\delta_{ij} - 2\partial_{(i}\xi_{j)} \right) \mathrm{d}\tilde{x}^i \mathrm{d}\tilde{x}^j \right],
\end{aligned} \tag{6.7}
$$

where we have dropped terms that are quadratic in ξ^i and defined $\xi_i' \equiv \partial_\eta \xi_i$. We apparently have introduced the metric perturbations $B_i = -\xi_i'$ and $\hat{E}_i = -\xi_i$. However, these are just fictitious **gauge modes** that can be removed by going back to the old coordinates.

As another example, consider a change in the time slicing, $\eta \mapsto \tilde{\eta} = \eta + \xi^0(\eta, \mathbf{x})$. The homogeneous density of the universe then gets perturbed,

$$
\rho(\eta) = \rho(\tilde{\eta} - \xi^0(\tilde{\eta}, \mathbf{x})) = \bar{\rho}(\tilde{\eta}) - \bar{\rho}' \xi^0 . \tag{6.8}
$$

Even in an unperturbed universe, a local change of the time coordinate can therefore introduce a fictitious density perturbation $\delta\rho = -\bar{\rho}' \xi^0$. Conversely, we can also remove a real perturbation in the energy density by choosing the hypersurface of constant time to coincide with the hypersurface of constant energy density. We then have $\delta\rho = 0$ although there are real inhomogeneities. Of course, the fluctuations haven't disappeared completely, but will manifest themselves as new perturbations in the metric.

These examples illustrate that we need a more physical way to identify true perturbations. One way to do this is to define perturbations in such a way that they don't change under a change of coordinates.

Coordinate transformations

Consider the following coordinate transformation

$$x^\mu(q) \;\mapsto\; \tilde{x}^\mu(q) \equiv x^\mu(q) + \xi^\mu(q), \quad \text{where} \quad \begin{array}{l} \xi^0 \equiv T, \\ \xi^i \equiv L^i = \partial^i L + \hat{L}^i. \end{array} \tag{6.9}$$

As indicated by the labels, the old and the new coordinates are evaluated at the *same* physical point q. As before, we assume that ξ^μ is small and can therefore be treated as a small perturbation. The function $T(\eta, \mathbf{x})$ defines the hypersurfaces of constant time in the new coordinates, while $L^i(\eta, \mathbf{x})$ determines the spatial coordinates on these hypersurfaces. We have split the spatial shift L^i into a scalar, L, and a divergenceless vector, $\hat{L}^i$.

We wish to understand how the metric perturbations transform under this change of coordinates. Before we get to this, however, we first look at a simpler problem: the transformation of a (Lorentz) scalar field $\phi(\eta, \mathbf{x})$, which may be the inflaton or the density of a fluid. Since we have only relabeled the coordinates, the value of the field at each point q must be unchanged

$$\phi(x^\mu) = \tilde{\phi}(\tilde{x}^\mu). \tag{6.10}$$

Writing both sides in terms of the background value of the field and its perturbation, we get

$$\begin{aligned} \bar{\phi}(\eta) + \delta\phi(x^\mu) &= \bar{\phi}(\tilde{\eta}) + \delta\tilde{\phi}(\tilde{x}^\mu) \\ &= \bar{\phi}(\eta + T) + \delta\tilde{\phi}(\tilde{x}^\mu). \end{aligned} \tag{6.11}$$

Note that the background solution $\bar{\phi}(\eta)$ is the same in both coordinates. Taylor expanding the functions on the right-hand side of (6.11), we find

$$\bar{\phi}(\eta) + \delta\phi(x^\mu) = \bar{\phi}(\eta) + \bar{\phi}'T + \delta\tilde{\phi}(x^\mu), \tag{6.12}$$

where we have dropped terms beyond linear order in the perturbations. The perturbation of the field therefore transforms as

$$\delta\tilde{\phi}(x^\mu) = \delta\phi(x^\mu) - \bar{\phi}'T(x^\mu). \tag{6.13}$$

Both sides in this expression are evaluated at the same coordinate values x^μ, which in general do *not* correspond to the same physical point. With that in mind, we will often drop the arguments and write the transformation rule as

$$\delta\tilde{\phi} = \delta\phi - \bar{\phi}'T. \tag{6.14}$$

Note that the perturbation of the field only depends on the choice of the temporal gauge.

To determine the transformation of the metric, we use that the spacetime interval is an invariant:

$$\mathrm{d}s^2 = g_{\mu\nu}(x)\mathrm{d}x^\mu\mathrm{d}x^\nu = \tilde{g}_{\alpha\beta}(\tilde{x})\mathrm{d}\tilde{x}^\alpha\mathrm{d}\tilde{x}^\beta \,, \tag{6.15}$$

where I have used a different set of dummy indices on both sides to make the next few lines clearer. Writing $\mathrm{d}\tilde{x}^\alpha = (\partial\tilde{x}^\alpha/\partial x^\mu)\mathrm{d}x^\mu$ (and similarly for $\mathrm{d}\tilde{x}^\beta$), we find

$$g_{\mu\nu}(x) = \frac{\partial\tilde{x}^\alpha}{\partial x^\mu}\frac{\partial\tilde{x}^\beta}{\partial x^\nu}\tilde{g}_{\alpha\beta}(\tilde{x}) \,. \tag{6.16}$$

This relates the metric in the old coordinates, $g_{\mu\nu}$, to the metric in the new coordinates, $\tilde{g}_{\alpha\beta}$. Given the specific transformations in (6.9), we have

$$\frac{\partial\tilde{x}^\alpha}{\partial x^\mu} = \begin{pmatrix} \partial\tilde{\eta}/\partial\eta & \partial\tilde{\eta}/\partial x^j \\ \partial\tilde{x}^i/\partial\eta & \partial\tilde{x}^i/\partial x^j \end{pmatrix} = \begin{pmatrix} 1+T' & \partial_j T \\ L^{i\prime} & \delta^i_j + \partial_j L^i \end{pmatrix} \,, \tag{6.17}$$

where α labels the rows and μ the columns. In the insert below, I will show what (6.16) implies for the transformation of the metric perturbations in (6.2). In terms of the SVT decomposition, we get

$$A \mapsto A - T' - \mathcal{H}T \,, \tag{6.18}$$

$$B \mapsto B + T - L' \,, \qquad\qquad \hat{B}_i \mapsto \hat{B}_i - \hat{L}'_i \,, \tag{6.19}$$

$$C \mapsto C - \mathcal{H}T - \frac{1}{3}\nabla^2 L \,, \tag{6.20}$$

$$E \mapsto E - L \,, \qquad\qquad \hat{E}_i \mapsto \hat{E}_i - \hat{L}_i \,, \qquad \hat{E}_{ij} \mapsto \hat{E}_{ij} \,. \tag{6.21}$$

Example I will work out the temporal component as an example and leave the rest as an exercise. Consider $\mu = \nu = 0$ in (6.16):

$$g_{00}(x) = \frac{\partial\tilde{x}^\alpha}{\partial\eta}\frac{\partial\tilde{x}^\beta}{\partial\eta}\tilde{g}_{\alpha\beta}(\tilde{x}) \,. \tag{6.22}$$

The only term that contributes to the right-hand side is the one with $\alpha = \beta = 0$. Consider for example $\alpha = 0$ and $\beta = i$. The off-diagonal component of the metric $\tilde{g}_{0i}$ is proportional to $\tilde{B}_i$, so it is a first-order perturbation. But $\partial\tilde{x}^i/\partial\eta$ is proportional to the first-order variable ξ^i, so the product is second order and can be neglected. A similar argument holds for $\alpha = i$ and $\beta = j$. Equation (6.22) therefore reduces to

$$g_{00}(x) = \left(\frac{\partial\tilde{\eta}}{\partial\eta}\right)^2 \tilde{g}_{00}(\tilde{x}) \,. \tag{6.23}$$

Substituting (6.9) and (6.2), we get

$$a^2(\eta)(1+2A) = (1+T')^2 a^2(\eta+T)(1+2\tilde{A})$$

$$= (1+2T'+\cdots)(a(\eta)+a'T+\cdots)^2(1+2\tilde{A})$$

$$= a^2(\eta)(1+2\mathcal{H}T+2T'+2\tilde{A}+\cdots) \,. \tag{6.24}$$

Hence, we find that, at first order, the metric perturbation A transforms as

$$A \mapsto \tilde{A} = A - T' - \mathcal{H}T, \tag{6.25}$$

which is indeed the result quoted in (6.18).

Exercise 6.1 Show that the other metric components transform as

$$B_i \mapsto \tilde{B}_i = B_i + \partial_i T - L_i', \tag{6.26}$$

$$E_{ij} \mapsto \tilde{E}_{ij} = E_{ij} - \mathcal{H}T\delta_{ij} - \partial_{(i}L_{j)}. \tag{6.27}$$

In terms of the SVT decomposition this leads to (6.19)–(6.21).

Gauge-invariant variables

One way to avoid the gauge problem is to define special combinations of the metric perturbations that do *not* transform under a change of coordinates. These are the so-called **Bardeen variables** [1]

$$\Psi \equiv A + \mathcal{H}(B - E') + (B - E')', \quad \hat{\Phi}_i \equiv \hat{B}_i - \hat{E}_i', \quad \hat{E}_{ij},$$

$$\Phi \equiv -C + \frac{1}{3}\nabla^2 E - \mathcal{H}(B - E'). \tag{6.28}$$

These gauge-invariant variables are the "real" spacetime perturbations since they cannot be removed by a gauge transformation.

Exercise 6.2 Show that Ψ, Φ and $\hat{\Phi}_i$ are gauge invariant. Explain why the tensor perturbation $\hat{E}_{ij}$ is gauge invariant at linear order.

Gauge fixing

An alternative solution to the gauge problem is to *fix the gauge* and keep track of *all* perturbations in both the metric and the matter. A convenient choice of gauge can often simplify the analysis and different gauges are used for different purposes. The following are popular gauges:

- **Newtonian gauge** We can use the freedom in the gauge functions T and L in (6.9) to set two of the four scalar metric perturbations to zero. The Newtonian gauge is defined by the choice

$$B = E = 0, \tag{6.29}$$

so that (6.2) becomes[6]

$$ds^2 = a^2(\eta) \left[-(1 + 2\Psi)d\eta^2 + (1 - 2\Phi)\delta_{ij}dx^i dx^j \right], \qquad (6.30)$$

where I have renamed the remaining two metric perturbations, $A \equiv \Psi$ and $C \equiv -\Phi$, to make contact with the Bardeen potentials in (6.28). One advantage of Newtonian gauge is that the metric is diagonal, which not only will simplify calculations, but also makes the physics transparent: hypersurfaces of constant time are orthogonal to the worldlines of observers at rest in the coordinates (since $B = 0$) and the induced geometry of the constant-time hypersurfaces is isotropic (since $E = 0$). Notice, moreover, the similarity of the metric (6.30) to the usual weak-field limit of GR (see Appendix A), with Ψ playing the role of the gravitational potential. Newtonian gauge will be our preferred gauge for studying structure formation, since the equations of motion will have the closest resemblance to the equations we found in the Newtonian analysis.

- **Spatially flat gauge**   A convenient gauge for computing inflationary perturbations is the spatially flat gauge, with

$$C = E = 0. \qquad (6.31)$$

In this gauge, we will be able to focus most directly on the fluctuations in the inflaton field $\delta\phi$ (see Chapter 8).

- **Synchronous gauge**   A historically important gauge is the synchronous gauge, with

$$A = B = 0. \qquad (6.32)$$

This gauge was introduced by Lifshitz in his pioneering work on cosmological perturbation theory [4] and played an important role in the early CMB codes. A disadvantage of the synchronous gauge is that it still contains spurious gauge degrees of freedom. These modes caused quite some confusion in the early history of cosmological perturbation theory and indeed were one of the motivations for Bardeen to develop his gauge-invariant approach [1].

6.1.2 Matter Perturbations

Next, we consider perturbations of the energy-momentum tensor.[7] We will find it convenient to define these perturbations for the tensor with mixed upper and lower indices:

---

[6] Beware of alternative conventions for the metric potentials in Newtonian gauge. Some authors swap Ψ and Φ, others use different signs. My convention is that of Ma and Bertschinger [2], which is also the one adopted by the Boltzmann code CLASS [3].

[7] I will sometimes refer to these as "matter perturbations," although I mean the perturbations in all species (including radiation).

$$T^0{}_0 \equiv -(\bar{\rho} + \delta\rho) \,,$$

$$T^i{}_0 \equiv -(\bar{\rho} + \bar{P})v^i \,, \qquad\qquad\qquad (6.33)$$

$$T^i{}_j \equiv (\bar{P} + \delta P)\,\delta^i_j + \Pi^i{}_j \,, \qquad \Pi^i{}_i \equiv 0 \,,$$

where v^i is the **bulk velocity** and $\Pi^i{}_j$ is the **anisotropic stress**. We will often write the **momentum density** as $q^i \equiv (\bar{\rho} + \bar{P})v^i$. If there are several contributions to the energy-momentum tensor (e.g. photons, baryons, dark matter, etc.), they are added, $T_{\mu\nu} = \sum_a T^{(a)}_{\mu\nu}$, so that

$$\delta\rho = \sum_a \delta\rho_a \,, \quad \delta P = \sum_a \delta P_a \,, \quad q^i = \sum_a q^i_{(a)} \,, \quad \Pi^{ij} = \sum_a \Pi^{ij}_{(a)} \,. \qquad (6.34)$$

We see that the perturbations in the density, pressure and anisotropic stress simply add. The velocities do *not* add, but the momentum densities do.

Exercise 6.3 Given $\bar{U}^\mu = a^{-1}(1,0,0,0)$ and $g_{\mu\nu}U^\mu U^\nu = -1$, show that the perturbed four-velocity must take the form

$$U^\mu = a^{-1}(1 - A, v^i) \,, \qquad\qquad\qquad (6.35)$$

$$U_\mu = a(-(1 + A), v_i + B_i) \,, \qquad\qquad\qquad (6.36)$$

where A and B_i are the metric perturbations defined in (6.2) and v^i is the bulk velocity. Substituting these results into the perturbed energy-momentum tensor of a fluid,

$$\delta T^\mu{}_\nu = (\delta\rho + \delta P)\bar{U}^\mu \bar{U}_\nu + (\bar{\rho} + \bar{P})(\delta U^\mu \bar{U}_\nu + \bar{U}^\mu \delta U_\nu) + \delta P \delta^\mu_\nu + \Pi^\mu{}_\nu \,, \quad (6.37)$$

with $U^\mu \Pi_{\mu\nu} = 0$, show that this leads to the same result as in (6.33). Show also that $T^0{}_i = (\bar{\rho} + \bar{P})(v_i + B_i)$.

As for the metric, we apply an SVT decomposition to the perturbations of the energy-momentum tensor: $\delta\rho$ and δP have scalar parts only, while v_i and q_i have scalar and vector parts,

$$v_i = \partial_i v + \hat{v}_i \,,$$
$$q_i = \partial_i q + \hat{q}_i \,. \qquad\qquad\qquad (6.38)$$

The scalar part of the velocity is sometimes written in terms of the **velocity divergence**, $\theta \equiv \partial_i v^i$. The anisotropic stress Π_{ij} has scalar, vector and tensor parts,

$$\Pi_{ij} = \partial_{\langle i}\partial_{j\rangle}\Pi + \partial_{(i}\hat{\Pi}_{j)} + \hat{\Pi}_{ij} \,. \qquad\qquad\qquad (6.39)$$

Often it is useful to work with the *rescaled* scalar anisotropic stress, which in Fourier space is defined by $\Pi_{ij}(\eta, \mathbf{k}) \subset -(\bar{\rho} + \bar{P})\hat{k}_{\langle i}\hat{k}_{j\rangle}\Pi(\eta, \mathbf{k})$ and corresponds to $\sigma = \frac{2}{3}\Pi$ in Ma and Bertschinger [2]. Finally, it is also convenient to write the density perturbations in terms of the dimensionless **density contrast**, $\delta \equiv \delta\rho/\rho$. Perturbation theory applies when the density contrast is smaller than unity.

In summary, scalar perturbations of the total matter are described by four perturbation variables, $(\delta, \delta P, v, \Pi)$. Similarly, the perturbations of distinct species $a = \gamma, \nu, c, b, \ldots$ are represented by $(\delta_a, \delta P_a, v_a, \Pi_a)$. Just like for the metric perturbations, however, these perturbations depend on the choice of coordinates.

Coordinate transformations

Under a coordinate transformation, $x^\mu \mapsto \tilde{x}^\mu$, the tensor $T^\mu{}_\nu$ transforms as

$$T^\mu{}_\nu(x) = \frac{\partial x^\mu}{\partial \tilde{x}^\alpha} \frac{\partial \tilde{x}^\beta}{\partial x^\nu} \tilde{T}^\alpha{}_\beta(\tilde{x}) \,. \tag{6.40}$$

The origin of this transformation law is explained in Section A.3.3. To evaluate (6.40), we require $\partial x^\mu/\partial \tilde{x}^\alpha$ which, considered as a matrix, is the inverse of $\partial \tilde{x}^\alpha/\partial x^\mu$, since

$$\frac{\partial x^\mu}{\partial \tilde{x}^\alpha} \frac{\partial \tilde{x}^\alpha}{\partial x^\nu} = \delta^\mu_\nu \,. \tag{6.41}$$

For the transformations in (6.9), the matrix $\partial \tilde{x}^\alpha/\partial x^\mu$ was given by (6.17):

$$\frac{\partial \tilde{x}^\alpha}{\partial x^\mu} = \begin{pmatrix} \partial\tilde\eta/\partial\eta & \partial\tilde\eta/\partial x^j \\ \partial\tilde x^i/\partial\eta & \partial\tilde x^i/\partial x^j \end{pmatrix} = \begin{pmatrix} 1 + T' & \partial_j T \\ L^{i\prime} & \delta^i_j + \partial_j L^i \end{pmatrix} . \tag{6.42}$$

To determine the corresponding inverse matrix, we make use of the fact that the inverse of a matrix of the form $\mathbf{1} + \varepsilon$, where $\mathbf{1}$ is the identity and ε is a small perturbation, is $\mathbf{1} - \varepsilon$ to first order in ε. It follows that

$$\frac{\partial x^\mu}{\partial \tilde{x}^\alpha} = \begin{pmatrix} \partial\eta/\partial\tilde\eta & \partial\eta/\partial\tilde x^j \\ \partial x^i/\partial\tilde\eta & \partial x^i/\partial\tilde x^j \end{pmatrix} = \begin{pmatrix} 1 - T' & -\partial_j T \\ -L^{i\prime} & \delta^i_j - \partial_j L^i \end{pmatrix} . \tag{6.43}$$

Substituting these results into (6.40), we can then determine how the perturbations of the energy-momentum tensor transform. I will derive the transformation of $\delta\rho$ as an example and leave the rest as an exercise. The results are

$$\delta\rho \mapsto \delta\rho - \bar{\rho}'T \,, \tag{6.44}$$

$$\delta P \mapsto \delta P - \bar{P}'T \,, \tag{6.45}$$

$$q_i \mapsto q_i + (\bar{\rho} + \bar{P})L'_i \,, \tag{6.46}$$

$$v_i \mapsto v_i + L'_i \,, \tag{6.47}$$

$$\Pi_{ij} \mapsto \Pi_{ij} \,. \tag{6.48}$$

Note that the density and pressure perturbations transform in the same way as the perturbation of the scalar field in (6.14). Of course, this had to be the case since ρ and P are both Lorentz scalars.

Example Applying (6.40) to the temporal component of the energy-momentum tensor, we get

$$T^0{}_0(x) = \frac{\partial \eta}{\partial \tilde{x}^\alpha} \frac{\partial \tilde{x}^\beta}{\partial \eta} \tilde{T}^\alpha{}_\beta(\tilde{x})$$

$$= \frac{\partial \eta}{\partial \tilde{\eta}} \frac{\partial \tilde{\eta}}{\partial \eta} \tilde{T}^0{}_0(\tilde{x}) + O(2). \tag{6.49}$$

Substituting $T^0{}_0 = -(\bar{\rho} + \delta\rho)$ on both sides, we get

$$\bar{\rho} + \delta\rho = (1 - T')(1 + T')\big(\bar{\rho}(\eta + T) + \delta\tilde{\rho}(x)\big)$$

$$= \bar{\rho}(\eta) + \bar{\rho}'T + \delta\tilde{\rho}(x) + O(2). \tag{6.50}$$

Solving this for $\delta\tilde{\rho}$ gives

$$\delta\tilde{\rho} = \delta\rho - \bar{\rho}'T, \tag{6.51}$$

which agrees with the result in (6.44).

Exercise 6.4 Derive (6.45)–(6.48).

Gauge-invariant variables

We see that the scalar and vector parts of the perturbations can be changed by a change of coordinates. As before, we can define specific combinations of variables for which these transformations cancel. There are various gauge-invariant quantities that can be formed from the metric and matter variables.

- One useful combination is

$$\bar{\rho}\Delta \equiv \delta\rho + \bar{\rho}'(v + B). \tag{6.52}$$

The quantity Δ is called the **comoving density contrast** because it reduces to the density contrast in the comoving gauge, with $v + B = 0$. As we will see below, in terms of the comoving density contrast the relativistic generalization of the Poisson equation takes the same form as in the Newtonian approximation,

$$\nabla^2 \Phi = 4\pi G a^2 \bar{\rho}\Delta, \tag{6.53}$$

where Φ is the Bardeen potential defined in (6.28).

- Two additional important gauge-invariant quantities are[8]

$$\zeta = -C + \frac{1}{3}\nabla^2 E + \mathcal{H}\frac{\delta\rho}{\bar{\rho}'}, \tag{6.54}$$

$$\mathcal{R} = -C + \frac{1}{3}\nabla^2 E - \mathcal{H}(v + B). \tag{6.55}$$

---

[8] Beware of alternative conventions for the overall signs of these quantities. Some authors also swap the meaning of $\mathcal{R}$ and ζ, so that their ζ is my $\mathcal{R}$.

These are called **curvature perturbations** because they reduce to the intrinsic curvature of the spatial slices in the uniform density and comoving gauges, respectively.

Exercise 6.5 Consider the induced metric on surfaces of constant time

$$\gamma_{ij} = a^2 \left[(1 + 2C)\delta_{ij} + 2E_{ij} \right].\tag{6.56}$$

Show that the intrinsic curvature on the spatial slices is

$$a^2 R_{(3)}[\gamma_{ij}] = -4\nabla^2 C + 2\partial_i \partial_j E^{ij}$$

$$= 4\nabla^2 \left(-C + \frac{1}{3}\nabla^2 E \right),\tag{6.57}$$

which explains why $-C + \frac{1}{3}\nabla^2 E$ is called the "curvature perturbation."

The three gauge-invariant perturbations Δ, ζ and $\mathcal{R}$ are not independent, but obey the following relation

$$\zeta = \mathcal{R} + \frac{\mathcal{H}}{\bar{\rho}'}\bar{\rho}\Delta.\tag{6.58}$$

We see from the Poisson equation (6.53) that the comoving density contrast vanishes on superhorizon scales, so that the two curvature perturbations become equal

$$\zeta \xrightarrow{\ k\ll\mathcal{H}\ } \mathcal{R}.\tag{6.59}$$

On large scales, we can therefore treat ζ and $\mathcal{R}$ interchangeably. A critical feature of the curvature perturbations is that they are constant on superhorizon scales if the matter perturbations are "adiabatic" (see Section 6.2.2). A nice way to describe the initial conditions is therefore in terms of ζ or $\mathcal{R}$.

More gauges

Above we used our gauge freedom to set two of the metric perturbations to zero. Alternatively, we can also define the gauge in the matter sector:

- **Uniform density gauge** We can use the freedom in the time slicing to set the total density perturbation to zero,

$$\delta\rho = 0.\tag{6.60}$$

The main scalar perturbation in this gauge is the curvature perturbation, $\delta g_{ij} = a^2(1 - 2\zeta)\delta_{ij}$.

- **Comoving gauge** Similarly, we can ask for the covariant velocity perturbation to vanish,

$$v + B = 0.\tag{6.61}$$

The main scalar perturbation in this gauge is then the curvature perturbation, $\delta g_{ij} = a^2(1 - 2\mathcal{R})\delta_{ij}$. This is another convenient gauge for the computation of the inflationary quantum fluctuations (see e.g. [5] and Problem 8.3).

There are different versions of the uniform density and comoving gauges depending on which of the metric fluctuations is set to zero.

6.1.3 Conservation Equations

So far, we haven't done very much. We defined the perturbations of the metric and the energy-momentum tensor, and discussed their gauge dependence. Our next task is to derive the evolution equations for these perturbations. The evolution of the metric is governed by the Einstein equation, $G_{\mu\nu} = 8\pi G\, T_{\mu\nu}$, while the evolution of the matter perturbations follows from the conservation of the energy-momentum tensor, $\nabla^\mu T_{\mu\nu} = 0$. Expanding these equations to linear order in the perturbations will give us the linearized equations of motion for the matter and metric perturbations.

It will be convenient to perform this analysis in a fixed gauge and we will take this to be the Newtonian gauge where the metric is

$$g_{\mu\nu} = a^2 \begin{pmatrix} -(1+2\Psi) & 0 \\ 0 & (1-2\Phi)\delta_{ij} \end{pmatrix}. \tag{6.62}$$

In this section, we will derive the equations of motion for the matter perturbations by linearizing $\nabla^\mu T_{\mu\nu} = 0$. If there is no energy and momentum transfer between the different components, then the species are separately conserved and we also have $\nabla^\mu T_{\mu\nu}^{(a)} = 0$. In that case, our results will apply to all species separately.

More explicitly, we have

$$\nabla_\mu T^\mu{}_\nu = 0$$
$$= \partial_\mu T^\mu{}_\nu + \Gamma^\mu_{\mu\alpha} T^\alpha{}_\nu - \Gamma^\alpha_{\mu\nu} T^\mu{}_\alpha. \tag{6.63}$$

To evaluate this, we need the perturbed connection coefficients. Using the metric (6.62), it is a straightforward (but tedious) exercise to show that

$$\Gamma^0_{00} = \mathcal{H} + \Psi', \tag{6.64}$$
$$\Gamma^0_{i0} = \partial_i \Psi, \tag{6.65}$$
$$\Gamma^i_{00} = \partial^i \Psi, \tag{6.66}$$
$$\Gamma^0_{ij} = \mathcal{H}\delta_{ij} - \left[\Phi' + 2\mathcal{H}(\Phi + \Psi)\right]\delta_{ij}, \tag{6.67}$$
$$\Gamma^i_{j0} = (\mathcal{H} - \Phi')\delta^i_j, \tag{6.68}$$
$$\Gamma^i_{jk} = -2\delta^i_{(j}\partial_{k)}\Phi + \delta_{jk}\partial^i\Phi. \tag{6.69}$$

Example Recall the general expression for the connection coefficients

$$\Gamma^{\alpha}_{\beta\gamma} = \frac{1}{2}g^{\alpha\rho}\left(\partial_{\beta}g_{\gamma\rho} + \partial_{\gamma}g_{\beta\rho} - \partial_{\rho}g_{\beta\gamma}\right). \tag{6.70}$$

To evaluate this, we need the inverse of the metric (6.62)

$$g^{\mu\nu} = a^{-2}\begin{pmatrix} -(1-2\Psi) & 0 \\ 0 & (1+2\Phi)\delta^{ij} \end{pmatrix}, \tag{6.71}$$

where we have used that to linear order $(1+2\Psi)^{-1} \approx (1-2\Psi)$. Applying this to Γ^{0}_{00}, we then get

$$\begin{aligned}
\Gamma^{0}_{00} &= \frac{1}{2}g^{0\rho}\left(\partial_{0}g_{0\rho} + \partial_{0}g_{0\rho} - \partial_{\rho}g_{00}\right) \\
&= \frac{1}{2}g^{00}\left(\partial_{0}g_{00} + \partial_{0}g_{00} - \partial_{0}g_{00}\right) \quad (\text{since } g^{0i} = 0) \\
&= \frac{1}{2}g^{00}\partial_{0}g_{00} \\
&= \frac{1}{2}a^{-2}(1-2\Psi)\,\partial_{0}\left[a^{2}(1+2\Psi)\right] \\
&= a^{-2}(1-2\Psi)\left[aa'(1+2\Psi) + a^{2}\Psi'\right] \\
&= \mathcal{H} + \Psi',
\end{aligned} \tag{6.72}$$

which is the result quoted in (6.64).

Exercise 6.6 Derive (6.65)–(6.69).

Substituting the perturbed connection coefficients into (6.63), we can derive the linearized evolution equations for the matter perturbations. The algebra is a bit involved and is mostly relegated to the box below. The results are summarized in the following paragraphs.

- **Continuity equation** Setting $\nu = 0$ in (6.63), we find

$$\boxed{\delta\rho' = -3\mathcal{H}(\delta\rho + \delta P) - \partial_{i}q^{i} + 3\Phi'(\bar{\rho} + \bar{P})}, \tag{6.73}$$

which is the continuity equation describing the evolution of the density perturbations. The first term on the right-hand side is just the dilution due to the background expansion, while the second term accounts for the local fluid flow. The last term (proportional to Φ') is a purely relativistic effect corresponding to the density changes caused by perturbations to the local expansion rate. This arises because we can think of $(1-\Phi)a$ as the "local scale factor" in the spatial part of the metric in Newtonian gauge. The first and last terms in (6.73) are therefore related.

Substituting $\delta\rho = \bar{\rho}\,\delta$ and $q^{i} = (\bar{\rho}+\bar{P})v^{i}$ into (6.73), we get an evolution equation for the density contrast

$$\boxed{\delta' = -\left(1 + \frac{\bar{P}}{\bar{\rho}}\right)(\mathbf{\nabla}\cdot\mathbf{v} - 3\Phi') - 3\mathcal{H}\left(\frac{\delta P}{\delta\rho} - \frac{\bar{P}}{\bar{\rho}}\right)\delta}\,. \tag{6.74}$$

In the limit $P \ll \rho$, we therefore recover the Newtonian equation (5.138) written in conformal time, $\delta' = -\mathbf{\nabla}\cdot\mathbf{v} + 3\Phi'$, but with a relativistic correction coming from the perturbed expansion rate. This correction is small on sub-Hubble scales and we reproduce the Newtonian limit.

- **Euler equation** Setting $\nu = i$ in (6.63), we get

$$\boxed{q_i' = -4\mathcal{H}q_i - (\bar{\rho} + \bar{P})\partial_i\Psi - \partial_i\delta P - \partial^j\Pi_{ij}}\,, \tag{6.75}$$

which is the relativistic version of the Euler equation. The first term on the right-hand side of (6.75) enforces the expected scaling of the momentum density, $q \propto a^{-4}$. The remaining terms are force terms due to gravity, pressure and anisotropic stress.

Substituting $q_i = (\bar{\rho} + \bar{P})v_i$ into (6.75), we get an evolution equation for the bulk velocity

$$\boxed{v_i' = -\left(\mathcal{H} + \frac{\bar{P}'}{\bar{\rho} + \bar{P}}\right)v_i - \frac{1}{\bar{\rho} + \bar{P}}\left(\partial_i\delta P + \partial^j\Pi_{ij}\right) - \partial_i\Psi}\,, \tag{6.76}$$

which should be compared to the Newtonian equation (5.139).

Derivation Consider first the $\nu = 0$ component of (6.63),

$$\partial_0 T^0{}_0 + \partial_i T^i{}_0 + \Gamma^\mu{}_{\mu 0}T^0{}_0 + \underbrace{\Gamma^\mu{}_{\mu i}T^i{}_0}_{O(2)} - \Gamma^0{}_{00}T^0{}_0 - \underbrace{\Gamma^0{}_{i0}T^i{}_0}_{O(2)} - \underbrace{\Gamma^i{}_{00}T^0{}_i}_{O(2)} - \Gamma^i{}_{j0}T^j{}_i = 0\,. \tag{6.77}$$

Substituting the perturbed energy-momentum tensor and connection coefficients gives

$$- (\bar{\rho} + \delta\rho)' - \partial_i q^i - (\mathcal{H} + \Psi' + 3\mathcal{H} - 3\Phi')(\bar{\rho} + \delta\rho)$$
$$+ (\mathcal{H} + \Psi')(\bar{\rho} + \delta\rho) - (\mathcal{H} - \Phi')\delta^i_j(\bar{P} + \delta P)\delta^j_i = 0\,, \tag{6.78}$$

and hence

$$\bar{\rho}' + \delta\rho' + \partial_i q^i + 3\mathcal{H}(\bar{\rho} + \delta\rho) - 3\bar{\rho}\Phi' + 3\mathcal{H}(\bar{P} + \delta P) - 3\bar{P}\Phi' = 0\,. \tag{6.79}$$

Writing the zeroth-order and first-order parts separately, we get

$$\bar{\rho}' = -3\mathcal{H}(\bar{\rho} + \bar{P})\,, \tag{6.80}$$
$$\delta\rho' = -3\mathcal{H}(\delta\rho + \delta P) - \partial_i q^i + 3\Phi'(\bar{\rho} + \bar{P})\,. \tag{6.81}$$

The zeroth-order equation (6.80) is simply the conservation of energy in the homogeneous background. The first-order equation (6.81) is the continuity equation for the density perturbation $\delta\rho$.

Next, consider the $\nu = i$ component of (6.63), $\partial_\mu T^\mu{}_i + \Gamma^\mu_{\mu\rho} T^\rho{}_i - \Gamma^\rho_{\mu i} T^\mu{}_\rho = 0$, and hence

$$\partial_0 T^0{}_i + \partial_j T^j{}_i + \Gamma^\mu_{\mu 0} T^0{}_i + \Gamma^\mu_{\mu j} T^j{}_i - \Gamma^0_{0i} T^0{}_0 - \Gamma^0_{ji} T^j{}_0 - \Gamma^j_{0i} T^0{}_j - \Gamma^j_{ki} T^k{}_j = 0 \ . \quad (6.82)$$

Substituting the perturbed energy-momentum tensor, with $T^0{}_i = +q_i$ and $T^i{}_0 = -q^i$, and the perturbed connection coefficients gives

$$q_i' + \partial_j \left[(\bar{P} + \delta P)\delta_i^j + \Pi^j{}_i \right] + 4\mathcal{H} q_i + (\partial_j \Psi - 3\partial_j \Phi)\bar{P}\delta_i^j + \partial_i \Psi \bar\rho$$
$$+ \mathcal{H}\delta_{ji} q^j - \mathcal{H}\delta_i^j q_j - \underbrace{\left(-2\delta^j_{(k}\partial_{i)}\Phi + \delta_{ki}\partial^j \Phi \right) \bar{P}\delta_j^k}_{-3\partial_i \Phi \bar{P}} = 0 \ . \quad (6.83)$$

Cleaning this up, we find

$$q_i' = -4\mathcal{H} q_i - (\bar\rho + \bar{P})\partial_i \Psi - \partial_i \delta P - \partial^j \Pi_{ij} \ , \quad (6.84)$$

which confirms the form of the Euler equation (6.75).

It is instructive to evaluate (6.74) and (6.76) for a few special cases:

- **Matter** Consider a non-relativistic fluid (i.e. matter), with $P_m = 0$ and $\Pi_m^{ij} = 0$. The continuity and Euler equations then simplify considerably

$$\delta_m' = -\boldsymbol{\nabla} \cdot \mathbf{v}_m + 3\Phi' \ , \quad (6.85)$$
$$\mathbf{v}_m' = -\mathcal{H}\mathbf{v}_m - \boldsymbol{\nabla}\Psi \ . \quad (6.86)$$

Each term in these equations should be rather intuitive. Combining the time derivative of (6.85) with the divergence of (6.86), we find

$$\delta_m'' + \mathcal{H}\delta_m' = \nabla^2 \Psi + 3(\Phi'' + \mathcal{H}\Phi') \ . \quad (6.87)$$
$$\quad\quad\quad \uparrow \quad\quad\quad\quad \uparrow$$
$$\quad\quad \text{friction} \quad \text{gravity}$$

Except for the terms arising from the time dependence of the potential, this is the same as the Newtonian equation for the evolution of matter perturbations. We will apply this equation to the clustering of dark matter perturbations.

- **Radiation** For a relativistic perfect fluid (i.e. radiation with no viscosity), we have $P_r = \frac{1}{3}\rho_r$ and $\Pi_r^{ij} = 0$. The continuity and Euler equations then become

$$\delta_r' = -\frac{4}{3}\boldsymbol{\nabla} \cdot \mathbf{v}_r + 4\Phi' \ , \quad (6.88)$$

$$\mathbf{v}_r' = -\frac{1}{4}\boldsymbol{\nabla}\delta_r - \boldsymbol{\nabla}\Psi \ . \quad (6.89)$$

Combining the time derivative of (6.88) with the divergence of (6.89), we get

$$\delta_r'' - \frac{1}{3}\nabla^2 \delta_r = \frac{4}{3}\nabla^2 \Psi + 4\Phi'' \ . \quad (6.90)$$
$$\quad\quad\quad \uparrow \quad\quad\quad\quad \uparrow$$
$$\quad\quad \text{pressure} \quad \text{gravity}$$

We see that the radiation perturbations don't experience the Hubble friction, but feel an additional pressure-induced force. Below, I will show how this equation allows for sound waves in the primordial plasma.

Summary

For future reference, let me collect here the linearized continuity and Euler equations:

$$\delta' = -\left(1 + \frac{\bar{P}}{\bar{\rho}}\right)(\partial_i v^i - 3\Phi') - 3\mathcal{H}\left(\frac{\delta P}{\delta \rho} - \frac{\bar{P}}{\bar{\rho}}\right)\delta\,, \tag{6.91}$$

$$v_i' = -\left(\mathcal{H} + \frac{\bar{P}'}{\bar{\rho} + \bar{P}}\right)v_i - \frac{1}{\bar{\rho} + \bar{P}}\left(\partial_i \delta P + \partial^j \Pi_{ij}\right) - \partial_i \Psi\,. \tag{6.92}$$

We typically work with these equations in Fourier space. For scalar fluctuations, we obtain the equations for the individual Fourier modes by the following replacements:

$$\partial_i \to ik_i\,,$$
$$v_i \to i\hat{k}_i v\,, \tag{6.93}$$
$$\Pi_{ij} \to -(\bar{\rho} + \bar{P})\hat{k}_{\langle i}\hat{k}_{j\rangle}\Pi\,,$$

where $\hat{k}_{\langle i}\hat{k}_{j\rangle} \equiv \hat{k}_i\hat{k}_j - \frac{1}{3}\delta_{ij}$. Note that we have absorbed factors of $k \equiv |\mathbf{k}|$ in v and Π—as well as rescaled the anisotropic stress by a factor of $(\bar{\rho} + \bar{P})$—so that $v(\eta, \mathbf{k})$ and $\Pi(\eta, \mathbf{k})$ are not quite the Fourier transforms of $v(\eta, \mathbf{x})$ and $\Pi(\eta, \mathbf{x})$. Equations (6.91) and (6.92) then lead to

$$\delta' = \left(1 + \frac{\bar{P}}{\bar{\rho}}\right)(kv + 3\Phi') - 3\mathcal{H}\left(\frac{\delta P}{\delta \rho} - \frac{\bar{P}}{\bar{\rho}}\right)\delta\,, \tag{6.94}$$

$$v' = -\left(\mathcal{H} + \frac{\bar{P}'}{\bar{\rho} + \bar{P}}\right)v - \frac{1}{\bar{\rho} + \bar{P}}k\,\delta P + \frac{2}{3}k\Pi - k\Psi\,. \tag{6.95}$$

These equations also apply separately for all components that aren't directly coupled to each other. For example, dark matter doesn't interact with other matter (except through gravity), so it satisfies its own conservation equations. In general, we will have separate evolution equations for each species with perturbations $(\delta_a, v_a, \delta P_a, \Pi_a)$. The gravitational coupling between the different species is captured by the expansion rate $\mathcal{H}$ and the metric potentials Φ and Ψ in (6.91) and (6.92).

Of course, *two* equations aren't sufficient to completely describe the evolution of the *four* perturbations $(\delta, v, \delta P, \Pi)$. To make progress, we either must make further simplifying assumptions or find additional evolution equations. We will do both.

- Sometimes we will assume a perfect fluid. Such a fluid is characterized by strong interactions which keep the pressure isotropic, $\Pi = 0$. For a barotropic fluid, with $P = P(\rho)$, the pressure perturbations can be written as $\delta P = c_s^2 \delta\rho$, where the sound speed is

$$c_s^2 \equiv \frac{\mathrm{d}P}{\mathrm{d}\rho} = \frac{\bar{P}'}{\bar{\rho}'} \, . \tag{6.96}$$

For a general fluid, the pressure may not just depend on the density and the sound speed is defined as

$$c_s^2 = \left(\frac{\partial P}{\partial \rho}\right)_S = \frac{\bar{P}'}{\bar{\rho}'} \, , \tag{6.97}$$

where the subscript indicates that the derivative is taken at fixed entropy and the second equality assumes that the expansion of the background is adiabatic (i.e. there is no entropy production). The perturbations are called **adiabatic** if

$$\delta P = c_s^2 \delta\rho = \frac{\bar{P}'}{\bar{\rho}'}\delta\rho \, . \tag{6.98}$$

Note that the perturbations in a barotropic fluid are necessarily adiabatic, but in general this need not be the case. If the perturbations are adiabatic, then they are described by only two independent variables, say δ and v, whose evolution is given by the continuity and Euler equations.

- Decoupled or weakly interacting species (e.g. neutrinos) cannot be described by a perfect fluid and the above simplifications for the anisotropic stress and the pressure perturbation do not apply. In that case, we cannot avoid solving the Boltzmann equation for the evolution of the perturbed distribution function f (see Appendix B).

- Decoupled cold dark matter is a peculiar case. It is collisionless and has a negligible velocity dispersion. It therefore behaves like a pressureless perfect fluid although it has no interactions and therefore isn't a fluid in the traditional sense [6].

6.1.4 Einstein Equations

We see from (6.91) and (6.92) that the evolution of the matter perturbations is influenced by the metric potentials Φ and Ψ. To close the system of equations, we therefore also need evolution equations for the metric perturbations. These follow from the Einstein equation, which we will analyze next.

Deriving the linearized Einstein equations in Newtonian gauge is conceptually straightforward, but algebraically a bit tedious. Once in your life you should do this computation by hand. After that you should use Mathematica or your favorite alternative to perform the algebra. On the book's website, you can find the Mathematica notebook CPT.nb which will allow you to reproduce all results in this section at the press of a button.

We require the perturbation to the Einstein tensor, $G_{\mu\nu} \equiv R_{\mu\nu} - \frac{1}{2}Rg_{\mu\nu}$, so we first need to calculate the perturbed Ricci tensor $R_{\mu\nu}$ and Ricci scalar R.

- **Ricci tensor** Recall that the Ricci tensor can be expressed in terms of the connection coefficients as

$$R_{\mu\nu} = \partial_\lambda \Gamma^\lambda_{\mu\nu} - \partial_\nu \Gamma^\lambda_{\mu\lambda} + \Gamma^\lambda_{\lambda\rho}\Gamma^\rho_{\mu\nu} - \Gamma^\rho_{\mu\lambda}\Gamma^\lambda_{\nu\rho}. \tag{6.99}$$

Substituting the perturbed connection coefficients (6.64)–(6.69), we find

$$R_{00} = -3\mathcal{H}' + \nabla^2\Psi + 3\mathcal{H}(\Phi' + \Psi') + 3\Phi'', \tag{6.100}$$

$$R_{0i} = 2\partial_i(\Phi' + \mathcal{H}\Psi), \tag{6.101}$$

$$R_{ij} = \left[\mathcal{H}' + 2\mathcal{H}^2 - \Phi'' + \nabla^2\Phi - 2(\mathcal{H}' + 2\mathcal{H}^2)(\Phi + \Psi)\right.$$
$$\left. - \mathcal{H}\Psi' - 5\mathcal{H}\Phi'\right]\delta_{ij} + \partial_i\partial_j(\Phi - \Psi). \tag{6.102}$$

I will derive R_{00} explicitly and leave the other components as an exercise.

Example The temporal component of the Ricci tensor is

$$R_{00} = \partial_\lambda \Gamma^\lambda_{00} - \partial_0 \Gamma^\lambda_{0\lambda} + \Gamma^\lambda_{\lambda\rho}\Gamma^\rho_{00} - \Gamma^\rho_{0\lambda}\Gamma^\lambda_{0\rho}. \tag{6.103}$$

The terms with $\lambda = 0$ cancel in the sum over λ, so we get

$$R_{00} = \partial_i \Gamma^i_{00} - \partial_0 \Gamma^i_{0i} + \Gamma^i_{i\rho}\Gamma^\rho_{00} - \Gamma^\rho_{0i}\Gamma^i_{0\rho}$$

$$= \partial_i \Gamma^i_{00} - \partial_0 \Gamma^i_{0i} + \Gamma^i_{i0}\Gamma^0_{00} + \underbrace{\Gamma^i_{ij}\Gamma^j_{00}}_{O(2)} - \underbrace{\Gamma^0_{0i}\Gamma^i_{00}}_{O(2)} - \Gamma^j_{0i}\Gamma^i_{0j}$$

$$= \nabla^2\Psi - 3\partial_0(\mathcal{H} - \Phi') + 3(\mathcal{H} + \Psi')(\mathcal{H} - \Phi') - (\mathcal{H} - \Phi')^2\delta^j_i\delta^i_j$$

$$= -3\mathcal{H}' + \nabla^2\Psi + 3\mathcal{H}(\Phi' + \Psi') + 3\Phi'', \tag{6.104}$$

which confirms the result (6.100).

Exercise 6.7 Derive (6.101) and (6.102). Verify the results using the Mathematica notebook provided on the book's website.

- **Ricci scalar** It is now relatively straightforward to compute the Ricci scalar

$$R = g^{00}R_{00} + 2\underbrace{g^{0i}R_{0i}}_{O(2)} + g^{ij}R_{ij}. \tag{6.105}$$

It follows that

$$a^2 R = -(1 - 2\Psi)R_{00} + (1 + 2\Phi)\delta^{ij}R_{ij}$$

$$= (1 - 2\Psi)\left[3\mathcal{H}' - \nabla^2\Psi - 3\mathcal{H}(\Phi' + \Psi') - 3\Phi''\right]$$
$$+ 3(1 + 2\Phi)\left[\mathcal{H}' + 2\mathcal{H}^2 - \Phi'' + \nabla^2\Phi - 2(\mathcal{H}' + 2\mathcal{H}^2)(\Phi + \Psi)\right.$$
$$\left. - \mathcal{H}\Psi' - 5\mathcal{H}\Phi'\right] + (1 + 2\Phi)\nabla^2(\Phi - \Psi). \tag{6.106}$$

Dropping nonlinear terms, this becomes

$$a^2 R = 6(\mathcal{H}' + \mathcal{H}^2)$$
$$- 2\nabla^2\Psi + 4\nabla^2\Phi - 12(\mathcal{H}' + \mathcal{H}^2)\Psi - 6\Phi'' - 6\mathcal{H}(\Psi' + 3\Phi'), \quad (6.107)$$

where the first line is the homogeneous part discussed in Chapter 2 and the second line is the linear correction.

- **Einstein equations** We are ready to compute the Einstein equation

$$G^\mu{}_\nu = 8\pi G T^\mu{}_\nu. \quad (6.108)$$

We chose to work with one index raised since it simplifies the form of the energy-momentum tensor. We will first consider the temporal part of the equation, $\mu = \nu = 0$. The relevant component of the Einstein tensor is

$$G^0{}_0 = g^{0\mu}G_{\mu 0} = g^{00}\left[R_{00} - \frac{1}{2}g_{00}R\right]$$
$$= -a^{-2}(1 - 2\Psi)R_{00} - \frac{1}{2}R, \quad (6.109)$$

where we have used that g^{0i} vanishes in Newtonian gauge and $g^{00}g_{00} = 1$ at linear order. Substituting (6.100) and (6.107), we find

$$\delta G^0{}_0 = -2a^{-2}\left[\nabla^2\Phi - 3\mathcal{H}(\Phi' + \mathcal{H}\Psi)\right], \quad (6.110)$$

and the temporal Einstein equation therefore gives

$$\boxed{\nabla^2\Phi - 3\mathcal{H}(\Phi' + \mathcal{H}\Psi) = 4\pi G a^2 \delta\rho}, \quad (6.111)$$

where $\delta\rho \equiv \sum_a \delta\rho_a$ is the *total* density perturbation. Equation (6.111) is the relativistic generalization of the Poisson equation. Inside the Hubble radius—i.e. for Fourier modes with $k \gg \mathcal{H}$—we have $|\nabla^2\Phi| \gg 3\mathcal{H}|\Phi' + \mathcal{H}\Psi|$, so that (6.111) reduces to $\nabla^2\Phi \approx 4\pi G a^2 \delta\rho$, which is the Poisson equation in the Newtonian limit. The GR corrections in (6.111) start to become important on scales comparable to the Hubble radius.

Next, we consider the purely spatial part of the Einstein equation. The relevant components of the Einstein tensor are

$$G^i{}_j = g^{ik}\left[R_{kj} - \frac{1}{2}g_{kj}R\right]$$
$$= a^{-2}(1 + 2\Phi)\delta^{ik}R_{kj} - \frac{1}{2}\delta^i_j R. \quad (6.112)$$

From (6.102), we see that most terms in R_{kj} are proportional to δ_{kj}. When contracted with δ^{ik} this leads to a myriad of terms proportional to δ^i_j. We don't want to deal with this mess. Instead, we will first focus on the *tracefree* part of $G^i{}_j$ which is sourced by the anisotropic stress $\Pi^i{}_j$. Using (6.102), this gives

$$\boxed{\partial_{\langle i}\partial_{j\rangle}(\Phi - \Psi) = 8\pi G a^2 \Pi_{ij}}. \quad (6.113)$$

With the replacements $\partial_{\langle i}\partial_{j\rangle} \to -k_{\langle i}k_{j\rangle}$ and $\Pi_{ij} = -(\bar{\rho} + \bar{P})\hat{k}_{\langle i}\hat{k}_{j\rangle}\Pi$, the corresponding equation in Fourier space reads

$$k^2(\Phi - \Psi) = 8\pi Ga^2(\bar{\rho} + \bar{P})\Pi\,, \tag{6.114}$$

where $\Pi \equiv \sum_a \Pi_a$. On large scales, dark matter and baryons can be described as perfect fluids with negligible anisotropic stress. Photons only start to develop an anisotropic stress component during the matter-dominated era when their energy density is subdominant. The only relevant source in (6.113) is therefore free-streaming neutrinos. To describe the neutrino-induced anisotropic stress requires going beyond the fluid approximation; see Appendix B. The effect is relatively small, so to the level of accuracy that we aspire to in this chapter it can be ignored. Equation (6.113) then implies $\Psi \approx \Phi$.

The time–space part of the Einstein equation is obtained in the same way. The relevant components of the Einstein tensor are

$$\begin{aligned}
G^0{}_i = g^{0\mu}G_{\mu i} &= g^{00}R_{0i} \\
&= -a^{-2}R_{0i} \\
&= -2a^{-2}\partial_i(\Phi' + \mathcal{H}\Psi)\,,
\end{aligned} \tag{6.115}$$

and together with $T^0{}_i = \partial_i q$, we find

$$\boxed{\Phi' + \mathcal{H}\Psi = -4\pi Ga^2 q}\,. \tag{6.116}$$

With this, the Poisson equation (6.111) can be written as

$$\boxed{\nabla^2\Phi = 4\pi Ga^2\bar{\rho}\Delta}\,, \tag{6.117}$$

where the comoving density contrast Δ was defined in (6.52). We see that in terms of the comoving density contrast the relativistic generalization of the Poisson equation takes the same form as in the Newtonian treatment, but is now valid on all scales.

Finally, we look at the trace of the space–space Einstein equation. It is a relatively straightforward exercise to show that this leads to the following evolution equation for the metric potential

$$\boxed{\Phi'' + \frac{1}{3}\nabla^2(\Psi - \Phi) + (2\mathcal{H}' + \mathcal{H}^2)\Psi + \mathcal{H}\Psi' + 2\mathcal{H}\Phi' = 4\pi Ga^2\delta P}\,, \tag{6.118}$$

where $\delta P \equiv \sum_a \delta P_a$ is the total pressure perturbation. Assuming $\Psi \approx \Phi$, this reduces to

$$\boxed{\Phi'' + 3\mathcal{H}\Phi' + (2\mathcal{H}' + \mathcal{H}^2)\Phi = 4\pi Ga^2\delta P}\,, \tag{6.119}$$

which becomes a closed equation for the evolution of Φ if we write $\delta P = c_s^2\delta\rho$ and use the Poisson equation to relate $\delta\rho$ to Φ.

Exercise 6.8 Derive (6.118).

Summary

Since the main results of this section are spread over several pages, it is probably useful to write them out in one place, so that they are easier to find later.

The linearized Einstein equations are

$$\nabla^2\Phi - 3\mathcal{H}(\Phi' + \mathcal{H}\Psi) = 4\pi G a^2 \delta\rho\,, \qquad (6.120)$$

$$-(\Phi' + \mathcal{H}\Psi) = 4\pi G a^2 q\,, \qquad (6.121)$$

$$\partial_{\langle i}\partial_{j\rangle}(\Phi - \Psi) = 8\pi G a^2 \Pi_{ij}\,, \qquad (6.122)$$

$$\Phi'' + \mathcal{H}\Psi' + 2\mathcal{H}\Phi' + \frac{1}{3}\nabla^2(\Psi - \Phi) + (2\mathcal{H}' + \mathcal{H}^2)\Psi = 4\pi G a^2 \delta P\,, \qquad (6.123)$$

where the sources on the right-hand side include a sum over all components. In the absence of anisotropic stress, equation (6.122) implies that the two metric potentials are equal, $\Psi \approx \Phi$, and (6.123) reduces to

$$\Phi'' + 3\mathcal{H}\Phi' + (2\mathcal{H}' + \mathcal{H}^2)\Phi = 4\pi G a^2 \delta P\,. \qquad (6.124)$$

Equations (6.120) and (6.121) can be combined into $\nabla^2\Phi = 4\pi G a^2 \bar\rho\Delta$, which is of the same form as the Newtonian Poisson equation, but is now valid on all scales.

Combing the Einstein equations with the conservation equations (6.91) and (6.92) leads to a closed system of equations (after specifying the equation of state and the sound speed of each fluid component). In the rest of this chapter, we will study the solutions of these equations for various situations of physical interest.

6.2 Initial Conditions

Before we can solve the above evolution equations, we must specify their initial conditions. At sufficiently early times, all scales of interest to current observations were outside of the Hubble radius. On such super-Hubble ("superhorizon") scales, the evolution of the perturbations becomes very simple, especially for adiabatic initial conditions (see below). In this section, I will describe the initial conditions for these superhorizon modes.

6.2.1 Superhorizon Limit

Consider the superhorizon limit of the continuity equations (6.85) and (6.88) for matter and radiation:

$$\delta_m' = 3\Phi', \tag{6.125}$$

$$\delta_r' = 4\Phi', \tag{6.126}$$

where the fact that the velocity terms can be dropped follows from the superhorizon limit of the Euler equations (6.86) and (6.89). Integrating these equations for photons, neutrinos, baryons and cold dark matter gives

$$
\begin{aligned}
\delta_\gamma &= 4\Phi + C_\gamma & \delta_\gamma &= 4\Phi + C_\gamma \\
\delta_\nu &= 4\Phi + C_\nu & \delta_\nu &= \delta_\gamma + S_\nu \\
\delta_c &= 3\Phi + C_c & \quad\rightarrow\quad & \delta_c = \tfrac{3}{4}\delta_\gamma + S_c \\
\delta_b &= 3\Phi + C_b & \delta_b &= \tfrac{3}{4}\delta_\gamma + S_b
\end{aligned}
\tag{6.127}
$$

where the parameters C_a are integration constants. Note that these constants, in general, are functions of the wavevector $\mathbf{k}$. The parameters S_a (for $a = \nu, c, b$) are called **isocurvature modes**.

A preferred set of initial conditions are given by the **adiabatic mode** with $S_\nu = S_c = S_b = 0$ and $C_\gamma \neq 0$. In that case, the initial values of all perturbations are determined by a single degree of freedom. As we will see in Chapter 8, adiabatic fluctuations are the natural prediction of the simplest inflationary models. Except for a few isolated comments, we will assume adiabatic initial conditions throughout the rest of this book. Besides being theoretically attractive, adiabatic initial conditions also seem to be preferred by observations (see Section 8.4.1).

Next, we look at the Einstein equation (6.120):

$$\nabla^2\Phi - 3\mathcal{H}(\Phi' + \mathcal{H}\Phi) = 4\pi G a^2\left(\bar{\rho}_\gamma\delta_\gamma + \bar{\rho}_\nu\delta_\nu + \bar{\rho}_c\delta_c + \bar{\rho}_b\delta_b\right), \tag{6.128}$$

where we have ignored the anisotropic stress of the neutrinos, so that $\Phi \approx \Psi$. On large scales, we can drop $\nabla^2\Phi$ and at early times the matter contributions are negligible. We then have

$$3\mathcal{H}(\Phi' + \mathcal{H}\Phi) = -4\pi G a^2\left(\bar{\rho}_\gamma\delta_\gamma + \bar{\rho}_\nu\delta_\nu\right). \tag{6.129}$$

Specializing to adiabatic initial conditions, we have $\delta_\nu = \delta_\gamma$ and $4\pi G a^2(\bar{\rho}_\gamma + \bar{\rho}_\nu) = 3\mathcal{H}^2/2$. Substituting $\mathcal{H} = 1/\eta$, we then get

$$\eta\Phi' + \Phi = -\frac{1}{2}\delta_\gamma. \tag{6.130}$$

Taking a time derivative, and using (6.126) to replace δ_γ', this becomes

$$\eta\Phi'' + 4\Phi' = 0. \tag{6.131}$$

The two solutions to this equation are the growing mode $\Phi = \text{const}$ and the decaying mode $\Phi \propto \eta^{-3}$. Focusing on the growing mode, $\Phi \equiv \Phi_i$, equation (6.130) then implies the following initial condition

$$\delta_\gamma(\eta_i, \mathbf{k}) = -2\Phi_i(\mathbf{k}) \,. \tag{6.132}$$

We see that the integration constant in (6.127) is fixed by the primordial value of the gravitational potential, $C_\gamma = -6\Phi_i$. For adiabatic perturbations, the fluctuations in all components are then related as

$$\boxed{\delta_\gamma = \delta_\nu = \frac{4}{3}\delta_c = \frac{4}{3}\delta_b = -2\Phi_i} \quad \text{(superhorizon)}\,, \tag{6.133}$$

so that the initial conditions of all fluctuations are specified by giving the value of the primordial potential Φ_i.

Gauge dependence   One important subtlety is the gauge dependence of the superhorizon dynamics. To illustrate this, consider the Poisson equation written in terms of the comoving density contrast

$$\Delta = -\frac{2}{3}\frac{k^2}{\mathcal{H}^2}\Phi \,. \tag{6.134}$$

We notice a few things from this. First, the comoving density contrast Δ is much smaller than δ on superhorizon scales, $k \ll \mathcal{H}$. Second, it is not constant on superhorizon scales, but evolves as $\Delta \propto \mathcal{H}^{-2} \propto \eta^2$. Should we be worried about this qualitative difference in the evolution of the density contrasts in the two different gauges? Isn't the density contrast something that we can measure? Not on superhorizon scales! On subhorizon scales, where we do all of our observations, we instead have

$$\frac{\Delta - \delta}{\Delta} = -\frac{3}{4}\frac{\mathcal{H}(\Phi' + \mathcal{H}\Phi)}{k^2\Phi} \xrightarrow{k \gg \mathcal{H}} 0 \,, \tag{6.135}$$

so that there is no difference between Δ and δ on small scales. This is a general feature: the gauge problem disappears for all measurable quantities on subhorizon scales.

Had we taken the anisotropic stress due to the neutrinos into account, we would instead have found (see Problem 6.2)

$$\delta_\gamma(\eta_i, \mathbf{k}) = -2\Psi_i(\mathbf{k}) = -2\Phi_i(\mathbf{k})\left(1 + \frac{2}{5}f_\nu\right)^{-1}, \tag{6.136}$$

where we have introduced the fractional neutrino density $f_\nu \equiv \bar{\rho}_\nu/(\bar{\rho}_\nu + \bar{\rho}_\gamma)$.

Exercise 6.9   Using the results from Chapter 3, show that

$$f_\nu \approx 0.41 \,, \tag{6.137}$$

so that $\Psi_i \approx 0.86\,\Phi_i$.

6.2.2 Adiabatic Perturbations

Given the special status of adiabatic perturbations, it is worth describing their properties in a bit more detail.

Adiabatic perturbations can be created by starting with a homogeneous universe and performing a *common, local shift in time* of all background quantities, $\eta \to \eta + \pi(\eta, \mathbf{x})$. After this time shift, some parts of the universe are "ahead" and others are "behind" in the evolution. As we will see in Chapter 8, quantum fluctuations during inflation produce precisely such a time shift in the background quantities. The induced pressure and density perturbations are

$$
\begin{aligned}
\delta P_a(\eta, \mathbf{x}) &= \bar{P}'_a \, \pi(\eta, \mathbf{x}) \\
\delta \rho_a(\eta, \mathbf{x}) &= \bar{\rho}'_a \, \pi(\eta, \mathbf{x})
\end{aligned}
\quad \Rightarrow \quad
\begin{aligned}
\frac{\delta P_a}{\delta \rho_a} &= \frac{\bar{P}'_a}{\bar{\rho}'_a}, \\
\frac{\delta \rho_a}{\delta \rho_b} &= \frac{\bar{\rho}'_a}{\bar{\rho}'_b},
\end{aligned}
\tag{6.138}
$$

where we have used that $\pi(\eta, \mathbf{x})$ is the *same* for all species a and b. The first relation on the right implies that

$$
\delta P_a = c_{s,a}^2 \delta \rho_a,
\tag{6.139}
$$

so that the perturbations are indeed adiabatic in the sense defined in (6.98). Using $\bar{\rho}'_a = -3\mathcal{H}(1 + w_a)\bar{\rho}_a$, the second relation on the right can be written as

$$
\frac{\delta_a}{1 + w_a} = \frac{\delta_b}{1 + w_b} \qquad \text{for all species } a \text{ and } b.
\tag{6.140}
$$

Components with the same equation of state therefore have the same fractional perturbations. Perturbations in matter ($w_m = 0$) and radiation ($w_r = \frac{1}{3}$) obey

$$
\boxed{\delta_m = \frac{3}{4}\delta_r},
\tag{6.141}
$$

which is the same relation as for the adiabatic mode in (6.127). An important corollary is that, for adiabatic fluctuations, the total density perturbation, $\delta \rho \equiv \sum_a \bar{\rho}_a \delta_a$, is dominated by the species that carries the dominant energy density $\bar{\rho}_a$, since all of the δ_a's are comparable.

Finally, we note that the time shift $\pi(\eta, \mathbf{x})$ used in the definition of adiabatic fluctuations has the interpretation of the **Goldstone boson** associated with the spontaneous breaking of time translation symmetry. This symmetry breaking arises because of the time dependence of the cosmological background. The language of spontaneous symmetry breaking has recently become very popular for describing cosmological perturbations and an effective field theory was constructed in terms of the Goldstone mode [7].

Exercise 6.10 Show that the thermodynamic relation $T\mathrm{d}S = \mathrm{d}U + P\mathrm{d}V$ implies

$$\frac{T\mathrm{d}S}{V} = \left(\rho_\nu - \frac{3}{4} r_\nu (\rho + P) \right) \left(\delta_\nu - \delta_\gamma \right)$$
$$+ \sum_{a=c,b} \left(\rho_a - r_a (\rho + P) \right) \left(\delta_a - \frac{3}{4} \delta_\gamma \right), \tag{6.142}$$

where $r_a \equiv n_a/n$ are the fractional number densities of the species. For adiabatic perturbations, this means that $\mathrm{d}S = 0$.

6.2.3 Isocurvature Perturbations*

The complement of adiabatic perturbations are isocurvature perturbations. For two components a and b, the (density) isocurvature perturbation is defined as

$$S_{ab} \equiv \frac{\delta_a}{1 + w_a} - \frac{\delta_b}{1 + w_b}. \tag{6.143}$$

We see from (6.140) that a nonzero value of this quantity measures the deviation from the adiabatic mode. Special cases of this isocurvature mode are given in (6.127). The name "isocurvature" was introduced because these perturbations can be chosen in such a way that the curvature perturbation ζ vanishes initially. In that case, an overdensity in one species compensates for an underdensity in another, resulting in no net curvature perturbation.

Let me illustrate this for the special case of the neutrino density isocurvature mode. From the definition of ζ in (6.54), we have

$$\zeta = \Phi - \frac{\delta\rho}{3(\rho + P)} = \Phi - \frac{f_\gamma \delta_\gamma + f_\nu \delta_\nu + f_b \delta_b + f_c \delta_c}{3 + f_\gamma + f_\nu}, \tag{6.144}$$

where we have used the continuity equation for $\bar\rho'$ and introduced $f_a \equiv \rho_a/\rho$. In the second equality, we used that $\sum_a f_a = 1$ to replace $f_b + f_c$ with $1 - (f_\gamma + f_\nu)$ in the denominator. At early times, we can ignore baryons and dark matter, $f_{b,c} \approx 0$, so that $f_\gamma \approx 1 - f_\nu$ and hence

$$\zeta = \Phi - \frac{\delta_\gamma}{4} + \frac{f_\nu(\delta_\gamma - \delta_\nu)}{4}$$
$$= -\frac{1}{4}\left(C_\gamma + f_\nu S_\nu \right), \tag{6.145}$$

where we used (6.127) in the final equality. The curvature perturbation therefore vanishes for $S_\nu = -C_\gamma/f_\nu$ and the relation between the density contrasts is

$$\delta_\gamma = 4\Phi - f_\nu S_\nu$$
$$= \delta_\nu - S_\nu = \frac{4}{3}\delta_c = \frac{4}{3}\delta_b. \tag{6.146}$$

In Problem 6.2, you will show that

$$S_\nu = \frac{15 + 4f_\nu}{f_\nu(1 - f_\nu)} \Phi_i \approx 68.8\, \Phi_i, \tag{6.147}$$

where the final equality is for $f_\nu \approx 0.41$. Equation (6.146) then implies that $\delta_\gamma = -24.2\,\Phi_i$ and $\delta_\nu = 44.6\,\Phi_i$.

While there is only one way for the fluctuations to be adiabatic, there are many different types of isocurvature perturbations. Beside the density isocurvature perturbations described so far, we can also have isocurvature modes in the velocity perturbations. Moreover, we can form arbitrary linear combinations of the different isocurvature modes and correlate them with the adiabatic mode. The general phenomenology of isocurvature perturbations is described in [8].

Fortunately, there are good reasons to favor adiabatic initial conditions. First of all, isocurvature fluctuations would produce distinctive signatures in the CMB temperature and polarization anisotropies [9] that have not been seen in the data. At most, isocurvature fluctuations can therefore be a subdominant component. Adiabatic perturbations are also attractive on theoretical grounds because they arise naturally in the simplest inflationary models (see Chapter 8) and are frozen on superhorizon scales (see Section 6.2.4). The latter property is important for the predictive power of inflation as it allows us to be agnostic about the uncertain physics of reheating. In contrast, the amplitude of primordial isocurvature perturbations is strongly model-dependent and sensitive to the post-inflationary evolution. In particular, isocurvature fluctuations can be washed out by thermal equilibrium: If all particle species are in thermal equilibrium after inflation and their local densities are determined by the temperature (with vanishing chemical potential) then the primordial perturbations are adiabatic. The existence of primordial isocurvature modes therefore requires at least one field to decay into some species whose abundance is not determined by thermal equilibrium (e.g. cold dark matter after decoupling) or respects some conserved quantum numbers, like baryon or lepton numbers. For instance, the neutrino density isocurvature mode discussed above could be due to spatial fluctuations in the chemical potential of neutrinos.

Given the complexity of models with isocurvature fluctuations and the fact that observations don't seem to require them, we will assume adiabatic initial conditions for the rest of this book.

6.2.4 Curvature Perturbations

Figure 6.3 shows the superhorizon evolution of the gravitational potential in the transition from radiation domination to matter domination. We see that the gravitational potential is a constant during the radiation and matter eras, but varies in between. This is a general feature: the potential is frozen on superhorizon scales during phases of the evolution where the equation of state is a constant, but evolves when the equation of state changes.

Exercise 6.11 Consider a universe dominated by a fluid with constant equation of state $w \neq -1$. Assuming adiabatic perturbations and vanishing anisotropic

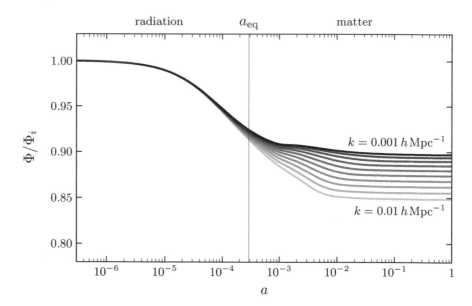

Fig. 6.3   Evolution of the gravitational potential during the transition from radiation to matter domination, ignoring the anisotropic stress from neutrinos.

stress, show that the gravitational potential obeys

$$\Phi'' + 3(1 + w)\mathcal{H}\Phi' - w\nabla^2\Phi = 0 \,. \qquad (6.148)$$

This shows that the growing mode is a constant on superhorizon scales.

It would be convenient to identify an alternative perturbation variable that stays constant on large scales even when the equation of state changes. This is particularly important for the transition from inflation to the radiation-dominated universe. Since the equation of state during reheating is unknown, it is crucial that we can track a quantity whose evolution doesn't depend on these details. This will allow us to match the predictions made by inflation to the fluctuations in the primordial plasma after inflation. For adiabatic initial conditions, the conserved perturbations are the curvature perturbations introduced in (6.54) and (6.55), which in Newtonian gauge read

$$\zeta = \Phi - \frac{\delta\rho}{3(\bar{\rho} + \bar{P})} \,, \qquad (6.149)$$

$$\mathcal{R} = \Phi - \mathcal{H}v \,, \qquad (6.150)$$

where we have used the continuity equation to replace $\bar{\rho}'$. To prove that ζ is indeed conserved on super-Hubble scales, we take a time derivative of (6.149):

$$\zeta' = \Phi' - \frac{\delta\rho'}{3(\bar{\rho} + \bar{P})} + \frac{(\bar{\rho}' + \bar{P}')}{3(\bar{\rho} + \bar{P})^2}\delta\rho \,. \qquad (6.151)$$

Using the continuity equations,

$$\bar{\rho}' = -3\mathcal{H}(\bar{\rho} + \bar{P}),$$
$$\delta\rho' = -3\mathcal{H}(\delta\rho + \delta P) - \partial_i q^i + 3\Phi'(\bar{\rho} + \bar{P}),$$

(6.152)

we get

$$(\bar{\rho} + \bar{P})\frac{\zeta'}{\mathcal{H}} = \left(\delta P - \frac{\bar{P}'}{\bar{\rho}'}\delta\rho\right) + \frac{1}{3}\frac{\partial_i q^i}{\mathcal{H}}.$$

(6.153)

For adiabatic perturbations, the term in brackets on the right-hand side vanishes, while $\partial_i q^i = \nabla^2 q$ is suppressed by a factor of $k^2/\mathcal{H}^2 \ll 1$ on large scales, so that $\zeta'/\mathcal{H} \xrightarrow{k \ll \mathcal{H}} O(k^2/\mathcal{H}^2) \approx 0$. We have therefore established the conservation of the curvature perturbation on superhorizon scales.

Exercise 6.12 Show that the comoving curvature perturbation satisfies

$$(\bar{\rho} + \bar{P})\frac{\mathcal{R}'}{\mathcal{H}} = \left(\delta P - \frac{\bar{P}'}{\bar{\rho}'}\delta\rho\right) + \frac{\bar{P}'}{\bar{\rho}'}\frac{\nabla^2\Phi}{4\pi G a^2},$$

(6.154)

and is therefore also conserved on superhorizon scales. Of course, this had to be true since we showed in (6.59) that $\mathcal{R}$ is equal to ζ on large scales.

Exercise 6.13 Consider a universe dominated by a fluid with constant equation of state. Show that the superhorizon limit of the curvature perturbation is

$$\mathcal{R} \xrightarrow{k \ll \mathcal{H}} \frac{5 + 3w}{3 + 3w}\Phi.$$

(6.155)

Use this to show that the amplitude of super-Hubble modes of Φ drops by a factor of 9/10 in the radiation-to-matter transition (as seen in Fig. 6.3).

6.2.5 Primordial Power Spectrum

Using the result of Exercise 6.13, the superhorizon value of the gravitational potential during the radiation era is

$$\Phi(\eta, \mathbf{k}) = \frac{2}{3}\mathcal{R}_i(\mathbf{k}) \quad \text{(superhorizon)},$$

(6.156)

where $\mathcal{R}_i(\mathbf{k}) \equiv \mathcal{R}(0, \mathbf{k})$ is the initial value of the conserved curvature perturbation. Ignoring the effect of neutrinos, so that $\Psi \approx \Phi$, and assuming adiabatic initial conditions, we then also have

$$\boxed{\delta_r = \frac{4}{3}\delta_m = -2\Phi = -\frac{4}{3}\mathcal{R}_i} \quad \text{(superhorizon)}.$$

(6.157)

The initial amplitudes of all fluctuations are determined by $\mathcal{R}_i$. We take the power spectrum of the primordial curvature perturbations to be

$$\mathcal{P}_{\mathcal{R}}(t_i, k) = A_s k^{n_s - 4},\tag{6.158}$$

where A_s and n_s are constants. In Chapter 8, we will show that inflation naturally predicts a nearly scale-invariant spectrum, with $n_s \approx 1$. In the following, we will discuss how these scale-invariant fluctuations evolve after they re-enter the horizon.

6.3 Growth of Matter Perturbations

In this section, I will describe the evolution of pressureless matter fluctuations which will be relevant for the clustering of dark matter and baryons (after decoupling). In the next section, I will then treat the evolution of photons and baryons before decoupling, where radiation pressure plays an important role.

As we have seen in Chapter 5, an important scale in structure formation is the horizon at matter–radiation equality. A mode entering the horizon at the time of equality has wavenumber $k_{eq} = \mathcal{H}_{eq}$. Modes with $k > k_{eq}$ enter the horizon during the radiation era, while modes with $k < k_{eq}$ enter during the matter era. As we saw in the previous chapter, this leads to a scale-dependent evolution of the matter perturbations (see Section 5.2.3). We will now derive this evolution again, connecting it more rigorously to the superhorizon limit of the fluctuations.

6.3.1 Evolution of the Potential

Let us begin with the evolution of the gravitational potential. For simplicity, we will neglect anisotropic stress and assume adiabatic perturbations. Analytic solutions for Φ can then be found when the modes are on superhorizon scales, or when either radiation or matter dominate the background (see Fig. 6.4).

When the universe is dominated by a component with a constant equation of state, the evolution equation is (see Exercise 6.11)

$$\Phi'' + 3(1 + w)\mathcal{H}\Phi' - w\nabla^2\Phi = 0.\tag{6.159}$$

This time we keep the gradient term. During matter domination, we have $w \approx 0$ and equation (6.159) takes the same form on subhorizon scales as on superhorizon scales. The solutions for the growing and decaying modes are

$$\boxed{\Phi(a, \mathbf{k}) = C_1(\mathbf{k}) + C_2(\mathbf{k})\, a^{-5/2}} \quad \text{(matter era)},\tag{6.160}$$

where we used $a \propto \eta^2$ to write the decaying mode in terms of the scale factor. The wavenumber-dependent amplitudes $C_1(\mathbf{k})$ and $C_2(\mathbf{k})$ are determined by the initial conditions and the evolution during the radiation era. We see that the growing

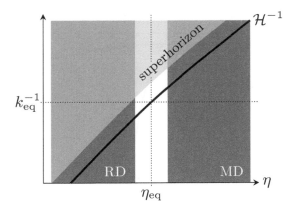

Fig. 6.4 Illustration of the three different regimes (shaded areas) in which analytic solutions can be found for the gravitational potential.

mode of the gravitational potential is a constant on all scales during the matter era. This is worth remembering.

Things are a bit more interesting during the radiation era. Using $w \approx 1/3$ in (6.159), we find that each Fourier mode now satisfies

$$\Phi'' + \frac{4}{\eta}\Phi' + \frac{1}{3}k^2\Phi = 0 \,. \tag{6.161}$$

Writing $\Phi = u(\varphi)/\varphi$, where $\varphi \equiv k\eta/\sqrt{3}$, we have

$$\frac{\mathrm{d}^2 u}{\mathrm{d}\varphi^2} + \frac{2}{\varphi}\frac{\mathrm{d}u}{\mathrm{d}\varphi} + \left(1 - \frac{2}{\varphi^2}\right)u = 0 \,. \tag{6.162}$$

The solutions of this equation are the spherical Bessel functions (see Appendix D)

$$j_1(\varphi) = +\frac{\sin\varphi}{\varphi^2} - \frac{\cos\varphi}{\varphi} = +\frac{\varphi}{3} + O(\varphi^3) \,, \tag{6.163}$$

$$n_1(\varphi) = -\frac{\cos\varphi}{\varphi^2} - \frac{\sin\varphi}{\varphi} = -\frac{1}{\varphi^2} + O(\varphi^0) \,. \tag{6.164}$$

Since $n_1(\varphi)$ blows up for small φ (early times), we reject that solution on the basis of initial conditions. The solution for the gravitational potential during the radiation era then is

$$\boxed{\Phi(\eta, \mathbf{k}) = 2\mathcal{R}_i \frac{\sin\varphi - \varphi\cos\varphi}{\varphi^3}} \quad \text{(radiation era)}\,, \tag{6.165}$$

where the overall normalization was determined by $\Phi(0, \mathbf{k}) = \frac{2}{3}\mathcal{R}(0, \mathbf{k}) \equiv \frac{2}{3}\mathcal{R}_i(\mathbf{k})$ (see Exercise 6.13). Notice that (6.165) is valid on all scales. Taking the limits $\varphi \ll 1$ and $\varphi \gg 1$ gives the solutions on superhorizon and subhorizon scales, respectively.

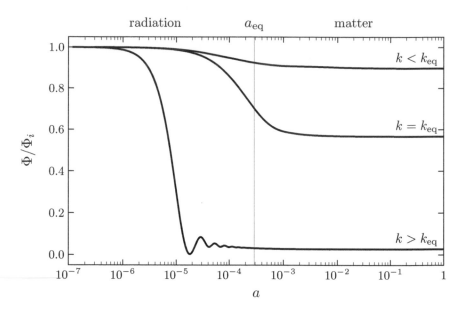

Numerical solutions for the linear evolution of the gravitational potential in a flat Einstein–de Sitter universe with $\Omega_b = 0.05$ and $\Omega_c = 0.95$. Shown are three representative Fourier modes with $k = \{0.001, 0.05, 1\}\,h\,\mathrm{Mpc}^{-1}$. On large scales, the gravitational potential is a constant except for a 10% drop in the transition from radiation to matter domination. After horizon entry a mode decays and oscillates during the radiation era, but stays constant during the matter era.

These are

$$
\Phi(\eta, \mathbf{k}) \approx
\begin{cases}
\dfrac{2}{3}\mathcal{R}_i & \text{(superhorizon)}, \\[2ex]
-6\mathcal{R}_i\,\dfrac{\cos\left(\frac{1}{\sqrt{3}}k\eta\right)}{(k\eta)^2} & \text{(subhorizon)}.
\end{cases}
\tag{6.166}
$$

During the radiation era, subhorizon modes of Φ therefore oscillate with a frequency $\frac{1}{\sqrt{3}}k$ and have an amplitude that decays as $\eta^{-2} \propto a^{-2}$.

Figure 6.5 shows numerical solutions for the evolution of the gravitational potential for three representative wavelengths. As predicted, the potential is constant when the modes are outside the horizon. Two of the modes enter the horizon during the radiation era, after which their amplitudes decrease as a^{-2}. The resulting amplitudes in the matter era are therefore strongly suppressed. During the matter era, the potential is a constant on all scales. The long-wavelength mode enters the horizon during the matter era, so its amplitude is only suppressed by the factor of 9/10 coming from the radiation-to-matter transition (see Section 6.2.4).

In Problem 6.1, you will derive an analytic solution for the superhorizon evolution of Φ:

$$\Phi(a, \mathbf{k}) = \frac{2\mathcal{R}_i}{30y^3}\left[16\sqrt{1+y} + 9y^3 + 2y^2 - 8y - 16\right] \quad \text{(superhorizon)}, \quad (6.167)$$

where $y \equiv a/a_{\text{eq}}$. It is easy to confirm that this solution reduces to a constant at both early times ($y \ll 1$) and late times ($y \gg 1$), and accounts for the 9/10 factor in the transition from the radiation era to the matter era.

Exercise 6.14 Consider the evolution equation (6.124) for the gravitational potential. Show that during the dark energy-dominated era, this equation can be written as

$$\frac{d^2\Phi}{da^2} + \frac{5}{a}\frac{d\Phi}{da} + \frac{3}{a^2}\Phi = 0, \quad (6.168)$$

where $a(\eta) = [1 + H_\Lambda(\eta_0 - \eta)]^{-1}$, with $H_\Lambda = \text{const}$. Note that this equation is valid on all scales. Show that the growing mode solution decays as $\Phi \propto a^{-1}$.

6.3.2 Clustering of Dark Matter

Next, I will describe the growth of dark matter perturbations. The evolution of subhorizon perturbations of a pressureless fluid was treated in the previous chapter using the Newtonian approximation. In the following, I will reproduce these results in the relativistic framework and extend them to superhorizon scales.

Matter era

Let us begin with the evolution during the matter era. Since matter is the dominant component, the Poisson equation reads

$$\nabla^2\Phi \approx 4\pi G a^2 \bar{\rho}_m \Delta_m . \quad (6.169)$$

The solution for the comoving density contrast Δ_m can therefore be obtained directly from our previous result for Φ. Using (6.160) and $a^2\bar{\rho}_m \propto a^{-1}$, we find

$$\Delta_m(a, \mathbf{k}) = -\frac{k^2\Phi}{4\pi G a^2 \bar{\rho}_m} = \tilde{C}_1(\mathbf{k})\, a + \tilde{C}_2(\mathbf{k})\, a^{-3/2} \quad \text{(matter era)}, \quad (6.170)$$

just as in the Newtonian treatment, but now valid on all scales. Notice that the growing mode of Δ_m evolves as a outside the horizon, while δ_m stays constant. Inside the horizon, $\delta_m \approx \Delta_m$ and the density contrasts in both gauges evolve as the scale factor a.

Dark energy era

It is straightforward to extend this to the dark energy-dominated era. Since dark energy has no density fluctuations, the Poisson equation is still of the form (6.169). To get an evolution equation for Δ_m, we use a neat trick. Since $a^2 \bar{\rho}_m \propto a^{-1}$, we have $\Phi \propto \Delta_m/a$. The Einstein equation (6.124) then implies

$$\partial_\eta^2(\Delta_m/a) + 3\mathcal{H}\partial_\eta(\Delta_m/a) + (2\mathcal{H}' + \mathcal{H}^2)(\Delta_m/a) = 0\,, \tag{6.171}$$

which rearranges to

$$\Delta_m'' + \mathcal{H}\Delta_m' + (\mathcal{H}' - \mathcal{H}^2)\Delta_m = 0\,. \tag{6.172}$$

Combining the Friedmann equations (2.156) and (2.157) for a universe with matter and dark energy gives

$$\begin{aligned}
2(\mathcal{H}' - \mathcal{H}^2) = (2\mathcal{H}' + \mathcal{H}^2) - 3\mathcal{H}^2 &= -8\pi G a^2 \bar{P} - 8\pi G a^2 \bar{\rho} \\
&= +8\pi G a^2 \bar{\rho}_\Lambda - 8\pi G a^2(\bar{\rho}_m + \bar{\rho}_\Lambda) \\
&= -8\pi G a^2 \bar{\rho}_m\,,
\end{aligned} \tag{6.173}$$

and the equation for the comoving density contrast becomes

$$\Delta_m'' + \mathcal{H}\Delta_m' - 4\pi G a^2 \bar{\rho}_m \Delta_m = 0\,. \tag{6.174}$$

This is similar to the Newtonian equation (5.62), but is now valid on all scales. In the dark energy-dominated regime, we have $\mathcal{H}^2 \gg 4\pi G a^2 \bar{\rho}_m$ and we can drop the last term in (6.174) to get

$$\Delta_m'' - \frac{1}{\eta}\Delta_m' \approx 0\,, \tag{6.175}$$

which has the following solution

$$\boxed{\Delta_m(a, \mathbf{k}) = \tilde{C}_1(\mathbf{k}) + \tilde{C}_2(\mathbf{k})a^{-2}} \quad \text{(dark energy era)}\,, \tag{6.176}$$

where we have used $\eta \propto a^{-1}$. This recovers the suppression of the growth of structure that we found in the previous chapter, but it now holds on all scales.

Exercise 6.15 Show that (6.176) also follows directly from the solution to (6.168).

Radiation era

During the radiation era, matter is a subdominant component and we cannot use the above trick to determine the evolution of matter perturbations from the Einstein equation for the gravitational potential. Instead, we must work with the continuity and Euler equations. In Section 6.1.3, we showed that these imply the following evolution equation for the matter density contrast

$$\delta_m'' + \mathcal{H}\delta_m' = \nabla^2\Phi + 3(\Phi'' + \mathcal{H}\Phi')\,, \tag{6.177}$$

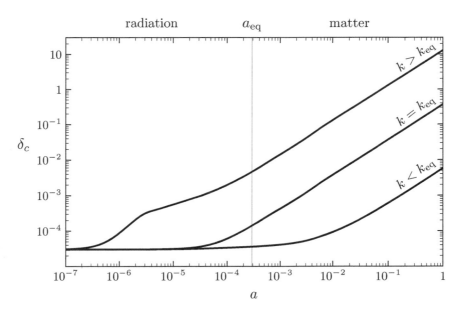

Fig. 6.6 Evolution of the density contrast of dark matter in a flat Einstein–de Sitter universe with $\Omega_b = 0.05$ and $\Omega_c = 0.95$. Shown are three representative Fourier modes with $k = \{0.005, 0.05, 5\}\, h\, \mathrm{Mpc}^{-1}$.

where $\Phi = \Phi_r + \Phi_m$ is sourced by both radiation and matter. The contribution from the radiation, Φ_r, is rapidly oscillating on subhorizon scales, while the contribution from matter, Φ_m, is a constant. The solution δ_m therefore inherits a "fast mode" sourced by Φ_r and a "slow mode" sourced by Φ_m. It turns out that the fast mode is suppressed by a factor of $(\mathcal{H}/k)^2$ relative to the slow mode [10]. This reflects the fact that the matter can't react to the fast change in the gravitational potential and effectively only evolves in response to the time-averaged potential. As a result, δ_m is sourced by Φ_m even deep in the radiation era. Using $\nabla^2\Phi \approx \nabla^2\Phi_m = 4\pi G a^2 \bar\rho_m \delta_m$ and $\Phi'' = \Phi' \approx 0$, we get

$$\delta_m'' + \mathcal{H}\delta_m' - 4\pi G a^2 \bar\rho_m \delta_m \approx 0, \tag{6.178}$$

where $4\pi G a^2 \bar\rho_m = \frac{3}{2}\Omega_m \mathcal{H}^2$. The conformal Hubble parameter for a universe filled with matter and radiation is

$$\frac{\mathcal{H}}{\mathcal{H}_0} = \frac{\Omega_m}{\sqrt{\Omega_r}}\frac{\sqrt{1+y}}{y}, \qquad y \equiv \frac{a}{a_{\mathrm{eq}}}, \tag{6.179}$$

where $a_{\mathrm{eq}} = \Omega_r/\Omega_m$ is the scale factor at matter–radiation equality. Using y as the time variable, equation (6.178) becomes the so-called **Mészáros equation**

$$\frac{\mathrm{d}^2\delta_m}{\mathrm{d}y^2} + \frac{2+3y}{2y(1+y)}\frac{\mathrm{d}\delta_m}{\mathrm{d}y} - \frac{3}{2y(1+y)}\delta_m = 0, \tag{6.180}$$

whose solutions are

$$
\delta_m \propto
\begin{cases}
1 + \dfrac{3}{2}\, y\,, \\[2ex]
\left(1 + \dfrac{3}{2}\, y\right) \ln\left(\dfrac{\sqrt{1+y}+1}{\sqrt{1+y}-1}\right) - 3\sqrt{1+y}\,.
\end{cases}
\tag{6.181}
$$

In the limit $y \ll 1$ (RD), the growing mode solution is $\delta_m \propto \ln y \propto \ln a$, i.e. the matter fluctuations only grow logarithmically in the radiation era. Significant growth of dark matter inhomogeneities only occurs when the universe becomes matter dominated. Indeed, in the limit $y \gg 1$ (MD), the growing mode solution is $\delta_m \propto y \propto a$.

Figure 6.6 shows numerical solutions for the evolution of the dark matter density contrast for three representative Fourier modes. Notice that the mode with $k > k_{\mathrm{eq}}$ receives a small boost in its amplitude at horizon crossing, before settling into the slow logarithmic growth predicted for the subhorizon evolution in the radiation era.

6.3.3 Matter Power Spectrum

We have derived enough of the clustering of matter fluctuations to explain the shape of the matter power spectrum at late times. We start at a time η_i when all modes of interest were still outside the Hubble radius. We take the initial power spectrum to be scale invariant, so that $k^3 \mathcal{P}_m(\eta_i, k) = \mathrm{const}$. We would like to see how the scale dependence of the spectrum evolves with time. (See also Section 5.2.3 for a complementary discussion.)

Consider first the time η_{eq}. Modes with $k < k_{\mathrm{eq}}$ are still outside the horizon and the spectrum for these scales must have the same shape as the initial spectrum (see Fig. 6.7). Modes with $k > k_{\mathrm{eq}}$ evolved as $\ln a$ during the time they spent inside the horizon in the radiation era. Their amplitude is therefore enhanced by a factor of $\ln(a_{\mathrm{eq}}/a_k)$, where a_k is the moment of horizon crossing of the mode k. Since $k = (aH)_k \propto a_k^{-1}$ during the radiation era, this gives the $(\ln k)^2$ scaling of the spectrum for $k > k_{\mathrm{eq}}$.

Next, we take a time η_0 after matter–radiation equality. All subhorizon modes grow as a and the amount by which the modes will have grown depends on when they entered the horizon. For $k < k_{\mathrm{eq}}$, we therefore have $k^3 \mathcal{P}_m \propto (a_0/a_k)^2 \propto k^4$, where we used that $k = (aH)_k \propto a_k^{-1/2}$ during the matter era. For $k > k_{\mathrm{eq}}$, we instead get $k^3 \mathcal{P}_m \propto (a_0/a_{\mathrm{eq}})^2 \ln(a_{\mathrm{eq}}/a_k)^2 \propto (\ln k)^2$ as before. Modes that entered the horizon in the radiation era all evolve in the same way during the matter era, so the spectrum is uniformly boosted and no additional scale dependence is generated for these scales. Finally, modes with $k < k_0$ are still outside the horizon and therefore the spectrum must have the same scaling as the initial spectrum, $k^3 \mathcal{P}_m \propto \mathrm{const}$. Combining these results, we have

$$\mathcal{P}_m(\eta_0, k) = \begin{cases} k^{-3} & k < k_0, \\ k & k_0 < k < k_{\rm eq}, \\ k^{-3}(\ln k)^2 & k > k_{\rm eq}. \end{cases} \qquad (6.182)$$

By definition, the regime $k < k_0$, with $k_0 = a_0 H_0 \approx 3 \times 10^{-4}\, h\,{\rm Mpc}^{-1}$, corresponds to superhorizon modes today which are unobservable. The scaling of the spectrum on subhorizon scales, $k > k_0$, is the same as that obtained in Section 5.2.3.

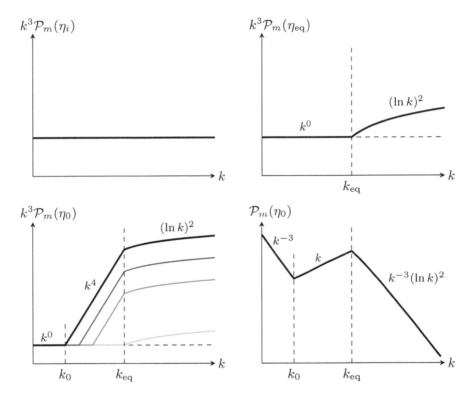

Fig. 6.7 Evolution of the matter power spectrum. *Top left*: The spectrum is scale invariant at some initial time η_i. *Top right*: At matter–radiation equality, the spectrum scales as $(\ln k)^2$ for $k > k_{\rm eq}$. *Bottom left*: At a later time η_0, the spectrum for $k_0 > k > k_{\rm eq}$ developed at k^4 scaling due to the subhorizon growth of the fluctuations. *Bottom right*: Final scalings for $\mathcal{P}_m(\eta_0, k)$. (Figure adapted from [11].)

6.4 Evolution of Photons and Baryons

Figure 6.8 shows the evolution of perturbations in dark matter, baryons and photons for a representative Fourier mode. Initially, the fluctuations in all components are of equal size (up to a factor of 4/3). Before decoupling, the baryons are tightly coupled to the photons and oscillate after horizon crossing. At the same time, the

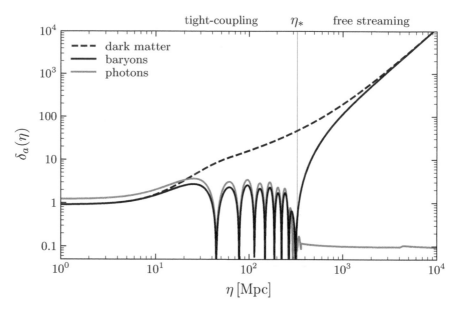

Fig. 6.8   Evolution of the density contrasts of dark matter (*dashed*), baryons (*black*) and photons (*gray*) for $k = 0.25\, h\, \mathrm{Mpc}^{-1}$. Note that around the time of decoupling the fluid approximation breaks down and the evolution is *not* captured by the equations described in this chapter. Instead, this figure was produced by solving the Boltzmann equations for the distribution functions of each species numerically.

dark matter experiences a slow logarithmic growth during the radiation era and a more rapid power law growth during the matter era. This explains why $\delta_c \gg \delta_b$ at decoupling. After decoupling, the baryons lose the pressure support of the photons and fall into the gravitational potential wells created by the dark matter. The density contrasts of dark matter and baryons eventually become equal again (see Exercise 5.2). Let me remark that it is crucial that the clustering of dark matter had a head start and then aided the growth of the baryon perturbations. Without the dark matter-assisted growth, the baryons wouldn't be able to cluster fast enough to explain the formation of galaxies.

In this section, we will study the evolution of photons and baryons prior to decoupling. During that time, the photons and baryons acted as a single fluid with pressure provided by the photons and an extra mass density by the baryons. To build up intuition, we will start by ignoring the baryons and study the photon perturbations alone.[9] After that, we will discuss how this dynamics is modified by the presence of baryons.

6.4.1 Radiation Fluctuations

Let us first consider the evolution of perturbations in the relativistic perfect fluid during the radiation era. Because the radiation component dominates, we use the

---

[9] For simplicity, we will ignore the photon anisotropic stress. In reality, this is only a good approximation because of the coupling to baryons.

same trick as before and consider the Poisson equation, which now reads

$$\nabla^2 \Phi = \frac{3}{2}\mathcal{H}^2 \Delta_r \,. \tag{6.183}$$

In Fourier space, we therefore have

$$\Delta_r(\eta, \mathbf{k}) = -\frac{2}{3}(k\eta)^2 \Phi(\eta, \mathbf{k})\,, \tag{6.184}$$

and, using our solution (6.165) for the gravitational potential during the radiation era, we find

$$\boxed{\Delta_r(\eta, \mathbf{k}) = -4\mathcal{R}_i \frac{\sin\varphi - \varphi\cos\varphi}{\varphi}} \quad \text{(radiation era)}\,, \tag{6.185}$$

where $\varphi \equiv k\eta/\sqrt{3}$. This solution is again valid on all scales. Taking the limits $\varphi \ll 1$ and $\varphi \gg 1$ gives the solutions on superhorizon and subhorizon scales:

$$\Delta_r(\eta, \mathbf{k}) \approx \begin{cases} -\dfrac{4}{9}\mathcal{R}_i(k\eta)^2 & \text{(superhorizon)}\,, \\[2mm] 4\mathcal{R}_i \cos\left(\dfrac{1}{\sqrt{3}}k\eta\right) & \text{(subhorizon)}\,. \end{cases} \tag{6.186}$$

We see that the comoving density contrast grows as $\Delta_r \propto \eta^2 \propto a^2$ on superhorizon scales, while it oscillates with a constant amplitude on subhorizon scales.

Exercise 6.16 Using

$$\delta_r = -\frac{2}{3}(k\eta)^2\Phi - 2\eta\,\Phi' - 2\Phi\,, \tag{6.187}$$

determine the solution for δ_r from (6.165). What are the superhorizon and subhorizon limits of the solution?

Next, we consider the evolution in the matter era. Since the radiation fluctuations are subdominant, we must use the continuity and Euler equations to follow their evolution. In Section 6.1.3, we showed that the radiation density contrast satisfies

$$\delta_r'' - \frac{1}{3}\nabla^2\delta_r = \frac{4}{3}\nabla^2\Phi + 4\Phi'' \quad \Leftarrow \quad \begin{cases} \delta_r' = -\dfrac{4}{3}\nabla\cdot\mathbf{v}_r + 4\Phi'\,, \\[3mm] \mathbf{v}_r' = -\dfrac{1}{4}\nabla\delta_r - \nabla\Phi\,, \end{cases} \tag{6.188}$$

where, for later convenience, I have also shown how the equation of motion for δ_r arises from the continuity and Euler equations. Recall that, during matter domination, the gravitational potential is a constant on all scales, so that

$$\delta_r'' - \frac{1}{3}\nabla^2\delta_r = \frac{4}{3}\nabla^2\Phi = \text{const.} \tag{6.189}$$

This is the equation of a harmonic oscillator with a constant driving force. The subhorizon fluctuations in the radiation density therefore oscillate with a constant amplitude around a shifted equilibrium point, $-4\Phi_0(\mathbf{k})$, where $\Phi_0(\mathbf{k})$ is the k-dependent amplitude of the gravitational potential in the matter era. This k-

dependence arises from the initial conditions and the nontrivial transfer function of the dark matter fluctuations. The solution for the density contrast is

$$\delta_r(\eta, \mathbf{k}) = C(\mathbf{k})\cos(\varphi) + D(\mathbf{k})\sin(\varphi) - 4\Phi_0(\mathbf{k}),\tag{6.190}$$

where $\varphi \equiv k\eta/\sqrt{3}$.

6.4.2 Photon–Baryon Fluid

So far, we have ignored the effects of baryons on the radiation fluid (except for their role in suppressing the photon anisotropic stress). Let us now fix this. We will work in the so-called **tight-coupling approximation**, where the coupling between photons and baryons is so strong that they can be treated as a single photon–baryon fluid with velocity $\mathbf{v}_b = \mathbf{v}_\gamma$. Only the combined momentum density of the photons and baryons is now conserved:

$$\begin{aligned}\mathbf{q} &= (\bar{\rho}_\gamma + \bar{P}_\gamma)\mathbf{v}_\gamma + (\bar{\rho}_b + \bar{P}_b)\mathbf{v}_b \\ &= \frac{4}{3}(1+R)\bar{\rho}_\gamma\mathbf{v}_\gamma, \qquad R \equiv \frac{3}{4}\frac{\bar{\rho}_b}{\bar{\rho}_\gamma} = 0.6\left(\frac{\Omega_b h^2}{0.02}\right)\left(\frac{a}{10^{-3}}\right).\end{aligned}\tag{6.191}$$

The fractional contribution from the baryons is characterized by the dimensionless parameter R. This parameter is small at early times, but grows linearly with $a(t)$ and becomes of order one around the time of recombination.

Recall the Euler equation for the evolution of the momentum density

$$\mathbf{q}' + 4\mathcal{H}\mathbf{q} = -(\bar{\rho} + \bar{P})\boldsymbol{\nabla}\Psi - \boldsymbol{\nabla}\delta P,\tag{6.192}$$

where $\Psi \approx \Phi$ in the absence of anisotropic stress. Substituting (6.191), the left-hand side can be written as

$$\mathbf{q}' + 4\mathcal{H}\mathbf{q} = \frac{\partial_\eta(a^4\mathbf{q})}{a^4} = \frac{4}{3}\bar{\rho}_\gamma\big[(1+R)\mathbf{v}_\gamma\big]',\tag{6.193}$$

where we have used that $a^4\bar{\rho}_\gamma$ is a constant. Inserting $\bar{\rho} + \bar{P} = \frac{4}{3}(1+R)\bar{\rho}_\gamma$ and $\delta P = \frac{1}{3}\bar{\rho}_\gamma\delta_\gamma$ on the right-hand side of (6.192), we get

$$-(\bar{\rho} + \bar{P})\boldsymbol{\nabla}\Psi - \boldsymbol{\nabla}\delta P = -\frac{4}{3}\bar{\rho}_\gamma\left[(1+R)\boldsymbol{\nabla}\Psi + \frac{1}{4}\boldsymbol{\nabla}\delta_\gamma\right].\tag{6.194}$$

Combining (6.193) and (6.194) then gives

$$\underset{\underset{\substack{\text{inertial}\\\text{mass}}}{\uparrow}}{\big[(1+R)\mathbf{v}_\gamma\big]'} = -\frac{1}{4}\boldsymbol{\nabla}\delta_\gamma - \underset{\underset{\substack{\text{gravitational}\\\text{mass}}}{\uparrow}}{(1+R)\boldsymbol{\nabla}\Psi}.\tag{6.195}$$

For $R = 0$, this reduces to the Euler equation for the radiation fluid shown in (6.188). The corrections to this equation are easy to understand: The coupling to the baryons adds extra "weight" to the photon–baryon fluid. This increases both the momentum density and the gravitational force term by a factor of $(1 + R)$.

The baryons, on the other hand, don't contribute to the pressure, so the pressure force term does *not* receive this factor.

Since the scattering of photons and baryons doesn't exchange energy, the continuity equation in (6.188) does not get modified and still reads

$$\delta_\gamma' = -\frac{4}{3} \boldsymbol{\nabla} \cdot \mathbf{v}_\gamma + 4\Phi' . \tag{6.196}$$

Combining the continuity and Euler equations, as before, we then find

$$\boxed{\delta_\gamma'' + \frac{R'}{1+R} \delta_\gamma' - \frac{1}{3(1+R)} \nabla^2 \delta_\gamma = \frac{4}{3} \nabla^2 \Psi + 4\Phi'' + \frac{4R'}{1+R} \Phi'} , \tag{6.197}$$

$$\underset{\text{pressure}}{\uparrow} \qquad\qquad \underset{\text{gravity}}{\uparrow}$$

which is the fundamental equation describing the evolution of density perturbations in the photon–baryon fluid.

From the pressure term in (6.197) we can read off the **sound speed** of the photon–baryon fluid

$$c_s^2 \equiv \frac{1}{3(1+R)} . \tag{6.198}$$

This is consistent with the definition of the adiabatic sound speed $c_s^2 = \delta P / \delta \rho$. To see this, we use $\delta P = \delta P_\gamma = \frac{1}{3} \delta \rho_\gamma$ and $\delta \rho = \delta \rho_\gamma + \delta \rho_b$, together with $\delta \rho_b = \frac{3}{4}(\bar{\rho}_b / \bar{\rho}_\gamma) \delta \rho_\gamma = R \, \delta \rho_\gamma$ (for adiabatic perturbations).

6.4.3 Cosmic Sound Waves

The dynamics associated with (6.197) is rather complex and will be discussed in detail in Chapter 7. Here, I will just give a brief sketch of the basic phenomenology and explain its observational consequences.

Acoustic oscillations

For simplicity, let us ignore the effects of gravity and consider solutions to the homogeneous equation:

$$\delta_\gamma'' + \frac{R'}{1+R} \delta_\gamma' - \frac{1}{3(1+R)} \nabla^2 \delta_\gamma = 0 . \tag{6.199}$$

There are two timescales in the problem: the expansion time and the oscillation time. On small scales, $k \gg \mathcal{H}/c_s$, the latter is much shorter than the former. A WKB approximation can then be used to obtain the solution to the damped harmonic oscillator equation (see Problem 7.4):

$$\delta_\gamma(\eta, \mathbf{k}) = C(\mathbf{k}) \frac{\cos(k r_s)}{(1+R)^{1/4}} + D(\mathbf{k}) \frac{\sin(k r_s)}{(1+R)^{1/4}} , \tag{6.200}$$

where $r_s(\eta) \equiv \int_0^\eta c_s(\eta') \, d\eta'$ is the **sound horizon** at the time η. At the time of photon decoupling, η_*, the sound horizon is $r_s(\eta_*) \approx 145 \, \mathrm{Mpc}$ (or 500 million

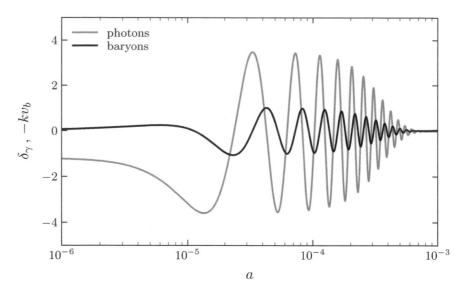

Fig. 6.9   Evolution of the density of photons (*gray*) and the velocity of baryons (*black*) for $k = 1\,h\,\mathrm{Mpc}^{-1}$. The damping of the oscillations arises from the diffusion of the photons on small scales.

light-years). The functions $C(\mathbf{k})$ and $D(\mathbf{k})$ are fixed by the superhorizon initial conditions. For adiabatic initial conditions, the superhorizon modes are time independent, so the sine solution is not excited, $D(\mathbf{k}) = 0$. The amplitude of the cosine solution is determined by the value of the perturbation at horizon crossing. An important feature of the solution is that all Fourier modes of a given wavelength reach the extrema of their oscillations at the same time, irrespective of the direction of the wavevector $\mathbf{k}$. This **phase coherence** allows for the constructive interference of many waves, which leaves an imprint in cosmological observables (see below).

A numerical solution for the acoustic oscillations in the photon–baryon fluid is shown in Fig. 6.9. An important feature that is not captured by our simplified analysis is the damping of the fluctuations, which arises because the photons diffuse out of regions of high density into regions of low density, washing out the density contrast. This so-called **diffusion damping** is especially relevant for small-scale fluctuations (high-k modes) and to describe it requires going beyond the tight-coupling approximation. A more accurate treatment must also include the time-dependent gravitational driving force that we have ignored. In Chapter 7, we will have fun with the subtle, but wonderful, physics of the sound waves in the primordial plasma.

CMB anisotropies

When we are looking at the CMB, we are seeing a snapshot of the primordial sound waves at the moment of photon decoupling. We see from the solution (6.200) that the oscillation frequency of the photon–baryon fluid depends on the wavenumber k, meaning that different modes decouple at different phases in their evolution and

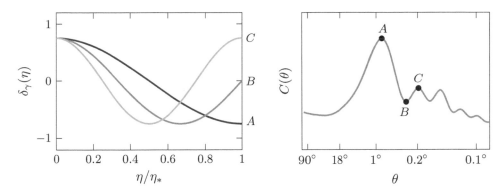

Fig. 6.10 Cartoon illustrating the origin of the peaks in the CMB power spectrum. *Left*: Fluctuations of different wavelengths are captured at different moments in their evolution and therefore have different amplitudes at decoupling. *Right*: Since the square of the amplitude determines the power on a given length scale, waves that are captured at an extremum (A or C) produce the peaks in the CMB spectrum, while waves that are captured with zero amplitude (B) produce the troughs.

therefore have different amplitudes at last-scattering (see Fig. 6.10). Evaluating the solution at decoupling, $\eta = \eta_*$, gives the k-dependent amplitude of the fluctuations at last-scattering. This amplitude has extrema at

$$k_n = n\pi/r_s(\eta_*)\,, \qquad (6.201)$$

with $n = 1, 2, \cdots$. After projecting the Fourier modes onto the spherical surface of last-scattering, these become the peaks in the angular power spectrum of the CMB anisotropies. Note that the power spectrum measures the square of the amplitude, $|\delta_\gamma|^2$, so the maxima and minima in δ_γ both become peaks in the power spectrum. The positions of the peaks are determined by a combination of the sound horizon at decoupling and the angular diameter distance to the surface of last-scattering. The heights of the peaks carry information about the amount of dark matter and baryons in the universe. The size of the damping on small scales depends on the density of neutrinos. This is how the measurements of the CMB anisotropies—like those by the WMAP and Planck satellites, as well as other ground-based experiments—have allowed precise measurements of the cosmological parameters and revolutionized the field of cosmology. It is an absolutely wonderful story, which I will describe in much more detail in Chapter 7.

Baryon acoustic oscillations*

The same oscillations that are observed in the CMB anisotropies also leave an imprint in the clustering of galaxies. This arises because photons and baryons oscillate together before decoupling. Like the photons, the different Fourier modes of the baryons decouple at different phases in their evolution. The initial conditions for the gravitational growth of the baryons therefore include the oscillations of the photon–baryon fluid. This oscillatory feature gets transferred to the gravitational

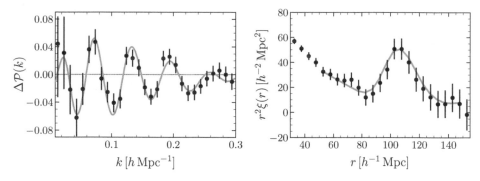

Fig. 6.11   Measurement of the BAO signal in Fourier space (*left*) and in position space (*right*). To accentuate the oscillatory feature, the left plot shows the deviation $\Delta\mathcal{P}(k)$ from the smooth spectrum. (Figure adapted from [14].)

potential and the matter fluctuations, but because baryons are only about 15% of the total matter density, the effect is smaller than in the CMB. Since correlations in the galaxy distribution are inherited from correlations in the matter density, these **baryon acoustic oscillations** (BAO) are imprinted in the galaxy power spectrum (see Fig. 6.11). In the following, I will give a simple analytic derivation of the BAO signal [12].

The time when the baryons are released from the drag of the photons is known as the **drag epoch**. Although photon and baryon decoupling are closely related, they are not the same. In fact, the baryons decouple slightly later than the photons at $z_d \approx 1060$ (compared to $z_* \approx 1090$). After the drag epoch, the baryons behave as pressureless matter and we can write the total matter density contrast as

$$\delta_m = f_b \delta_b + (1 - f_b)\delta_c \,, \tag{6.202}$$

where $f_b \equiv \rho_b/\rho_m$ is the fractional baryon density. Near decoupling, the solution for the total matter density is given by (6.181):

$$\delta_m(\eta, \mathbf{k}) = C_1(\mathbf{k})D_1(y) + C_2(\mathbf{k})D_2(y) \,, \tag{6.203}$$

where $y(\eta) \equiv a(\eta)/a_{\rm eq}$ and

$$
\begin{aligned}
D_1(y) &= \frac{2}{3} + y \approx y \,, \\
D_2(y) &= D_1(y) \ln\left(\frac{\sqrt{1+y}+1}{\sqrt{1+y}-1}\right) - 2\sqrt{1+y} \approx \frac{8}{45}y^{-3/2} \,.
\end{aligned}
\tag{6.204}
$$

The growing mode solution at late times is $\delta_m \approx C_1(\mathbf{k})\,y$. We would like to write this solution in terms of its initial conditions at baryon decoupling. To do this, we match δ_m and its derivative δ'_m at the end of the drag epoch. The matching conditions are

$$C_1(\mathbf{k})D_1(\eta_d) + C_2(\mathbf{k})D_2(\eta_d) = f_b\delta_b(\eta_d) + (1 - f_b)\delta_c(\eta_d)\,,$$
$$C_1(\mathbf{k})D_1'(\eta_d) + C_2(\mathbf{k})D_2'(\eta_d) = f_b\delta_b'(\eta_d) + (1 - f_b)\delta_c'(\eta_d)\,,$$

(6.205)

where we have used (6.202) on the right-hand side. Solving this for $C_1(\mathbf{k})$, and only keeping the baryonic contribution which carries the oscillatory feature, we find

$$C_1(\mathbf{k}) = f_b\frac{D_2'}{D_1D_2' - D_1'D_2}\left[\delta_b(\eta_d, \mathbf{k}) - \frac{D_2}{D_2'}\delta_b'(\eta_d, \mathbf{k})\right]\,,$$

(6.206)

where all functions of time are evaluated at η_d and the primes are derivatives with respect to η, not y. Note that $\partial_\eta f = \mathcal{H}y\,\partial_y f$. Using $y_d \gg 1$, and replacing all growth functions D_i by their asymptotic expressions, we get

$$\delta_m(\eta, \mathbf{k}) = C_1(\mathbf{k})\,y \approx \frac{3f_b}{5}\left[\delta_b(\eta_d, \mathbf{k}) + \frac{2}{3}(k\eta_d)\,v_b(\eta_d, \mathbf{k})\right]\frac{a}{a_d}\,,$$

(6.207)

where we have used that $\delta_b' = kv_b$. On scales that are smaller than the horizon at the drag epoch, we have $k\eta_d \gg 1$ and the second term, proportional to the velocity v_b, dominates. This is called the **velocity overshoot**. When the baryons are released at the drag epoch, they move according to their velocity and generate a new density perturbation. This is the dominant effect on small scales.

As long as the tight-coupling approximation holds, we have $v_b = v_\gamma = \frac{3}{4}\delta_\gamma'/k$. The cosine contribution in the solution (6.200) for δ_γ then leads to a dominant sine contribution in the solution (6.207) for δ_m. A recent measurement of the BAO signal is shown in Fig. 6.11. It is rather amazing how this signal arising from primordial sound waves in the early universe is so clearly visible to the clustering of galaxies billions of years later. In position space, the BAO feature maps to a peak around $110\,h^{-1}\mathrm{Mpc}$. The position of this peak is an important "standard ruler." It depends on the sound horizon at the drag epoch and the angular diameter distance to the observed objects. Measurements of the BAO help to break degeneracies in the CMB data and have therefore played an important role in measurements of the cosmological parameters (see Chapter 7).

6.5 Gravitational Waves

A new era of science was initiated on September 14, 2015, when the Laser Interferometer Gravitational-Wave Observatory (LIGO) detected the first **gravitational waves** (GWs) from the merger of two black holes. This detection marked the beginning of multi-messenger astronomy and, in the future, gravitational waves may also provide a new window into the early universe. In this section, I will collect some mathematical background on cosmological GWs that will be useful for our discussions in Chapters 7 and 8. Further details can be found in [15].

Gravitational waves are tensor perturbations to the spatial metric,

$$ds^2 = a^2(\eta) \left[-d\eta^2 + (\delta_{ij} + h_{ij})dx^i dx^j \right] . \tag{6.208}$$

Since the perturbation h_{ij} is symmetric ($h_{ij} = h_{ji}$), transverse ($\partial_i h_{ij} = 0$) and traceless ($h_{ii} = 0$), it contains $6 - 3 - 1 = 2$ independent modes (corresponding to the two polarizations of the gravitational wave). To make this more explicit, we write the Fourier modes of h_{ij} as

$$h_{ij}(\eta, \mathbf{k}) = \sum_{\lambda=+,\times} h_\lambda(\eta, \mathbf{k}) \, \epsilon_{ij}^\lambda(\hat{\mathbf{k}}) , \tag{6.209}$$

where $\epsilon_{ij}^\lambda(\hat{\mathbf{k}})$ are two independent polarization tensors and $h_\lambda(\eta, \mathbf{k})$ are the corresponding mode functions. The polarization tensors can be taken to be real and satisfy $\epsilon_{ij}^\lambda(\hat{\mathbf{k}}) = \epsilon_{ij}^\lambda(-\hat{\mathbf{k}})$, so that $h_{ij}(\eta, \mathbf{x})$ is real if $h_\lambda^*(\eta, \mathbf{k}) = h_\lambda(\eta, -\mathbf{k})$. The polarization tensors are symmetric ($\epsilon_{ij}^\lambda = \epsilon_{ji}^\lambda$), transverse ($\hat{k}_i \epsilon_{ij}^\lambda = 0$) and traceless ($\epsilon_{ii}^\lambda = 0$). Our normalization for the polarization basis will be

$$\sum_{i,j} \epsilon_{ij}^\lambda \epsilon_{ij}^{\lambda'} = 2\delta^{\lambda\lambda'} , \tag{6.210}$$

but other conventions (with different numerical constants on the right-hand side) can be found in the literature. Explicitly, the polarization tensors can be written as

$$\begin{aligned} \epsilon_{ij}^+(\hat{\mathbf{k}}) &\equiv \hat{m}_i \hat{m}_j - \hat{n}_i \hat{n}_j , \\ \epsilon_{ij}^\times(\hat{\mathbf{k}}) &\equiv \hat{m}_i \hat{m}_j + \hat{n}_i \hat{n}_j , \end{aligned} \tag{6.211}$$

where $\hat{\mathbf{m}}$ and $\hat{\mathbf{n}}$ are two unit vectors that are orthogonal to $\hat{\mathbf{k}}$ and to each other. For a gravitational wave with a wavevector pointing in the z-direction, i.e. $\hat{\mathbf{k}} = (0, 0, 1)$, we can choose $\hat{\mathbf{m}} \equiv \hat{\mathbf{x}}$ and $\hat{\mathbf{n}} \equiv \hat{\mathbf{y}}$, so that

$$h_{ij} = h_+ \begin{pmatrix} 1 & 0 & 0 \\ 0 & -1 & 0 \\ 0 & 0 & 0 \end{pmatrix} + h_\times \begin{pmatrix} 0 & 1 & 0 \\ 1 & 0 & 0 \\ 0 & 0 & 0 \end{pmatrix} = \begin{pmatrix} h_+ & h_\times & 0 \\ h_\times & -h_+ & 0 \\ 0 & 0 & 0 \end{pmatrix} . \tag{6.212}$$

Sometimes, it is useful to work in the so-called **helicity basis**, where

$$\begin{aligned} \epsilon_{ij}^{\pm 2} &\equiv (\epsilon_{ij}^+ \pm i\epsilon_{ij}^\times)/2 , \\ h_{\pm 2} &\equiv h_+ \mp ih_\times . \end{aligned} \tag{6.213}$$

Under a rotation by an angle ψ around the wavevector $\mathbf{k}$, we have $h_{\pm 2} \mapsto e^{\pm 2i\psi} h_{\pm 2}$, which shows that the modes $h_{\pm 2}$ describe states of helicity $+2$ and -2 (also called "right-handed" and "left-handed" polarization, respectively).

In Problem 6.4, you will show that the linearized Einstein equation implies the following evolution equation for the tensor perturbations:

$$h_{ij}'' + 2\mathcal{H}h_{ij}' - \nabla^2 h_{ij} = 16\pi G a^2 \hat{\Pi}_{ij}, \qquad (6.214)$$

where $\hat{\Pi}_{ij}$ is the transverse, traceless part of the anisotropic stress. In the absence of a source, $\hat{\Pi}_{ij} = 0$, this equation describes the free propagation of gravitational waves in an expanding universe. It is convenient to remove the Hubble friction term by defining $f_{ij} \equiv a(\eta)h_{ij}$. Each polarization mode then satisfies

$$f_\lambda'' + \left(k^2 - \frac{a''}{a} \right) f_\lambda = 0. \qquad (6.215)$$

Let us solve this equation for a generic scale factor, $a(\eta) \propto \eta^\beta$, which includes radiation domination ($\beta = 1$), matter domination ($\beta = 2$) and inflation ($\beta \approx -1$) as special cases. The solution is

$$h_\lambda(\eta, \mathbf{k}) = \frac{C_\lambda(\mathbf{k})}{a(\eta)} \eta\, j_{\beta-1}(k\eta) + \frac{D_\lambda(\mathbf{k})}{a(\eta)} \eta\, y_{\beta-1}(k\eta), \qquad (6.216)$$

where $j_\beta(x)$ and $y_\beta(x)$ are spherical Bessel functions (see Appendix D) and the constants $C_\lambda(\mathbf{k})$ and $D_\lambda(\mathbf{k})$ must be fixed by the initial conditions.

Note that, for a power-law scale factor, we have $a''/a \propto \mathcal{H}^2$. On sub-Hubble scales, $k \gg \mathcal{H}$, we can therefore drop the a''/a term in (6.215) and the solution is

$$h_\lambda(\eta, \mathbf{k}) = \frac{C_\lambda(\mathbf{k})}{a(\eta)} e^{ik\eta} + \frac{D_\lambda(\mathbf{k})}{a(\eta)} e^{-ik\eta} \quad (\text{for } k \gg \mathcal{H}), \qquad (6.217)$$

which you can check also follows from the $k\eta \gg 1$ limit of (6.216). On sub-Hubble scales, we therefore get the expected plane wave solutions with an amplitude that decays as a^{-1}. On super-Hubble scales, $k \ll \mathcal{H}$, we instead drop the k^2 term and the solution becomes

$$h_\lambda(\eta, \mathbf{k}) = C_\lambda(\mathbf{k}) + D_\lambda(\mathbf{k}) \int \frac{\mathrm{d}\eta}{a^2(\eta)} \quad (\text{for } k \ll \mathcal{H}). \qquad (6.218)$$

The second term decays with the expansion of the universe, so that the growing mode of the gravitational wave is a constant on super-Hubble scales.[10]

In cosmology, we are interested in the power spectrum of a stochastic background of gravitational waves, $\mathcal{P}_h(k) \equiv (2\pi^2/k^3)\Delta_h^2(k)$. We define this power spectrum, so that the variance is

$$\sum_{i,j} \langle h_{ij}^2(\mathbf{x}) \rangle = \int \mathrm{d}\ln k\, \Delta_h^2(k), \qquad (6.219)$$

---

[10] In a radiation-dominated universe, a''/a vanishes identically, so that the solution (6.218) doesn't apply. In that case, the super-Hubble solution is the $k\eta \ll 1$ limit of (6.217), which again has a decaying mode and a constant mode.

where we have suppressed the time dependence. The power spectrum for the individual polarization modes then is

$$\langle h_\lambda(\mathbf{k}) h_{\lambda'}(\mathbf{k}') \rangle = \frac{2\pi^2}{k^3} \frac{\Delta_h^2(k)}{4} \times (2\pi)^3 \delta_{\mathrm{D}}(\mathbf{k} + \mathbf{k}') \, \delta_{\lambda\lambda'} \,, \tag{6.220}$$

where the factor of 4 follows from our normalization of the polarization basis in (6.210). In Chapter 8, we will show how quantum fluctuations during inflation produce a scale-invariant spectrum of primordial GWs, while in Chapter 7, we will explain how these GWs affect the anisotropies in the CMB.

6.6 Summary

In this chapter, we have developed the foundations of cosmological perturbation theory. By linearizing the Einstein equations, we derived the evolution equations for small fluctuations and solved them for various special cases.

Focusing on scalar perturbations, the most general perturbation of the spacetime metric is

$$g_{\mu\nu} = a^2 \begin{pmatrix} -(1 + 2A) & \partial_i B \\ \partial_i B & (1 + 2C)\delta_{ij} + 2\partial_{\langle i}\partial_{j\rangle} E \end{pmatrix}, \tag{6.221}$$

where $\partial_{\langle i}\partial_{j\rangle} = \partial_i \partial_j - \frac{1}{3}\nabla^2$. The perturbed energy-momentum tensor is

$$\begin{aligned}
T^0{}_0 &\equiv -(\bar\rho + \delta\rho)\,, \\
T^i{}_0 &\equiv -(\bar\rho + \bar P)\,\partial^i v\,, \\
T^0{}_i &\equiv (\bar\rho + \bar P)\,\partial_i(v + B)\,, \\
T^i{}_j &\equiv (\bar P + \delta P)\delta^i_j + \partial^{\langle i}\partial_{j\rangle}\Pi\,,
\end{aligned} \tag{6.222}$$

and we often work in terms of the momentum density $q \equiv (\bar\rho + \bar P)v$ and the density contrast $\delta \equiv \delta\rho/\rho$. Coordinate transformations can change the values of these perturbation variables. The problem of unphysical gauge modes can be addressed by working with combinations of perturbations that are invariant under coordinate transformations. Important gauge-invariant variables are the Bardeen potential and the comoving density contrast

$$\Psi = A + \mathcal{H}(B - E') + (B - E')'\,, \tag{6.223}$$

$$\Phi = -C + \frac{1}{3}\nabla^2 E - \mathcal{H}(B - E')\,, \tag{6.224}$$

$$\Delta = \delta + \frac{\bar\rho'}{\bar\rho}(v + B)\,, \tag{6.225}$$

as well as the two curvature perturbations

$$\zeta = -C + \frac{1}{3}\nabla^2 E + \mathcal{H}\frac{\delta\rho}{\bar{\rho}'}, \tag{6.226}$$

$$\mathcal{R} = -C + \frac{1}{3}\nabla^2 E - \mathcal{H}(v + B). \tag{6.227}$$

On superhorizon scales, ζ and $\mathcal{R}$ are equal and time independent if the matter perturbations are adiabatic. Perturbations are called adiabatic if

$$\delta P = c_s^2 \delta\rho = \frac{\bar{P}'}{\bar{\rho}'}\delta\rho, \tag{6.228}$$

and fluctuations in the densities of matter and radiation obey

$$\delta_m = \frac{3}{4}\delta_r. \tag{6.229}$$

The initial conditions of our universe are well described by adiabatic perturbations.

An alternative to the gauge-invariant formalism is to work in a fixed gauge, track the evolution of all fluctuations and compute observables. Unphysical gauge modes must cancel in physical answers. Popular gauges are

- Newtonian: $B = E = 0$ • Constant density: $\delta\rho = B = 0$
- Spatially flat: $C = E = 0$ • Comoving: $v = B = 0$
- Synchronous: $A = B = 0$

We derived the equations of motion for the perturbations in Newtonian gauge. The linearized Einstein equations are

$$\nabla^2\Phi - 3\mathcal{H}(\Phi' + \mathcal{H}\Psi) = 4\pi G a^2 \delta\rho, \tag{6.230}$$

$$-(\Phi' + \mathcal{H}\Psi) = 4\pi G a^2 q, \tag{6.231}$$

$$\partial_{\langle i}\partial_{j\rangle}(\Phi - \Psi) = 8\pi G a^2 \Pi_{ij}, \tag{6.232}$$

$$\Phi'' + \mathcal{H}\Psi' + 2\mathcal{H}\Phi' + \frac{1}{3}\nabla^2(\Psi - \Phi) + (2\mathcal{H}' + \mathcal{H}^2)\Psi = 4\pi G a^2 \delta P, \tag{6.233}$$

where the sources on the right-hand side include a sum over all components. In the absence of anisotropic stress, equation (6.232) implies that the two metric potentials are equal, $\Psi \approx \Phi$, and (6.233) reduces to

$$\Phi'' + 3\mathcal{H}\Phi' + (2\mathcal{H}' + \mathcal{H}^2)\Phi = 4\pi G a^2 \delta P. \tag{6.234}$$

Equations (6.230) and (6.231) can be combined into

$$\nabla^2\Phi = 4\pi G a^2 \bar{\rho}\Delta, \tag{6.235}$$

which is of the same form as the Newtonian Poisson equation, but is now valid on all scales.

		Radiation era	Matter era
Φ	$k < \mathcal{H}$	const	const
	$k > \mathcal{H}$	$a^{-2}\cos(k\eta/\sqrt{3})$	const
Δ_r	$k < \mathcal{H}$	a^2	a
	$k > \mathcal{H}$	$\cos(k\eta/\sqrt{3})$	$\cos(k\eta/\sqrt{3}) + \text{const}$
Δ_m	$k < \mathcal{H}$	a^2	a
	$k > \mathcal{H}$	$\ln a$	a

Table 6.1 Summary of the evolution of cosmological perturbations

The conservation of the energy-momentum tensor gives the continuity and Euler equations

$$\delta' = -(1+w)(\boldsymbol{\nabla}\cdot\mathbf{v} - 3\Phi') - 3\mathcal{H}(c_s^2 - w)\delta\,, \tag{6.236}$$

$$\mathbf{v}' = -\mathcal{H}(1 - 3w)\mathbf{v} - \frac{c_s^2}{1+w}\boldsymbol{\nabla}\delta - \boldsymbol{\nabla}\Phi\,, \tag{6.237}$$

where $c_s^2 = \delta P/\delta\rho$ is the sound speed of the fluid and $w = \bar{P}/\bar{\rho}$ is its equation of state. These equations apply separately for every non-interacting fluid.

We discussed the solutions to the above perturbation equations in various special limits. Table 6.1 summarizes the results for Φ, Δ_r and Δ_m. The density contrast δ is equal to Δ on subhorizon scales, but differs on superhorizon scales, where it is a constant.

Before decoupling, photons and baryons can be treated as a single tightly-coupled fluid. The density contrast of the photon–baryon fluid evolves as

$$\delta_\gamma'' + \frac{R'}{1+R}\,\delta_\gamma' - \frac{1}{3(1+R)}\nabla^2\delta_\gamma = \frac{4}{3}\nabla^2\Psi + 4\Phi'' + \frac{4R'}{1+R}\Phi'\,, \tag{6.238}$$

where $R \equiv \frac{3}{4}\bar{\rho}_b/\bar{\rho}_\gamma \propto a(t)$. This is the equation of a harmonic oscillator with a gravitational driving force, which plays an important role in the physics of the CMB anisotropies (see Chapter 7).

Further Reading

Cosmological perturbation theory is an important subject that is treated in every cosmology textbook. In addition, there are many good dedicated reviews. Two influential reviews are by Kodama and Sasaki [16] and Mukhanov, Feldman and Brandenberger [17]. A more recent alternative by Malik and Wands [18] includes results up to second order in perturbation theory. A must-read is the classic by Ma and Bertschinger [2] which presents the linearized evolution equations in both synchronous and Newtonian gauge. This paper has become the industry standard, so that many papers follow the notation and conventions of Ma and Bertschinger. You may also want to try the lecture notes of Lesgourgues [19], Brandenberger [20] and Piattella [21]. A systematic treatment of isocurvature perturbations can be found in [8]. A description of the CMB anisotropies that is roughly at the level of this chapter, but contains further details, can be found in [22, 23]. In addition, Wayne Hu's website (http://background.uchicago.edu/~whu/) contains many helpful animations. Finally, my analytic treatment of the BAO feature was based on the paper by Hu and Sugiyama [12].

Problems

6.1 Superhorizon evolution of the gravitational potential

In this problem, you will derive an analytic solution for the superhorizon evolution of the gravitational potential that remains valid in the transition from the radiation era to the matter era.

The starting point is the following Einstein equation

$$\nabla^2\Phi - 3\mathcal{H}(\Phi' + \mathcal{H}\Phi) = 4\pi G a^2 \delta\rho\,,$$

where $\delta\rho = \sum_a \delta\rho_a$. On superhorizon scales, we can drop $\nabla^2\Phi$ and interpret this as an evolution equation for Φ.

1. Assuming adiabatic perturbations, show that the superhorizon evolution of the potential in a universe with matter and radiation satisfies

$$y\frac{\mathrm{d}\Phi}{\mathrm{d}y} + \Phi = -\frac{4 + 3y}{6(1 + y)}\delta_m\,,$$

where $y \equiv a/a_{\mathrm{eq}}$ and δ_m is the density contrast of the matter perturbations. Use $\delta_m' = 3\Phi'$ to write this as a closed equation for Φ.

2. Consider the following change of variables

$$u \equiv \frac{y^3\Phi}{\sqrt{1 + y}}\,,$$

and show that the above evolution equation becomes

$$\frac{\mathrm{d}^2u}{\mathrm{d}y^2} + \left[-\frac{2}{y} + \frac{3}{2(1 + y)} - \frac{3}{4 + 3y}\right]\frac{\mathrm{d}u}{\mathrm{d}y} = 0\,.$$

By integrating this equation, show that the solution for the gravitational potential is

$$\Phi(\eta) = \frac{\Phi_i}{10y^3}\left[16\sqrt{1+y} + 9y^3 + 2y^2 - 8y - 16\right],$$

where Φ_i is the primordial value of the potential. Note that for $y \to \infty$, this gives $\Phi \to \frac{9}{10}\Phi_i$, confirming our result for the change in the gravitational potential during the transition from the radiation era to the matter era.

6.2 Superhorizon initial conditions

In this problem, you will derive the superhorizon initial conditions for perturbations, accounting for the anisotropic stress due to neutrinos.

Consider the following Einstein equations during the radiation era:

$$k^2\Phi + 3\mathcal{H}(\Phi' + \mathcal{H}\Psi) = -4\pi G a^2(\bar{\rho}_\gamma\delta_\gamma + \bar{\rho}_\nu\delta_\nu), \tag{1}$$

$$k^2(\Phi - \Psi) = 8\pi G a^2(\bar{\rho}_\nu + \bar{P}_\nu)\Pi_\nu, \tag{2}$$

where Π_ν is the (rescaled) neutrino-induced anisotropic stress. In the tight-coupling approximation, we have $\Pi_\nu' \approx -\frac{2}{5}kv_\nu$ (see Section B.3.3).

1. Specializing to the case of adiabatic initial conditions, show that the superhorizon limit of (1) implies

$$\eta\Phi'' + 3\Phi' + \Psi' = 0.$$

2. Taking multiple time derivatives of (2) show that Φ satisfies

$$\eta^3\Phi'''' + 12\eta^2\Phi''' + 4\left(9 + \frac{2}{5}f_\nu\right)\eta\Phi'' + 8\left(3 + \frac{2}{5}f_\nu\right)\Phi' = 0,$$

where $f_\nu \equiv \bar{\rho}_\nu/(\bar{\rho}_\gamma + \bar{\rho}_\nu)$. Determine the four solutions of this equation.

3. Focussing on the growing mode, $\Phi = \Phi_i$, show that

$$\Psi = \Phi_i\left(1 + \frac{2}{5}f_\nu\right)^{-1} \approx 0.86\,\Phi_i$$

and $\delta_\gamma = -2\Psi \approx -1.72\,\Phi_i$.

4. Repeating the analysis for the neutrino density isocurvature mode (see Section 6.2.3), show that

$$\Psi = -2\Phi_i, \quad S_\nu \equiv \delta_\nu - \delta_\gamma = \frac{15 + 4f_\nu}{f_\nu(1 - f_\nu)}\Phi_i \approx 68.8\,\Phi_i,$$

and $\delta_\gamma \approx -24.2\,\Phi_i$.

6.3 Perturbation theory in Mathematica

Download the Mathematica notebook `CPT.nb` from the book's website. The notebook derives the Einstein equations for scalar perturbations. The goal of this problem is for you to become familiar with the notebook, so that you can later use it for your own calculations.

1. Read through the notebook and make sure you understand its basic operations.

2. Now use the notebook to confirm the conservation equations:

$$\delta\rho' = -3\mathcal{H}(\delta\rho + \delta P) + 3\Phi'(\bar{\rho} + \bar{P}) - \partial_i q^i \,,$$
$$q_i' = -4\mathcal{H}q_i - (\bar{\rho} + \bar{P})\partial_i\Phi - \partial_i\delta P \,,$$

where $q^i = (\bar{\rho} + \bar{P})v^i$.

Hint: To determine say the $\nu = 0$ component of $\nabla_\mu T^\mu{}_\nu = 0$, write $\partial_\mu T^\mu{}_0 = -\Gamma^\mu_{\mu\lambda}T^\lambda{}_0 + \Gamma^\lambda_{\mu 0}T^\mu{}_\lambda$ and compute the two sides separately.

6.4 Gravitational waves

In this problem, you will compute the evolution equation for gravitational waves in a cosmological background. You will first do this by hand and then confirm your result in Mathematica.

Consider the line element

$$\mathrm{d}s^2 = a^2(\eta)\left[-\,\mathrm{d}\eta^2 + (\delta_{ij} + h_{ij})\mathrm{d}x^i\mathrm{d}x^j\right],$$

where h_{ij} is symmetric, trace-free and transverse. To linear order in h_{ij}, the nonzero Christoffel symbols are

$$\Gamma^0_{00} = \mathcal{H}\,,$$
$$\Gamma^0_{ij} = \mathcal{H}\delta_{ij} + \mathcal{H}h_{ij} + \frac{1}{2}h'_{ij}\,,$$
$$\Gamma^i_{j0} = \mathcal{H}\delta^i_j + \frac{1}{2}\delta^{il}h'_{lj}\,,$$
$$\Gamma^i_{jk} = \frac{1}{2}\left(\partial_j h^i{}_k + \partial_k h^i{}_j - \partial^i h_{jk}\right).$$

If you are feeling energetic, you may try to verify these expressions.

1. Show that the spatial part to the perturbed Einstein tensor is

$$\delta G_{ij} = \frac{1}{2}\left(h''_{ij} - \nabla^2 h_{ij} + 2\mathcal{H}h'_{ij} - 2h_{ij}(2\mathcal{H}' + \mathcal{H}^2)\right).$$

Hint: Convince yourself that the Ricci scalar has no tensor perturbations at first order.

2. Combine the previous result with the perturbation to the stress tensor, $\delta T_{ij} = a^2(\bar{P}h_{ij} + \hat{\Pi}_{ij})$, to show that the perturbed Einstein equation (in Fourier space) is

$$h''_{ij} + 2\mathcal{H}h'_{ij} + k^2 h_{ij} = 16\pi G a^2 \hat{\Pi}_{ij}\,.$$

Assuming vanishing anisotropic stress, $\hat{\Pi}_{ij} = 0$, discuss the solutions of this equation on scales that are larger and smaller than the Hubble radius.

3. Derive the above results using Mathematica.

   Hint: Consider a single gravitational wave with a wavevector **k** pointing in the z-direction and write h_{ij} in terms of its two polarization modes. Derive everything in terms of the functions h_+ and $h_\times$.

7 Cosmic Microwave Background

Observations of the CMB have played a pivotal role in establishing the standard cosmological model. They have given us a detailed understanding of the geometry and composition of the universe, and provided the first evidence that the primordial fluctuations originated from quantum fluctuations during a period of inflation. What makes the CMB such a powerful cosmological probe is the fact that the fluctuations were captured when they were still small and therefore accurately described by linear perturbation theory. Unlike the nonlinear structure formation described in Chapter 5, which has many sources of uncertainty, the physics of the CMB can be understood from first principles. Small fluctuations in the primordial plasma evolve under a well-defined set of equations, allowing very accurate predictions of the expected CMB temperature anisotropies.

Figure 7.1 is the stunning image of the universe when it was only 370 000 years old. It shows variations of the intensity of the CMB photons across the sky, which reflect perturbations in the density at the time of photon decoupling. The CMB temperature fluctuations are analyzed statistically by measuring the correlations between hot and cold spots as a function of their angular separation. The result is the angular power spectrum shown in Fig. 7.2. The figure also shows a fit of the

Fig. 7.1 Temperature fluctuations in the cosmic microwave background as measured by the WMAP satellite [1]. A color version of the higher-resolution Planck map [2] can be found on the book's website (but isn't reproduced here because it doesn't print well in grayscale).

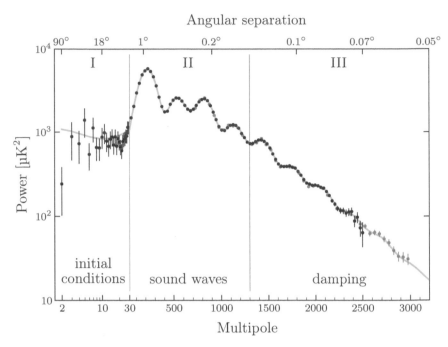

Fig. 7.2 Power spectrum of the CMB anisotropies measured by the Planck satellite (*black*) [2] and the South Pole Telescope (*gray*) [3]. For low multipoles (large angular scales), the spectrum is determined by the primordial initial conditions alone. For intermediate multipoles, the imprint of sound waves in the photon–baryon fluid is seen, while for large multipoles (small angular scales) damping becomes important. (To avoid clutter, I have not added relevant data from the Atacama Cosmology Telescope [4].)

theoretical prediction for the CMB spectrum to the data. The agreement between the theory and the data is remarkable.

It is useful to separate the CMB spectrum into three different regimes, which in Fig. 7.2 are labeled as I, II and III:

- Region I: Large-angle correlations are sourced by fluctuations that were still outside of the horizon at recombination. These fluctuations did not evolve before photon decoupling and are therefore a direct probe of the initial conditions.

- Region II: Perturbations with shorter wavelengths entered the horizon before recombination. As we saw in Section 6.4.3, inside the horizon the perturbations in the tightly-coupled photon–baryon fluid propagate as sound waves supported by the large photon pressure. The oscillation frequency of these waves is a function of their wavelength, so that different modes are captured at different moments in their evolution when the CMB was released at photon decoupling. This is the origin of the oscillatory pattern seen in the angular power spectrum.

- Region III: On small scales, the random diffusion of the photons can erase the density contrast in the plasma, leading to a damping of the wave amplitudes.

This suppresses the amplitude of the CMB power spectrum on small angular scales (large multipole moments).

In this chapter, I will provide a (hopefully) pedagogical introduction to the physics of the CMB anisotropies.[1] The goal is to understand how the cosmological parameters determine the shape of the power spectrum, so that we can understand how the CMB measurements constrain these parameters. To develop intuition, I will present the evolution of fluctuations in a hydrodynamic approximation. This is a good approximation as long as photons and baryons can be treated as a single tightly-coupled fluid, but breaks down below the mean free path of the photons. A fully self-consistent treatment of the CMB anisotropies requires the heavy machinery of the Boltzmann equation and will be given in Appendix B.

7.1 Anisotropies in the First Light

The largest anisotropy in the CMB is a temperature dipole of magnitude $3.36\,\mathrm{mK}$, which is due to the motion of the Solar System with respect to the rest frame of the microwave background. As you will show in Problem 7.1, this motion creates a Doppler shift in the energies of the observed photons and hence a temperature anisotropy of the form

$$\frac{\delta T(\hat{\mathbf{n}})}{T} = \hat{\mathbf{n}} \cdot \mathbf{v} = v\cos\theta\,, \tag{7.1}$$

where $\hat{\mathbf{n}}$ is a unit vector pointing in the observer's line of sight, which is opposite to the momentum of the incoming radiation $\hat{\mathbf{p}}$, and $\mathbf{v}$ is the velocity of the observer (see Fig. 7.3). As expected, the temperature is higher if we move towards the radiation ($\hat{\mathbf{n}} \cdot \mathbf{v} = v$) and smaller if we move away from it ($\hat{\mathbf{n}} \cdot \mathbf{v} = -v$). Fitting this dipolar anisotropy to the data, we find that the speed of the Solar System relative to the CMB rest frame is $v = 368\,\mathrm{km/s}$.[2] After subtracting the dipole, we are left with the cosmological signal shown in Fig. 7.1. We are interested in the statistical correlations of these primordial temperature fluctuations.

---

[1] I am immensely grateful to Anthony Challinor and Eiichiro Komatsu for a very careful reading of this chapter and for many important corrections.

[2] We believe that the inferred speed has three main components: (1) the orbital velocity of the Solar System with respect to the center of our Galaxy; (2) the velocity of our Galaxy with respect to the center-of-mass of the Local Group of galaxies; and (3) the velocity of the Local Group with respect to the CMB rest frame. The first two components can be measured independently, leading to a velocity of about 307 km/s, but pointing almost exactly in the opposite direction as the velocity inferred from the CMB dipole [5]. Subtracting this known contribution, the measured bulk flow of the Local Group is 626 ± 30 km/s [6]. Part of this (about 220 km/s) can be accounted for by the pull of the nearby Virgo cluster of galaxies. Explaining the rest of the bulk motion of the Local Group remains an open problem (but see [7]).

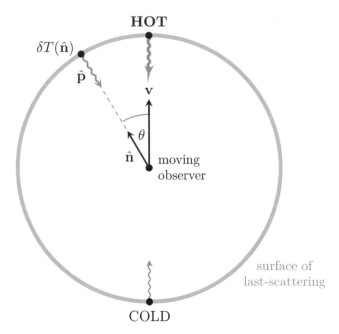

HOT

COLD

Fig. 7.3   The motion of the Solar System relative to the CMB rest frame produces a dipolar pattern in the observed CMB temperature.

7.1.1 Angular Power Spectrum

Let $T(\hat{\mathbf{n}}) \equiv \bar{T}_0 \left[1 + \Theta(\hat{\mathbf{n}})\right]$ be the measured CMB temperature in a direction $\hat{\mathbf{n}}$ in the sky, where $\bar{T}_0$ is the average CMB temperature today and we have introduced the fractional temperature fluctuation $\Theta(\hat{\mathbf{n}}) \equiv \delta T(\hat{\mathbf{n}})/\bar{T}_0$. Comparing the temperatures at two distinct points $\hat{\mathbf{n}}$ and $\hat{\mathbf{n}}'$ (see Fig. 7.4) gives the two-point correlation function

$$C(\theta) \equiv \langle \Theta(\hat{\mathbf{n}})\Theta(\hat{\mathbf{n}}') \rangle , \tag{7.2}$$

where $\cos\theta \equiv \hat{\mathbf{n}} \cdot \hat{\mathbf{n}}'$. As before, the angle brackets denote an average over an ensemble of universes. Of course, our universe is only one member of this ensemble, but we can estimate the ensemble average by dividing the CMB into independent patches and averaging over the correlations in each patch. For large-angle correlations, the number of independent patches is small and this estimate will have a large cosmic variance (see below).

Given that we observe fluctuations on the spherical surface of last-scattering, it is convenient to expand the temperature field in spherical harmonics

$$\Theta(\hat{\mathbf{n}}) = \sum_{l=2}^{\infty} \sum_{m=-l}^{l} a_{lm} Y_{lm}(\hat{\mathbf{n}}) , \tag{7.3}$$

where the expansion coefficients a_{lm} are called **multipole moments**. A few relevant mathematical properties of the spherical harmonics are reviewed in Appendix D.

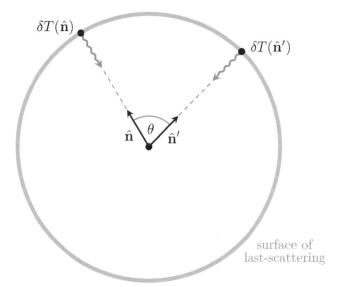

Fig. 7.4 Illustration of the two-point correlations between temperature fluctuations on the surface of last-scattering.

There are various phase conventions for the Y_{lm}; I will adopt $Y_{lm}^* = (-1)^m Y_{l,-m}$, so that $a_{lm}^* = (-1)^m a_{l,-m}$ for a real field $\Theta(\hat{\mathbf{n}})$. The two-point function of the multipole moments is defined as

$$\langle a_{lm} a_{l'm'}^* \rangle = C_l \, \delta_{ll'} \delta_{mm'} \,, \tag{7.4}$$

where C_l is the **angular power spectrum** and the Kronecker deltas are a consequence of statistical isotropy. The angular power spectrum is the harmonic space equivalent of the two-point correlation function in real space. Indeed, substituting (7.3) into (7.2), we get

$$\begin{aligned} C(\theta) &= \langle \Theta(\hat{\mathbf{n}})\Theta(\hat{\mathbf{n}}') \rangle \\ &= \sum_{lm} \sum_{l'm'} \langle a_{lm} a_{l'm'}^* \rangle Y_{lm}(\hat{\mathbf{n}}) Y_{l'm'}^*(\hat{\mathbf{n}}') \\ &= \sum_l C_l \sum_m Y_{lm}(\hat{\mathbf{n}}) Y_{l'm'}^*(\hat{\mathbf{n}}') \\ &= \sum_l \frac{2l+1}{4\pi} C_l \, P_l(\cos\theta) \,, \end{aligned} \tag{7.5}$$

where $P_l(\cos\theta)$ are Legendre polynomials (see Appendix D). We see that the moments of the angular power spectrum appear as coefficients in an expansion of $C(\theta)$ in terms of Legendre polynomials. Using the orthogonality of the Legendre polynomials, we can also write

$$C_l = 2\pi \int_{-1}^{1} \mathrm{d}\cos\theta \, C(\theta) P_l(\cos\theta) \,. \tag{7.6}$$

The information contained in the C_l's is therefore completely equivalent to that of the function $C(\theta)$.

The variance of the temperature anisotropy field is

$$C(0) = \sum_l \frac{2l+1}{4\pi} C_l \approx \int d\ln l \, \frac{(2l+1)l C_l}{4\pi} \approx \int d\ln l \, \frac{l(l+1)C_l}{2\pi}, \qquad (7.7)$$

where the final equality holds for $l \gg 1$. The power per logarithmic interval in l is

$$\Delta_T^2 \equiv \frac{l(l+1)}{2\pi} C_l \, \bar{T}_0^2. \qquad (7.8)$$

We usually plot the CMB power spectrum as Δ_T^2, which on large scales (small multipoles) will be independent of l if the primordial fluctuations are scale invariant (see Section 7.3.2). Note that our Δ_T^2 is the same as $\mathcal{D}_l$ in the Planck papers.

Finally, let us return to the issue of **cosmic variance**. For fixed l, we have $2l+1$ different a_{lm}'s, allowing for $2l+1$ independent estimates of the true C_l's. Imagine that we have made a full-sky, noise-free observation of the temperature field $\Theta(\hat{\mathbf{n}})$ and extracted its multipole moments a_{lm}. An **estimator** for C_l is

$$\hat{C}_l \equiv \frac{1}{2l+1} \sum_m |a_{lm}|^2. \qquad (7.9)$$

This estimator is unbiased in the sense that $\langle \hat{C}_l \rangle = C_l$. However, the estimator has a nonzero variance corresponding to an irreducible error in our determination of the true power spectrum

$$\frac{\Delta C_l}{C_l} \equiv \frac{\sqrt{\langle (C_l - \hat{C}_l)^2 \rangle}}{C_l} = \sqrt{\frac{2}{2l+1}}. \qquad (7.10)$$

As expected, this cosmic variance is largest for small l, corresponding to large scales. This explains why the error bars in Fig. 7.2, which contain both cosmic variance and instrumental noise, are largest for small multipoles. Current measurements from Planck are dominated by cosmic variance up to $l \sim 2000$.

Exercise 7.1 Use Wick's theorem,

$$\langle a_{lm} a_{lm}^* a_{lm'} a_{lm'}^* \rangle = \langle a_{lm} a_{lm}^* \rangle \langle a_{lm'} a_{lm'}^* \rangle + \langle a_{lm} a_{lm'} \rangle \langle a_{lm}^* a_{lm'}^* \rangle$$
$$+ \langle a_{lm} a_{lm'}^* \rangle \langle a_{lm}^* a_{lm'} \rangle, \qquad (7.11)$$

to derive (7.10).

7.1.2 A Road Map

Our main goal in this chapter is to understand how the observed power spectrum of CMB anisotropies is related to the spectrum of initial curvature perturbations:

$$\Delta_{\mathcal{R}}^2(k) \equiv \frac{k^3}{2\pi^2} \mathcal{P}_{\mathcal{R}}(k) \quad \longrightarrow \quad C_l = 4\pi \int d\ln k \, \Theta_l^2(k) \, \Delta_{\mathcal{R}}^2(k).$$

The transfer function $\Theta_l(k)$ that maps $\Delta_{\mathcal{R}}^2(k)$ to C_l captures the evolution of the fluctuations in the primordial plasma, the free streaming of the photons after decoupling and the projection of the anisotropies onto the sky. Since the primordial power spectrum is rather featureless, it is this transfer function that leads to all the nontrivial structure in the CMB power spectrum.

The following schematic illustrates the main steps involved in the computation of the transfer function, and hence the CMB power spectrum:

$$
\mathcal{R}(0, \mathbf{k}) \xrightarrow{\text{evolution (§7.4)}} \begin{pmatrix} \delta_\gamma \\ \Psi \\ v_b \end{pmatrix}_{\eta_*} \xrightarrow[\text{projection (§7.3)}]{\text{free streaming (§7.2)}} \delta T(\hat{\mathbf{n}})
$$

We will first show how the free streaming of photons after decoupling relates the observed temperature fluctuations to fluctuations on the surface of last-scattering. We will see that the CMB anisotropies depend on the fluctuations in the photon density (δ_γ), the gravitational potential (Ψ) and the baryon velocity (v_b) at the time of decoupling (η_*). We will also explain that the projection of these inhomogeneities onto the observer's sky leads to nontrivial angular variations. Finally, we will link the fluctuations at last-scattering to the initial curvature perturbations ($\mathcal{R}$), which requires us to follow the evolution of coupled perturbations from early times until decoupling. Since the evolution is linear, we can study each Fourier mode separately, and the final anisotropy spectrum is obtained by summing over many Fourier modes weighted by the spectrum of the initial conditions.

7.2 Photons in a Clumpy Universe

We begin with the free streaming of photons after decoupling. As the photons travel through the inhomogeneous universe they gain or lose energy, which will affect the observed temperature anisotropies. In this section, we will compute this effect.

7.2.1 Gravitational Redshift

The evolution of the photons after decoupling is governed by the geodesic equation (see Chapter 2 and Appendix A)

$$
\frac{\mathrm{d}P^\mu}{\mathrm{d}\lambda} = -\Gamma_{\nu\rho}^\mu P^\nu P^\rho, \tag{7.12}
$$

where the parameter λ is defined such that $P^\mu = \mathrm{d}x^\mu/\mathrm{d}\lambda$. We will work in the Newtonian gauge where the line element with scalar fluctuations is

$$
\mathrm{d}s^2 = a^2(\eta) \left[-(1 + 2\Psi)\mathrm{d}\eta^2 + (1 - 2\Phi)\delta_{ij}\mathrm{d}x^i\mathrm{d}x^j \right]. \tag{7.13}
$$

The case of tensor perturbations will be explored in Problem 7.6.

It is convenient to write the left-hand side of the geodesic equation as

$$\frac{\mathrm{d}P^\mu}{\mathrm{d}\lambda} = \frac{\mathrm{d}\eta}{\mathrm{d}\lambda}\frac{\mathrm{d}P^\mu}{\mathrm{d}\eta} = P^0\frac{\mathrm{d}P^\mu}{\mathrm{d}\eta}\,. \tag{7.14}$$

To determine the evolution of the photon energy, we consider the time component of (7.12):

$$\begin{aligned}
\frac{\mathrm{d}P^0}{\mathrm{d}\eta} &= -\Gamma^0_{\nu\rho}\frac{P^\nu P^\rho}{P^0} \\[2mm]
&= -\Gamma^0_{00}P^0 - 2\Gamma^0_{0i}P^i - \Gamma^0_{ij}\frac{P^i P^j}{P^0} \\[2mm]
&= -(\mathcal{H}+\Psi')P^0 - 2\partial_i\Psi P^i - [\mathcal{H}-\Phi'-2\mathcal{H}(\Phi+\Psi)]\delta_{ij}\frac{P^i P^j}{P_0}\,.
\end{aligned} \tag{7.15}$$

Before we continue, we have to address a small subtlety that we have so far brushed under the carpet. The four-momentum components $P^\mu = (P^0, P^i)$ are defined in the coordinate frame, while what an observer actually measures as the photon energy and momentum are the components $P^{\hat\mu} = (E, P^{\hat\imath})$ in their *local inertial frame*. In a perturbed universe, the two are not the same. To relate these two sets of momentum components, we use that

$$\begin{aligned}
\eta_{\hat\mu\hat\nu}P^{\hat\mu}P^{\hat\nu} &= g_{\mu\nu}P^\mu P^\nu\,, \\
-E^2 + \delta_{ij}P^{\hat\imath}P^{\hat\jmath} &= g_{00}(P^0)^2 + g_{ij}P^i P^j\,.
\end{aligned} \tag{7.16}$$

The energy and momentum in the local inertial frame can therefore be written as[3]

$$\begin{aligned}
E &= \sqrt{-g_{00}}\,P^0\,, \\
p^2 &\equiv g_{ij}P^i P^j = \delta_{ij}P^{\hat\imath}P^{\hat\jmath}\,,
\end{aligned} \tag{7.17}$$

and hence we have

$$\begin{aligned}
P^0 &= \frac{E}{\sqrt{-g_{00}}} = \frac{E}{\sqrt{a^2(1+2\Psi)}} = \frac{E}{a}(1-\Psi)\,, \\
P^i &= \frac{E}{\sqrt{g_{ii}}}\hat p^i = \frac{E}{\sqrt{a^2(1-2\Phi)}}\hat p^i = \frac{E}{a}(1+\Phi)\hat p^i\,,
\end{aligned} \tag{7.18}$$

where $\hat p^i$ is the unit vector in the photon's direction of propagation. Substituting (7.18) into (7.15), we get

$$\begin{aligned}
\frac{\mathrm{d}}{\mathrm{d}\eta}\left(\frac{E}{a}(1-\Psi)\right) = & -(\mathcal{H}+\Psi')\frac{E}{a}(1-\Psi) - 2\partial_i\Psi\frac{E}{a}(1+\Phi)\hat p^i \\
& - [\mathcal{H}-\Phi'-2\mathcal{H}(\Phi+\Psi)]\frac{E}{a}\frac{(1+\Phi)^2}{(1-\Psi)}\,,
\end{aligned} \tag{7.19}$$

---

[3] The condition in (7.16) only fixes the relation between the components in the two frames up to a Lorentz transformation. In the following, we assume that the observer is at rest and that their spatial basis vectors are aligned with the spatial coordinate directions.

which, at first order, cleans up rather nicely

$$\boxed{\frac{1}{E}\frac{\mathrm{d}E}{\mathrm{d}\eta} = -\mathcal{H} + \Phi' - \hat{p}^i \partial_i \Psi}\,,\tag{7.20}$$

where we have used that

$$\frac{\mathrm{d}\Psi}{\mathrm{d}\eta} = \frac{\partial\Psi}{\partial\eta} + \frac{\mathrm{d}x^i}{\mathrm{d}\eta}\partial_i\Psi \equiv \Psi' + \hat{p}^i\partial_i\Psi\,.\tag{7.21}$$

Each term on the right-hand side of (7.20) has a clear physical interpretation: The first term describes the redshifting of the photon energy due to the expansion of the universe, $E \propto a^{-1}$. To understand the origin of the second term, we note that the metric potential Φ can be viewed as a local perturbation of the scale factor, $\tilde{a}(\eta, \mathbf{x}) = a(\eta)(1 - \Phi)$. The Φ' term in (7.20) then simply captures the fact that the photon energy now decreases as $E \propto \tilde{a}^{-1}$. The third term, proportional to $\partial_i\Psi$, describes the gravitational redshift (or blueshift) as the photon travels out of (or falls into) a gravitational potential well. Using (7.21), we can write this term as

$$\hat{p}^i\partial_i\Psi = \frac{\mathrm{d}\Psi}{\mathrm{d}\eta} - \frac{\partial\Psi}{\partial\eta} \equiv \frac{\mathrm{d}\Psi}{\mathrm{d}\eta} - \Psi'\,,\tag{7.22}$$

so that (7.20) becomes

$$\boxed{\frac{\mathrm{d}\ln(aE)}{\mathrm{d}\eta} = -\frac{\mathrm{d}\Psi}{\mathrm{d}\eta} + \Phi' + \Psi'}\,.\tag{7.23}$$

This equation determines how the inhomogeneities in the spacetime affect the photon energy, beyond the usual redshifting due to the expansion of the universe.

7.2.2 Line-of-Sight Solution

By integrating the geodesic equation (7.23), we can relate the energy of a CMB photon at decoupling to its energy today, when it enters our detectors. We will assume for simplicity that recombination was nearly instantaneous, so that all photons were released at the same time η_*. This assumption is justified by looking at the time evolution of the **visibility function** (see Section 3.2.5), which we now define in terms of conformal time:

$$g(\eta) \equiv \frac{\mathrm{d}}{\mathrm{d}\eta}e^{-\tau} = -\tau'e^{-\tau}\,,\tag{7.24}$$

where τ is the optical depth. Recall that the visibility function describes the probability that a photon last scattered in the interval $[\eta, \eta + \mathrm{d}\eta]$. Figure 7.5 shows that the visibility function is sharply peaked at $\eta_* \approx 373\,000\,\mathrm{yrs}$ (or $z_* \approx 1090$), so that it is a reasonable approximation to take the CMB photons to be emitted at a fixed time η_*, the moment of last-scattering.

Letting our location be the origin of the coordinates, $\mathbf{x}_0 \equiv 0$, the photons in a direction $\hat{\mathbf{n}}$ were emitted at the point $\mathbf{x}_* \equiv \chi_*\hat{\mathbf{n}}$, where $\chi_* = \eta_0 - \eta_*$ is the distance to the last-scattering surface (in a flat universe). Integrating (7.23) from the time of emission η_* to the time of observation η_0, we then get

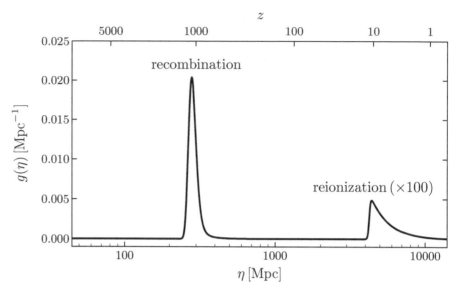

Fig. 7.5 Visibility function $g(\eta)$ as a function of conformal time η (or redshift z). We see that the function is sharply peaked near hydrogen recombination and has a second peak at reionization. The reionization peak has been multiplied by a factor of 100 to make it visible in the plot.

$$\ln(aE)_0 = \ln(aE)_* - (\Psi_0 - \Psi_*) + \int_{\eta_*}^{\eta_0} \mathrm{d}\eta \, (\Phi' + \Psi') . \tag{7.25}$$

After decoupling, the photon distribution function maintains the same shape, because all photons simply move along geodesics. Since the Bose–Einstein distribution is a function of E/T, this means that the effective photon temperature satisfies $T \propto E$ (see Chapter 3). This allows us to relate the evolution of the perturbed photon energy to the temperature anisotropy:

$$aE \propto a\bar{T} \left(1 + \frac{\delta T}{\bar{T}} \right), \tag{7.26}$$

where $\bar{T}(\eta)$ is the mean temperature. Taylor expanding the logarithms in (7.25) to first order in $\delta T/\bar{T}$, and keeping in mind that $a_0\bar{T}_0 = a_*\bar{T}_*$, we find

$$\boxed{\left.\frac{\delta T}{\bar{T}}\right|_0 = \left.\frac{\delta T}{\bar{T}}\right|_* + \Psi_* + \int_{\eta_*}^{\eta_0} \mathrm{d}\eta \, (\Phi' + \Psi')}, \tag{7.27}$$

where we have dropped the value of the gravitational potential at our location, Ψ_0, since it only contributes to the $l = 0$ monopole and is hence unobservable.

7.2.3 Fluctuations at Last-Scattering

We see from (7.27) that the observed temperature fluctuations today, δT_0, depend on the intrinsic temperature fluctuations at the moment of last-scattering, δT_*. The latter have two distinct sources:

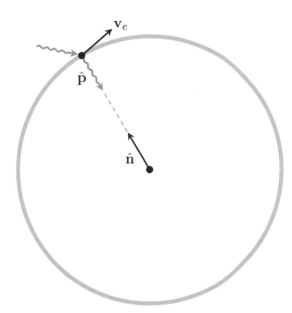

Fig. 7.6   The motion of electrons at the surface of last-scattering produces an additional temperature anisotropy.

- First, there are temperature fluctuations associated to fluctuations in the photon density at recombination; since $\rho_\gamma \propto T^4$, the fractional temperature fluctuations are $\delta T/T \subset \delta_\gamma/4$. Regions of enhanced photon density are slightly hotter than the average, while regions of lower density are slightly colder.

- Second, there are temperature fluctuations associated to the fluctuations in the bulk velocity of the electrons at recombination. When the photons scatter off these electrons, they receive an additional Doppler shift leading to temperature fluctuations of the form $\delta T/T \subset \hat{\mathbf{p}} \cdot \mathbf{v}_e$, where $\hat{\mathbf{p}}$ is a unit vector associated with the momentum of the scattered photon (see Fig. 7.6).[4] A formal derivation of this result—using the Boltzmann equation—will be given in Appendix B. However, the form of the answer is rather intuitive; in particular, notice the similarity of this formula with (7.1), the Doppler-induced temperature anisotropy due to our motion relative to the CMB rest frame.

  A photon observed in the direction $\hat{\mathbf{n}}$ has $\hat{\mathbf{p}} = -\hat{\mathbf{n}}$. Moreover, since the electrons are strongly interacting with the baryons, we can use $\mathbf{v}_e = \mathbf{v}_b$ and write $\delta T/T \subset -\hat{\mathbf{n}} \cdot \mathbf{v}_b$.

---

[4] We can also think of this temperature fluctuation as arising because the photons are emitted with different peculiar velocities at each point on the last-scattering surface. The projection of these velocities onto the line-of-sight describes the Doppler shift of the photon energy.

Combining these two sources of temperature fluctuations, we get

$$\left.\frac{\delta T}{\bar{T}}\right|_* = \left(\frac{1}{4}\delta_\gamma - \hat{\mathbf{n}} \cdot \mathbf{v}_b\right)_*, \tag{7.28}$$

where the subscript $*$ indicates that a quantity is evaluated at the time η_* and the position $\mathbf{x}_*$ on the surface of last-scattering.

Substituting (7.28) into (7.27), we get

$$\frac{\delta T}{\bar{T}}(\hat{\mathbf{n}}) = \left(\frac{1}{4}\delta_\gamma + \Psi\right)_* - (\hat{\mathbf{n}} \cdot \mathbf{v}_b)_* + \int_{\eta_*}^{\eta_0} d\eta\,(\Phi' + \Psi'), \tag{7.29}$$

$$\begin{array}{ccc} \uparrow & \uparrow & \uparrow \\ \text{SW} & \text{Doppler} & \text{ISW} \end{array}$$

where we have dropped the subscript 0 on the observed $\delta T/\bar{T}$ to avoid clutter. The answer in (7.29) has been separated into three contributions:

- **SW** The first term is the so-called *Sachs–Wolfe term* [8]. It combines the intrinsic temperature fluctuations associated to the photon density fluctuations at last-scattering, $\delta_\gamma/4$, with the induced temperature perturbation Ψ_* arising from the gravitational redshift of the photons. In our conventions, $\Psi < 0$ corresponds to an overdensity, so that negative Ψ_* leads to a temperature decrement, as expected since a photon loses energy when climbing out of a potential well.

- **Doppler** The second term is the *Doppler term* that we discussed before. Since the baryon velocity vanishes on superhorizon scales, this contribution is subdominant on large scales. We will find that the oscillations in the Doppler term are out of phase with the oscillations in the Sachs–Wolfe term, so that adding the Doppler contribution reduces the contrast between the peaks and troughs in the spectrum (see the left panel of Fig. 7.7).

- **ISW** The last term describes the additional gravitational redshift due to the evolution of the metric potentials along the line-of-sight. We call this the *integrated Sachs–Wolfe (ISW) effect*. If the gravitational potential is a constant, then the photons gain energy when moving into a high-density region, but lose the same amount of energy when leaving the region. The net effect would be zero. A nonzero ISW effect therefore requires that the potential changes with time. As we have seen in Chapter 6, for most of the history of the universe—namely, when the universe was matter dominated—the gravitational potential was constant and therefore didn't contribute to the ISW term. At early times, however, the residual amount of radiation gives $\Phi' \neq 0$, which results in a nonzero *early ISW effect*. Adding this contribution raises the height of the first peak of the spectrum (see the right panel of Fig. 7.7). Similarly, at late times, dark energy becomes relevant, leading to a *late ISW effect*, which adds power for low l.

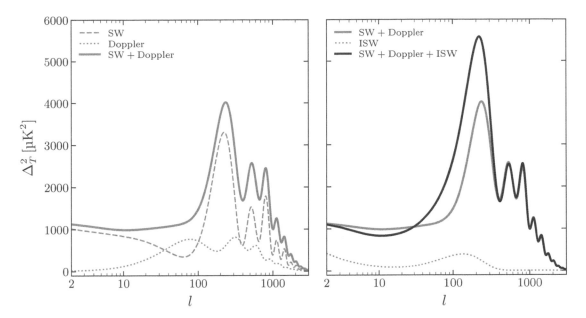

Illustration of the different contributions to the CMB power spectrum. The left panel shows the effect of adding the Doppler term to the Sachs–Wolfe term, while the right panel displays the effect of adding the ISW term.

7.3 Anisotropies from Inhomogeneities

The line-of-sight solution (7.29) shows how the observed CMB temperature fluctuations in a given direction are related to fluctuations at the surface of last-scattering in that direction. In this section, I will describe how the inhomogeneities in the primordial plasma give rise to correlations between the temperatures in different directions.

7.3.1 Spatial-to-Angular Projection

As we have seen in the previous chapters, the evolution of the primordial fluctuations is best studied in Fourier space, because the individual Fourier modes evolve independently. These Fourier modes describe plane wave solutions in the primordial plasma and the three-dimensional density field can be found by adding many of these plane waves with different amplitudes, wavelengths and phases. Figure 7.8 illustrates how even a single Fourier mode gives rise to angular variations in the CMB temperature on the surface of last-scattering. We will now study the mathematical details of this projection effect.

Let us for a moment ignore the ISW contribution. The temperature fluctuation in the direction $\hat{\mathbf{n}}$ is then directly related to the fluctuations in the perturbations

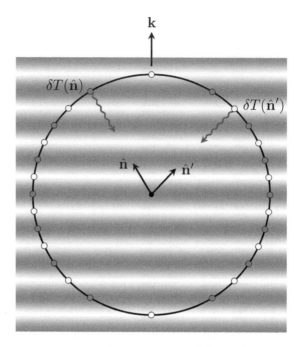

Fig. 7.8 Illustration of the temperature anisotropy created by a single plane wave inhomogeneity at recombination.

at the point $\mathbf{x}_* = \chi_* \hat{\mathbf{n}}$ on the surface of last-scattering. For the reason that we just described, it is useful to express the latter in terms of its Fourier components. For scalar fluctuations, we have $\mathbf{v}_b = i\hat{\mathbf{k}} v_b$ and hence the line-of-sight solution (7.29) can be written as

$$\Theta(\hat{\mathbf{n}}) \equiv \frac{\delta T}{\bar{T}}(\hat{\mathbf{n}}) = \int \frac{\mathrm{d}^3 k}{(2\pi)^3} \, e^{i\mathbf{k}\cdot(\chi_*\hat{\mathbf{n}})} \Big[F(\eta_*, \mathbf{k}) - i(\hat{\mathbf{k}}\cdot\hat{\mathbf{n}}) \, G(\eta_*, \mathbf{k}) \Big] , \qquad (7.30)$$

where $F \equiv \frac{1}{4}\delta_\gamma + \Psi$ and $G \equiv v_b$. Since the evolution is linear, it is convenient to factor out the initial curvature perturbation $\mathcal{R}_i(\mathbf{k}) \equiv \mathcal{R}(0, \mathbf{k})$ and write

$$\Theta(\hat{\mathbf{n}}) = \int \frac{\mathrm{d}^3 k}{(2\pi)^3} \, e^{i\mathbf{k}\cdot(\chi_*\hat{\mathbf{n}})} \Big[F_*(k) - i(\hat{\mathbf{k}}\cdot\hat{\mathbf{n}}) \, G_*(k) \Big] \mathcal{R}_i(\mathbf{k}) , \qquad (7.31)$$

where $F_*(k) \equiv F(\eta_*, \mathbf{k})/\mathcal{R}_i(\mathbf{k})$ and $G_*(k) \equiv G(\eta_*, \mathbf{k})/\mathcal{R}_i(\mathbf{k})$ are the transfer functions for the Sachs–Wolfe and Doppler terms. Note that these transfer functions only depend on the magnitude of the wavevector, $k \equiv |\mathbf{k}|$, while the initial perturbations are a function of its direction. We then use the Legendre expansion of the exponential

$$e^{i\mathbf{k}\cdot(\chi_*\hat{\mathbf{n}})} = \sum_l i^l (2l+1) j_l(k\chi_*) P_l(\hat{\mathbf{k}}\cdot\hat{\mathbf{n}}) , \qquad (7.32)$$

where j_l are the spherical Bessel functions (see Appendix D). The factor of $i(\hat{\mathbf{k}}\cdot\hat{\mathbf{n}})$ of the Doppler term leads to

$$i(\hat{\mathbf{k}} \cdot \hat{\mathbf{n}}) \, e^{i\chi_* \mathbf{k} \cdot \hat{\mathbf{n}}} = \frac{\mathrm{d}}{\mathrm{d}(k\chi_*)} e^{i(k\chi_*)\hat{\mathbf{k}} \cdot \hat{\mathbf{n}}} = \sum_l i^l (2l+1) j_l'(k\chi_*) P_l(\hat{\mathbf{k}} \cdot \hat{\mathbf{n}}), \qquad (7.33)$$

where the prime denotes a derivative with respect to the argument of the spherical Bessel function. Substituting (7.32) and (7.33) into (7.31), we get

$$\Theta(\hat{\mathbf{n}}) = \sum_l i^l (2l+1) \int \frac{\mathrm{d}^3 k}{(2\pi)^3} \, \Theta_l(k) \, \mathcal{R}_i(\mathbf{k}) P_l(\hat{\mathbf{k}} \cdot \hat{\mathbf{n}}), \qquad (7.34)$$

where we have defined the transfer function

$$\Theta_l(k) \equiv F_*(k) j_l(k\chi_*) - G_*(k) j_l'(k\chi_*). \qquad (7.35)$$

Inserting (7.34) into the two-point function (7.2), and using

$$\langle \mathcal{R}_i(\mathbf{k}) \mathcal{R}_i(\mathbf{k}') \rangle = \frac{2\pi^2}{k^3} \Delta_{\mathcal{R}}^2(k) \, (2\pi)^3 \delta_{\mathrm{D}}(\mathbf{k} + \mathbf{k}'), \qquad (7.36)$$

$$P_l(-\hat{\mathbf{k}} \cdot \hat{\mathbf{n}}) = (-1)^l P_l(\hat{\mathbf{k}} \cdot \hat{\mathbf{n}}), \qquad (7.37)$$

$$\int \mathrm{d}\hat{\mathbf{k}} \, P_l(\hat{\mathbf{k}} \cdot \hat{\mathbf{n}}) P_{l'}(\hat{\mathbf{k}} \cdot \hat{\mathbf{n}}') = \frac{4\pi}{2l+1} P_l(\hat{\mathbf{n}} \cdot \hat{\mathbf{n}}') \delta_{ll'}, \qquad (7.38)$$

we find

$$\langle \Theta(\hat{\mathbf{n}}) \Theta(\hat{\mathbf{n}}') \rangle = \sum_l \frac{2l+1}{4\pi} \left[4\pi \int \mathrm{d}\ln k \, \Theta_l^2(k) \Delta_{\mathcal{R}}^2(k) \right] P_l(\hat{\mathbf{n}} \cdot \hat{\mathbf{n}}'). \qquad (7.39)$$

Comparing this to (7.5), we identify the angular power spectrum as

$$\boxed{C_l = 4\pi \int \mathrm{d}\ln k \, \Theta_l^2(k) \Delta_{\mathcal{R}}^2(k)}. \qquad (7.40)$$

We see that the CMB power spectrum is determined by the power spectrum of the primordial curvature perturbations, $\Delta_{\mathcal{R}}^2(k)$, and the transfer function $\Theta_l(k)$. The latter describes both the evolution until decoupling and the projection onto the surface of last-scattering.

Exercise 7.2 Using the addition theorem

$$P_l(\hat{\mathbf{k}} \cdot \hat{\mathbf{n}}) = \sum_{|m| \leq l} \frac{4\pi}{2l+1} Y_{lm}^*(\hat{\mathbf{k}}) Y_{lm}(\hat{\mathbf{n}}), \qquad (7.41)$$

show that the line-of-sight solution (7.34) can be written in the form (7.3), with

$$a_{lm} = 4\pi i^l \int \frac{\mathrm{d}^3 k}{(2\pi)^3} \, \Theta_l(k) \, \mathcal{R}_i(\mathbf{k}) \, Y_{lm}^*(\hat{\mathbf{k}}). \qquad (7.42)$$

Show that substituting this into (7.4) gives the result in (7.40).

Including the ISW contribution, and repeating the same steps, we find that the complete transfer function is

$$
\Theta_l(k) = F_*(k)\, j_l(k\chi_*) - G_*(k)\, j_l'(k\chi_*) + \int_{\eta_*}^{\eta_0} d\eta\, (\Phi' + \Psi')\, j_l(k\chi)\,, \tag{7.43}
$$

where $\chi(\eta) = \eta_0 - \eta$. Substituting (7.43) into (7.40), in principle leads to six terms: the power spectrum of the SW term, C_l^{SW}, that of the Doppler term, C_l^{D}, that of the ISW term, C_l^{ISW}, and three cross spectra. In practice, the cross spectra are small, although the cross correlation between the SW and ISW terms plays a non-negligible role in the height of the first peak.

7.3.2 Large Scales: Sachs–Wolfe Effect

The low multipoles of the CMB spectrum ($l < 100$) are created by superhorizon fluctuations at recombination. From Fig. 7.7, we see that this regime is dominated by the Sachs–Wolfe term. Let us evaluate it for adiabatic initial conditions.

Since decoupling occurs during the matter era, the superhorizon limit of the fluctuations implies $-2\Psi \approx -2\Phi \approx \delta = \delta_m = \frac{3}{4}\delta_\gamma$ and the observed CMB temperature fluctuations on large scales therefore are

$$
\Theta(\hat{\mathbf{n}}) \approx \left(\frac{1}{4}\delta_\gamma + \Psi\right)_* = \frac{1}{3}\Psi_* = \frac{1}{5}\mathcal{R}_i\,, \tag{7.44}
$$

where the final equality follows from (6.155). Two points are worth noting: First, since there has been no evolution on large scales, this limit of the CMB spectrum directly probes the initial conditions. Second, the gravitational redshift has won over the intrinsic temperature fluctuations. This means that an overdensity at last-scattering ($\delta_{\gamma,*} > 0$), corresponding to a potential well ($\Psi_* < 0$), leads to a cold spot in the CMB map ($\Theta < 0$). Conversely, a hot spot corresponds to an underdensity at last-scattering.

The transfer function (7.43) for the large-scale Sachs–Wolfe term then simply is

$$
\Theta_l^{\mathrm{SW}}(k) = \frac{1}{5} j_l(k\chi_*)\,, \tag{7.45}
$$

and the power spectrum (7.40) becomes

$$
C_l^{\mathrm{SW}} = \frac{4\pi}{25} \int_0^\infty d\ln k\, \Delta_{\mathcal{R}}^2(k)\, j_l^2(k\chi_*)\,. \tag{7.46}
$$

For a primordial spectrum of the power law form, $\Delta_{\mathcal{R}}^2(k) = A_{\mathrm{s}}(k/k_0)^{n_{\mathrm{s}}-1}$, the integral can be evaluated analytically and we find

$$
C_l^{\mathrm{SW}} = \frac{4\pi}{25} A_{\mathrm{s}}(k_0\chi_*)^{1-n_{\mathrm{s}}}\, 2^{n_{\mathrm{s}}-4}\, \pi\, \frac{\Gamma(3-n_{\mathrm{s}})}{\Gamma^2(4-\frac{n_{\mathrm{s}}}{2})}\, \frac{\Gamma(l + \frac{n_{\mathrm{s}}-1}{2})}{\Gamma(l + 2 - \frac{n_{\mathrm{s}}-1}{2})}\,. \tag{7.47}
$$

For $n_{\mathrm{s}} = 1$, the first ratio of gamma functions becomes $\Gamma(2)/\Gamma^2(3/2) = 4/\pi$, while the second ratio is $[l(l+1)]^{-1}$. Moreover, the scale dependence from $(k_0\chi_*)^{1-n_{\mathrm{s}}}$

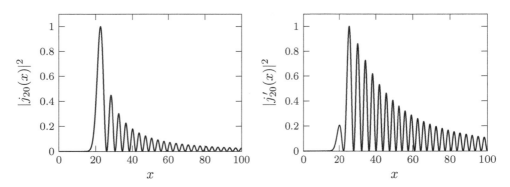

Fig. 7.9   Illustration of the spherical Bessel function $j_l(x)$ and its derivative for $l = 20$.

disappears, as expected for a spectrum with no intrinsic scale, and the power spectrum becomes

$$\frac{l(l+1)}{2\pi} C_l^{\text{SW}} = \frac{A_{\text{s}}}{25}. \tag{7.48}$$

We see that a scale-invariant primordial spectrum, $k^3 \mathcal{P}_{\mathcal{R}}(k) = \text{const}$, corresponds to the combination $l(l+1)C_l$ being a constant, independent of the multipole moment l. The amplitude of the large-scale CMB spectrum is a direct measure of the amplitude A_{s} of the primordial fluctuations.

7.3.3 Small Scales: Sound Waves

The larger multipoles in the CMB spectrum ($l > 100$) are sourced by small-scale fluctuations whose nontrivial evolution on sub-Hubble scales needs to be accounted for. We will discuss this in detail in the next section. Here, we just briefly comment on the projection effect for these scales.

For large values of l, the spherical Bessel function $j_l(x)$ is peaked near $x \approx l$ (cf. Fig. 7.9) and therefore acts like a delta function in the integral (7.40). The derivative of the Bessel function $j_l'(x)$ is not as sharply peaked at $x \approx l$, so the projection from wavenumber k to multipole l is less sharp for the Doppler term.[5] We will indicate this by writing $x \sim l$, rather than $x \approx l$. The Sachs–Wolfe and Doppler contributions to the power spectrum can therefore be written as

$$\frac{l(l+1)}{2\pi} C_l^{\text{SW}} \sim \left. F_*^2(k) \, \Delta_{\mathcal{R}}^2(k) \right|_{k \approx l/\chi_*}, \tag{7.49}$$

$$\frac{l(l+1)}{2\pi} C_l^{\text{D}} \sim \left. G_*^2(k) \, \Delta_{\mathcal{R}}^2(k) \right|_{k \sim l/\chi_*}, \tag{7.50}$$

---

[5] In fact, the contribution from $x = l$ actually vanishes in the Doppler integral; see (7.53). The reason for this can be understood geometrically. For a plane wave traveling in the z-direction (see Fig. 7.8), the condition $x = l$ corresponds to lines-of-sight in the xy-plane. The Doppler effect for such lines-of-sight is zero because they are perpendicular to the velocity vectors at the last-scattering surface, $\mathbf{v}_b = i\hat{\mathbf{z}} v_b$, so that $\mathbf{v}_b \cdot \hat{\mathbf{n}} = 0$. Moving away from the xy-plane we start picking up nonzero Doppler effects, but at $l < x$. Thanks to Eiichiro Komatsu for pointing this out to me.

where $F_*(k)$ and $G_*(k)$ are the transfer functions defined in (7.31). This is derived more rigorously in Exercise 7.3. For scale-invariant initial conditions, $\Delta_{\mathcal{R}}^2(k) = $ const, the power spectra C_l^{SW} and C_l^{D} are therefore determined simply by the squares of the transfer functions evaluated at $k = l/\chi_*$. Oscillations in k become oscillations in l.

Exercise 7.3 For large l, the spherical Bessel functions can be approximated as

$$j_l(x) \to \begin{cases} \cos b \cos\left[\nu(\tan b - b) - \pi/4\right]/\nu\sqrt{\sin b} & x > \nu, \\ 0 & x < \nu, \end{cases} \qquad (7.51)$$

where $\nu \equiv l + 1/2$ and $\cos b \equiv \nu/x$, with $0 \le b \le \pi/2$. Use this to show that the Sachs–Wolfe and Doppler contributions to the power spectrum can be written as

$$\frac{l(l+1)}{2\pi} C_l^{\mathrm{SW}} = \int_1^\infty \frac{\mathrm{d}\beta}{\beta^2\sqrt{\beta^2-1}} F_*^2\left(\frac{l\beta}{\chi_*}\right) \Delta_{\mathcal{R}}^2\left(\frac{l\beta}{\chi_*}\right), \qquad (7.52)$$

$$\frac{l(l+1)}{2\pi} C_l^{\mathrm{D}} = \int_1^\infty \frac{\mathrm{d}\beta}{\beta^4}\sqrt{\beta^2-1}\, G_*^2\left(\frac{l\beta}{\chi_*}\right) \Delta_{\mathcal{R}}^2\left(\frac{l\beta}{\chi_*}\right), \qquad (7.53)$$

where we see explicitly that the integrand of the Doppler contribution vanishes for $\beta = 1$. Given the transfer functions F_* and G_*, these integrals are easy to evaluate numerically. Most of the contribution comes from $\beta \sim 1$, so the integrals can be performed with a finite cutoff, say $\beta = 5$, without losing much accuracy [9].

7.4 Primordial Sound Waves

To complete the derivation of the CMB power spectrum, we have to compute the transfer functions describing the evolution before recombination:

$$\mathcal{R}(0, \mathbf{k}) \xrightarrow{\text{transfer function}} \begin{pmatrix} \delta_\gamma \\ \Psi \\ v_b \end{pmatrix}_{\eta_*}$$

A numerical solution of the transfer function of the photon density contrast δ_γ is shown in Fig. 7.10. Before we study the rather complex physics of the transfer function in detail, it will be helpful to first discuss the fundamental scales of the problem. This will tell us when particular effects become important.

- **Sound horizon** Sound waves can only propagate if their wavelengths are smaller than the sound horizon

$$r_s(\eta) \equiv \int_0^\eta \mathrm{d}\tilde\eta\, c_s(\tilde\eta). \qquad (7.54)$$

In Problem 7.2, you will show that the comoving sound horizon at decoupling is $r_{s,*} \approx 145\,\mathrm{Mpc}$. The effects of sound waves will only be important for modes

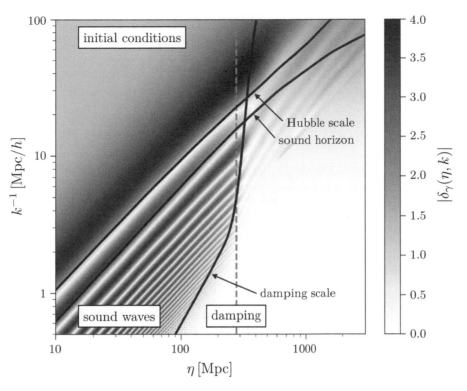

Fig. 7.10 Transfer function of the photon density contrast as a function of k and η, computed with CLASS [10]. Illustrated are also the evolution of the Hubble scale, the sound horizon and the damping scale, and the moment of recombination (dashed line). We see that sound waves propagate below the sound horizon and are suppressed below the damping scale.

with $kr_{s,*} > 1$. Since $l \sim k\chi_*$, with $\chi_* \approx 14\,\mathrm{Gpc}$, this implies that sound waves should affect the CMB spectrum for $l > 100$. This is indeed what is seen in the left panel of Fig. 7.7.

- **Damping scale** On small scales, photon diffusion leads to a damping of the waves. Essentially, high-energy photons in hot regions can diffuse into cold regions, thereby erasing the temperature difference. This diffusion can be modeled as a random walk, with the step size being the photons' mean free path $\ell_\gamma = 1/(an_e\sigma_T)$. In a time interval $\Delta\eta$, the number of scatterings is $N = \Delta\eta/\ell_\gamma$ and the mean-squared distance explored by the random walk is $\langle \Delta r^2 \rangle = N\ell_\gamma^2 = \ell_\gamma\Delta\eta$. Integrating this from $\eta = 0$ to a time η, we get the squared diffusion length

$$ r_D^2(\eta) \sim \int_0^\eta \mathrm{d}\tilde\eta\, \ell_\gamma(\tilde\eta)\,. \tag{7.55} $$

We see that the diffusion length is roughly the geometric mean between the age of the universe in conformal time and the mean free path of the photons. In Problem 7.3, you will show that the diffusion length at decoupling is $r_{D,*} \approx 7\,\mathrm{Mpc}$. Moreover, we will see in Section 7.4.4 that on scales smaller than the diffusion length, $kr_{D,*} > 1$, the fluctuations are suppressed by a factor of $e^{-2(kr_{D,*})^2}$. In harmonic space, the damping scale becomes $l_D \sim \chi_*/\sqrt{2}r_{D,*} \sim 1400$, so we expect the CMB fluctuations to be exponentially damped for $l > 1400$. In Section 7.4.4, we will improve this estimate by including an additional damping due to the finite surface of last-scattering.

7.4.1 Photon–Baryon Dynamics

We will now study the evolution of the photon–baryon fluid in more detail. We already gave a rough description of this in Section 6.4. Here, we will re-derive the equations describing the dynamics of the photon–baryon fluid, but present the derivation in a way that will allow us to extend it beyond the tight-coupling limit (see Section 7.4.4).

Hydrodynamic equations

At linear order, Thomson scattering does not exchange energy, so the continuity equations for the photons and baryons still apply in their original forms

$$\delta_\gamma' = \frac{4}{3}kv_\gamma + 4\Phi', \tag{7.56}$$

$$\delta_b' = kv_b + 3\Phi'. \tag{7.57}$$

Momentum, on the other hand, is exchanged by the scattering, so the Euler equations are modified. In particular, the Euler equation for the photons becomes

$$v_\gamma' + \frac{1}{4}k\delta_\gamma - \frac{2}{3}k\Pi_\gamma + k\Psi = -\Gamma(v_\gamma - v_b), \tag{7.58}$$

where the source term on the right-hand side—sometimes called the **drag term**—describes the momentum transfer between the photons and baryons. A formal derivation of this equation requires the Boltzmann equation and is given in Appendix B. The answer, however, is rather intuitive: the momentum transfer is proportional to the scattering rate, Γ, and to the difference between the photon and baryon velocities, $v_{\gamma-b} \equiv v_\gamma - v_b$ (also referred to as the **slip velocity**). As expected, the scattering tries to make the photons move together with the baryons, so that $v_\gamma \to v_b$ for large Γ.

The Euler equation for the baryons is

$$v_b' + \mathcal{H}v_b + k\Psi = \frac{\Gamma}{R}(v_\gamma - v_b), \tag{7.59}$$

where the form of the right-hand side is fixed by the requirement that the combined momentum density of the photons and baryons must be conserved. The factor of $1/R$ in the drag term in (7.59) simply arises because the momentum density of the baryons is smaller than that of the photons by a factor of $R \equiv \frac{3}{4}\bar\rho_b/\bar\rho_\gamma$.

Tight-coupling limit

We first study these equations in the limit of large scattering rate. Let us begin by rearranging (7.59) as

$$v_{\gamma-b} = \frac{R}{\Gamma} \left[v_b' + \mathcal{H} v_b + k\Psi \right] . \tag{7.60}$$

To lowest order in Γ^{-1}, the right-hand side is suppressed and the slip velocity is small, $v_{\gamma-b} \approx 0$, or $v_b \approx v_\gamma$. However, since $v_{\gamma-b}$ is multiplied by Γ in the Euler equations, we need the next-to-leading order solution for $v_{\gamma-b}$. We obtain this by substituting $v_b \approx v_\gamma$ into the right-hand side of (7.60):

$$v_{\gamma-b} = \frac{R}{\Gamma} \left[v_\gamma' + \mathcal{H} v_\gamma + k\Psi \right] . \tag{7.61}$$

Using this in (7.58), we get

$$v_\gamma' + \frac{R\mathcal{H}}{1+R} v_\gamma + \frac{1}{4(1+R)} k\delta_\gamma + k\Psi = 0 , \tag{7.62}$$

where we have dropped the photon anisotropic stress Π_γ because it is suppressed for large Γ (see below). Combining this with the continuity equation (7.56), we find a second-order equation for the photon density contrast

$$\boxed{\delta_\gamma'' + \frac{R'}{1+R} \delta_\gamma' + k^2 c_s^2 \delta_\gamma = -\frac{4}{3} k^2 \Psi + 4\Phi'' + \frac{4R'}{1+R} \Phi'} , \tag{7.63}$$

where we have used that $R\mathcal{H} = R'$ (since $R \propto a$) and introduced the sound speed $c_s^2 = [3(1+R)]^{-1}$. This is the same as the result (6.197) obtained in Section 6.4.

7.4.2 High-Frequency Solution

In the rest of this section, we will study the solutions to the master equation (7.63). In full generality, the equation does not have an analytic solution, but we can make progress by studying special limits.

We first consider the high-frequency (or short-wavelength) limit, $k \gg \mathcal{H}$. Following Weinberg [9], we split the solution into "fast" and "slow" modes:

$$f = f^{\text{fast}} + f^{\text{slow}} , \tag{7.64}$$

where $f \equiv \{\delta_\gamma, \Psi, \Phi\}$. The fast modes evolve on a scale k^{-1}, so that $\partial_\eta f^{\text{fast}} \sim k f^{\text{fast}}$, while the slow modes vary on the Hubble timescale $\mathcal{H}^{-1}$, so that $\partial_\eta f^{\text{slow}} \sim \mathcal{H} f^{\text{slow}}$. Dark matter fluctuations do not have any fast modes (because the dark matter is pressureless). The fast mode of the gravitational potential is therefore sourced only by the radiation fluctuations. From the Poisson equation, we have

$$- k^2 \Phi^{\text{fast}} = 4\pi G a^2 \, \bar\rho_\gamma \delta_\gamma^{\text{fast}} \lesssim \mathcal{H}^2 \delta_\gamma^{\text{fast}} , \tag{7.65}$$

so that the gravitational potential is suppressed by a factor of $\mathcal{H}^2/k^2 \ll 1$ in the high-frequency limit. The fast mode of the photon density contrast therefore obeys the *homogeneous* equation

$$\delta_\gamma'' + \frac{R'}{1+R}\delta_\gamma' + k^2 c_s^2 \delta_\gamma = 0 \quad \text{(fast mode)}. \qquad (7.66)$$

This equation can be solved in a **WKB approximation** (see Problem 7.4). Consider the ansatz

$$\delta_\gamma = A(\eta) \exp\left(\pm ik \int_0^\eta c_s(\tilde{\eta})\, \mathrm{d}\tilde{\eta}\right) \equiv A(\eta) e^{\pm i\varphi(\eta)}, \qquad (7.67)$$

where the function $A(\eta)$ is slowly varying on the timescale of the oscillations. Taking time derivatives of (7.67), we get

$$\delta_\gamma' = A' e^{\pm i\varphi} \pm ikc_s A e^{\pm i\varphi} \approx \pm ikc_s \delta_\gamma, \qquad (7.68)$$

$$\begin{aligned}
\delta_\gamma'' &= A'' e^{\pm i\varphi} \pm 2ikc_s A' e^{\pm i\varphi} \pm ikc_s' A e^{i\varphi} - k^2 c_s^2 A e^{i\varphi} \\
&\approx \left(\pm 2ikc_s \frac{A'}{A} \mp \frac{ikc_s}{2}\frac{R'}{1+R} - k^2 c_s^2\right)\delta_\gamma,
\end{aligned} \qquad (7.69)$$

where we used that $A' \ll ikc_s A$, $A'' \ll ikc_s A'$ and

$$c_s' = \frac{\mathrm{d}}{\mathrm{d}\eta}\left(\frac{1}{\sqrt{3(1+R)}}\right) = -\frac{1}{2}c_s\frac{R'}{1+R}. \qquad (7.70)$$

Substituting (7.68) and (7.69) into (7.66) then gives

$$\frac{A'}{A} = -\frac{1}{4}\frac{R'}{1+R} \quad \Rightarrow \quad A(\eta) \propto (1+R)^{-1/4}, \qquad (7.71)$$

and the solution of (7.66) takes the form

$$\delta_\gamma^{\text{fast}} = (1+R)^{-1/4}\left[C\cos\varphi + D\sin\varphi\right]. \qquad (7.72)$$

The integration constants C and D are fixed by the superhorizon initial conditions. For adiabatic initial conditions, we expect $C \gg D$ and the oscillating part of the solution is a pure cosine.

To complete the solution, we have to add the slow mode of the photon density contrast. In this case, we can drop all time derivative terms in (7.63) and the solution becomes

$$\delta_\gamma^{\text{slow}} = -4(1+R)\Psi^{\text{slow}}, \qquad (7.73)$$

where Ψ^{slow} is now also sourced by the dark matter perturbations. The full WKB solution then is

$$\boxed{\delta_\gamma = (1+R)^{-1/4}\left[C\cos\varphi + D\sin\varphi\right] - 4(1+R)\Psi}. \qquad (7.74)$$

We see that the gravitational force term has shifted the zero-point of the oscillations. Since the gravitational potential decays on sub-Hubble scales during the radiation era, the size of the zero-point shift will depend on the time that the mode has spent

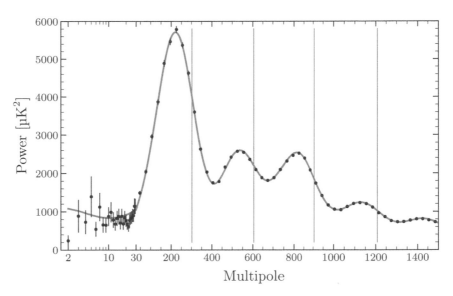

Fig. 7.11 Comparison of the prediction (7.76) for the peak locations (vertical lines) with the observed CMB spectrum. The mismatch shows that the CMB peak locations are *not* captured well by the solution in the high-frequency limit.

inside the horizon during the radiation era (which is dictated by the wavelength of the fluctuation).

The transfer function for the Sachs–Wolfe term then is

$$F_*(k) \equiv \frac{\frac{1}{4}\delta_\gamma(\eta_*,\mathbf{k}) + \Psi(\eta_*,\mathbf{k})}{\mathcal{R}_i(\mathbf{k})}$$
$$= \frac{1}{4}(1+R)^{-1/4}\frac{C(\mathbf{k})\cos\varphi_* + D(\mathbf{k})\sin\varphi_*}{\mathcal{R}_i(\mathbf{k})} - R\frac{\Psi(\eta_*,\mathbf{k})}{\mathcal{R}_i(\mathbf{k})}, \qquad (7.75)$$

with $\varphi_* = k r_{s,*}$. In Section 7.3.3, we have seen that the angular power spectrum is given roughly by $|F_*(k)|^2$, with $k \approx l/\chi_*$. Since $C \gg D$, we therefore expect the peaks of the spectrum to be at

$$l_n = n\pi\frac{\chi_*}{r_{s,*}} \approx n\pi\frac{14\,\mathrm{Gpc}}{145\,\mathrm{Mpc}} \approx n \times 302, \qquad (7.76)$$

where $n \in \mathbb{Z}$. However, this prediction does *not* match the measured locations of the first few peaks, which are $l_1 = 220.0 \pm 0.5$, $l_2 = 537.5 \pm 0.7$ and $l_3 = 810.8 \pm 0.7$ [11]. A graphical illustration of this mismatch is given in Fig. 7.11. While the prediction (7.76) works reasonably well for the peak spacing, the predicted peak locations do not coincide with the true locations, and the typical shift δl_n is about one fourth of the oscillation period. To determine the precise peak locations, we need a more accurate solution and a more careful treatment of the spatial-to-angular projection.

7.4.3 Semi-Analytic Solution*

Unfortunately, there is no analytic solution for the dynamics of the gravitationally-driven photon–baryon fluid that is valid at all times and for all wavelengths. However, Weinberg found an interesting semi-analytic solution that uses an interpolation between two analytic solutions for $k \ll k_{\rm eq}$ and $k \gg k_{\rm eq}$. The following is a summary of the very detailed analysis in Weinberg's book [9]. Even this summary is still somewhat technical and could be skipped on a first reading. An alternative semi-analytic solution due to Hu and Sugiyama [12] is discussed in Problem 7.5.

- Modes with $k < k_{\rm eq}$ are outside the horizon during the radiation era and only enter the horizon during the matter era (if at all). The superhorizon evolution of the fluctuations is known analytically. Moreover, once the universe becomes matter dominated, an analytic solution is possible on all scales. By matching this solution to the superhorizon solution, we obtain an analytic solution for $k < k_{\rm eq}$ with the correct initial conditions (see Fig. 7.12). For realistic cosmological parameters, the time of decoupling, η_*, is relatively close to the time of equality, $\eta_{\rm eq}$, so the asymptotic limit described here is not really reached by the time of last-scattering. Assuming $\eta_* \gg \eta_{\rm eq}$ is nevertheless still useful because it motivates an ansatz for the interpolation function. This function will then apply at η_*.

- Modes with $k > k_{\rm eq}$ instead enter the horizon during the radiation era. As long as the universe is dominated by radiation, an analytic solution exists for all wavenumbers. Moreover, once the mode is deep inside the horizon, an analytic solution is possible that remains valid even when matter becomes important. By matching the two solutions in the region of overlap—i.e. for subhorizon modes in the radiation era—we obtain an analytic solution for $k > k_{\rm eq}$ with the correct initial conditions (see Fig. 7.12).

In the following, I will first derive these two analytic solutions and then describe Weinberg's interpolation.

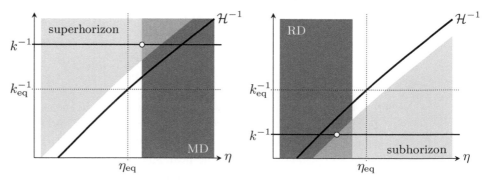

Fig. 7.12   Illustration of the matching of analytic solutions for modes with $k < k_{\rm eq}$ (*left*) and $k > k_{\rm eq}$ (*right*).

Long-wavelength solution

We first consider modes with $k < k_{\text{eq}}$, which enter the horizon only after matter–radiation equality. The superhorizon evolution of adiabatic fluctuations was derived in Section 6.2 (see also Problem 6.1). There we found that the superhorizon limit of the photon density contrast in the matter era is

$$\delta_\gamma = \frac{4}{3}\delta_c = -\frac{8}{3}\Phi \quad (k < \mathcal{H} < k_{\text{eq}}), \tag{7.77}$$

where we used that $\delta_c = -2\Phi$. In terms of the curvature perturbation $\mathcal{R}_i(\mathbf{k})$, we have

$$\Psi \approx \Phi = \frac{3}{5}\mathcal{R}_i, \quad \delta_\gamma = -\frac{8}{5}\mathcal{R}_i \qquad (k < \mathcal{H} < k_{\text{eq}}). \tag{7.78}$$

We will use this as an initial condition for the modes entering the horizon during the matter era.

Initially, the effect of the baryons is still negligible, $R \ll 1$, and the oscillator equation (7.63) becomes

$$\delta_\gamma'' + \frac{1}{3}k^2\delta_\gamma = -\frac{4}{3}k^2\Psi, \tag{7.79}$$

where we used that Φ is a constant during the matter era. Note that this equation is valid on all scales. Its solution is

$$\delta_\gamma = C\cos\varphi + D\sin\varphi - 4\Psi, \tag{7.80}$$

where $\varphi = k\eta/\sqrt{3}$. Matching the $\varphi \ll 1$ limit of (7.80) to the superhorizon limit (7.78), we get

$$\delta_\gamma = \frac{4}{5}\mathcal{R}_i\big[\cos\varphi - 3\big]. \tag{7.81}$$

Eventually, the baryon-to-photon ratio R becomes large and cannot be ignored anymore. For modes with $k > \mathcal{H}$, this happens when they are already deep inside the horizon. The WKB solution (7.74) then applies and we have

$$\boxed{\delta_\gamma = \frac{4}{5}\mathcal{R}_i\left[(1+R)^{-1/4}\cos\varphi - 3(1+R)\right]} \quad (k < k_{\text{eq}}), \tag{7.82}$$

where $\varphi = k\int \mathrm{d}\eta/\sqrt{3(1+R)}$. This solution is smoothly connected to the solution in (7.81).

Short-wavelength solution

Next, we consider modes with $k > k_{\text{eq}}$. These mode entered the Hubble radius during the radiation era. We therefore first derive an analytic solution in a radiation-dominated universe (valid on all scales) and then match the high-frequency limit of this solution to the WKB solution (7.74) which remains valid even after matter–radiation equality.

During the radiation era, we have $R \ll 1$ and the effects of the baryons can be ignored. This simplifies the evolution equation (7.63) to

$$\delta_\gamma'' + \frac{1}{3}k^2\delta_\gamma = -\frac{4}{3}k^2\Psi + 4\Phi'', \tag{7.83}$$

where the time dependence of the metric potentials must now be accounted for. Defining $d_\gamma \equiv \delta_\gamma - 4\Phi$ and $\varphi \equiv k\eta/\sqrt{3}$, we get

$$d_\gamma'' + d_\gamma = -4(\Phi + \Psi), \tag{7.84}$$

where the prime is now a derivative with respect to φ. The solution to this equation is

$$d_\gamma = C\cos\varphi + D\sin\varphi - 4\int_0^\varphi d\tilde\varphi \sin(\varphi - \tilde\varphi)\,(\Phi + \Psi)(\tilde\varphi), \tag{7.85}$$

where $C = -4\mathcal{R}_i$ and $D = 0$ for adiabatic initial conditions.

Exercise 7.4 Let $S_1 = \cos\varphi$ and $S_2 = \sin\varphi$ be the two homogeneous solutions of (7.84). Using

$$G(\varphi, \tilde\varphi) = \frac{S_1(\tilde\varphi)S_2(\varphi) - S_1(\varphi)S_2(\tilde\varphi)}{S_1(\varphi)S_2'(\varphi) - S_1'(\varphi)S_2(\varphi)}, \tag{7.86}$$

confirm the form of the Green's function in (7.85).

Using the trigonometric identity $\sin(\varphi - \tilde\varphi) = \sin\varphi\cos\tilde\varphi - \cos\varphi\sin\tilde\varphi$, the result in (7.85) can be written as

$$d_\gamma = (C + \Delta C)\cos\varphi + \Delta D\sin\varphi, \tag{7.87}$$

where

$$\Delta C = 4\int_0^\varphi d\tilde\varphi\,\sin\tilde\varphi\,(\Phi + \Psi)(\tilde\varphi), \tag{7.88}$$

$$\Delta D = -4\int_0^\varphi d\tilde\varphi\,\cos\tilde\varphi\,(\Phi + \Psi)(\tilde\varphi). \tag{7.89}$$

To evaluate this, we need the solutions for the metric potentials $\Phi(\eta)$ and $\Psi(\eta)$. These were found in Section 6.3.1. Ignoring the anisotropic stress due to neutrinos, we obtained

$$\Phi = \Psi = 2\mathcal{R}_i\,\frac{\sin\varphi - \varphi\cos\varphi}{\varphi^3}. \tag{7.90}$$

Substituting this into (7.88) and (7.89), we get

$$\Delta C = 8\mathcal{R}_i \left(1 - \frac{\sin^2 \varphi}{\varphi^2} \right), \tag{7.91}$$

$$\Delta D = 8\mathcal{R}_i \frac{\cos \varphi \sin \varphi - \varphi}{\varphi^2}, \tag{7.92}$$

and equation (7.87) becomes

$$d_\gamma = 4\mathcal{R}_i \left(\cos \varphi - \frac{2 \sin \varphi}{\varphi} \right). \tag{7.93}$$

This is a remarkably simple solution for the fluctuations in the radiation era, valid on all scales. We see that the solution is only a pure cosine in the high-frequency limit, $\varphi \gg 1$. For intermediate wavelengths, however, the gravitational evolution has led to an admixture from the sine solution.

To extend this solution into the matter era, we match the $\varphi \gg 1$ limit of (7.93), $d_\gamma \to \delta_\gamma \approx 4\mathcal{R}_i \cos \varphi$, to the WKB solution (7.74). The resulting solution is

$$\delta_\gamma = 4\mathcal{R}_i (1 + R)^{-1/4} \cos \varphi - 4(1 + R)\Psi. \tag{7.94}$$

To complete the analysis, we still need to relate the potential Ψ to the initial perturbation $\mathcal{R}_i$. In the matter era, we have

$$\Psi \approx -\frac{3}{2} \frac{\mathcal{H}^2}{k^2} \delta_c, \tag{7.95}$$

so Ψ can be obtained from the solution for the dark matter density contrast δ_c. In the box below, I show that

$$\delta_c \approx -1.54 \, \mathcal{R}_i \ln \left(0.15 \frac{k}{k_{\mathrm{eq}}} \right) (k_{\mathrm{eq}}\eta)^2 \qquad (k > k_{\mathrm{eq}}), \tag{7.96}$$

and (7.95) therefore becomes

$$\Psi \approx \frac{3}{5} \mathcal{R}_i \frac{\ln(0.15\, k/k_{\mathrm{eq}})}{(0.31\, k/k_{\mathrm{eq}})^2}. \tag{7.97}$$

Feeding this into (7.94), we obtain the solution for the photon density contrast in the matter era:

$$\boxed{\delta_\gamma = \frac{4}{5}\mathcal{R}_i \left[5(1 + R)^{-1/4} \cos \varphi - 3(1 + R) \frac{\ln(0.15\, k/k_{\mathrm{eq}})}{(0.31\, k/k_{\mathrm{eq}})^2} \right]} \qquad (k > k_{\mathrm{eq}}). \tag{7.98}$$

Compared to the long-wavelength solution (7.82), the amplitude of the oscillations is enhanced by a factor of 5 and the shift of the zero-point is modulated by a nontrivial transfer function.

Derivation   In the following, we derive the dark matter transfer function for modes with $k > k_{\rm eq}$. The continuity and Euler equations for pressureless dark matter are

$$\delta_c' = kv_c + 3\Phi' , \tag{7.99}$$

$$v_c' + \mathcal{H}v_c = -k\Psi . \tag{7.100}$$

Substituting $kv_c = (\delta_c - 3\Phi)'$ into (7.100), we find

$$(ad_c')' = -3a\Psi , \tag{7.101}$$

where we have defined $d_c \equiv \delta_c - 3\Phi$ and primes now denote derivatives with respect to $\varphi \equiv k\eta/\sqrt{3}$. Integrating (7.101), we get

$$d_c(\varphi) = -3 \int_0^\varphi \frac{\mathrm{d}\tilde{\varphi}}{a} \int_0^{\tilde{\varphi}} a\Psi(\bar{\varphi}) \, \mathrm{d}\bar{\varphi} . \tag{7.102}$$

Note that this result is exact and valid for all k. During the radiation era, the gravitational potential is determined by the radiation and Ψ can be treated as an external source as far as the evolution of the dark matter is concerned. In this regime, we can use the solution (7.90) and $a \propto \eta \propto \varphi$ to evaluate the integral in (7.102):

$$\begin{aligned}
d_c(\varphi) &= -3 \int_0^\varphi \frac{\mathrm{d}\tilde{\varphi}}{\tilde{\varphi}} \int_0^{\tilde{\varphi}} \bar{\varphi} \, \Psi(\bar{\varphi}) \, \mathrm{d}\bar{\varphi} \\
&= -6\mathcal{R}_i \int_0^\varphi \frac{\mathrm{d}\tilde{\varphi}}{\tilde{\varphi}} \int_0^{\tilde{\varphi}} \frac{\sin\bar{\varphi} - \bar{\varphi}\cos\bar{\varphi}}{\bar{\varphi}^2} \, \mathrm{d}\bar{\varphi} \\
&= -6\mathcal{R}_i \int_0^\varphi \frac{\mathrm{d}\tilde{\varphi}}{\tilde{\varphi}} \left(1 - \frac{\sin\tilde{\varphi}}{\tilde{\varphi}} \right) + d_c(0) ,
\end{aligned} \tag{7.103}$$

where the integration constant is fixed by the superhorizon limit, $d_c(0) = -3\mathcal{R}_i$. Writing $\tilde{\varphi}^{-2}\sin\tilde{\varphi}$ as $-\mathrm{d}\tilde{\varphi}^{-1}/\mathrm{d}\tilde{\varphi}\sin\tilde{\varphi}$ and integrating by parts, we find

$$d_c(\varphi) = 6\mathcal{R}_i \left(\mathrm{Ci}(\varphi) - \frac{\sin\varphi}{\varphi} - \ln\varphi \right)\Big|_0^\varphi - 3\mathcal{R}_i , \tag{7.104}$$

where $\mathrm{Ci}(\varphi)$ is the "cosine integral"

$$\mathrm{Ci}(\varphi) \equiv \int_\varphi^\infty \frac{\cos\tilde{\varphi}}{\tilde{\varphi}} \mathrm{d}\tilde{\varphi} . \tag{7.105}$$

The superhorizon limit of the cosine integral is

$$\lim_{\varphi \to 0} \mathrm{Ci}(\varphi) = \gamma_E + \ln\varphi + O(\varphi^2) , \tag{7.106}$$

where $\gamma_E = 0.5772\ldots$ is the Euler–Mascheroni constant, while the high-frequency limit, $\varphi \to \infty$, vanishes by definition. The high-frequency limit of the solution $d_c(\varphi)$ therefore is

$$d_c(\varphi) \xrightarrow{\varphi \gg 1} \delta_c(\varphi) = 6\mathcal{R}_i \left(\frac{1}{2} - \gamma_E - \ln\varphi + O(\varphi^{-1}) \right) , \tag{7.107}$$

which describes the evolution after the mode entered the horizon, but before matter–radiation equality.

We would like to compare this solution to the solution (6.181) for matter fluctuations on sub-Hubble scales (but *not* restricted to the radiation era):

$$\delta_c = C_1 \left(1 + \frac{3}{2}y\right) + C_2 \left[\left(1 + \frac{3}{2}y\right)\ln\left(\frac{\sqrt{1+y}+1}{\sqrt{1+y}-1}\right) - 3\sqrt{1+y}\right], \quad (7.108)$$

where $y \equiv a/a_{\mathrm{eq}}$. In the limit $y \ll 1$, this solution gives

$$\delta_c \xrightarrow{y \ll 1} (C_1 - 3C_2) - C_2 \ln(y/4) + O(y)$$
$$= \left[C_1 - 3C_2 + C_2 \ln\left(\frac{2}{\sqrt{3}(\sqrt{2}-1)}\frac{k}{k_{\mathrm{eq}}}\right)\right] - C_2 \ln\varphi, \quad (7.109)$$

where we have used (2.162) to relate $y \propto \eta$ to φ and introduced $k_{\mathrm{eq}} = \eta_{\mathrm{eq}}^{-1}$. This is consistent with (7.107) if

$$C_1 = 6\mathcal{R}_i \left[\frac{7}{2} - \gamma_E - \ln\left(\frac{2}{\sqrt{3}(\sqrt{2}-1)}\frac{k}{k_{\mathrm{eq}}}\right)\right] \approx -6\mathcal{R}_i \ln\left(0.15\frac{k}{k_{\mathrm{eq}}}\right), \quad (7.110)$$

$$C_2 = 6\mathcal{R}_i. \quad (7.111)$$

During the matter-dominated era ($y \gg 1$), the second term in (7.108) is a decaying mode and can be dropped. We then find

$$\delta_c \xrightarrow{y \gg 1} \frac{3}{2}C_1 y = -9\mathcal{R}_i \ln\left(0.15\frac{k}{k_{\mathrm{eq}}}\right)(k_{\mathrm{eq}}\eta)^2(\sqrt{2}-1)^2,$$

$$\approx \boxed{-1.54\,\mathcal{R}_i \ln\left(0.15\frac{k}{k_{\mathrm{eq}}}\right)(k_{\mathrm{eq}}\eta)^2}, \quad (7.112)$$

where we used (2.162) for $y \propto \eta^2$ in the matter era. This solution has the $\ln k$ scaling that we found in Section 6.3.2, but now includes the precise numerical factors that are required to relate δ_c to $\mathcal{R}_i$.

Solution at last-scattering

The analytic solutions (7.82) and (7.98) imply the following limits of the Sachs–Wolfe transfer function at last-scattering:

$$F_*(k) \equiv \frac{(\frac{1}{4}\delta_\gamma + \Psi)_*}{\mathcal{R}_i}$$

$$= \begin{cases} \dfrac{1}{5}\left[\dfrac{1}{(1+R_*)^{1/4}}\cos(kr_{s,*}) - 3R_*\right] & k < k_{\mathrm{eq}}, \\[3mm] \dfrac{1}{5}\left[\dfrac{5}{(1+R_*)^{1/4}}\cos(kr_{s,*}) - 3R_*\dfrac{\ln(0.15\,k/k_{\mathrm{eq}})}{(0.31\,k/k_{\mathrm{eq}})^2}\right] & k > k_{\mathrm{eq}}. \end{cases} \quad (7.113)$$

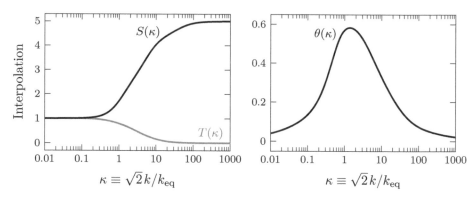

Fig. 7.13 Illustration of the interpolation functions appearing in the solution (7.114).

We would now like to find a solution that interpolates between these two limits and is therefore valid for modes with $k \sim k_{\text{eq}}$. Motivated by the form of the solutions in (7.113), Weinberg proposed the following ansatz

$$F_*(k) = \frac{1}{5}\left[\frac{S(k)}{(1+R_*)^{1/4}}\cos[kr_{s,*}+\theta(k)] - 3R_*\,T(k)\right],\qquad(7.114)$$

where $T(k)$, $S(k)$ and $\theta(k)$ are momentum-dependent interpolation functions, with the limits

$$k \ll k_{\text{eq}}:\qquad S \to 1,\quad T \to 1,\quad \theta \to 0,$$

$$k \gg k_{\text{eq}}:\qquad S \to 5,\quad T \to \frac{\ln(0.15\,k/k_{\text{eq}})}{(0.31\,k/k_{\text{eq}})^2},\quad \theta \to 0.\qquad(7.115)$$

As described by Weinberg, the interpolation between the limits $k \ll k_{\text{eq}}$ and $k \gg k_{\text{eq}}$ can almost be done "by hand," with any smooth interpolation giving reasonable results for the CMB spectrum. For more accurate results, the interpolation functions have to be found numerically by solving the coupled equations of the photon–baryon fluid and the dark matter. The result is summarized by the following fitting functions [9][6]

$$S(\kappa) \equiv \left[\frac{1+(1.209\kappa)^2+(0.5116\kappa)^4+\sqrt{5}(0.1657\kappa)^6}{1+(0.9459\kappa)^2+(0.4249\kappa)^4+(0.167\kappa)^6}\right]^2,$$

$$T(\kappa) \equiv \frac{\ln[1+(0.124\kappa)^2]}{(0.124\kappa)^2}\left[\frac{1+(1.257\kappa)^2+(0.4452\kappa)^4+(0.2197\kappa)^6}{1+(1.606\kappa)^2+(0.8568\kappa)^4+(0.3927\kappa)^6}\right]^{1/2},\qquad(7.116)$$

$$\theta(\kappa) \equiv \left[\frac{(1.1547\kappa)^2+(0.5986\kappa)^4+(0.2578\kappa)^6}{1+(1.723\kappa)^2+(0.8707\kappa)^4+(0.4581\kappa)^6+(0.2204\kappa)^8}\right]^{1/2},$$

---

[6] The formula for $\theta(\kappa)$ corrects a misprint in the first edition of Weinberg's book. Thanks to Eiichiro Komatsu for pointing this out to me.

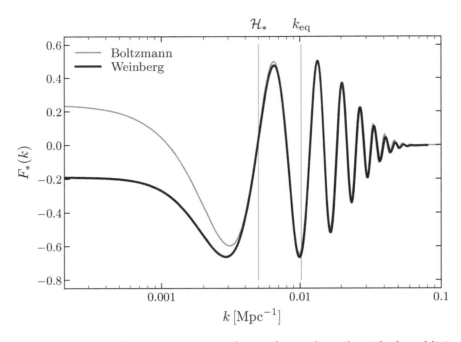

Fig. 7.14 Comparison between Weinberg's semi-analytic solution (7.114), with the additional damping factor discussed in Section 7.4.4, and the exact result from the Boltzmann code CLASS. The disagreement for small k is expected since Weinberg's solution only holds for modes that are inside the horizon at last-scattering, $k \gg \mathcal{H}_*$.

where $\kappa \equiv \sqrt{2}\,k/k_{\mathrm{eq}}$. Plots of these fitting functions are given in Fig. 7.13. Note that it takes quite long for these interpolation functions to reach the high-frequency limit, especially $S(\kappa)$.

The function $\theta(\kappa)$ leads to shifts in the CMB peak positions relative to the estimates of the high-frequency solution given in Section 7.4.1.[7] As Weinberg showed, the agreement with the observed peak positions is now "almost embarrassingly good" [9]. Indeed, the comparison between Weinberg's SW transfer function (7.114) and the exact numerical result in Fig. 7.14 shows perfect agreement of the oscillatory part of the transfer function. The disagreement at small k is expected since Weinberg's solution only applies to subhorizon modes at last-scattering. In this low-frequency limit, we should take $R_* \to 0$ in (7.114) since sound waves are not important on these large scales. This reproduces the correct Sachs–Wolfe limit of Section 7.3.2.

---

[7] Understanding this phase shift analytically is possible, but quite nontrivial. As shown in [11], the phase shift $\theta(\kappa)$ has multiple different origins, such as the breakdown of tight coupling, neutrino free-streaming and a transient evolution of the gravitational potential. Moreover, a significant part of the mismatch in the predicted peak locations (7.76) comes from the fact that the $l \sim k\chi_*$ mapping from Fourier modes to spherical harmonics is not very accurate [11]. This accounts for about half of the shifts seen in Fig. 7.11, and the rest is captured by the nontrivial phase shift $\theta(\kappa)$.

The solution above does not contain the effects of neutrinos. As we have seen in Chapter 6 (see Problem 6.2), the anisotropic stress carried by the neutrinos has a small effect on the superhorizon initial conditions for the metric potentials. Moreover, it can be shown that the free-streaming neutrinos lead to a small phase shift in the high-frequency limit of the solution [13, 14], $\theta(k \gg k_{\text{eq}}) \to 0.063\pi$. These effects have to be taken into account in order to match the precision of current experiments, but aren't essential to understand the main qualitative features of the CMB spectrum.

From the solution for the photon density contrast, we can infer the baryon velocity and hence the transfer function for the Doppler contribution. Using the continuity equation for the photons, which during the matter era reads $\delta'_\gamma = \frac{4}{3}kv_\gamma$, we obtain $G_*(k) = (v_b)_*/\mathcal{R}_i \approx (v_\gamma)_*/\mathcal{R}_i = 3(\delta'_\gamma)_*/(4\mathcal{R}_i k)$. Taking the derivative of the solution for δ_γ with respect to conformal time, we find

$$\boxed{G_*(k) = -\frac{\sqrt{3}}{5}\frac{S(k)}{(1+R_*)^{3/4}}\sin[kr_{s,*} + \theta(k)]}.$$ (7.117)

We see that the oscillations in the Doppler term are exactly out of phase with the oscillations in the photon density.

7.4.4 Small-Scale Damping

We have seen in plots of the CMB spectrum that the power is suppressed for large multipoles ($l > 1000$). This damping is not yet included in our solution. There are two sources of damping: First, photon diffusion between hot and cold regions erases temperature differences on small scales. Second, the surface of last-scattering has a finite thickness. Averaging over this finite width suppresses short-wavelength fluctuations. In the following, we will derive these two effects in detail.

Silk damping

Since the effect of photon diffusion is only important deep inside the Hubble radius, we can ignore the effects of expansion; e.g. $\mathcal{H}v_b \ll v'_b$ in the Euler equation for the baryons (7.59). We can also drop the gravitational forcing terms, since $-k^2\Phi \sim \mathcal{H}^2\delta$, so that $k\Psi$ is small relative to the photon pressure term, $k\delta_\gamma$, and the change in the baryon momentum density, Rv'_b. The Euler equations for the photons and baryons then are

$$v'_\gamma + \frac{1}{4}k\delta_\gamma - \frac{2}{3}k\Pi_\gamma = -\Gamma v_{\gamma-b},$$ (7.118)

$$Rv'_b = +\Gamma v_{\gamma-b},$$ (7.119)

where we have restored the photon anisotropic stress Π_γ. Above we studied these equations at leading order in an expansion in $k/\Gamma = k\ell_\gamma < 1$. Now, we will extend this treatment to include the next-to-leading order correction, which will capture the damping of the fluctuations.

Combining the two Euler equations, we get an equation for the slip velocity

$$v'_{\gamma-b} = - \left(\frac{1+R}{R}\right) \Gamma v_{\gamma-b} - \frac{1}{4}k\delta_\gamma + \frac{2}{3}k\Pi_\gamma \,. \tag{7.120}$$

In the tight-coupling limit, we can drop the photon anisotropic stress and the time variation of the slip velocity is small. We then have

$$v_{\gamma-b} \approx -\frac{R}{1+R}\frac{k}{\Gamma}\frac{\delta_\gamma}{4} \,. \tag{7.121}$$

This equation has an interesting physical interpretation: the slip velocity characterizes the bulk velocity of the photons in the rest frame of the baryons and is therefore associated with an energy flux in that frame. Moreover, since $\rho_\gamma \propto T_\gamma^4$, the right-hand side of (7.121) can be written as $-\ell_\gamma \nabla T_\gamma / \bar{T}_\gamma$, which means that it describes the energy flux coming from a gradient in the photon temperature. A finite slip velocity is therefore associated with **thermal conduction**.

Adding the two Euler equations, we find

$$Rv'_b + v'_\gamma + \frac{1}{4}k\delta_\gamma - \frac{2}{3}k\Pi_\gamma = 0 \,. \tag{7.122}$$

Using the conduction equation (7.121), the time derivative of the baryon velocity can be written as

$$v'_b = v'_\gamma - v'_{\gamma-b}$$
$$\approx v'_\gamma + \frac{R}{1+R}\frac{k}{\Gamma}\frac{\delta'_\gamma}{4} \,, \tag{7.123}$$

where we have ignored time derivatives of R and Γ on sub-Hubble scales. Equation (7.122) then becomes

$$(1+R)v'_\gamma = -\frac{1}{4}k\delta_\gamma - \frac{R^2}{4(1+R)}\frac{k}{\Gamma}\delta'_\gamma + \frac{2}{3}k\Pi_\gamma \,. \tag{7.124}$$

In Appendix B, we show that the photon anisotropic stress is related to the gradient of the bulk velocity (see Section B.3.1)

$$\Pi_\gamma \approx -\frac{4}{9}\frac{k}{\Gamma}v_\gamma \,, \tag{7.125}$$

which leads to an effective **photon viscosity**. Equation (7.124) then reads

$$(1+R)v'_\gamma = -\frac{1}{4}k\delta_\gamma - \frac{1}{4}\left(\frac{8}{9} + \frac{R^2}{1+R}\right)\frac{k}{\Gamma}\delta'_\gamma \,, \tag{7.126}$$

where we have used the continuity equation on sub-Hubble scales, $\delta'_\gamma = \frac{4}{3}kv_\gamma$. Taking a time derivative of the continuity equation, $\delta''_\gamma = \frac{4}{3}kv'_\gamma$, and substituting (7.126) then gives

$$\boxed{\delta''_\gamma + \frac{k^2}{3(1+R)}\frac{1}{\Gamma}\left(\frac{8}{9} + \frac{R^2}{1+R}\right)\delta'_\gamma + k^2 c_s^2 \delta_\gamma = 0} \,. \tag{7.127}$$

We see that going beyond leading order in the tight-coupling approximation has introduced a friction term proportional to δ'_γ. As the derivation shows, this friction is a combination of thermal conduction and photon viscosity. Notice that the size of the friction term depends on the wavenumber k. The associated damping of the fluctuations is called **Silk damping** [15].

As in Section 7.4.2, we can solve (7.127) in a WKB approximation. Consider the ansatz

$$\delta_\gamma \propto \exp\left(i \int_0^\eta \omega(\tilde\eta)\, d\tilde\eta\right), \quad \text{with} \quad \omega' \ll \omega^2. \tag{7.128}$$

Taking time derivatives of this ansatz, we get

$$\delta_\gamma' = i\omega\delta_\gamma, \tag{7.129}$$

$$\delta_\gamma'' = (-\omega^2 + i\omega')\delta_\gamma \approx -\omega^2\delta_\gamma, \tag{7.130}$$

and, using this in (7.127), we find

$$(k^2 c_s^2 - \omega^2) + \frac{k^2}{3(1+R)}\frac{1}{\Gamma}\left(\frac{8}{9} + \frac{R^2}{1+R}\right)i\omega = 0. \tag{7.131}$$

Substituting $\omega = kc_s + \delta\omega$, and expanding in small $\delta\omega$, gives

$$\delta\omega = i\frac{k^2}{6(1+R)}\frac{1}{\Gamma}\left(\frac{8}{9} + \frac{R^2}{1+R}\right). \tag{7.132}$$

Since this correction is imaginary, the solution (7.128) receives an exponential suppression

$$\boxed{\delta_\gamma \propto e^{-k^2/k_S^2}\exp\left(ik\int c_s(\tilde\eta)\, d\tilde\eta\right)}, \tag{7.133}$$

where the damping wavenumber is

$$\boxed{k_S^{-2}(\eta) \equiv \int_0^\eta \frac{d\tilde\eta}{6(1+R)\Gamma(\tilde\eta)}\left(\frac{8}{9} + \frac{R^2}{1+R}\right)}. \tag{7.134}$$

Including polarization corrections to the scattering would give the same result with $8/9 \to 16/15$ [16]. We see that diffusion damping can be modeled by an e^{-k^2/k_S^2} envelope to the oscillatory part of the solutions discussed in the previous section. This is therefore incorporated into Weinberg's solutions (7.114) and (7.117) by replacing the transfer function $S(k)$ with $e^{-k^2/k_{S,*}^2}S(k)$, where $k_{S,*}$ is given by (7.134) evaluated at last-scattering. For standard cosmological parameters, a good estimate of the Silk damping scale is (see Problem 7.3)

$$k_{S,*}^{-1} \approx 7.2\,\text{Mpc}. \tag{7.135}$$

This corresponds to damping in the CMB power spectrum for multipoles larger than $l_S \sim k_{S,*}\chi_*/\sqrt2 \sim 1370$.

Landau damping

So far, we have assumed that recombination was an instantaneous process, corresponding to the visibility function being a perfect delta function. As we see from Fig. 7.5, however, in reality the visibility function has a finite width. For short-wavelength fluctuations this has to be taken into account.

Near its maximum, the visibility function can be approximated by a Gaussian

$$g(\eta) = \frac{\exp[-(\eta - \eta_*)^2/2\sigma_g^2]}{\sqrt{2\pi}\sigma_g}, \tag{7.136}$$

where $\sigma_g \approx 15.5\,\mathrm{Mpc}$. The fluctuations must now be averaged over this finite width of the visibility function. This averaging makes little difference for slowly evolving fluctuations which can therefore be evaluated at the mean time of decoupling η_*. However, the amplitude of the rapidly oscillating part of the solution can be significantly reduced by the averaging. This effect is sometimes called **Landau damping** because it is similar to the damping that occurs for oscillations with a spread of frequencies.

In Appendix B, we derive the line-of-sight solution in the case of a nontrivial visibility function; see Section B.2. For the Sachs–Wolfe term, we find

$$\Theta_l^{\mathrm{SW}}(k) = \int_0^{\eta_0} \mathrm{d}\eta\, g(\eta) \left(\frac{1}{4}\delta_\gamma + \Psi\right) j_l(k\chi), \tag{7.137}$$

where $\chi(\eta) = \eta_0 - \eta$. Note that this reduces to (7.43) in the case of instantaneous recombination, with $g(\eta) = \delta_{\mathrm{D}}(\eta - \eta_*)$. Over the small range of support of the visibility function, the Bessel function can be treated as time independent and be pulled out of the integral. The remaining integral then is

$$\int_0^{\eta_0} \mathrm{d}\eta\, g(\eta) \cos\left[\int_0^\eta \omega(\tilde{\eta})\, \mathrm{d}\tilde{\eta}\right] \approx \int_0^{\eta_0} \mathrm{d}\eta\, g(\eta) \cos\left[\int_0^{\eta_*} \omega(\tilde{\eta})\, \mathrm{d}\tilde{\eta} + \omega_*(\eta - \eta_*)\right]$$

$$\approx \exp(-\omega_*^2\sigma_g^2/2) \cos\left[\int_0^{\eta_*} \omega(\tilde{\eta})\, \mathrm{d}\tilde{\eta}\right], \tag{7.138}$$

where $\omega_* = \omega(\eta_*) = kc_{s,*}$. We see that the averaging has induced an exponential damping factor $\exp(-k^2/k_{L,*}^2)$, with a scale

$$k_{L,*}^{-1} = \frac{c_{s,*}\sigma_g}{\sqrt{2}} = \frac{\sigma_g}{\sqrt{6(1 + R_*)}} \approx 5.0\,\mathrm{Mpc}. \tag{7.139}$$

Since the functional dependence of this effect is the same as for Silk damping, we can combine the two. To obtain the effective damping scale, the Silk and Landau damping scales must be added in quadrature

$$k_{D,*}^{-1} = \sqrt{k_{S,*}^{-2} + k_{L,*}^{-2}} \approx 8.8\,\mathrm{Mpc}. \tag{7.140}$$

The multipole moment corresponding to this damping scale is $l_D \sim k_{D,*}\chi_*/\sqrt{2} \sim 1125$, which agrees with the observed CMB spectrum in Fig. 7.2.

7.4.5 Summary of Results

Let us take stock and summarize the results of this section. Weinberg's semi-analytic solutions for the Sachs–Wolfe and Doppler transfer functions are

$$F_*(k) = \frac{1}{5}\left[e^{-k^2/k^2_{D,*}}\frac{S(k)}{(1+R_*)^{1/4}}\cos[kr_{s,*}+\theta(k)] - 3R_*T(k)\right], \qquad (7.141)$$

$$G_*(k) = -\frac{\sqrt{3}}{5}e^{-k^2/k^2_{D,*}}\frac{S(k)}{(1+R_*)^{3/4}}\sin[kr_{s,*}+\theta(k)], \qquad (7.142)$$

where the interpolation functions $S(k)$, $T(k)$ and $\theta(k)$ are defined in (7.116) and plotted in Fig. 7.13. Using the result of Exercise 7.3, the power spectrum can be written as

$$\frac{l(l+1)}{2\pi}C_l = \int_1^\infty \frac{d\beta}{\beta^2\sqrt{\beta^2-1}}\left[F_*^2\left(\frac{l\beta}{\chi_*}\right) + \frac{\beta^2-1}{\beta^2}G_*^2\left(\frac{l\beta}{\chi_*}\right)\right]\Delta_{\mathcal{R}}^2\left(\frac{l\beta}{\chi_*}\right). \qquad (7.143)$$

The integral converges rapidly and accurate results are obtained with a relatively low cutoff at $\beta = 5$ [9]. Since most of the support of the integral is near $\beta = 1$, the power spectrum is the sum of the squares of the transfer functions of the Sachs–Wolfe and Doppler transfer functions, evaluated at $k \approx l/\chi_*$. The full spectrum must also include the ISW contribution and the cross correlations between the (I)SW and Doppler contributions.

Figure 7.15 shows the different contributions to the CMB power spectrum. We see that most of the shape of the spectrum comes from the Sachs–Wolfe term, with the Doppler term slightly reducing the contrast between the peaks and troughs. The ISW effect has a significant effect on the height of the first peak. In the next

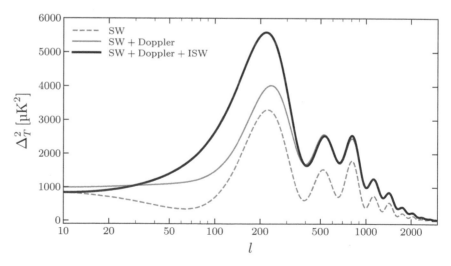

Fig. 7.15 Illustration of the different contributions to the CMB power spectrum. Variations of the cosmological parameters have distinct effects on the shape of the spectrum.

section, we will discuss in some detail how the shape of the spectrum depends on the cosmological parameters.

7.5 Understanding the Power Spectrum

The ΛCDM model is defined by six parameters, which we take to be

$$\{A_\mathrm{s}, n_\mathrm{s}, \omega_b, \omega_m, \Omega_\Lambda, \tau\}\,.$$

The first two parameters, A_s and n_s, describe the spectrum of the initial curvature perturbations. The remaining four parameters determine the evolution of the fluctuations as well as their projection onto the sky. The physical baryon and matter densities are given by the parameters $\omega_b \equiv \Omega_b h^2$ and $\omega_m \equiv \Omega_m h^2$, where h is the dimensionless Hubble constant. The fractional density in the cosmological constant is Ω_Λ. The photon density is fixed by the measured CMB temperature $T_0 = 2.725\,\mathrm{K}$, and is therefore not a free parameter. Neutrinos are treated as massless, so that their density is fixed in terms of the photon density. A nonzero neutrino mass is easy to incorporate and can be constrained by its effects on the gravitationally lensed spectrum. The Hubble constant H_0 (or h) is not included as a free parameter because we are assuming a flat universe, so that $h = \sqrt{\omega_m/(1 - \Omega_\Lambda)}$. The final parameter τ is the integrated optical depth to recombination, which is nonzero because the light of the first stars reionized the universe. The measured values of the base parameters of the ΛCDM cosmology are given in Table 7.1.

Table 7.1 Observed values of the parameters of the LCDM cosmology [17]

Parameter	Meaning	Value
$10^9 A_\mathrm{s}$	scalar amplitude	2.098 ± 0.023
n_s	scalar spectral index	0.965 ± 0.004
$100\,\omega_b$	physical baryon density	2.237 ± 0.015
$100\,\omega_m$	physical matter density	14.30 ± 0.11
Ω_Λ	dark energy density	0.685 ± 0.007
τ	optical depth	0.054 ± 0.007

In the following, I will explain how these parameters determine the shape of the CMB spectrum. I will also present the effects of a few extensions of the minimal ΛCDM model.

7.5.1 Peak Locations

The key characteristics of the CMB spectrum are the positions and heights of the acoustic peaks. To obtain a qualitative understanding for the shape of the spectrum, it is therefore essential that we understand how these features depend on the cosmological parameters.

The angular scale of the acoustic peaks is given by $\theta_s \equiv r_{s,*}/d_{A,*}$, where $r_{s,*}$ is the sound horizon at decoupling and $d_{A,*}$ is the comoving angular diameter distance to the last-scattering surface (see Section 2.2.3). So far in this chapter, we have assumed a flat universe where $d_{A,*} = \chi_*$. The sound horizon depends on pre-recombination physics, so in the ΛCDM model it is determined by ω_b and ω_m alone (for fixed radiation density). The distance to last-scattering, on the other hand, depends on the geometry (Ω_k, H_0) and the energy content (Ω_Λ, Ω_m) after recombination. Breaking the degeneracy between pre- and post-recombination physics requires additional data sets (e.g. BAO data or supernovae).

7.5.2 Peak Heights

The peak heights are determined by four distinct effects:

1. **Diffusion damping** Recall that the diffusion length is roughly the geometric mean of the photon mean free path and the age of the universe at recombination. The former depends on the baryon density (ω_b), while the latter is a function of the total matter density (ω_m).

2. **Early ISW effect** As we explained in Section 7.2.3, the residual radiation density at recombination leads to a time dependence in the potential, $\Phi' \neq 0$, which affects the gravitational redshift of the CMB photons. Since Φ is small on small scales, the main effect is on the first peak of the spectrum.[8] Increasing the relative amount of radiation (e.g. by decreasing the matter density) raises the height of the first peak.

3. **Baryon loading** Baryons add extra weight to the fluid which displaces the zero-point of the acoustic oscillations. As I will show below, this "baryon loading" changes the relative heights of the odd and even peaks in the CMB spectrum. The heights of alternating peaks are therefore a measure of the baryon density.

4. **Radiation driving** Changes in the relative amount of matter and radiation affect the peak heights through an effect called "radiation driving." As we will see, reducing the amount of matter increases the heights of the first few peaks. Since the radiation density is fixed by the CMB temperature, this provides a way to measure the total matter density of the universe.

The last two effects are somewhat subtle and deserve further explanation.

Baryon loading

Baryons contribute to both the inertial and gravitational mass of the photon–baryon fluid, $m_{\text{eff}} \propto 1 + R$, but not to the pressure. Increasing the baryon density therefore

---

[8] Another important effect is that the early ISW effect continues to increase the power even after decoupling. This is the reason that the first peak not only becomes taller, but also fatter, extending to lower multipoles; see Fig. 7.15.

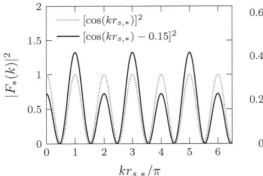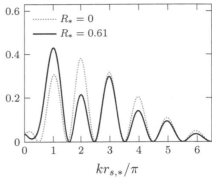

Fig. 7.16 Illustration of the effect of baryons on the Sachs–Wolfe transfer function. The left panel shows the idealized case of pure cosine oscillations without damping. We see that a shift in the zero-point of the oscillations produces alternating peak heights for the square of the transfer function. The same is seen for Weinberg's solution (7.114) in the right panel, but the Silk damping and the WKB factor $(1 + R)^{-1/4}$ mask the effect on the third and fifth peaks.

changes the balance of pressure and gravity in the evolution of the photon–baryon fluid.

One characteristic effect of this "baryon loading" is the shift of the zero-point of the oscillations, described by the $-3RT(k)$ term in (7.114). Since the CMB power spectrum depends on the *square* of the transfer function, this shift of the zero-point leads to differences in the heights of alternating peaks (see Fig. 7.16). The effect is clearly seen in Weinberg's solution for the Sachs–Wolfe transfer function, although the impact on the heights of the third and fifth peaks is masked by the damping of the fluctuations.

The time dependence of the baryon-to-photon ratio R affects the amplitude of the acoustic oscillations through the WKB factor $(1+R)^{-1/4}$ in (7.114). This effect can also be understood through an analogy with classical mechanics, where the ratio of the energy $E = \frac{1}{2} m_{\mathrm{eff}} \omega^2 A^2$ to the frequency ω of an oscillator is an adiabatic invariant. In our case, the evolution of R induces a slow evolution of the sound speed and hence the oscillation frequency $\omega \propto (1 + R)^{-1/2}$. Combining this with $m_{\mathrm{eff}} \propto (1 + R)$, the constraint $m_{\mathrm{eff}} \omega A^2 = \mathrm{const}$ leads to $A \propto (1 + R)^{-1/4}$.

Radiation driving

As we have seen in Section 6.3.1, the gravitational potential decays when a mode enters the horizon during the radiation era. This leads to a boost in the amplitude of the fluctuations which is captured by the transfer function $S(k)$ in (7.114). Intuitively, this can be understood as follows: the moment at which the potential starts to decay coincides with the compression of the photon–baryon fluid, since both begin at horizon crossing. The potential therefore assists the first compression of the fluid through a near-resonant driving force, but doesn't resist the subsequent rarefaction of the fluid (see Fig. 7.17). The amplitude of the acoustic oscillations

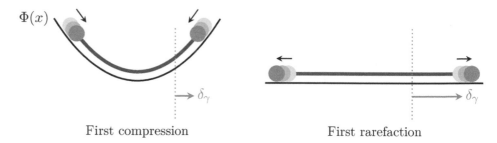

First compression   First rarefaction

Fig. 7.17 Cartoon illustrating the radiation driving effect. The photon–baryon fluid oscillates in the presence of a gravitational potential. At the first compression, the gravitational potential decays due to the large radiation pressure. While the potential therefore assists the compression of the fluid, it doesn't resist the subsequent expansion. The amplitude of the oscillations therefore increases.

therefore increases. In addition, the time-dependent potential $\Phi(\eta)$ induces a time dilation effect in the force term, cf. (7.63), which further increases the amplitude of the photon fluctuations. Heuristically, an overdensity (with negative Φ) induces a "stretching" of the space as seen by the form of the metric (7.13). As the potential decays, the space "re-contracts" and the photons are blueshifted. The combination of the above two effects boosts the amplitude of short-wavelength fluctuations by a factor of about 5. This effect is clearly seen in Weinberg's solution for the Sachs–Wolfe transfer function, cf. Fig. 7.18.

Only modes that enter the sound horizon during the radiation-dominated epoch experience this "radiation driving." We therefore expect the amplitude of the peaks in the CMB to increase as we go from low multipoles to high multipoles, with the

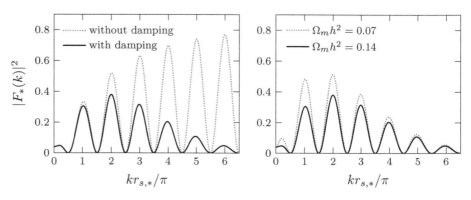

Fig. 7.18 Illustration of the effect of matter on the Sachs–Wolfe transfer function (7.114), with and without diffusion damping. The result without damping clearly shows the boost of short-wavelength modes due to the radiation driving effect (left panel). Reducing the matter density pushes the moment of matter–radiation equality to later times (so that k_{eq} is smaller) and hence enhances the effect (right panel).

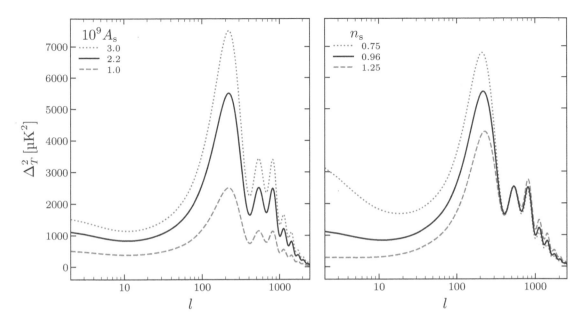

Fig. 7.19   Variation of the CMB power spectrum with changes of the amplitude A_s (*left*) and the tilt n_s (*right*) of the primordial power spectrum.

point of the transition depending of the relative amount of matter and radiation in the universe.

7.5.3 LCDM Cosmology

We are now ready to explain how the CMB spectrum depends on the base parameters of the ΛCDM model, $\{A_s, n_s, \omega_b, \omega_m, \Omega_\Lambda, \tau\}$.

Initial conditions

Our computations of the CMB spectrum assume a power law ansatz for the power spectrum of primordial curvature perturbations

$$\Delta_{\mathcal{R}}^2(k) = A_s \left(\frac{k}{k_0}\right)^{n_s - 1}, \tag{7.144}$$

where $k_0 = 0.05\,\mathrm{Mpc}^{-1}$ is an arbitrarily chosen pivot scale. Since the evolution is linear, it is easy to understand how changes in the amplitude A_s and the tilt n_s of the initial spectrum affect the observed anisotropy spectrum. Variations of A_s simply rescale the overall amplitude of the spectrum, while changing the value of n_s changes the relative amount of power on large and small scales (see Fig. 7.19).

Arguably, one of the most remarkable discoveries coming from observations of the CMB is that the primordial spectrum is close to scale invariant, but not exactly:

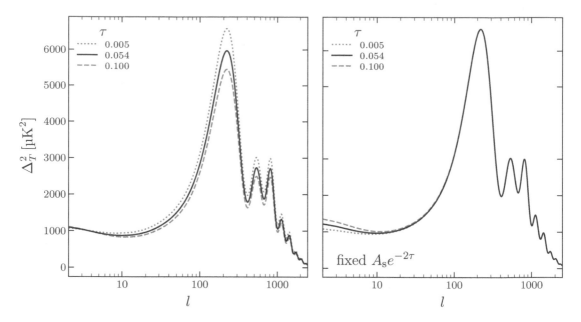

Fig. 7.20 Effect of reionization on the CMB power spectrum. In the right panel, the amplitude A_s has been adjusted to keep the combination $A_\mathrm{s} e^{-2\tau}$ fixed.

$n_\mathrm{s} = 0.965 \pm 0.004$ [17]. As we will see in Chapter 8, this deviation from perfect scale invariance is a prediction of inflation, where it arises from the slow decay of the inflationary energy density. The measurement of the parameter n_s is our first measurement of the dynamics during inflation. We will have much more to say about this in the next chapter.

Reionization

There is an effect that we haven't yet mentioned that also affects the overall amplitude of the spectrum. The ultraviolet light from the first stars reionized the universe, so that the CMB photons have a small probability to scatter off the free electrons in the late universe. To quantify this, consider a photon in a direction $\hat{\mathbf{n}}$, with temperature $T(\hat{\mathbf{n}}) = \bar{T}[1 + \Theta(\hat{\mathbf{n}})]$. The probability of no scattering to occur is $e^{-\tau}$, while the probability of rescattering is $1 - e^{-\tau}$. The temperature of the rescattered photons will be the temperature $\bar{T}$ of the equilibrated ionized regions. The observed temperature today then is

$$T_0(\hat{\mathbf{n}}) = \bar{T}_0[1 + \Theta(\hat{\mathbf{n}})]e^{-\tau} + \bar{T}_0(1 - e^{-\tau}) = \bar{T}_0(1 + \Theta(\hat{\mathbf{n}})e^{-\tau}). \qquad (7.145)$$

We see that the observed anisotropies are suppressed by a factor of $e^{-\tau}$ and the power spectrum becomes

$$C_l \to C_l \, e^{-2\tau}. \qquad (7.146)$$

The effect only occurs on scales that are smaller than the horizon at reionization, corresponding to multipoles $l > 10$. Since the power spectrum is uniformly reduced by a factor of $e^{-2\tau}$, the impact of reionization is degenerate with a change of the amplitude of the initial fluctuations A_s; observations only constrain the combination $A_s e^{-2\tau}$ (see Fig. 7.20). To break this degeneracy requires an independent measurement of the optical depth τ. This is provided by CMB polarization (see Section 7.6).

Baryons

A change in the baryon density ω_b has three main effects:

1. It changes the sound speed and hence leads to a different sound horizon at recombination,

$$r_{s,*} = \int_0^{a_*} \frac{\mathrm{d}\ln a}{aH} c_s(a). \tag{7.147}$$

Associated with this change in $r_{s,*}$ is a change in the angular scale of the sound horizon $\theta_s = r_{s,*}/\chi_*$, where χ_* is the comoving distance to the last-scattering surface

$$\chi_* = \int_{a_*}^1 \frac{\mathrm{d}\ln a}{aH}. \tag{7.148}$$

The value of θ_s determines the peak locations of the CMB spectrum.

2. It changes the zero-point of the acoustic oscillations, enhancing the difference in the heights of odd and even peaks in the spectrum (see Section 7.5.2).

3. It affects the damping of the fluctuations. Reducing the baryon density reduces the tight coupling between the photons and baryons and hence increases the damping effect.

The effect on the peak locations is degenerate with other parameters (such as the dark energy density Ω_Λ) that change the distance to last-scattering χ_* and hence the observed angular scale θ_s. In the left panel of Fig. 7.21, I have therefore plotted the spectrum for different values of ω_b, but fixed θ_s. This removes the effect on the peak locations and highlights the effect on the peak amplitudes. We see that increasing the baryon density raises the first peak and suppresses the second peak, as expected from our discussion on baryon loading. For the higher peaks, the Silk damping and the WKB factor $(1 + R)^{-1/4}$ mask the effect. Looking closely at the spectrum for $l > 1000$, we also see the enhanced damping of the fluctuations for reduced baryon density.

The observations of the CMB lead to $\omega_b = 0.02237 \pm 0.00015$ or $\Omega_b = 0.0493 \pm 0.0006$ [17]. As we have seen in Section 3.2.4, this value of the baryon density is consistent with the value required by BBN.

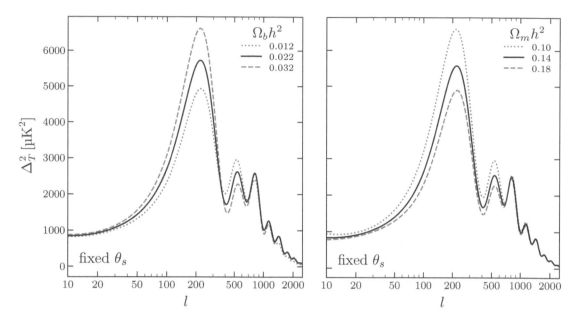

Fig. 7.21 Effects of varying the baryon density $\Omega_b h^2$ (*left*) and the matter density $\Omega_m h^2$ (*right*) at fixed angular scale of the sound horizon θ_s.

Dark matter

A change in the matter density ω_m has four main effects:

1. It changes the sound horizon at recombination through a change in the evolution of the Hubble rate; cf. (7.147). As we described before, it also affects the distance χ_* to the last-scattering surface. The combination of these two effects determines the angular scale of the sound horizon θ_s and hence the peak locations.

2. It shifts the time of matter–radiation equality. Reducing the matter density shifts the moment of equality to a later time which enhances the radiation driving effect described in Section 7.5.2 and hence increases the peak heights.

3. It determines the early ISW effect by changing the relative amount of matter and radiation at the time of recombination. Reducing the matter density increases the early ISW effect, which leads to a boost in the height of the first peak.

4. It affects the diffusion scale by changing the time to recombination.

In the right panel of Fig. 7.21, I have plotted the CMB spectrum for different values of ω_m. The angular scale of the sound horizon θ_s has again been kept fixed. As expected, reducing the matter density leads to a boost in the amplitudes of the first few peaks of the spectrum. The effect saturates for the higher peaks which therefore aren't affected by changes in the matter density.

The observations of the CMB give $\omega_m = 0.1430 \pm 0.0011$ or $\Omega_m = 0.3153 \pm 0.0073$ [17]. Comparing this to the measurement of the baryon density, we conclude

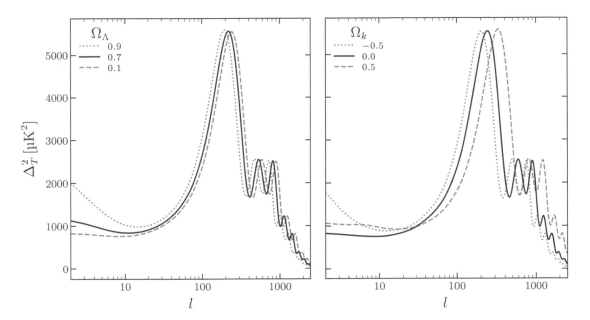

Fig. 7.22 Variation of the CMB power spectrum with dark energy density (*left*) and spatial curvature (*right*), keeping $\Omega_b h^2$ and $\Omega_m h^2$ fixed.

that the universe must have non-baryonic dark matter. The measured amount of the dark matter is consistent with the requirements from structure formation discussed in Chapters 5 and 6.

Dark energy

Changing the dark energy density Ω_Λ affects the CMB spectrum mainly through its effect on the distance to the last-scattering surface. The latter depends on the Hubble rate after recombination, which in a flat universe is

$$H^2(a) = H_0^2 \left(\Omega_m a^{-3} + \Omega_\Lambda\right). \tag{7.149}$$

Since the matter density is constrained by its effect on the peak heights, this allows us to measure the dark energy density. Increasing Ω_Λ at fixed $\Omega_m h^2$ increases the Hubble rate and hence decreases the distance to last-scattering, so that the peaks move to larger angular scales (lower l). This is indeed what is seen in Fig. 7.22. Since dark energy affects the CMB mostly through its effect on the distance to last-scattering, it is degenerate with any other parameter that also changes this distance, such as spatial curvature. Breaking this **geometric degeneracy** requires either external data sets or a measurement of CMB lensing. Combining measurements of the CMB and BAO gives $\Omega_\Lambda = 0.685 \pm 0.007$ [17].

As we can see from Fig. 7.22, increasing the amount of dark energy also leads to an increase in the power on large scales (low l). This arises from the late ISW effect.

The onset of dark energy induces a decay of the gravitational potential, which reduces the net gravitational redshift in traversing the large-scale time-varying gravitational potentials and hence increases the large-scale CMB temperature fluctuations.

Curvature

Throughout this chapter, we have assumed a spatially flat universe. It is interesting to relax this assumption and ask how the CMB can test if the universe is really flat. The effect of spatial curvature on the CMB power spectrum is easy to understand. For all reasonable deviations from a flat universe, spatial curvature is negligible at early times and therefore mainly affects the CMB through a projection effect. (Large deviations from spatial curvature would also alter the early ISW effect.) Physical scales on the last-scattering surface are now related to the observed angular scales by the (comoving) angular diameter distance $d_{A,*}$. For small deviations from a flat universe, we have

$$
d_{A,*} =
\begin{cases}
R_0 \sin(\chi_*/R_0) \approx \chi_* \left(1 - \dfrac{\chi_*^2}{6R_0^2}\right) & \text{(positively curved)}, \\[4mm]
R_0 \sinh(\chi_*/R_0) \approx \chi_* \left(1 + \dfrac{\chi_*^2}{6R_0^2}\right) & \text{(negatively curved)}.
\end{cases}
\tag{7.150}
$$

We see that the angular diameter distance increases for a negatively curved universe. The peaks in the CMB spectrum therefore move to smaller scales (larger multipoles). Conversely, in a positively curved universe the peaks would move to larger scales (smaller multipoles). This is indeed what is seen in Fig. 7.22. (Recall that $\Omega_k < 0$ for a positively curved universe.) The figure also illustrates the significant effect that curvature has on the low-l spectrum through its impact on the late ISW effect.

Hubble constant

Our parameterization of the ΛCDM model in (7.144) did not include the Hubble constant. Assuming a flat universe, the Friedmann equation relates the Hubble constant to the parameters Ω_m and Ω_Λ, so that it becomes a derived parameter. We could also have used the Hubble constant as a free parameter and made the dark energy density a derived parameter. In that case, the Hubble rate in (7.149) would be written as

$$
H^2(a) \propto \left(\omega_m a^{-3} + h^2 - \omega_m\right).
\tag{7.151}
$$

Constraining ω_m through its effect on the peak heights (and/or external data sets) then allows the CMB to measure the Hubble constant h. These observations find $h = 0.674 \pm 0.005$ [17]. Interestingly, this is somewhat in tension with local measurements of the Hubble constant using supernovae, which give $h = 0.730 \pm 0.010$ [18]. This so-called "Hubble tension" remains an important open problem of the ΛCDM concordance model.

7.5.4 Beyond LCDM

It is also interesting to study extensions beyond the ΛCDM cosmology. I will discuss two well-motivated examples: extra relativistic species and tensor modes.

Extra relativistic species

In the ΛCDM model, the radiation density is fixed in terms of the CMB temperature (with the contribution from neutrinos determined by the Standard Model). However, extra light species or non-standard neutrino properties can lead to extra radiation density.

The total energy density in relativistic species is often defined as

$$\rho_r = \left[1 + \frac{7}{8} \left(\frac{4}{11} \right)^{4/3} N_{\text{eff}} \right] \rho_\gamma \,, \tag{7.152}$$

where ρ_γ is the energy density of photons. The parameter N_{eff} is referred to as the "effective number of neutrinos," although there may be contributions to N_{eff} that have nothing to do with neutrinos. The Standard Model predicts $N_{\text{eff}} = 3.046$ and the current constraint from the Planck satellite is $N_{\text{eff}} = 2.99 \pm 0.17$ [17]. Deviations from the standard value may arise if the neutrinos have non-standard properties or if new physics at high energies leads to additional weakly coupled light species (see Problem 3.5).

The main effect of adding extra radiation to the early universe is to increase the damping of the CMB spectrum (see Fig. 7.23). Increasing N_{eff} increases H_*, the expansion rate at recombination. This would change both the damping scale θ_D and the acoustic scale θ_s, with the ratio scaling as

$$\frac{\theta_D}{\theta_s} = \frac{1}{r_{s,*} k_D} \propto \frac{1}{H_*^{-1} H_*^{1/2}} = H_*^{1/2} \,. \tag{7.153}$$

Since θ_s is measured very accurately by the peak locations, we need to keep it fixed. This can be done, for example, by simultaneously increasing the Hubble constant H_0. Increasing N_{eff} (and hence H_*) at fixed θ_s then implies larger θ_D, so that the damping kicks in at larger scales, reducing the power in the damping tail (see Fig. 7.23). By accurately measuring the small-scale CMB anisotropies, observations can therefore put a constraint on the number of relativistic species at recombination.

The main limiting factor in these measurements is a degeneracy with the primordial helium fraction $Y_P \equiv 4n_{\text{He}}/n_b$. At fixed ω_b, increasing Y_P decreases the number density of free electrons, which increases the diffusion length and hence reduces the power in the damping tail. The parameters Y_P and N_{eff} are therefore anti-correlated. This degeneracy is broken by the phase shift induced by free-streaming relativistic species [13, 14].

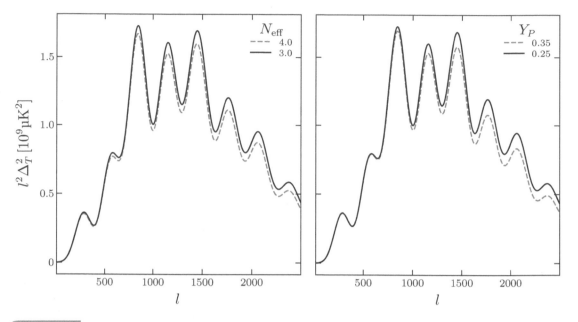

Fig. 7.23 Variation of the CMB temperature power spectrum as a function of $N_{\rm eff}$ (*left*) and Y_P (*right*) for fixed angular size of the sound horizon θ_s. The spectra have been multiplied by a factor of l^2 to emphasize the effects at large l.

As shown in Problem 3.5, light particles that decoupled before the QCD phase transition contribute at the percent level to the radiation density of the universe. Although such a signal is an order of magnitude smaller than the sensitivity of current CMB observations, it is within reach of the next generation of CMB experiments [19].

Gravitational waves

In Problem 7.6, you will explore the impact of tensor metric fluctuations, $\delta g_{ij} = a^2 h_{ij}$, on the CMB temperature fluctuations. The presence of these gravitational waves creates temperature anisotropies at last-scattering, as can be determined from the geodesic equation for the photons. Working through Problem 7.6, you will find that the line-of-sight solution is

$$\Theta^{(t)}(\hat{\mathbf{n}}) = -\frac{1}{2} \int_{\eta_*}^{\eta_0} \mathrm{d}\eta\, h'_{ij} \hat{n}^i \hat{n}^j \,. \tag{7.154}$$

Moreover, you will show that the tensor-induced power spectrum is

$$C_l^{(t)} = 4\pi \int \mathrm{d}\ln k \left| \Theta_l^{(t)}(k) \right|^2 \Delta_h^2(k) \,, \tag{7.155}$$

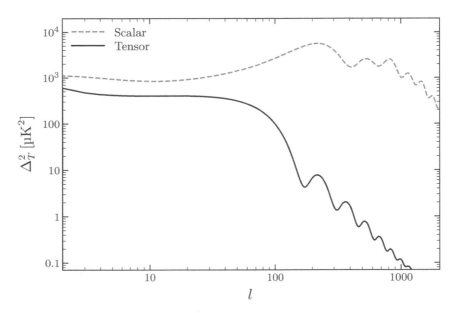

Fig. 7.24 Illustration of the tensor contribution to the CMB temperature power spectrum (for $r = 1$). The decay of the signal for $l > 100$ corresponds to the redshifting of the gravitational wave amplitude inside the horizon.

where $\Delta_h^2(k)$ is the primordial power spectrum of the tensor modes (as defined in Section 6.5). You will derive an explicit form for the transfer function $\Theta_l^{(t)}(k)$ in Problem 7.6. A scale-invariant tensor spectrum, $\Delta_h^2(k) \approx$ const, leads to $l(l + 1)C_l^{(t)} \approx$ const on large scales. On small scales, we expect the signal to be suppressed since the amplitude of gravitational waves decays inside the horizon. This is indeed confirmed by the numerical result in Fig. 7.24.

The measurement of the large-scale CMB temperature anisotropy spectrum constrains the tensor amplitude to $r < 0.1$ [17], where $r \equiv \Delta_h^2(k_0)/\Delta_\mathcal{R}^2(k_0)$ is the tensor-to-scalar ratio. Stronger constraints on r require the measurement of CMB polarization, especially its B-mode type. We will discuss this in the next section.

7.6 A Glimpse at CMB Polarization*

Measurements of the polarization of the CMB are at the forefront of observational cosmology. In this section, I will give an introduction to the physics of CMB polarization. This is a somewhat technical subject and I will not give all mathematical details. Instead my goal is to provide an intuitive understanding of the E/B decomposition of the polarization signal and to explain why B-modes are a unique signature of primordial tensor modes.

7.6.1 Polarization from Scattering

It is easy to see why scattering produces polarization. Consider the scattering of radiation by a free electron as illustrated in Fig. 7.25. Imagine that the angle between the incoming and outgoing rays is 90°; e.g. let the incoming and outgoing rays be in the directions $-\hat{\mathbf{x}}$ and $+\hat{\mathbf{z}}$, respectively. Even if the incoming radiation is unpolarized, the outgoing radiation will become polarized. This is because the polarization direction must be transverse to the direction of propagation. The component of the initial polarization that is parallel to the direction of the outgoing radiation will therefore not survive the scattering. For the scattering shown in the left panel of Fig. 7.25 all of the intensity along the z-axis is blocked and only that along the y-axis is transmitted. The net result is polarization in the direction $\hat{\mathbf{y}}$.

Of course, it is not enough to study just one incoming ray and a very special scattering angle. In general, radiation will come from all directions and scatter

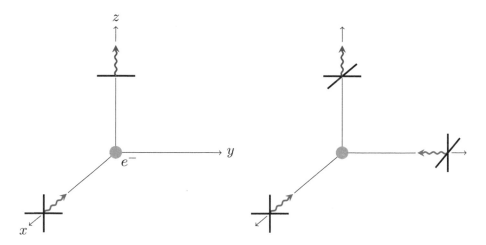

Fig. 7.25 Each scattering event creates polarization (*left*), but no net polarization is generated if the incident radiation is isotropic (*right*). The left figure shows the scattering of unpolarized radiation coming from a single direction. Only the component of the electric field orthogonal to the outgoing direction survives the scattering. The right figure shows the scattering of two rays of an isotropic radiation field. Each scattered ray becomes polarized, but the sum of all outgoing rays is unpolarized.

into all directions. If the incident radiation is isotropic, then the net result after scattering will still be unpolarized radiation. To see this, consider adding a ray of unpolarized radiation in the direction $-\hat{\mathbf{y}}$ as in the right panel in Fig. 7.25. After scattering in the direction $\hat{\mathbf{z}}$, it will produce outgoing polarization in the direction $\hat{\mathbf{x}}$. The sum of the polarized radiation created by the two incoming rays will be unpolarized radiation. To create a net polarization, we require anisotropy in the incoming radiation.

Exercise 7.5 Show with a sketch that the scattering of a dipolar anisotropy also does not lead to a net polarization.

Figure 7.26 shows that polarization is generated if the incident radiation has a quadrupolar anisotropy. The intensity of the radiation coming from the directions $\pm\hat{\mathbf{y}}$ is now larger than that coming from $\pm\hat{\mathbf{x}}$. The outgoing radiation therefore will be polarized in the $\hat{\mathbf{x}}$-direction. It is useful to remember that the polarization direction of the outgoing radiation is aligned with the hot regions.

The polarization pattern observed on the sky will be the result of the scattering of the temperature inhomogeneities in the primordial plasma. As always, these inhomogeneities are best described in a Fourier decomposition, with each Fourier mode describing a plane wave perturbation. It is therefore useful to first determine the polarization generated by a single plane wave. We will do this more formally below, but the basic result can be understand from the cartoons introduced in this section. Consider a plane wave photon density perturbation propagating in a direction transverse to the line-of-sight (see Fig. 7.27). The plane wave corresponds

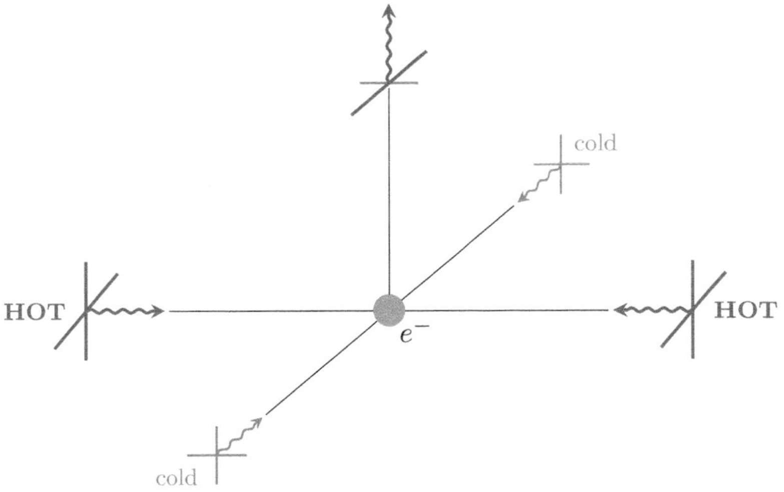

Fig. 7.26 A net polarization is generated if the incident radiation has a quadrupolar anisotropy. The polarization direction of the outgoing radiation is aligned with the hot regions.

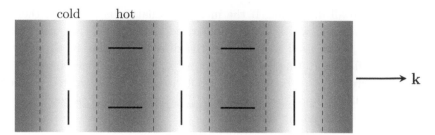

cold   hot

Fig. 7.27 Illustration of E-mode polarization created by a plane wave density perturbation. The sketched polarization pattern is for radiation scattering out of the plane of the paper. We see that the polarization directions are parallel or perpendicular to the wavevector **k**.

to hot and cold modulations of the photon temperature. Using our earlier assertion that the polarization is aligned with the hot regions, we predict the polarization pattern shown in Fig. 7.27. Importantly, the polarization directions are either parallel or perpendicular to the wavevector **k**. We call such a polarization pattern an **E-mode**.

Rotating all polarization directions by 45 degrees would give the so-called **B-mode** polarization (see Fig. 7.28). By symmetry, this type of polarization pattern *cannot* be generated by scalar (density) fluctuations. Instead, the B-mode pattern is a key signature of gravitational waves in the early universe. We will discuss this in detail in Section 7.6.5.

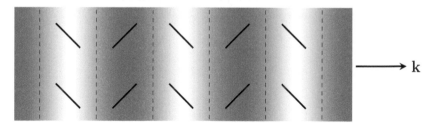

Fig. 7.28 Illustration of B-mode polarization created by a plane wave. The polarization directions are rotated by 45° relative to the E-mode pattern shown in Fig. 7.27. The azimuthal symmetry of scalar density fluctuations (i.e. rotational symmetry around the wavevector **k**) forbids such a polarization pattern.

7.6.2 Statistics of CMB Polarization

To go beyond the simple cartoon description of CMB polarization we need to introduce a bit of formalism.

Stokes parameters

Consider a monochromatic electromagnetic plane wave propagating in the z-direction. The electric field of the wave is

$$\mathbf{E}(z,t) = \text{Re}\left[(E_x\hat{\mathbf{x}} + E_y\hat{\mathbf{y}})\,e^{i(kz-\omega t)}\right],\qquad(7.156)$$

where $\omega = kc$. We can define the complex amplitudes as $E_x = |E_x|e^{i\varphi_x}$ and $E_y = |E_y|e^{i\varphi_y}$. At a given location, which we can take to be $z = 0$, we then have

$$\mathbf{E}(t) = |E_x|\cos(\omega t)\,\hat{\mathbf{x}} + |E_y|\cos(\omega t - \varphi)\,\hat{\mathbf{y}},\qquad(7.157)$$

where we have defined $\varphi \equiv \varphi_y - \varphi_x$ and chosen the origin of time so that $\varphi_x \equiv 0$. In general, the field vector in (7.157) traces out an ellipse in the xy-plane. For $\varphi = 0$ (or π), the x- and y-components of the field oscillate in phase (or anti-phase) and the ellipse degenerates into a line. We say that the wave is *linearly polarized*. For $\varphi = \pm\pi/2$, the wave is *circularly polarized*.

The polarization of electromagnetic radiation is then defined by the **Stokes parameters**

$$\begin{aligned}
I &\equiv |E_x|^2 + |E_y|^2,\\
Q &\equiv |E_x|^2 - |E_y|^2,\\
U &\equiv 2|E_x||E_y|\cos\varphi,\\
V &\equiv 2|E_x||E_y|\sin\varphi,
\end{aligned}\qquad(7.158)$$

where I measures the intensity of the radiation, while Q, U, V describe its polarization state. The expressions in (7.158) describe fully polarized radiation, with $I^2 = Q^2 + U^2 + V^2$, which the CMB is certainly not; the polarization amplitude of the CMB is less than 10% of the temperature anisotropy. Recall that, in SI units, the energy density of an electromagnetic wave is $\rho = \varepsilon_0|\mathbf{E}|^2$ and its intensity is $\mathcal{I} = \rho c = c\varepsilon_0|\mathbf{E}|^2$. Averaged over many oscillation periods, we have

$$\langle\mathcal{I}\rangle = c\varepsilon_0\langle|\mathbf{E}|^2\rangle = \tfrac{1}{2}c\varepsilon_0(|E_x|^2 + |E_y|^2) = \tfrac{1}{2}c\varepsilon_0 I.\qquad(7.159)$$

In units with $c = \varepsilon_0 \equiv 1$, we therefore get $I = 2\langle\mathcal{I}\rangle$, so that the Stokes parameter I indeed is a measure of the time-averaged intensity. (Sometimes the Stokes parameters are defined with an alternative normalization, so that $I = \langle\mathcal{I}\rangle$. I will follow what seems to be the standard convention in the CMB literature.)

The Stokes parameter V vanishes for linear polarization (with $\varphi = 0, \pi$). Since circular polarization usually isn't produced in the early universe, we will set $V \equiv 0$ in the following. The remaining Stokes parameters Q and U then describe the

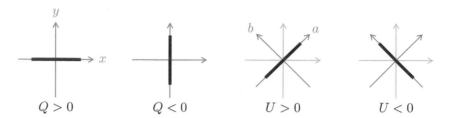

Fig. 7.29   Graphical illustration of the Stokes Q and U parameters for linear polarization.

linear polarization of the radiation. The parameter Q is the difference between the intensity along the x-axis, $|E_x|^2$, and that along the y-axis, $|E_y|^2$. Radiation that is polarized along the x-axis has $Q > 0$, while radiation that is polarized along the y-axis has $Q < 0$. Similarly, the parameter U can be thought of as the difference in the intensity along two coordinate axes a and b that are rotated counterclockwise by 45° (see Fig. 7.29):

$$U = |E_a|^2 - |E_b|^2$$
$$= 2\,\mathrm{Re}[E_x E_y^*]\,, \qquad\qquad (7.160)$$

where we have used that $E_{x,y} = (E_a \pm iE_b)/\sqrt{2}$. Radiation that is polarized along the a-axis has $U > 0$, while radiation polarized along the b-axis has $U < 0$.

An obvious problem with these definitions is that they depend on our arbitrary choice of coordinates. Indeed, rotating the x- and y-axes counterclockwise by an angle ϕ changes the components of the electric field and hence the Stokes parameters:

$$\begin{pmatrix} E_x' \\ E_y' \end{pmatrix} = \begin{pmatrix} \cos\phi & \sin\phi \\ -\sin\phi & \cos\phi \end{pmatrix} \begin{pmatrix} E_x \\ E_y \end{pmatrix},$$

$$\begin{pmatrix} Q' \\ U' \end{pmatrix} = \begin{pmatrix} \cos 2\phi & \sin 2\phi \\ -\sin 2\phi & \cos 2\phi \end{pmatrix} \begin{pmatrix} Q \\ U \end{pmatrix}. \qquad (7.161)$$

The transformation of the Stokes parameters can also be written in the following compact form

$$Q' \pm iU' = e^{\mp 2i\phi}(Q \pm iU)\,, \qquad\qquad (7.162)$$

which shows that polarization transforms like a **spin-2 field**. To avoid this ambiguity, we will introduce the coordinate-independent E- and B-modes of the polarization field.

Decomposition into E- and B-modes

On scales smaller than a few degrees, the sky can be approximated as flat. This **flat-sky limit** greatly simplifies the mathematical treatment of CMB polarization, so we will use it in the following. Let $\boldsymbol{\theta} = (\theta_x, \theta_y)$ be the position on a flat region of the sky and define the symmetric and trace-free polarization tensor as

$$P_{ab}(\boldsymbol{\theta}) \equiv \begin{pmatrix} Q(\boldsymbol{\theta}) & U(\boldsymbol{\theta}) \\ U(\boldsymbol{\theta}) & -Q(\boldsymbol{\theta}) \end{pmatrix}. \tag{7.163}$$

Being made out of the Stokes parameters, this polarization tensor of course also transforms under a rotation of the coordinates. However, by taking spatial derivatives of P_{ab} we can define two quantities that are invariant under rotations

$$\begin{aligned} \nabla^2 E &\equiv \partial_a \partial_b P_{ab} \,, \\ \nabla^2 B &\equiv \epsilon_{ac} \partial_b \partial_c P_{ab} \,, \end{aligned} \tag{7.164}$$

where ϵ_{ab} is the antisymmetric tensor. In fact, these definitions of E- and B-modes even hold beyond the flat-sky limit if we promote the derivatives to covariant derivatives on the sphere. Writing (7.164) out in terms of the Stokes parameters, we get

$$\begin{aligned} \nabla^2 E &\equiv (\partial_x^2 - \partial_y^2)Q + 2\partial_x \partial_y U \,, \\ \nabla^2 B &\equiv (\partial_x^2 - \partial_y^2)U - 2\partial_x \partial_y Q \,. \end{aligned} \tag{7.165}$$

We note that the relation between the E- and B-modes and the Stokes parameters is *non-local*.

The E/B decomposition of the polarization is similar to the decomposition of a vector field into a gradient of a scalar function and the curl of a vector

$$\mathbf{V} = \boldsymbol{\nabla} G - \boldsymbol{\nabla} \times \mathbf{C} \,, \tag{7.166}$$

where $\boldsymbol{\nabla} \cdot \mathbf{C} = 0$. Taking the divergence and the curl of the vector, we can isolate the two components

$$\begin{aligned} \nabla^2 G &= \boldsymbol{\nabla} \cdot \mathbf{V} \,, \\ \nabla^2 \mathbf{C} &= \boldsymbol{\nabla} \times \mathbf{V} \,, \end{aligned} \tag{7.167}$$

which indeed looks very similar to (7.164). This decomposition of a vector field into a curl-free gradient part and a divergence-free curl part also explains the names E-mode and B-mode: in electrostatics, the electric field has vanishing curl, $\boldsymbol{\nabla} \times \mathbf{E} = 0$, while a magnetic field has zero divergence, $\boldsymbol{\nabla} \cdot \mathbf{B} = 0$.

Polarization power spectra

The CMB map is now described by the temperature T and the polarization modes E and B, which we collectively call $X \equiv \{T, E, B\}$. The main benefit of working in the flat-sky limit is that we can use ordinary Fourier transforms instead of spherical harmonics to describe the anisotropies:

$$X(\boldsymbol{l}) = \int \mathrm{d}^2\boldsymbol{\theta} \, X(\boldsymbol{\theta}) \, e^{-i\boldsymbol{l}\cdot\boldsymbol{\theta}} \,, \tag{7.168}$$

where l is the two-dimensional wavevector conjugate to the position $\boldsymbol{\theta}$. The relations in (7.165) imply that the Fourier components of the E- and B-modes satisfy

$$E(\boldsymbol{l}) = \frac{(l_x^2 - l_y^2)Q(\boldsymbol{l}) + 2l_x l_y U(\boldsymbol{l})}{l_x^2 + l_y^2},$$

$$B(\boldsymbol{l}) = \frac{(l_x^2 - l_y^2)U(\boldsymbol{l}) - 2l_x l_y Q(\boldsymbol{l})}{l_x^2 + l_y^2}. \tag{7.169}$$

You should confirm that although $Q(\boldsymbol{l})$ and $U(\boldsymbol{l})$ transform under a rotation of the θ_x- and θ_y-axes, $E(\boldsymbol{l})$ and $B(\boldsymbol{l})$ are invariant. The power spectra of the fluctuations are then defined in the usual way as

$$\langle X(\boldsymbol{l})Y(\boldsymbol{l}')\rangle = (2\pi)^2 \delta_{\mathrm{D}}(\boldsymbol{l} + \boldsymbol{l}')\, C_l^{XY}, \tag{7.170}$$

where $X, Y = \{T, E, B\}$. In principle, this corresponds to six different spectra. However, in practice, only four of these spectra are expected to be nonzero. To see this, consider a reflection about the x-axis (called a parity inversion). This leads to

$$\theta_y \to -\theta_y, \quad Q \to Q, \quad U \to -U, \quad l_x \to l_x, \quad l_y \to -l_y. \tag{7.171}$$

From (7.169) we see that the E-mode is invariant under this reflection, while the B-mode changes sign:

$$E(\boldsymbol{l}) \to E(\boldsymbol{l}), \quad B(\boldsymbol{l}) \to -B(\boldsymbol{l}). \tag{7.172}$$

The field E is therefore a parity-even scalar field (like the temperature T), while B is a parity-odd pseudoscalar field. If the physics of the primordial universe is parity conserving, then we expect $C_l^{\mathrm{TB}} = C_l^{\mathrm{EB}} = 0$.

On the full sky, we have to use spherical harmonics to describe the polarization field and generalize our definitions of E- and B-modes. I will just sketch the logic, leaving the somewhat complicated mathematical details to [20, 21].

We have seen that polarization is a spin-2 field, cf. (7.162). We therefore can't use ordinary scalar spherical harmonics to describe its decomposition into multipole moments. Instead, we must use so-called **spin-weighted spherical harmonics** [20, 21]. The polarization field can then be written as

$$Q(\hat{\mathbf{n}}) \pm iU(\hat{\mathbf{n}}) = \sum_{lm} a_{\pm 2, lm}\, {}_{\pm 2}Y_{lm}(\hat{\mathbf{n}}), \tag{7.173}$$

where the explicit forms of the spin-2 spherical harmonics ${}_{\pm 2}Y_{lm}(\hat{\mathbf{n}})$ can be found in [20, 21]. The multipole coefficients $a_{2,lm}$ and $a_{-2,lm}$ can be combined into parity-even and parity-odd combinations:

$$a_{lm}^E \equiv -(a_{2,lm} + a_{-2,lm})/2,$$

$$a_{lm}^B \equiv -(a_{2,lm} - a_{-2,lm})/2i, \tag{7.174}$$

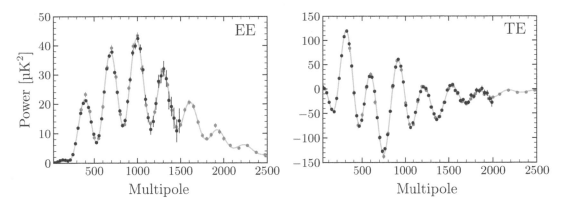

Fig. 7.30   E-mode power spectrum (*left*) and its cross correlation with the temperature fluctuations (*right*) measured by the Planck satellite (*black*) [2] and the Atacama Cosmology Telescope (*gray*) [23]. The curves are *not* fits to the data, but instead are predictions given the parameters inferred from the fit to the temperature power spectrum. (To avoid clutter, I have not added relevant data from the South Pole Telescope [24].)

which are the multipole coefficients of the E- and B-modes:

$$
\begin{aligned}
E(\hat{\mathbf{n}}) &= \sum_{lm} a_{E,lm} Y_{lm}(\hat{\mathbf{n}})\,, \\
B(\hat{\mathbf{n}}) &= \sum_{lm} a_{B,lm} Y_{lm}(\hat{\mathbf{n}})\,.
\end{aligned}
\tag{7.175}
$$

The polarization power spectra are then defined in the usual way:

$$
\begin{aligned}
\langle a_{lm} a^{E*}_{l'm'} \rangle &= C_l^{\mathrm{TE}} \delta_{ll'} \delta_{mm'}\,, \\
\langle a^{E}_{lm} a^{E*}_{l'm'} \rangle &= C_l^{\mathrm{EE}} \delta_{ll'} \delta_{mm'}\,, \\
\langle a^{B}_{lm} a^{B*}_{l'm'} \rangle &= C_l^{\mathrm{BB}} \delta_{ll'} \delta_{mm'}\,.
\end{aligned}
\tag{7.176}
$$

CMB polarization was first detected in 2002 by the Degree Angular Scale Interferometer (DASI) [22]. It has since been measured with increasing precision by the WMAP and Planck satellites, as well as a number of ground-based experiments. Figure 7.30 shows the TE and EE spectra measured by Planck [2] and ACT [23].

7.6.3 Visualizing E- and B-modes

Having introduced the formal definitions of E- and B-modes, it is useful to relate them back to the pictures we gave in Section 7.6.1. We return to the flat-sky approximation. For simplicity, we align the x-axis of our coordinates with the position vector $\boldsymbol{\theta}$, and let ϕ_l be the angle of the wavevector $\boldsymbol{l} = l(\cos\phi_l, \sin\phi_l)$. The Fourier modes in (7.169) then are

$$
\begin{aligned}
E(\boldsymbol{l}) &= +Q(\boldsymbol{l}) \cos(2\phi_l) + U(\boldsymbol{l}) \sin(2\phi_l)\,, \\
B(\boldsymbol{l}) &= -Q(\boldsymbol{l}) \sin(2\phi_l) + U(\boldsymbol{l}) \cos(2\phi_l)\,.
\end{aligned}
\tag{7.177}
$$

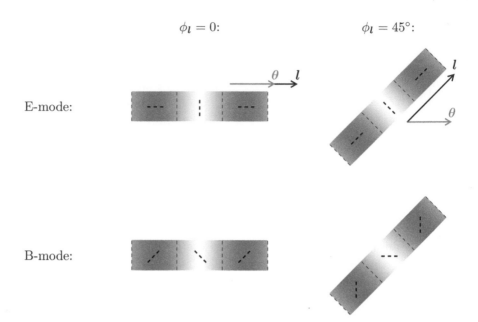

$\phi_l = 0:$ $\phi_l = 45°:$

E-mode:

B-mode:

Fig. 7.31 Polarization patterns corresponding to a pure E-mode (*top*) and a pure B-mode (*bottom*) for two different orientations of the plane wave. We see that for the E-mode the polarization directions are either parallel or perpendicular to the wavevector, while for the B-mode they are tilted by 45°.

Inverting these expressions lets us write the Stokes parameters in terms of the parameters E and B:

$$Q(\boldsymbol{l}) = E(\boldsymbol{l})\cos(2\phi_l) - B(\boldsymbol{l})\sin(2\phi_l)\,,$$
$$U(\boldsymbol{l}) = E(\boldsymbol{l})\sin(2\phi_l) + B(\boldsymbol{l})\cos(2\phi_l)\,. \tag{7.178}$$

The polarization generated by a single plane wave then is

$$Q(\boldsymbol{\theta}) = \mathrm{Re}\big[\,(E(\boldsymbol{l})\cos(2\phi_l) - B(\boldsymbol{l})\sin(2\phi_l))\,e^{i\boldsymbol{l}\cdot\boldsymbol{\theta}}\big]\,,$$
$$U(\boldsymbol{\theta}) = \mathrm{Re}\big[\,(E(\boldsymbol{l})\sin(2\phi_l) + B(\boldsymbol{l})\cos(2\phi_l))\,e^{i\boldsymbol{l}\cdot\boldsymbol{\theta}}\big]\,, \tag{7.179}$$

where the factor of $e^{i\boldsymbol{l}\cdot\boldsymbol{\theta}}$ describes the modulation of the polarization amplitude induced by the plane wave perturbation. From this expression, we can infer the polarization patterns corresponding to pure E- and B-modes created by a plane wave perturbation.

For a pure E-mode, we have

$$Q(\boldsymbol{\theta}) = \mathrm{Re}\big[E(\boldsymbol{l})\cos(2\phi_l)\,e^{i\boldsymbol{l}\cdot\boldsymbol{\theta}}\big]\,,$$
$$U(\boldsymbol{\theta}) = \mathrm{Re}\big[E(\boldsymbol{l})\sin(2\phi_l)\,e^{i\boldsymbol{l}\cdot\boldsymbol{\theta}}\big]\,. \tag{7.180}$$

In the top of Fig. 7.31, I show the corresponding polarization patterns for two representative orientations of the wavevector $\boldsymbol{l}$. In both cases, the polarization directions are parallel or perpendicular to the wavevector. This defining feature

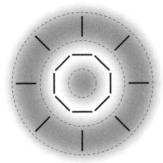 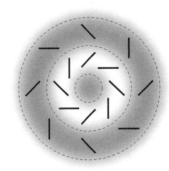

Fig. 7.32   E- and B-mode patterns created by a radial wave in the xy-plane for radiation in the z-direction (out of the page). Note that when reflected about a line going through the center, the E-mode pattern remains unchanged, while the B-mode pattern changes. (Figure adapted from [25].)

of E-mode polarization holds for any orientation of the plane wave perturbation. Another way to diagnose the presence of the E-mode is to rotate the coordinate system so that the x'-axis is aligned with the direction l. The Stokes U parameter will then vanish and the polarization will be pure Q. This is another coordinate-independent way to define the E-mode.

For a pure B-mode, we have

$$
\begin{aligned}
Q(\boldsymbol{\theta}) &= -\mathrm{Re}\big[B(\boldsymbol{l})\sin(2\phi_l)\,e^{i\boldsymbol{l}\cdot\boldsymbol{\theta}}\big], \\
U(\boldsymbol{\theta}) &= +\mathrm{Re}\big[B(\boldsymbol{l})\cos(2\phi_l)\,e^{i\boldsymbol{l}\cdot\boldsymbol{\theta}}\big].
\end{aligned}
\tag{7.181}
$$

In the bottom of Fig. 7.31, I present the corresponding polarization patterns for the same two orientations of the wavevector l as before. The polarization directions are now tilted by $45°$ with respect to the wavevector, which is the defining characteristic of B-mode polarization. Rotating the coordinates so that the x'-axis is aligned with l, the Stokes Q parameter would vanish and the polarization is pure U.

More complex polarization patterns arise from the superposition of the polarization patterns created by plane wave perturbations. In Fig. 7.32, we consider the polarization generated by a radial wave in the xy-plane. This wave can be thought of as a superposition of plane waves with equal phases and amplitudes, but different azimuthal angles ϕ_l. You should convince yourself that a superposition of the polarization patterns shown in Fig. 7.31 (for varying ϕ_l) produces the polarization shown in Fig. 7.32. We see that the E-mode pattern alternates between being radial and tangential, while the B-mode polarization has a characteristic "swirl" pattern.

7.6.4 E-modes from Scalars

We will now repeat the discussion in Section 7.6.1 with more equations and a larger degree of generality. We will mostly follow the excellent treatment in Dodelson's

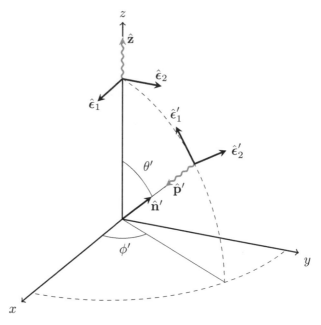

Fig. 7.33   Geometry and coordinates for incoming radiation from the direction $\hat{\mathbf{n}}' = -\hat{\mathbf{p}}'$ and outgoing radiation in the direction $\hat{\mathbf{z}}$.

book [26]. In this section, we will assume that the primordial fluctuations are sourced by adiabatic scalar fluctuations and show that these do *not* produce B-modes. The case of primordial tensor fluctuations is treated in the next section.

Thomson scattering revisited

In Section 7.6.1, I showed graphically how unpolarized radiation becomes polarized by scattering. Mathematically, this is described by the dependence of the scattering cross section on the polarization of the incoming and the outgoing radiation. If the polarization direction of the incoming radiation is $\hat{\epsilon}'$ and that of the outgoing radiation is $\hat{\epsilon}$, then the differential cross section is

$$\frac{\mathrm{d}\sigma}{\mathrm{d}\Omega} = \frac{3\sigma_T}{8\pi}|\hat{\epsilon} \cdot \hat{\epsilon}'|^2 . \tag{7.182}$$

The dependence on $\hat{\epsilon} \cdot \hat{\epsilon}'$ enforces that only the component of the incoming polarization that is orthogonal to the direction of the outgoing radiation is transmitted.

Without loss of generality, we can take the direction of the outgoing radiation to be $\hat{\mathbf{z}}$ and choose $\hat{\epsilon}_1 \equiv \hat{\mathbf{x}}$ and $\hat{\epsilon}_2 \equiv \hat{\mathbf{y}}$ to be an orthogonal basis of polarization vectors (see Fig. 7.33). The Q-type polarization of the outgoing radiation then is

$$Q(\hat{\mathbf{z}}) = A \int \mathrm{d}\Omega' \, f(\hat{\mathbf{n}}') \sum_{j=1}^{2} \left(|\hat{\mathbf{x}} \cdot \hat{\epsilon}_j'|^2 - |\hat{\mathbf{y}} \cdot \hat{\epsilon}_j'|^2 \right) , \tag{7.183}$$

where $f(\hat{\mathbf{n}}')$ is the distribution function of the incoming radiation and A is the overall normalization (which would be fixed by the intensity of the incoming radiation). We have assumed that the incident radiation is unpolarized, so $f(\hat{\mathbf{n}}')$ doesn't depend on $\hat{\boldsymbol{\epsilon}}'_j$. To evaluate the inner products in (7.183), we write the polarization basis of the incoming radiation in Cartesian coordinates

$$
\begin{aligned}
\hat{\boldsymbol{\epsilon}}'_1 &\equiv -\hat{\boldsymbol{\theta}}' = (-\cos\theta'\cos\phi', -\cos\theta'\sin\phi', \sin\theta') \,, \\
\hat{\boldsymbol{\epsilon}}'_2 &\equiv \hat{\boldsymbol{\phi}}' = (-\sin\phi', \cos\phi', 0) \,.
\end{aligned}
\tag{7.184}
$$

We then get

$$
\begin{aligned}
Q(\hat{\mathbf{z}}) &= A \int d\Omega'\, f(\hat{\mathbf{n}}') \left[\cos^2\theta'\cos^2\phi' + \sin^2\phi' - \cos^2\theta'\sin^2\phi' - \cos^2\phi'\right] \\
&= -A \int d\Omega'\, f(\hat{\mathbf{n}}')\, \sin^2\theta'\cos 2\phi' \,.
\end{aligned}
\tag{7.185}
$$

The combination of angles in this integral is proportional to a sum of $l = 2$ spherical harmonics, $\sin^2\theta'\cos 2\phi' \propto [Y_{22} + Y_{2,-2}](\hat{\mathbf{n}}')$; see (D.6) in Appendix D. By orthogonality of the spherical harmonics, the integral will then pick out the $l = 2, m = \pm 2$ components of $f(\hat{\mathbf{n}}')$. We have therefore derived the crucial result that a nonzero Q will only be produced if the incident radiation has a nonzero quadrupole moment. We have seen this in pictures before, but it is nice to have it drop out of the equations.

The U-type polarization of the outgoing radiation is defined with respect to the polarization basis $\hat{\boldsymbol{\epsilon}}_a = (\hat{\mathbf{x}} + \hat{\mathbf{y}})/\sqrt{2}$ and $\hat{\boldsymbol{\epsilon}}_b = (\hat{\mathbf{x}} - \hat{\mathbf{y}})/\sqrt{2}$. We then find

$$
\begin{aligned}
U(\hat{\mathbf{z}}) &= \frac{A}{2} \int d\Omega'\, f(\hat{\mathbf{n}}') \sum_{j=1}^{2} \left(|(\hat{\mathbf{x}} + \hat{\mathbf{y}}) \cdot \hat{\boldsymbol{\epsilon}}'_j|^2 - |(\hat{\mathbf{x}} - \hat{\mathbf{y}}) \cdot \hat{\boldsymbol{\epsilon}}'_j|^2\right) \\
&= -A \int d\Omega'\, f(\hat{\mathbf{n}}')\, \sin^2\theta'\sin 2\phi' \,,
\end{aligned}
\tag{7.186}
$$

where we have substituted (7.184) to obtain the result in the second line. The angular dependence of the integrand is now proportional to the difference of $l = 2$ spherical harmonics, $\sin^2\theta'\sin 2\phi' \propto [Y_{22} - Y_{2,-2}](\hat{\mathbf{n}}')$, so that again only the quadrupole moment of $f(\hat{\mathbf{n}}')$ produces polarization.

Polarization from a single plane wave

We would like to predict the polarization signals generated by the scattering of an inhomogeneous distribution of photons. We first write the perturbed photon distribution function as

$$
\begin{aligned}
f(\hat{\mathbf{n}}', \mathbf{x}) &= \exp\left(\frac{E}{\bar{T}[1 + \Theta(\hat{\mathbf{n}}', \mathbf{x})]} - 1\right)^{-1} \\
&\approx \bar{f} - \frac{d\bar{f}}{d\ln E}\, \Theta(\hat{\mathbf{n}}', \mathbf{x}) \,.
\end{aligned}
\tag{7.187}
$$

As always, it is convenient to describe the spatial dependence of the fluctuations in Fourier space, $\Theta(\hat{\mathbf{n}}', \mathbf{x}) \to \Theta(\hat{\mathbf{n}}', \mathbf{k})$. The fluctuations created by scalar density perturbations have an axial symmetry. This means that the perturbations in the direction $\hat{\mathbf{n}}'$ only depend on magnitude of the wavevector k and the angle defined by $\mu' = \hat{\mathbf{k}} \cdot \hat{\mathbf{n}}'$. We then write the temperature perturbation as an expansion in Legendre polynomials

$$\Theta(k, \mu') = \sum_l i^l (2l + 1)\, \Theta_l(k) P_l(\mu')\,. \tag{7.188}$$

Only the $l = 2$ quadrupole moment of this expansion leads to polarization. Substituting this quadrupole into (7.185) and (7.186), we get

$$Q(\hat{\mathbf{z}}) \propto \Theta_2(k) \int_0^\pi d\theta' \sin\theta' \int_0^{2\pi} d\phi'\, P_2(\mu') \sin^2\theta' \cos 2\phi'\,,$$
$$U(\hat{\mathbf{z}}) \propto \Theta_2(k) \int_0^\pi d\theta' \sin\theta' \int_0^{2\pi} d\phi'\, P_2(\mu') \sin^2\theta' \sin 2\phi'\,. \tag{7.189}$$

To evaluate the angular integrals, we must first write μ' in terms of θ' and ϕ'. Following Dodelson [26], we will do this in three steps:

1. First, we take the wavevector to be parallel to the x-axis, $\mathbf{k} \parallel \hat{\mathbf{x}}$. This gives

$$\mu' = \hat{\mathbf{k}} \cdot \hat{\mathbf{n}}' = \hat{\mathbf{x}} \cdot \hat{\mathbf{n}}' = \hat{n}'_x$$
$$= \sin\theta' \cos\phi'\,, \tag{7.190}$$

and hence

$$P_2(\mu') = \frac{3}{2}(\sin\theta' \cos\phi')^2 - \frac{1}{2}\,. \tag{7.191}$$

Only the first term of the Legendre polynomial survives the angular integration in (7.189) and we get

$$Q(\hat{\mathbf{z}}, \mathbf{k} \parallel \hat{\mathbf{x}}) \propto \frac{4\pi}{5} \Theta_2(k)\,,$$
$$U(\hat{\mathbf{z}}, \mathbf{k} \parallel \hat{\mathbf{x}}) \propto 0\,. \tag{7.192}$$

This is consistent with what we found before: the polarization created by density perturbations has vanishing Stokes parameter U when the x-axis is aligned with the wavevector of the plane wave.

2. Next, we take the wavevector to lie in the xz-plane, $\mathbf{k} \perp \hat{\mathbf{y}}$. Writing

$$\hat{\mathbf{k}} = (\sin\theta_k\,, 0\,, \cos\theta_k)\,, \tag{7.193}$$

we get

$$\mu' = \hat{\mathbf{k}} \cdot \hat{\mathbf{n}}' = \sin\theta_k \hat{n}'_x + \cos\theta_k \hat{n}'_z$$
$$= \sin\theta_k \sin\theta' \cos\phi' + \cos\theta_k \cos\theta'\,, \tag{7.194}$$

and hence

$$P_2(\mu') = \frac{3}{2}\sin^2\theta_k (\sin\theta' \cos\phi')^2 + \cdots\,, \tag{7.195}$$

where the ellipses stand for terms that lead to vanishing contributions when integrated against $\cos 2\phi'$ or $\sin 2\phi'$. The only difference with the relevant term in (7.191) is the extra factor of $\sin^2 \theta_k$. We therefore get

$$Q(\hat{\mathbf{z}}, \mathbf{k} \perp \hat{\mathbf{y}}) \propto \frac{4\pi}{5} \sin^2 \theta_k \Theta_2(k),$$
$$U(\hat{\mathbf{z}}, \mathbf{k} \perp \hat{\mathbf{y}}) \propto 0.$$
(7.196)

Projected onto the xy-plane the direction of the wavevector is still aligned with the x-axis, so the Stokes parameter U still vanishes.

3. Finally, we consider a wavevector pointing in an arbitrary direction. Its components are

$$\hat{\mathbf{k}} = (\sin \theta_k \cos \phi_k,\ \sin \theta_k \sin \phi_k,\ \cos \theta_k).$$
(7.197)

I will leave it to you to show that this leads to the following Stokes parameters:

$$Q(\hat{\mathbf{z}}, \mathbf{k}) \propto \frac{4\pi}{5} \sin^2 \theta_k \cos 2\phi_k\, \Theta_2(k),$$
$$U(\hat{\mathbf{z}}, \mathbf{k}) \propto \frac{4\pi}{5} \sin^2 \theta_k \sin 2\phi_k\, \Theta_2(k).$$
(7.198)

Comparison with (7.178) reveals that this is indeed a pure E-mode:

$$E(\hat{\mathbf{z}}, \mathbf{k}) \propto \frac{4\pi}{5} \sin^2 \theta_k\, \Theta_2(k),$$
$$B(\hat{\mathbf{z}}, \mathbf{k}) \propto 0,$$
(7.199)

where we have implicitly used that l is the projection of $\mathbf{k}$ onto the plane of the sky to identify ϕ_k with ϕ_l.

Exercise 7.6 Derive the result in (7.198).

Ultimately, we need the result not just for the specific direction $\hat{\mathbf{z}}$, but for a general line-of-sight $\hat{\mathbf{n}}$. To obtain this, we perform the rotation $\hat{\mathbf{z}} \to -\hat{\mathbf{n}}$, with the minus sign appearing because $\hat{\mathbf{n}}$ is the inverse of the direction of the momentum of the incoming photons, $\hat{\mathbf{n}} = -\hat{\mathbf{p}}$. Since the E-mode is invariant under this rotation, we must simply replace $\cos \theta_k$ with $-\hat{\mathbf{n}} \cdot \hat{\mathbf{k}}$ and (7.199) becomes

$$E(\hat{\mathbf{n}}, \mathbf{k}) \propto \frac{4\pi}{5} \left[1 - (\hat{\mathbf{n}} \cdot \hat{\mathbf{k}})^2 \right] \Theta_2(k),$$
$$B(\hat{\mathbf{n}}, \mathbf{k}) \propto 0.$$
(7.200)

This result assumes the flat-sky approximation and hence only applies to directions $\hat{\mathbf{n}}$ near the z-axis. The general result is much more complicated, but the same conclusion holds: no B-modes are generated by scalar density fluctuations.

Quadrupole from velocity gradients

We have seen that polarization is generated by the scattering of a quadrupolar anisotropy. However, in the tight-coupling regime, this anisotropy is erased by the frequent scattering of the photons. Polarization is therefore only generated near recombination when anisotropies grow by free streaming. Since the electron density drops sharply at recombination there is only a short time window in which the effect produces significant polarization of the CMB.

In the case of density perturbations, the quadrupole moment is produced by velocity gradients in the photon–baryon fluid. In Appendix B, I will derive this explicitly using the Boltzmann equation. Here, I will give a more heuristic explanation of the effect [27]. Consider scattering at a position $\mathbf{x}_0$. The scattered photons came from a position $\mathbf{x} = \mathbf{x}_0 + \ell_\gamma \hat{\mathbf{n}}'$, where $\ell_\gamma = \Gamma^{-1}$ is the mean free path and $\hat{\mathbf{n}}'$ is the direction of the incoming photon. The velocity of the photon–baryon fluid at that position was

$$\mathbf{v}_b(\mathbf{x}) \approx \mathbf{v}_b(\mathbf{x}_0) + \ell_\gamma\, \hat{\mathbf{n}}' \cdot \boldsymbol{\nabla}\, \mathbf{v}_b(\mathbf{x}_0)\,. \tag{7.201}$$

The temperature seen by the scatterer at $\mathbf{x}_0$ is Doppler shifted and hence given by

$$\Theta(\mathbf{x}_0, \hat{\mathbf{n}}') \equiv \frac{\delta T(\mathbf{x}_0, \hat{\mathbf{n}}')}{\bar{T}} = \hat{\mathbf{n}}' \cdot [\mathbf{v}_b(\mathbf{x}) - \mathbf{v}_b(\mathbf{x}_0)] \approx \ell_\gamma \hat{n}'_i \hat{n}'_j \partial_i (v_b)_j\,. \tag{7.202}$$

Being quadratic in $\hat{\mathbf{n}}'$, the induced temperature anisotropy indeed has the required quadrupole component Θ_2. The Fourier transform of (7.202) is

$$\Theta(\mathbf{k}, \hat{\mathbf{n}}') = \int \mathrm{d}^3 x\, e^{-i\mathbf{k}\cdot\mathbf{x}}\, \Theta(\mathbf{x}, \hat{\mathbf{n}}') = \ell_\gamma \hat{n}'_i \hat{n}'_j (ik^i)(i\hat{k}^j v_b)$$

$$= -\frac{k}{\Gamma} v_b\, (\hat{\mathbf{k}} \cdot \hat{\mathbf{n}}')^2 \subset -\frac{2}{3}\frac{k}{\Gamma} v_b\, P_2(\mu')\,, \tag{7.203}$$

from which we can read off $\Theta_2(k)$.

E-mode spectrum

Let $\mathbf{x}_0$ now be a point on the surface of last-scattering. Using (7.200) and (7.203), the E-mode polarization created by a plane wave perturbation then is

$$E(\hat{\mathbf{n}}, \mathbf{k}) \propto \left[1 - (\hat{\mathbf{n}} \cdot \hat{\mathbf{k}})^2\right] \Theta_2(k)\, e^{i\mathbf{k}\cdot\mathbf{x}_0}$$

$$\propto \frac{k}{\Gamma} v_b(\eta_*, \mathbf{k})(1 - \mu^2)\, e^{ik\eta_0\mu}\,, \tag{7.204}$$

where we have included the factor of $e^{i\mathbf{k}\cdot\mathbf{x}_0}$ to describe the modulation of the amplitude as a function of the position $\mathbf{x}_0 = (\eta_0 - \eta_*)\hat{\mathbf{n}} \approx \eta_0 \hat{\mathbf{n}}$. Next, we expand the exponential in spherical Bessel functions and extract the multipole moment $E_l(\mathbf{k})$. The details are given in Appendix B, where we find that

$$E_l(\mathbf{k}) \propto \frac{k}{\Gamma} v_b(\eta_*, \mathbf{k}) \sqrt{\frac{(l+2)!}{(l-2)!}} \frac{j_l(k\eta_0)}{(k\eta_0)^2} . \tag{7.205}$$

For large l, the square root becomes l^2 and the polarization transfer function is

$$\boxed{\Theta_{E,l}(k) \equiv \frac{E_l(\mathbf{k})}{\mathcal{R}_i(\mathbf{k})} \propto \frac{k}{\Gamma} G_*(k) \frac{l^2}{(k\eta_0)^2} j_l(k\eta_0)} , \tag{7.206}$$

where $G_*(k) \equiv v_b(\eta_*, \mathbf{k})/\mathcal{R}_i(\mathbf{k})$ is the transfer function of the photon–baryon velocity. The E-mode power spectrum then is

$$C_l^{\mathrm{EE}} = 4\pi \int_0^\infty \mathrm{d}\ln k \; \Theta_{E,l}^2(k) \, \Delta_{\mathcal{R}}^2(k) . \tag{7.207}$$

A more exact treatment would derive $\Theta_{E,l}(k)$ from the Boltzmann equation (see Appendix B).

Figure 7.34 shows the predicted E-mode spectrum and compares it to the temperature power spectrum. Since the E-mode signal is created by the gradient of the fluid velocity, its oscillations are expected to be out of phase with the oscillations in the temperature spectrum; recall that $\delta_\gamma' = -\frac{4}{3}\nabla \cdot \mathbf{v}_b$. This is indeed seen in Fig. 7.34: the peaks of the E-mode spectrum coincide with the troughs of the

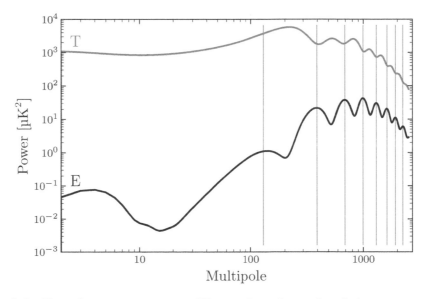

Fig. 7.34   Plot of the E-mode power spectrum. We see that the peaks of the spectrum are aligned with the troughs of the temperature power spectrum. The peak at $l < 10$ is due to the rescattering of the CMB photons after reionization.

temperature power spectrum. The polarization spectrum also has a pronounced peak at $l < 10$, which is due to the rescattering of the CMB photons after reionization. The size of this reionization peak depends on the optical depth τ, so that measurements of the low-l polarization help to break the degeneracy between τ and $A_{\rm s}$ that we found for the temperature fluctuations (see Section 7.5.3).

7.6.5 B-modes from Tensors

We have seen that primordial density fluctuations cannot produce a B-mode polarization signal. As I will now show, such a signal is therefore a unique discovery channel for primordial gravitational waves. The search for primordial B-modes is one of the most active areas of observational cosmology.

Above we found that the scattering of incident radiation with temperature anisotropies $\Theta(\hat{\mathbf{n}}')$ creates the following polarization for outgoing radiation near the z-axis:

$$
\begin{aligned}
Q(\mathbf{z}) &\propto \int d\Omega'\, \Theta(\hat{\mathbf{n}}')\, \sin^2\theta' \cos 2\phi' \,, \\
U(\mathbf{z}) &\propto \int d\Omega'\, \Theta(\hat{\mathbf{n}}')\, \sin^2\theta' \sin 2\phi' \,.
\end{aligned}
\tag{7.208}
$$

We also showed that only the quadrupole component of $\Theta(\hat{\mathbf{n}}')$ produces polarization. I will now show how such a quadrupolar anisotropy is generated by gravitational waves.

Quadrupole from gravitational waves

Consider a gravitational wave propagating in the z-direction. The perturbation of the spatial metric is

$$
\delta g_{ij} = a^2 h_{ij} = a^2 \begin{pmatrix} h_+ & h_\times & 0 \\ h_\times & -h_+ & 0 \\ 0 & 0 & 0 \end{pmatrix}.
\tag{7.209}
$$

The presence of this perturbation affects the free streaming of the photons. Using (7.209) in the geodesic equation gives (see Problem 7.6)

$$
\begin{aligned}
\frac{d\Theta^{(t)}}{d\eta} = \frac{d\ln(aE)}{d\eta} &= -\frac{1}{2}\frac{\partial h_{ij}}{\partial \eta}(\hat{\mathbf{n}}')^i(\hat{\mathbf{n}}')^j \\
&= -\frac{1}{2}\frac{\partial h_+}{\partial \eta}\left[(\hat{n}_x')^2 - (\hat{n}_y')^2\right] - \frac{\partial h_\times}{\partial \eta}\hat{n}_x'\hat{n}_y' \\
&= -\frac{1}{2}\sin^2\theta'\left(\frac{\partial h_+}{\partial \eta}\cos 2\phi' + \frac{\partial h_\times}{\partial \eta}\sin 2\phi'\right),
\end{aligned}
\tag{7.210}
$$

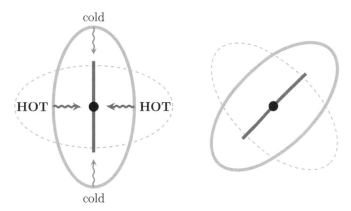

cold

HOT 〜〜〜 ● 〜〜〜 HOT

cold

Fig. 7.35 Cartoon illustrating that the anisotropic stretching and compressing of space by a gravitational wave creates a temperature quadrupole and hence leads to CMB polarization. The two polarizations of the gravitational wave produce polarization of the CMB photons with a relative angle of 45°. This is why gravitational waves produce both E- and B-modes, while density perturbations create only E-modes.

where we have written the direction of the incoming photon as

$$\hat{\mathbf{n}}' = (\sin\theta'\cos\phi',\ \sin\theta'\sin\phi',\ \cos\theta')\,. \tag{7.211}$$

The angular dependence in (7.210) induces a quadrupole moment in the radiation field and the scattering of this tensor-induced temperature anisotropy will then lead to polarization.

To understand this intuitively, consider how a gravitational wave propagating in the z-direction distorts space in the xy-plane (see Fig. 7.35). The key characteristic of gravitational waves is that these distortions are anisotropic, stretching space in one direction and compressing it in an orthogonal direction. Along the expanding direction, the photons are redshifted and become colder, while in the contracting direction, the photons are blueshifted and become hotter. This produces precisely the temperature quadrupole required to create CMB polarization. Since the gravitational wave has the two polarization modes rotated by 45°, we get two polarization patterns with a relative orientation of 45°. This is the fundamental reason why gravitational waves produce both E- and B-modes. In the following, I will prove this more rigorously.

B-modes from gravitational waves

Let us first focus on the $\times$ polarization of the gravitational wave. The induced temperature fluctuations then have the following angular dependence:

$$\Theta_{\times}^{(t)}(\hat{\mathbf{n}}',\mathbf{k}\parallel\hat{\mathbf{z}}) \propto \frac{1}{2}\sin^2\theta'\sin 2\phi' = \hat{n}'_x\hat{n}'_y\,. \tag{7.212}$$

This is the result for a wavevector pointing in the z-direction, but we need the result for a more general wavevector $\mathbf{k}$. To show that gravitational waves produce B-modes, it suffices to consider the special case

$$\hat{\mathbf{k}} = (\sin\theta_k\,,\, 0\,,\, \cos\theta_k) \perp \hat{\mathbf{y}}\,, \tag{7.213}$$

where θ_k is the angle between $\hat{\mathbf{k}}$ and the inverse of the line-of-sight $-\hat{\mathbf{n}}$. The projection of this wavevector onto the sky is parallel to the x-axis ($\phi_l = 0$), so it should lead to vanishing U if the polarization is a pure E-mode; cf. (7.180). Conversely, finding $U \neq 0$ for this wavevector would prove that a nonzero B-mode was generated.

We first need to transform the result in (7.212), so that it gives the temperature anisotropy induced by the wavevector in (7.213). We achieve this by rotating the coordinate system by an angle $-\theta_k$ around the y-axis. This gives

$$\begin{aligned}
\hat{n}'_x &\to \cos\theta_k\,\hat{n}'_x - \sin\theta_k\,\hat{n}'_z\,,\\
\hat{n}'_y &\to \hat{n}'_y\,,
\end{aligned} \tag{7.214}$$

and hence

$$\Theta^{(t)}_\times(\hat{\mathbf{n}}',\mathbf{k}\perp\hat{\mathbf{y}}) \propto \cos\theta_k\,\hat{n}'_x\hat{n}'_y - \sin\theta_k\,\hat{n}'_y\hat{n}'_z$$

$$\propto \cos\theta_k\,[\sin^2\theta'\sin 2\phi'] - \sin\theta_k\,[\sin\theta'\cos\theta'\sin\phi']\,. \tag{7.215}$$

The second term leads to a vanishing contribution in (7.208), while the first term gives

$$Q_\times(\hat{\mathbf{z}},\mathbf{k}\perp\hat{\mathbf{y}}) = 0\,, \tag{7.216}$$

$$U_\times(\hat{\mathbf{z}},\mathbf{k}\perp\hat{\mathbf{y}}) \propto \cos\theta_k \int_{-1}^{1} d\cos\theta'\,\sin^4\theta' \int_0^{2\pi} d\phi'\,\sin^2(2\phi')$$

$$= \frac{16\pi}{15}\cos\theta_k\,. \tag{7.217}$$

Except for a plane wave that is oriented precisely orthogonal to the line-of-sight ($\theta_k = 90°$), we have found a nonzero Stokes parameter U. Gravitational waves therefore indeed produce B-mode polarization.

Exercise 7.7 Repeating the above analysis for the $+$ polarization of the gravitational wave, show that

$$Q_+(\hat{\mathbf{z}},\mathbf{k}\perp\hat{\mathbf{y}}) \propto \frac{4\pi}{15}(\cos^2\theta_k + 1)\,, \tag{7.218}$$

$$U_+(\hat{\mathbf{z}},\mathbf{k}\perp\hat{\mathbf{y}}) = 0\,. \tag{7.219}$$

We see that this polarization signal has vanishing U parameter and is hence an E-mode.

In summary, we have found that gravitational waves produce both E- and B-modes, while density fluctuations only create E-modes.

B-mode spectrum

The solid line in Fig. 7.36 shows the B-mode spectrum expected for a scale-invariant spectrum of primordial gravitational waves. The damping of the signal for $l > 100$ is caused by the finite thickness of the last-scattering surface [31], while the peak at $l < 10$ is again due to the rescattering of the CMB photons after reionization. The oscillations in the spectrum are analogous to the acoustic oscillations in the angular power spectrum from scalar perturbations, but their frequency is set by the speed of light and not the sound speed of the photon–baryon fluid. The peaks of the spectrum are therefore at $l_n \approx n\pi\chi_*/\eta_*$. Note that the amplitude of the signal is *not* predicted, but depends on the unknown tensor-to-scalar ratio r. Recent measurements of CMB polarization imply the upper bound $r < 0.035$ [32].

Although scalar fluctuations don't produce B-mode polarization at the moment of last-scattering, they *do* create B-modes at later times via gravitational lensing. As the CMB photons travel through the large-scale structure of the universe they get deflected and E-mode polarization gets converted into B-modes. The resulting lensing-induced B-mode power spectrum is shown as the dashed line in Fig. 7.36. The signal has recently been detected by BICEP, SPTpol and Polarbear. While the lensing B-modes provide an interesting consistency check, they are a foreground in the search for primordial B-modes.

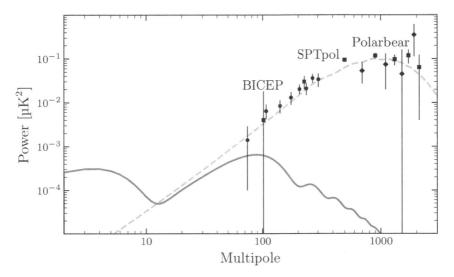

Fig. 7.36 Current measurements of the B-mode spectrum [28–30]. The measurements have detected the lensing-induced B-modes (dashed line) but are not yet sensitive enough to reveal the primordial B-mode signal (solid line). Shown is the primordial B-mode signal corresponding to a tensor-to-scalar ratio of $r = 0.01$.

7.7 Summary

In this chapter, we have studied the physics of the CMB anisotropies. We related the observed temperature fluctuations $\delta T(\hat{\mathbf{n}})$ to the primordial curvature perturbations $\mathcal{R}_i(\mathbf{k}) \equiv \mathcal{R}(0, \mathbf{k})$, and computed the two-point correlation function, $\langle \delta T(\hat{\mathbf{n}}) \delta T(\hat{\mathbf{n}}') \rangle$ given the power spectrum $\langle \mathcal{R}_i(\mathbf{k}) \mathcal{R}_i(\mathbf{k}') \rangle$. The key steps involved in the calculation are summarized by the following schematic:

$$
\mathcal{R}_i(\mathbf{k}) \xrightarrow{\text{evolution (§7.4)}} \begin{pmatrix} \delta_\gamma \\ \Psi \\ v_b \end{pmatrix}_{\eta_*} \xrightarrow[\text{projection (§7.3)}]{\text{free streaming (§7.2)}} \delta T(\hat{\mathbf{n}}) \,.
$$

We worked through this in reverse order:

- *Free streaming* By integrating the geodesic equation along the line-of-sight, we obtained a relation between the observed temperature fluctuations in a particular direction and the corresponding fluctuations on the surface of last-scattering

$$
\frac{\delta T}{\bar{T}}(\hat{\mathbf{n}}) = \left(\frac{1}{4}\delta_\gamma + \Psi \right)_* - (\hat{\mathbf{n}} \cdot \mathbf{v}_b)_* + \int_{\eta_*}^{\eta_0} \mathrm{d}\eta \, (\Phi' + \Psi') \,, \tag{7.220}
$$

where the subscript $*$ indicates quantities evaluated at last-scattering. The three terms on the right-hand side of (7.220) are the Sachs–Wolfe (SW) term, the Doppler term and the integrated Sachs–Wolfe (ISW) term, respectively.

- *Projection* We then wrote the temperature anisotropies as a sum over spherical harmonics

$$
\frac{\delta T}{\bar{T}}(\hat{\mathbf{n}}) = \sum_{lm} a_{lm} Y_{lm}(\hat{\mathbf{n}}) \,. \tag{7.221}
$$

Ignoring the ISW contribution, the multipole moments arising from the line-of-sight solution (7.220) are

$$
a_{lm} = 4\pi i^l \int \frac{\mathrm{d}^3 k}{(2\pi)^3} \underbrace{\left[F_*(k)\, j_l(k\chi_*) - G_*(k)\, j_l'(k\chi_*) \right]}_{\equiv\, \Theta_l(k)} \mathcal{R}_i(\mathbf{k})\, Y_{lm}^*(\hat{\mathbf{k}}) \,, \tag{7.222}
$$

where $F_*(k) \equiv (\frac{1}{4}\delta_\gamma + \Psi)(\eta_*, \mathbf{k})/\mathcal{R}_i(\mathbf{k})$ and $G_*(k) \equiv v_b(\eta_*, \mathbf{k})/\mathcal{R}_i(\mathbf{k})$ are the transfer functions for the Sachs–Wolfe and Doppler terms, respectively. The two-point function of the multipole moments, $\langle a_{lm} a_{l'm'}^* \rangle = C_l\, \delta_{ll'}\, \delta_{mm'}$, defines the angular power spectrum C_l. For the solution (7.222), the power spectrum is given by

$$
C_l = 4\pi \int \mathrm{d}\ln k \, \Theta_l^2(k)\, \Delta_{\mathcal{R}}^2(k) \,. \tag{7.223}
$$

We see that the power spectrum is determined by the power spectrum of the primordial curvature perturbations, $\Delta_{\mathcal{R}}^2(k)$, and the transfer function, $\Theta_l(k)$.

Table 7.2 Effects determining the shape of the CMB power spectrum and their dependence on the parameters of the LCDM model (adapted from [33])

	Effect	Relevant quantity	Parameter
(1)	Peak locations	$\theta_s = \dfrac{r_{s,*}}{d_{A,*}}$	$\leftarrow \omega_m, \omega_b$ $\leftarrow \Omega_\Lambda, \omega_m$
(2)	Odd/even peak amplitudes	R_*	ω_b
(3)	Overall peak amplitude	a_{eq}	ω_m
(4)	Damping envelope	$\theta_D = \dfrac{r_{D,*}}{d_{A,*}}$	$\leftarrow \omega_m, \omega_b$ $\leftarrow \Omega_\Lambda, \omega_m$
(5)	Global amplitude	$\Delta_{\mathcal{R}}$	A_s
(6)	Global tilt	$\mathrm{d}\ln \Delta_{\mathcal{R}}^2/\mathrm{d}\ln k$	n_s
(7)	Amplitude for $l > 10$	τ	τ
(8)	Tilt for $l < 20$	$a_{\Lambda m}$	Ω_Λ

The latter describes both the evolution until decoupling and the projection onto the surface of last-scattering.

- *Evolution* To determine the transfer functions $F_*(k)$ and $G_*(k)$, we studied the evolution of fluctuations in the primordial plasma. As long as photons and baryons are tightly coupled, they can be treated as a single fluid. Fluctuations in the photon density then satisfy the equation of a forced harmonic oscillator

$$\delta_\gamma'' + \frac{R'}{1+R}\delta_\gamma' + c_s^2 k^2 \delta_\gamma = -\frac{4}{3}k^2\Psi + 4\Phi'' + \frac{4R'}{1+R}\Phi', \tag{7.224}$$

where $R \equiv \frac{3}{4}\bar{\rho}_b/\bar{\rho}_\gamma$ and $c_s^2 \equiv [3(1+R)]^{-1}$. This describes the propagation of sound waves in the plasma. On small scales, the time dependence of the metric potentials can be ignored and that of the baryon-to-photon ratio $R(t)$ can be accounted for in a WKB approximation. The high-frequency solution of (7.224) then is

$$\delta_\gamma = (1+R)^{-1/4}\Big[C\cos(kr_s) + D\sin(kr_s)\Big] - 4(1+3R)\Psi, \tag{7.225}$$

where $r_s = \int \mathrm{d}\eta/\sqrt{3(1+R)}$ is the sound horizon. The constants C and D must be determined by matching to the superhorizon limit, with $C \gg D$ for adiabatic initial conditions. We also discussed a more accurate semi-analytic solution to (7.224) due to Weinberg [9]:

$$\delta_\gamma = \frac{4\mathcal{R}_i}{5}\left[\frac{S(k)}{(1+R)^{1/4}}\cos(kr_s + \theta(k)) - (1+3R)\,T(k)\right], \tag{7.226}$$

where $S(k)$, $T(k)$ and $\theta(k)$ are defined as interpolation functions in (7.116). Finally, we included photon viscosity which becomes important on scales less than the diffusion length of the photons and leads to an exponential damping of the fluctuations.

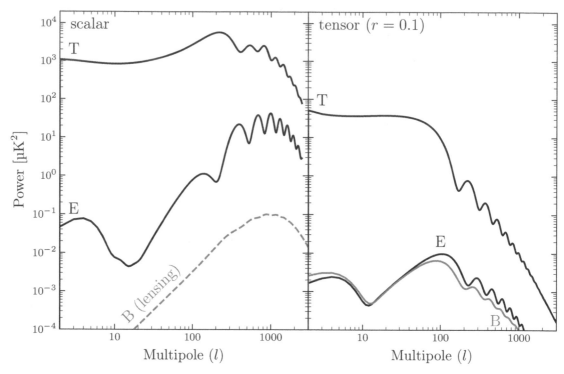

Fig. 7.37 Summary of the CMB spectra arising from primordial scalar perturbations (*left*) and tensor perturbations (*right*). To avoid clutter, I didn't plot the TE cross spectra.

We explained how the shape of the CMB power spectrum depends on the ΛCDM parameters $\{A_{\rm s}, n_{\rm s}, \omega_b, \omega_m, \Omega_\Lambda, \tau\}$. The main effects are summarized in Table 7.2. Three length scales are imprinted in the spectrum:

- $r_{s,*}$: The peak locations depend on the sound horizon at last-scattering.
- k_D^{-1}: The damping of the spectrum is determined by the diffusion scale.
- $d_{A,*}$: The map to angular scales depends on the distance to last-scattering.

The sound horizon and the diffusion scale are fixed by pre-recombination physics and hence in the ΛCDM model depend only on ω_b and ω_m (for fixed radiation density). These parameters also determine the peak heights through baryon loading (ω_b) and radiation driving (ω_m). The angular diameter distance to last-scattering is sensitive to the geometry (Ω_k, H_0) and the energy content (Ω_m, Ω_Λ) of the late universe.

Finally, I gave a brief introduction to CMB polarization. I showed that polarization is generated by scattering if the temperature anisotropies seen by the electrons have a quadrupole moment. The signal can be separated into a curl-free E-mode and a divergenceless B-mode. We proved that density fluctuations only produce E-modes, making the B-mode signal a unique signature of primordial gravitational waves (see Fig. 7.37). As we will discuss in the next chapter, these gravitational waves carry critical information about the physics of inflation.

Further Reading

There are many fantastic lecture notes on CMB physics. The ones I am most familiar with are [33–36]. The presentation in this chapter was inspired by the lectures of Anthony Challinor [37] and Eiichiro Komatsu [38], and I am grateful to Anthony and Eiichiro for enlightening me about many subtle aspects of this topic. Wayne Hu's thesis [39] is a masterpiece with a lot of very useful pedagogical material. Dodelson's book is also especially good for this part of cosmology [26]. CMB polarization is treated in [40–43] and a classic review of CMB lensing is [44]. All numerical plots in this chapter were produced with the Boltzmann code CLASS [10].

Problems

7.1 CMB dipole

The CMB is assumed to be isotropic with a temperature T in an inertial frame S. Consider another inertial frame S', moving with velocity v with respect to S. The energy E and three-momentum $\mathbf{p}$ of a particle in the two frames are related by a Lorentz transformation

$$E = \gamma(E' + \mathbf{v} \cdot \mathbf{p}'),$$

where $\gamma = 1/\sqrt{1 - v^2/c^2}$. A photon has $E = pc$.

1. Show that the CMB will also appear thermal in S', but with an anisotropic temperature

$$T'(\theta') = \frac{T}{\gamma(1 - (v/c)\cos\theta')} = T\left(1 + \frac{v}{c}\cos\theta' + O(v^2/c^2)\right),$$

where θ' is the angle between the velocity $\mathbf{v}$ and the direction of the incoming photon $\hat{\mathbf{n}} = -\hat{\mathbf{p}}'$.

2. Let T'_+ and T'_- be the maximum and minimum temperatures seen in the frame S'. Show that

$$T = \sqrt{T'_+ T'_-}\,.$$

The observed CMB has $T'_+ - T'_- \approx 6.5\,\mathrm{mK}$ and $T = \sqrt{T'_+ T'_-} \approx 2.7\,\mathrm{K}$. How fast are we traveling with respect to the universe's preferred inertial frame?

3. Consider now the cosmic neutrino background (CνB). The neutrinos have a small mass and so $E^2 = p^2 c^2 + m^2 c^4$. Today, the neutrinos are traveling at non-relativistic speeds. Show that the CνB will not be thermal, even at a fixed angle.

7.2 Sound horizon

Show that the sound horizon can be written as

$$r_s(\eta) = \frac{2}{3k_{\mathrm{eq}}} \sqrt{\frac{6}{R_{\mathrm{eq}}}} \ln\left(\frac{\sqrt{1 + R(\eta)} + \sqrt{R(\eta) + R_{\mathrm{eq}}}}{1 + \sqrt{R_{\mathrm{eq}}}} \right),$$

where $k_{\mathrm{eq}} = \mathcal{H}(\eta_{\mathrm{eq}})$ and $R_{\mathrm{eq}} = R(\eta_{\mathrm{eq}})$. Determine the size of the sound horizon at last-scattering, using the best-fit values of the cosmological parameters. What CMB multipole moment does this scale correspond to?

7.3 Damping scale

In this problem, you will estimate the Silk damping scale in the pre-recombination limit (i.e. assuming that all electrons are free).

1. Ignoring helium recombination, show that

$$\frac{n_e \sigma_T}{H} = 0.069 \, \Omega_b h \frac{H_0}{H a^3},$$

where $\sigma_T = 0.665 \times 10^{-24} \, \mathrm{cm}^2$ is the Thomson cross section. How would this result be changed by the presence of helium?

2. Setting $R = 0$, show that the Silk damping scale is

$$k_S^{-2}(\eta) = \frac{8}{45} \int_0^\eta \frac{\mathrm{d}\tilde{\eta}}{n_e \sigma_T a(\tilde{\eta})}$$

$$= (5.9 \, \mathrm{Mpc})^2 \left(\frac{a}{0.001}\right)^{5/2} f(a/a_{\mathrm{eq}}) \left(\frac{\Omega_b h^2}{0.022}\right)^{-1} \left(\frac{\Omega_m h^2}{0.143}\right)^{-1/2},$$

where

$$f(y) \equiv 5\sqrt{1 + 1/y} - \frac{20}{3}(1 + 1/y)^{3/2} + \frac{8}{3}\left[(1 + 1/y)^{5/2} - 1/y^{5/2}\right].$$

Plot your result as a function of the scale factor a and determine the approximate damping scale at recombination. Is the true damping scale larger or smaller than this?

7.4 WKB solution

Consider the equation of a damped harmonic oscillator

$$u'' + \gamma(\eta)u' + \omega^2(\eta)u = 0,$$

where $\gamma > 0$. We assume that the frequency is slowly varying, in the sense that $\omega' \ll \omega^2$.

1. Let $u \equiv fv$ and find the function $f(\eta)$ that removes the damping term. Show that the oscillator equation then becomes

$$v'' + \left(\omega^2 - \frac{1}{2}(\gamma' + \gamma^2) \right) v = 0 \,,$$

which reduces to $v'' + \omega^2 v = 0$ for $\omega^2 \gg \{\gamma', \gamma^2\}$.

2. Consider the ansatz $v = e^{i\delta}$, with $|\delta''| \ll (\delta')^2$. Show that the solution of the oscillator equation is

$$v \propto \omega^{-1/2} \exp\left(\pm i \int^{\eta} \omega \, d\tilde{\eta} \right) .$$

3. Apply these results to the equation

$$\delta_\gamma'' + \frac{R'}{1+R} \delta_\gamma' + \frac{k^2}{3(1+R)} \delta_\gamma = 0 \,.$$

Show that the amplitude of the oscillations scales as $\delta_\gamma \propto (1+R)^{-1/4}$.

7.5   Hu–Sugiyama solution

In this problem, you will derive a semi-analytic solution for the oscillations in the photon–baryon fluid.

1. Show that equation (7.63) can be written as

$$\left[\frac{d^2}{d\eta^2} + \frac{R'}{1+R} \frac{d}{d\eta} + k^2 c_s^2 \right] d_\gamma = -\frac{4k^2}{3} \left(\frac{\Phi}{1+R} + \Psi \right),$$

where $d_\gamma \equiv \delta_\gamma - 4\Phi$ and $c_s^2 \equiv [3(1+R)]^{-1}$. In Problem 7.4, we derived the homogeneous solution of this equation in the WKB approximation. The solution of the inhomogeneous equation can be written as

$$d_\gamma \subset -\frac{4k^2}{3} \int_0^{\eta} d\tilde{\eta} \left[\frac{\Phi(\tilde{\eta})}{1+R(\tilde{\eta})} + \Psi(\tilde{\eta}) \right] G(\eta, \tilde{\eta}; k) \,,$$

where $G(\eta, \tilde{\eta}; k)$ is the Green's function. Assuming adiabatic initial conditions, show that the complete solution is [12]

$$d_\gamma = -6\Phi_i \frac{\cos(kr_s)}{(1+R)^{1/4}}$$
$$-\frac{4k}{\sqrt{3}} \int_0^{\eta} d\tilde{\eta} \left[\frac{\Phi}{1+R} + \Psi \right] \sqrt{1+R} \, \sin[k(r_s(\eta) - r_s(\tilde{\eta}))] \,,$$

where $r_s(\eta)$ is the sound horizon.

2. In the radiation-dominated era, the baryon-to-photon ratio is small, $R \ll 1$, and the solution for $\Phi(\eta) \approx \Psi(\eta)$ is known explicitly (see Section 6.3.1). Determine the solution for d_γ in this regime.

7.6   Temperature anisotropies from tensor modes

In this problem, you will derive the temperature anisotropies induced by primordial tensor modes, i.e. metric perturbations of the form

$$ds^2 = a^2(\eta)\Big[-d\eta^2 + (\delta_{ij} + h_{ij})dx^i dx^j\Big],$$

where h_{ij} is transverse and traceless.

1. By considering the geodesic equation in the presence of the tensor perturbations, show that the energy of a photon evolves as

$$\frac{d\ln(aE)}{d\eta} = -\frac{1}{2}h'_{ij}\,\hat{p}^i\hat{p}^j\,,$$

   where $\hat{p}^i$ is a unit vector in the photon's direction of propagation.

   Hint: To linear order in h_{ij}, the relevant connection coefficients are $\Gamma^0_{00} = \mathcal{H}$ and $\Gamma^0_{ij} = \mathcal{H}\delta_{ij} + \mathcal{H}h_{ij} + \frac{1}{2}h'_{ij}$.

2. Assuming instantaneous decoupling, show that the line-of-sight solution for the induced temperature anisotropy is

$$\Theta^{(t)}(\hat{\mathbf{n}}) = -\frac{1}{2}\int_{\eta_*}^{\eta_0} d\eta \int \frac{d^3k}{(2\pi)^3}\, h'_{ij}(\eta,\mathbf{k})\,\hat{n}^i\hat{n}^j\, e^{ik\chi(\eta)\,\hat{\mathbf{k}}\cdot\hat{\mathbf{n}}}\,, \qquad (1)$$

   where $\chi(\eta) \equiv \eta_0 - \eta$ in a flat universe.

3. Consider a single gravitational wave with wavevector $\mathbf{k}$ pointing in the z-direction and write h_{ij} in terms of its two polarization modes,

$$h_{ij} = \begin{pmatrix} h_+ & h_\times & 0 \\ h_\times & -h_+ & 0 \\ 0 & 0 & 0 \end{pmatrix}.$$

   Show that

$$h'_{ij}\hat{n}^i\hat{n}^j = \sin^2\theta\left(h'_+\cos 2\phi + h'_\times \sin 2\phi\right),$$

   where θ and ϕ are the angles of $\hat{\mathbf{n}}$ in polar coordinates.

4. It is convenient to transform to the helicity basis, where $h_{\pm 2} \equiv h_+ \mp ih_\times$ (see Section 6.5). Show that the integrand in (1) then becomes

$$\mathcal{I}(\eta, k\hat{\mathbf{z}}) \equiv -\sqrt{\frac{2\pi}{15}} \sum_{\lambda=\pm 2} h'_\lambda(\eta, k\hat{\mathbf{z}})\, Y_{2,\lambda}(\hat{\mathbf{n}})\, e^{ik\chi\cos\theta}\,, \qquad (2)$$

   where $\cos\theta \equiv \hat{\mathbf{z}}\cdot\hat{\mathbf{n}}$.

5. Using the Rayleigh plane-wave expansion of the exponential, it can be shown that

$$Y_{2,\pm 2}(\hat{\mathbf{n}})\, e^{ik\chi\cos\theta} = -\sqrt{\frac{15}{8}} \sum_l \alpha_l\, \frac{j_l(k\chi)}{(k\chi)^2}\, Y_{l,\pm 2}(\hat{\mathbf{n}})\,,$$

   where $\alpha_l \equiv i^l\sqrt{2l+1}\sqrt{(l+2)!/(l-2)!}$.

So far, this is still for the special Fourier mode $\mathbf{k} = k\hat{\mathbf{z}}$. The contribution from a general Fourier mode $\mathbf{k}$ is obtained by rotating the result in (2), which gives

$$\mathcal{I}(\eta, \mathbf{k}) \equiv \sqrt{\frac{\pi}{2}} \sum_{\lambda=\pm 2} h'_\lambda(\eta, \mathbf{k}) \sum_{lm} \alpha_l \frac{j_l(k\chi)}{(k\chi)^2} D^l_{m\lambda}(\hat{\mathbf{k}}) Y_{lm}(\hat{\mathbf{n}}), \qquad (3)$$

where $D^l_{mm'}$ are Wigner matrices which satisfy the following orthogonality condition

$$\int d\hat{\mathbf{k}} \, D^l_{m\lambda}(\hat{\mathbf{k}}) D^{l'}_{m'\lambda}(\hat{\mathbf{k}}) = \frac{4\pi}{2l+1} \delta_{ll'} \delta_{mm'}.$$

Given the result in (3), show that the angular power spectrum of the tensor-induced anisotropies is

$$C_l^{(t)} = 4\pi \int d\ln k \left| \Theta_l^{(t)}(k) \right|^2 \Delta_h^2(k),$$

where $\Delta_h^2(k)$ is the primordial tensor power spectrum and $\Theta_l^{(t)}(k)$ is the transfer function. Show that the latter can be written as

$$\Theta_l^{(t)}(k) = \frac{1}{4} \sqrt{\frac{(l+2)!}{(l-2)!}} \int_{\eta_*}^{\eta_0} d\eta \, h'_{\perp 2}(\eta, k) \frac{j_l[k(\eta_0 - \eta)]}{[k(\eta_0 - \eta)]^2},$$

where $h_{\pm 2}(\eta, k) \equiv h_{\pm 2}(\eta, \mathbf{k})/h_{\pm 2}(0, \mathbf{k})$.

6. Discuss qualitatively the shape of the spectrum. In particular, show that a scale-invariant primordial spectrum leads to $l(l+1)C_l^{(t)} \approx$ const on large scales.

   Hint: To estimate the spectrum on large larges, take the shear of the gravitational waves to be $h'_\lambda(\eta, \mathbf{k}) \approx -\delta_D(\eta - 1/k) h_\lambda(\mathbf{k})$. In reality, the shear vanishes outside the horizon (where the tensor modes are frozen) and oscillates with a decaying amplitude inside the horizon. Here, this behaviour is being modeled by an impulsive shear at horizon entry.

7.7 CLASS

Download and install the CLASS software:

$$\text{https://lesgourg.github.io/class\_public/class.html}$$

Use it to reproduce some of the figures in this chapter.

Note: The website contains tutorials and example Python notebooks that will teach you the basic usage of CLASS.

8 Quantum Initial Conditions

One of the most remarkable features of inflation is that it provides a natural mechanism for creating the primordial density fluctuations that seeded the structure in the universe. Recall that the evolution of the inflaton field $\phi(t)$ governs the energy density of the early universe $\rho(t)$ and, hence, controls the end of inflation (see Fig. 8.1). Essentially, the field ϕ plays the role of a "clock" telling us the amount of inflationary expansion still remaining. However, by the uncertainty principle, arbitrarily precise timing is impossible in quantum mechanics. Instead, quantum-mechanical clocks necessarily have some variance, so the inflaton will have spatially varying fluctuations $\delta\phi(t, \mathbf{x})$. The time at which inflation ends will therefore vary across space. Although these fluctuations are a small quantum effect, they trigger large differences in the local expansion histories, which lead to significant differences in the local densities after inflation, $\delta\rho(t, \mathbf{x})$. Regions of space that end inflation first will start diluting and therefore end up having smaller densities, while regions that inflate for a longer time will have higher densities. It is worth emphasizing that the theory was not engineered to produce these fluctuations, but that their origin is instead a natural consequence of treating inflation quantum mechanically.

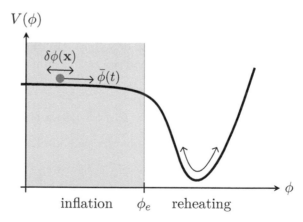

Fig. 8.1 Quantum fluctuations around the classical background evolution affect the duration of inflation. Regions acquiring negative fluctuations remain potential-dominated longer, and hence inflate longer, than regions with positive fluctuations. Different parts of the universe therefore undergo slightly different evolution, which induces variations in the density after inflation.

In this chapter, we will derive the spectrum of quantum fluctuations produced during inflation. We will begin, in Section 8.1, by deriving the classical equations of motion of the inflaton fluctuations in spatially flat gauge. We will find that each Fourier mode satisfies the equation of a harmonic oscillator with a time-dependent frequency. In Section 8.2, we will show how quantum zero-point fluctuations in these oscillators lead to a power spectrum of inflaton fluctuations. In Section 8.3, we will explain how these fluctuations are converted into density fluctuations after inflation, becoming the seeds for all structure in the universe. We will derive the forms of the primordial scalar and tensor power spectra, and relate these predictions to the shape of the inflaton potential in slow-roll models. In Section 8.4, we will present current observational constraints on inflationary models and comment on the prospects for future tests.

For pedagogical reasons, the presentation will focus on single-field slow-roll models of inflation. It is important to appreciate, however, that the landscape of inflationary models is much larger and there are many different mechanisms that can produce the inflationary expansion of the spacetime (see e.g. [1] for a review). An interesting way of describing a large class of models is in terms of an effective theory of the inflationary *fluctuations*. Given a suitable Hubble parameter $H(t)$ (whose origin doesn't have to be specified), we can construct an effective theory for the associated Goldstone boson of spontaneously broken time translations. On superhorizon scales, this Goldstone boson is closely related to the primordial curvature perturbations and adiabatic density fluctuations. By focusing directly on the inflationary fluctuations, this **EFT of inflation** provides an efficient way of describing the predictions of inflation [2]. The quantization procedure described in this chapter for slow-roll inflation applies straightforwardly in this broader context.

8.1 Inflationary Perturbations

The dynamics of the inflaton field is determined by the following action

$$S = \int d^4x \sqrt{-g} \left[-\frac{1}{2} g^{\mu\nu} \partial_\mu \phi \partial_\nu \phi - V(\phi) \right], \tag{8.1}$$

where $g \equiv \det g_{\mu\nu}$ (see Appendix A). The coupling to the metric will lead to a nontrivial mixing between the inflaton fluctuations $\delta\phi$ and the metric fluctuations $\delta g_{\mu\nu}$. The form of this mixing is gauge dependent. For example, in comoving gauge the inflaton fluctuations vanish, $\delta\phi \equiv 0$, and all fluctuations are in the metric. To make contact with the intuitive picture of inflaton fluctuations shown in Fig. 8.1, however, it will instead be convenient to work in the spatially flat gauge, where the line element (with scalar fluctuations) takes the form

$$ds^2 = a^2(\eta) \left[-(1 + 2A) d\eta^2 + 2\partial_i B \, dx^i d\eta + \delta_{ij} dx^i dx^j \right]. \tag{8.2}$$

As we will see, the metric fluctuations A and B are related to the inflaton fluctuations $\delta\phi$ through the Einstein equations, leaving a single degree of freedom to describe the inflationary fluctuations. We will begin by deriving the classical equation of motion for $\delta\phi$ and then discuss its quantization.

8.1.1 Equation of Motion

Varying of the action (8.1) leads to the Klein–Gordon equation for the inflaton field,

$$\frac{1}{\sqrt{-g}}\partial_\mu(\sqrt{-g}g^{\mu\nu}\partial_\nu\phi) = V_{,\phi}\,, \tag{8.3}$$

where $V_{,\phi} \equiv dV/d\phi$.

Exercise 8.1 Show that (8.3) follows from (8.1).

At first order in the fluctuations, the components of the inverse metric are

$$
\begin{aligned}
g^{00} &= -a^{-2}(1-2A)\,,\\
g^{0i} &= a^{-2}\partial^i B\,,\\
g^{ij} &= a^{-2}\delta^{ij}\,,
\end{aligned}
\tag{8.4}
$$

and $\sqrt{-g} = a^4(1+A)$. Substituting the perturbed field $\phi = \bar\phi+\delta\phi$ and the perturbed metric (8.2) into the equation of motion (8.3), we get

$$\delta\phi'' + 2\mathcal{H}\delta\phi' - \nabla^2\delta\phi = \left(A' + \nabla^2 B\right)\bar\phi' - 2a^2 V_{,\phi}A - a^2 V_{,\phi\phi}\delta\phi\,, \tag{8.5}$$

where $V_{,\phi\phi} \equiv d^2V/d\phi^2$.

Exercise 8.2 Show that (8.5) follows from (8.3).

Next, we want to use the Einstein equations to eliminate the metric perturbations A and B in favor of $\delta\phi$. For this, we must select the constraint equations that don't involve time derivatives of the metric perturbations. After some work (shown in the box below), one finds

$$A = \varepsilon\, \frac{\mathcal{H}}{\bar\phi'}\,\delta\phi\,, \tag{8.6}$$

$$\nabla^2 B = -\varepsilon\, \frac{\mathcal{H}}{\bar\phi'}\left(\delta\phi' + (\delta - \varepsilon)\,\mathcal{H}\delta\phi\right), \tag{8.7}$$

where we have introduced the slow-roll parameters

$$\varepsilon \equiv -\frac{\dot H}{H^2} = 1 - \frac{\mathcal{H}'}{\mathcal{H}^2} = 4\pi G\,\frac{(\bar\phi')^2}{\mathcal{H}^2}\,, \tag{8.8}$$

$$\delta \equiv -\frac{\ddot{\bar\phi}}{H\dot{\bar\phi}} = 1 - \frac{\bar\phi''}{\mathcal{H}\bar\phi'}\,. \tag{8.9}$$

Despite the appearance of the slow-roll parameters, we have *not* made any slow-roll approximation in (8.6) and (8.7). It is nevertheless worth highlighting that the metric perturbations are proportional to the slow-roll parameters, so that the mixing with the inflaton fluctuations vanishes at leading order in the slow-roll limit, $\varepsilon, \delta \to 0$. This is a special feature of spatially flat gauge. In a general gauge, this mixing would be order one even in the slow-roll limit. Even in spatially flat gauge, the leading slow-roll correction will play an important role on superhorizon scales, so for now we won't take any limit.

Derivation The full derivation of (8.6) and (8.7) is a bit tedious, so I will just show some of the intermediate steps and encourage you to check the rest. First, we consider the $0i$ Einstein equation. It takes the form

$$\delta G^0{}_i = -\frac{2\mathcal{H}}{a^2}\partial_i A = 8\pi G\, \delta T^0{}_i = -8\pi G\, \frac{\bar{\phi}'}{a^2}\partial_i \delta\phi\,, \tag{8.10}$$

where the final equality follows from the perturbed energy-momentum tensor of a scalar field (4.53). This implies the result in (8.6):

$$A = 4\pi G\, \frac{\bar{\phi}'}{\mathcal{H}}\,\delta\phi = \varepsilon\frac{\mathcal{H}}{\bar{\phi}'}\,\delta\phi\,, \tag{8.11}$$

where we used (8.8) in the final equality. Next, we look at the 00 Einstein equation,

$$\delta G^0{}_0 = \frac{2\mathcal{H}}{a^2}(3\mathcal{H}A + \nabla^2 B) = 8\pi G\, \delta T^0{}_0$$

$$= -8\pi G\left[a^{-2}\left(\bar{\phi}'\,\delta\phi' - (\bar{\phi}')^2 A\right) + V_{,\phi}\,\delta\phi\right]\,, \tag{8.12}$$

where the final equality again follows from (4.53). Using the background equation of motion (in conformal time),

$$a^2 V_{,\phi} = -\bar{\phi}'' - 2\mathcal{H}\bar{\phi}'\,, \tag{8.13}$$

and the solution (8.11) for A, we then obtain the result (8.7).

Exercise 8.3 Fill in the gaps in the derivation of equations (8.6) and (8.7).

The rest is straightforward (but somewhat lengthy) algebra. I will show you the key intermediate results and let you reproduce them as a useful exercise. Substituting (8.6) and (8.7) into (8.5) leads to a closed form equation for the inflaton fluctuations:

$$\delta\phi'' + 2\mathcal{H}\delta\phi' - \nabla^2\delta\phi = \left[2\varepsilon(3 + \varepsilon - 2\delta) - \frac{a^2 V_{,\phi\phi}}{\mathcal{H}^2}\right]\mathcal{H}^2\delta\phi\,. \tag{8.14}$$

Note that the terms proportional to $\delta\phi'$ have canceled in $A' + \nabla^2 B$, so that the mixing with the metric fluctuation only contributes to the *effective mass* of the inflaton fluctuations. Using the background equation of motion to write $V_{,\phi\phi}$ in terms of the slow-roll parameters, we find

$$\delta\phi'' + 2\mathcal{H}\delta\phi' - \nabla^2\delta\phi = \left[(3 + 2\varepsilon - \delta)(\varepsilon - \delta) - \frac{\delta'}{\mathcal{H}}\right]\mathcal{H}^2\delta\phi. \tag{8.15}$$

This still looks complicated, but we are not done yet.

The friction term $2\mathcal{H}\delta\phi'$ can be eliminated by defining the variable

$$f \equiv a\delta\phi, \tag{8.16}$$

so that

$$\delta\phi'' + 2\mathcal{H}\delta\phi' = \frac{1}{a}\left(f'' - (2 - \varepsilon)\mathcal{H}^2 f\right). \tag{8.17}$$

Collecting all the terms, we find that what looked like a mess becomes a beautiful equation, the so-called **Mukhanov–Sasaki equation**:

$$\boxed{f'' + \left(k^2 - \frac{z''}{z}\right)f = 0, \quad \text{where} \quad z \equiv \frac{a\bar{\phi}'}{\mathcal{H}}}. \tag{8.18}$$

Let me emphasize how remarkable (8.18) is: it contains the coupling between matter and metric fluctuations, does *not* make any slow-roll approximation, and is valid on all scales. It will be our master equation for the rest of the chapter.

Exercise 8.4 Fill in the gaps in the derivation of equation (8.18).

Hint: Show that

$$-\frac{a^2 V_{,\phi\phi}}{\mathcal{H}^2} = (\delta - 3)(\varepsilon + \delta) - \frac{\delta'}{\mathcal{H}}, \tag{8.19}$$

$$\frac{z''}{z} = \left[2 + 2\varepsilon - 3\delta + (2\varepsilon - \delta)(\varepsilon - \delta) - \frac{\delta'}{\mathcal{H}}\right]\mathcal{H}^2. \tag{8.20}$$

As we will see in the next section, to quantize the theory, the equation of motion will not be sufficient, but we need to know the quadratic action from which it arises. Given (8.18), this action is fixed up to an overall normalization

$$S_2 = \frac{N}{2}\int d\eta d^3x \left[(f')^2 - (\nabla f)^2 + \frac{z''}{z}f^2\right]. \tag{8.21}$$

However, the conjugate momentum $\pi \equiv \delta S_2/\delta f' = Nf'$ depends on the unfixed constant N, and the size of quantum fluctuations will not be fully specified, unless we determine N. To fix the normalization, we could substitute the perturbed inflaton field and the perturbed metric into the inflaton action (8.1) and expand to second order in $\delta\phi$, using the constraints (8.6) and (8.7) to replace the metric perturbations. This doesn't sound like fun. Fortunately, there is a simpler way to find N. Consider the early-time limit, in which the perturbations are deep inside the horizon and the metric perturbations and the potential are dynamically irrelevant in any reasonable gauge. In this limit, the quadratic action for the inflaton fluctuations is simply given by the kinetic term in the *unperturbed* spacetime

$$S_2^{(k \gg \mathcal{H})} \approx \frac{1}{2} \int d\eta\, d^3x\, a^2 \left[(\delta\phi')^2 - (\nabla\delta\phi)^2 \right]$$
$$\approx \frac{1}{2} \int d\eta\, d^3x \left[(f')^2 - (\nabla f)^2 \right]. \tag{8.22}$$

This shows that $N = 1$ in (8.21) and the conjugate momentum is $\pi = f'$.

8.1.2 From Micro to Macro

We note that (8.18) is the equation of a harmonic oscillator with a time-dependent frequency

$$\omega^2(\eta, k) \equiv k^2 - \frac{z''}{z} . \tag{8.23}$$

During slow-roll inflation, H and $d\bar{\phi}/dt$ (and hence $\bar{\phi}'/\mathcal{H}$) are approximately constant and

$$\frac{z''}{z} \approx \frac{a''}{a} \approx 2\mathcal{H}^2 , \tag{8.24}$$

where we used that $a' = a^2 H$. We see that the inverse of z''/z is a measure of the comoving Hubble radius $\mathcal{H}^{-1}$, which for simplicity I will often just refer to as the "horizon" (but see the discussion in Section 4.2).

- At early times, $\mathcal{H}^{-1}$ will be large and all modes are inside the horizon. Taking the limit $k^2 \gg |z''/z|$, equation (8.18) reduces to

$$f'' + k^2 f = 0 \qquad \text{(subhorizon)}. \tag{8.25}$$

  This is the equation of motion of a harmonic oscillator with a fixed frequency, $\omega = k$, which has the solutions $f \propto e^{\pm ik\eta}$. As we will see below, the amplitude of these oscillations will experience quantum mechanical zero-point fluctuations. We will study this in the next section.

- As the comoving horizon shrinks, modes will eventually exit the horizon and their evolution changes drastically (see Fig. 8.2). This can be seen by the frequency in (8.23) becoming imaginary on superhorizon scales. In physical coordinates, this evolution corresponds to microscopic scales being stretched outside of the constant Hubble radius H^{-1} to become macroscopic cosmological fluctuations. Taking the limit $k^2 \ll |z''/z|$, equation (8.18) now reads

$$f'' - \frac{z''}{z} f = 0 \qquad \text{(superhorizon)}. \tag{8.26}$$

  This has the growing mode solution $f \propto z$ and the decaying mode solution $f \propto z^{-2}$. As I will now show, the growing mode corresponds to a frozen curvature perturbation.

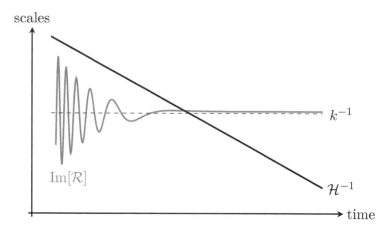

Fig. 8.2   Evolution of a single Fourier mode of the comoving curvature perturbation. The mode oscillates on subhorizon scales and freezes to a constant value on superhorizon scales.

Recall the definition of the comoving curvature perturbation (6.55), which in spatially flat gauge becomes

$$\mathcal{R} = -\mathcal{H}(v + B)\,, \tag{8.27}$$

where the peculiar velocity was defined through the perturbed momentum density $\delta T^i{}_0 = -(\bar{\rho} + \bar{P})\partial^i v$. Given the energy-momentum tensor of a scalar field in (4.53), we infer that

$$
\begin{aligned}
\delta T^i{}_0 &= g^{i\mu}\partial_\mu\phi\,\partial_0\phi = g^{i0}(\bar{\phi}')^2 + g^{ij}\bar{\phi}'\partial_j\delta\phi \\
&= a^{-2}(\bar{\phi}')^2\partial^i(B + \delta\phi/\bar{\phi}')\,.
\end{aligned} \tag{8.28}
$$

Since $\bar{\rho} + \bar{P} = a^{-2}(\bar{\phi}')^2$, we have $\delta T^i{}_0 = -a^{-2}(\bar{\phi}')^2\partial^i v$, so that $v + B = -\delta\phi/\bar{\phi}'$ and hence

$$\boxed{\mathcal{R} = \frac{\mathcal{H}}{\bar{\phi}'}\delta\phi = \frac{f}{z}} \xrightarrow{k \ll \mathcal{H}} \text{const}\,, \tag{8.29}$$

which proves that the growing mode of $\mathcal{R}$ is indeed frozen on large scales. Notice that (8.29) takes the form $\mathcal{R} = H\delta t$, confirming the intuition that the curvature perturbation is induced by the time delay to the end of inflation. Moreover, since $\mathcal{R}$ stays well-defined after inflation, while $\delta\phi$ loses its meaning, it is the natural variable to track in the transition from inflation to the hot Big Bang.

8.2 Quantum Fluctuations

As we have just seen, the Fourier modes of the field fluctuations satisfy the equation of motion of a harmonic oscillator. In this section, I will describe the quantization of these oscillators.

8.2.1 Quantum Harmonic Oscillators

> The career of a young theoretical physicist consists of treating the harmonic oscillator in ever-increasing levels of abstraction.   *Sidney Coleman*

To avoid getting distracted by Fourier indices and delta functions, we first study the simpler problem of a one-dimensional harmonic oscillator in quantum mechanics. For example, consider a mass m attached to a spring with spring constant κ. Let q be the deviation from the equilibrium point, whose equation of motion is $\ddot{q} = -\omega^2 q$, where $\omega^2 \equiv \sqrt{\kappa/m}$ is the oscillation frequency. The quantization of the oscillator is a textbook problem in quantum mechanics, which I will now briefly review.

The Hamiltonian of the oscillator is

$$H = \frac{1}{2}p^2 + \frac{1}{2}\omega^2 q^2 \,, \tag{8.30}$$

where we have rescaled the oscillator amplitude to set $m \equiv 1$ and introduced the conjugate momentum $p \equiv \dot{q}$. To solve the theory means finding the spectrum of energy eigenstates of this Hamiltonian. First, we promote the classical variables q, p to quantum operators $\hat{q}$, $\hat{p}$ and impose the canonical commutation relation

$$[\hat{q}, \hat{p}] \equiv \hat{q}\hat{p} - \hat{p}\hat{q} = i\hbar \,, \tag{8.31}$$

where $\hbar = 1.05 \times 10^{-34}\,\mathrm{J\,s}$ is the reduced Planck constant. An elegant way to solve for the spectrum of states is to express $\hat{q}$ and $\hat{p}$ in terms of annihilation and creation operators:

$$\begin{aligned}
\hat{q} &= \sqrt{\frac{\hbar}{2\omega}}(\hat{a} + \hat{a}^\dagger) \,, \\
\hat{p} &= -i\sqrt{\frac{\hbar\omega}{2}}(\hat{a} - \hat{a}^\dagger) \,,
\end{aligned} \tag{8.32}$$

where the commutation relation (8.31) is satisfied if

$$[\hat{a}, \hat{a}^\dagger] = 1 \,. \tag{8.33}$$

The Hamiltonian (8.30) can then be written as

$$\begin{aligned}
\hat{H} &= \frac{1}{2}\hat{p}^2 + \frac{1}{2}\omega^2 \hat{q}^2 \\
&= \frac{1}{2}\hbar\omega\left(\hat{a}\hat{a}^\dagger + \hat{a}^\dagger\hat{a}\right) = \hbar\omega\left(\hat{a}^\dagger\hat{a} + \frac{1}{2}\right).
\end{aligned} \tag{8.34}$$

The ground state of the oscillator is defined by

$$\hat{a}|0\rangle = 0, \tag{8.35}$$

and has energy $E_0 \equiv \langle 0|\hat{H}|0\rangle = \frac{1}{2}\hbar\omega$. Excited states are constructed by repeated application of the creation operator

$$|n\rangle = \frac{1}{\sqrt{n!}}(\hat{a}^\dagger)^n|0\rangle, \tag{8.36}$$

and have energy $E_n \equiv \langle n|\hat{H}|n\rangle = \hbar\omega(n + \frac{1}{2})$.

The expectation value of the position operator $\hat{q}$ in the ground state $|0\rangle$ vanishes,

$$\langle \hat{q} \rangle \equiv \langle 0|\hat{q}|0\rangle$$

$$= \sqrt{\frac{\hbar}{2\omega}}\,\langle 0|\overbrace{\hat{a} + \hat{a}^\dagger}|0\rangle$$

$$= 0, \tag{8.37}$$

because $\hat{a}$ annihilates $|0\rangle$ when acting on it from the left, and $\hat{a}^\dagger$ annihilates $\langle 0|$ when acting on it from the right. However, the expectation value of the square of the position operator receives finite **zero-point fluctuations**:

$$\langle |\hat{q}|^2 \rangle \equiv \langle 0|\hat{q}^\dagger \hat{q}|0\rangle$$

$$= \frac{\hbar}{2\omega}\langle 0|\overbrace{(\hat{a}^\dagger + \hat{a})}(\overbrace{\hat{a} + \hat{a}^\dagger})|0\rangle$$

$$= \frac{\hbar}{2\omega}\langle 0|\hat{a}\hat{a}^\dagger|0\rangle$$

$$= \frac{\hbar}{2\omega}\langle 0|[\hat{a}, \hat{a}^\dagger]|0\rangle$$

$$= \frac{\hbar}{2\omega}. \tag{8.38}$$

These zero-point fluctuations play an important role in many areas of physics and, in particular, seem to be the origin of all structure in the universe.

Uncertainty principle A heuristic way to derive the result (8.38) is from the uncertainty principle

$$\Delta\hat{q}\,\Delta\hat{p} = \frac{\hbar}{2}, \tag{8.39}$$

where $\Delta\hat{q}$ and $\Delta\hat{p}$ are the uncertainties in the position and momentum of the oscillator, respectively. The energy of the quantum harmonic oscillator due to these quantum fluctuations then is

$$E = \frac{\hbar^2}{8(\Delta\hat{q})^2} + \frac{1}{2}\omega^2(\Delta\hat{q})^2. \tag{8.40}$$

This energy is minimized for

$$\frac{\mathrm{d}E}{\mathrm{d}(\Delta\hat{q})^2} = -\frac{\hbar^2}{8(\Delta\hat{q})^4} + \frac{1}{2}\omega^2 = 0 \quad \Rightarrow \quad \Delta\hat{q} = \sqrt{\frac{\hbar}{2\omega}}, \tag{8.41}$$

reproducing our previous result (8.38). Substituting this back into (8.40) gives $E_0 = \frac{1}{2}\hbar\omega$, the zero-point energy of the oscillator in the ground state.

The only subtlety we will encounter when we apply this to inflation will be that the frequency of the oscillators is time dependent, $\omega = \omega(t)$. It is most convenient to describe this time dependence in the **Heisenberg picture** where operators vary in time, while states are time independent. An operator $\hat{O}$ in the Heisenberg picture satisfies the following evolution equation:

$$\frac{\mathrm{d}\hat{O}}{\mathrm{d}t} = \frac{i}{\hbar}[\hat{H}, \hat{O}], \tag{8.42}$$

where $\hat{H}$ is the Hamiltonian.

Exercise 8.5 Applying (8.42) to the operators $\hat{q}$ and $\hat{p}$, show that

$$\frac{\mathrm{d}^2\hat{q}}{\mathrm{d}t^2} + \omega^2\hat{q} = 0, \tag{8.43}$$

i.e. the quantum operator $\hat{q}$ obeys the same equation of motion as the classical variable q.

We now write the position operator as

$$\boxed{\hat{q}(t) = q(t)\,\hat{a}(t_i) + q^*(t)\,\hat{a}^\dagger(t_i)}, \tag{8.44}$$

where the complex mode function $q(t)$ satisfies $\ddot{q} + \omega^2(t)q = 0$. The annihilation operator $\hat{a}(t_i)$—and hence the associated vacuum state $|0\rangle$—depends on the time t_i at which it is defined. In the inflationary application, there will be a preferred moment at which we define the vacuum state, namely the beginning of inflation.

Substituting (8.44) into (8.31), we get

$$[\hat{q}, \hat{p}] = (q\dot{q}^* - \dot{q}q^*)[\hat{a}, \hat{a}^\dagger] = i\hbar. \tag{8.45}$$

In order for this to give $[\hat{a}, \hat{a}^\dagger] = 1$, we require the mode function to obey the normalization

$$\boxed{q\dot{q}^* - \dot{q}q^* = i\hbar}. \tag{8.46}$$

To completely fix the mode function and hence give a unique meaning to the operator $\hat{a}(t_i)$ and the corresponding vacuum state $|0\rangle$, we need to impose a second condition on the solution $q(t)$.

As we have seen in (8.25), at early times, the inflaton fluctuations satisfy the equation of motion of a harmonic oscillator with a fixed frequency. To model the situation during inflation, we therefore consider a harmonic oscillator whose frequency initially is a constant and then slowly develops a time dependence. We take the initial condition to be the ground state of the fixed-frequency oscillator (which will include the zero-point fluctuations discussed above) and then use the equation of motion with a time-dependent frequency to evolve these fluctuations forward in time.

To see how a preferred mode function arises for a harmonic oscillator with a fixed frequency, let us write the most general solution as

$$q(t) = r(t)e^{is(t)} \, , \tag{8.47}$$

where $r(t)$ and $s(t)$ are real functions of time, and (8.46) implies

$$\dot{s} = -\frac{\hbar}{2r^2} \, . \tag{8.48}$$

We would like to determine the solution that minimizes the vacuum expectation value of the Hamiltonian

$$\langle 0|\hat{H}|0\rangle = \frac{1}{2}\left(|\dot{q}|^2 + \omega^2|q|^2\right)$$

$$= \frac{1}{2}\left(\dot{r}^2 + r^2\dot{s}^2 + \omega^2 r^2\right) . \tag{8.49}$$

Substituting (8.48) into (8.49), we get

$$\langle 0|\hat{H}|0\rangle = \frac{1}{2}\left(\dot{r}^2 + \frac{\hbar^2}{4r^2} + \omega^2 r^2\right) . \tag{8.50}$$

This is minimized for $\dot{r} = 0$ and

$$0 = \frac{\mathrm{d}}{\mathrm{d}r^2}\left(\frac{\hbar^2}{4r^2} + \omega^2 r^2\right) = -\frac{\hbar^2}{4(r^2)^2} + \omega^2 \quad \Rightarrow \quad r = \sqrt{\frac{\hbar}{2\omega}} \, . \tag{8.51}$$

Substituting this back into (8.48), we find

$$\dot{s} = -\omega \quad \Rightarrow \quad s = -\omega t + \mathrm{const} \, . \tag{8.52}$$

Up to an irrelevant phase, we therefore get

$$\boxed{q(t) = \sqrt{\frac{\hbar}{2\omega}}e^{-i\omega t}} \quad (\omega = \mathrm{const}) \, . \tag{8.53}$$

We see that the ground state corresponds to the **positive-frequency solution**, $e^{-i\omega t}$, while adding any amount of the negative-frequency solution, $e^{+i\omega t}$, would raise the energy.

Using the solution (8.53) as an initial condition at $t = t_i$ uniquely fixes the solution to $\ddot{q} = -\omega^2(t)q$. Of course, the explicit solution depends on $\omega(t)$, which in the case of inflation is determined by the evolution of the inflationary background. Repeating the manipulations that led to (8.38), we get that the variance of the position operator is simply the square of the mode function:

$$\langle |\hat{q}|^2 \rangle = |q(t)|^2 \,.$$ (8.54)

This describes how the initial zero-point fluctuations, specified in (8.38), evolve in time.

8.2.2 Inflationary Vacuum Fluctuations

> With the new cosmology the universe must have been started off in some very simple way. What, then, becomes of the initial conditions required by dynamical theory? Plainly there cannot be any, or they must be trivial. We are left in a situation which would be untenable with the old mechanics. If the universe were simply the motion which follows from a given scheme of equations of motion with trivial initial conditions, it could not contain the complexity we observe. Quantum mechanics provides an escape from the difficulty. It enables us to ascribe the complexity to the quantum jumps, lying outside the scheme of equations of motion. The quantum jumps now form the uncalculable part of natural phenomena, to replace the initial conditions of the old mechanistic view.   *Paul Dirac* (in 1939)

We are now ready to apply the quantization procedure to the field fluctuations during inflation. For simplicity, we will work in the *slow-roll approximation*, taking the background spacetime to be de Sitter space, with $H = $ const and $a = -(H\eta)^{-1}$. The Mukhanov–Sasaki equation (8.18) then takes the form

$$f_{\mathbf{k}}'' + \left(k^2 - \frac{2}{\eta^2} \right) f_{\mathbf{k}} = 0 \,.$$ (8.55)

In spatially flat gauge, this is indeed a good approximation before horizon crossing. On superhorizon scales, however, slow-roll corrections are the leading effect. For example, ignoring the time dependence of the expansion rate will lead to a fictitious evolution of the curvature perturbation $\mathcal{R}$ on large scales. To minimize this error, we will evaluate all fluctuations at the moment of horizon crossing, $k = aH$, where we match them to the conserved curvature perturbation. Including a time-dependent expansion rate, $H(t)$, in the horizon crossing condition then extends our results to a quasi-de Sitter background. We also include the weak time dependence in the conversion from $\delta\phi$ to $\mathcal{R}$ in (8.29). In turns out that this simple prescription gives the right answer, as we will prove later by a more rigorous treatment.

Canonical quantization

The quantization of the field fluctuations is essentially the same as for the harmonic oscillator. We first promote the field $f(\eta, \mathbf{x})$ and its conjugate momentum $\pi(\eta, \mathbf{x})$ to quantum operators $\hat{f}(\eta, \mathbf{x})$ and $\hat{\pi}(\eta, \mathbf{x})$, and then impose

$$[\hat{f}(\eta, \mathbf{x}), \hat{\pi}(\eta, \mathbf{x}')] = i\delta_{\mathrm{D}}(\mathbf{x} - \mathbf{x}'),$$ (8.56)

where we have set $\hbar \equiv 1$. This is the field theory equivalent of (8.31). The delta function enforces *locality*: modes at different points in space are independent and the corresponding operators therefore commute. In Fourier space, we find

$$[\hat{f}_{\mathbf{k}}(\eta), \hat{\pi}_{\mathbf{k}'}(\eta)] = \int \mathrm{d}^3 x \int \mathrm{d}^3 x' \underbrace{[\hat{f}(\eta, \mathbf{x}), \hat{\pi}(\eta, \mathbf{x}')]}_{i\delta_{\mathrm{D}}(\mathbf{x} - \mathbf{x}')} e^{-i\mathbf{k}\cdot\mathbf{x}} e^{-i\mathbf{k}'\cdot\mathbf{x}'}$$

$$= i \int \mathrm{d}^3 x\, e^{-i(\mathbf{k}+\mathbf{k}')\cdot\mathbf{x}}$$

$$= i(2\pi)^3 \delta_{\mathrm{D}}(\mathbf{k} + \mathbf{k}'),$$ (8.57)

where the delta function implies that modes with different wavelengths commute.

As before, we are working in the Heisenberg picture where operators vary in time, while states are time independent. The operator solution $\hat{f}_{\mathbf{k}}(\eta)$ is determined by two initial conditions $\hat{f}_{\mathbf{k}}(\eta_i)$ and $\hat{\pi}_{\mathbf{k}}(\eta_i) = \partial_\eta \hat{f}_{\mathbf{k}}(\eta_i)$, and since the evolution equation is linear, the solution is linear in these operators. It is convenient to trade $\hat{f}_{\mathbf{k}}(\eta_i)$ and $\hat{\pi}_{\mathbf{k}}(\eta_i)$ for a single *non-Hermitian* operator $\hat{a}_{\mathbf{k}}(\eta_i)$, so that

$$\boxed{\hat{f}_{\mathbf{k}}(\eta) = f_k(\eta)\hat{a}_{\mathbf{k}} + f_k^*(\eta)\hat{a}_{-\mathbf{k}}^\dagger},$$ (8.58)

where the complex mode function $f_k(\eta)$ satisfies the classical equation of motion. This is the same as the mode expansion in (8.44) with some Fourier labels added. Since the operator $\hat{f}(\eta, \mathbf{x})$ is Hermitian, its Fourier transform must satisfy $\hat{f}_{\mathbf{k}}(\eta)^\dagger = \hat{f}_{-\mathbf{k}}(\eta)$. This explains the $-\mathbf{k}$ on $\hat{a}_{-\mathbf{k}}^\dagger$. We can then write $\hat{f}(\eta, \mathbf{x})$ as

$$\hat{f}(\eta, \mathbf{x}) = \int \frac{\mathrm{d}^3 k}{(2\pi)^3} \hat{f}_{\mathbf{k}}(\eta) e^{i\mathbf{k}\cdot\mathbf{x}}$$
$$= \int \frac{\mathrm{d}^3 k}{(2\pi)^3} \left[\hat{a}_{\mathbf{k}} f_k(\eta) e^{i\mathbf{k}\cdot\mathbf{x}} + \hat{a}_{\mathbf{k}}^\dagger f_k^*(\eta) e^{-i\mathbf{k}\cdot\mathbf{x}}\right],$$ (8.59)

which is manifestly Hermitian. As indicated by the notation, the mode function $f_k(\eta)$ depends only on the magnitude of the wavevector, $k \equiv |\mathbf{k}|$.[1]

As in (8.46), we choose the normalization of the mode functions, so that

$$f_k f_k'^* - f_k' f_k^* \equiv i.$$ (8.60)

Substituting (8.58) into (8.57), we then get

$$[\hat{a}_{\mathbf{k}}, \hat{a}_{\mathbf{k}'}^\dagger] = (2\pi)^3 \delta_{\mathrm{D}}(\mathbf{k} - \mathbf{k}'),$$ (8.61)

which is the field theory generalization of (8.33).

---

[1] Since the frequency $\omega_k(\eta) \equiv k^2 - 2/\eta^2$ in (8.55) depends only on the magnitude $k \equiv |\mathbf{k}|$, the evolution does not depend on direction. The constant operators $\hat{a}_{\mathbf{k}}$ and $\hat{a}_{\mathbf{k}}^\dagger$, on the other hand, define initial conditions which may depend on direction.

Choice of vacuum

The vacuum state is defined in the usual way

$$\hat{a}_{\mathbf{k}}(\eta_i)|0\rangle = 0. \tag{8.62}$$

However, because the Hamiltonian is time dependent, the choice of the vacuum state depends on the time η_i at which it is defined. It is natural to define that vacuum as the ground state of the Hamiltonian at the beginning of inflation. Formally, we take the limit $\eta_i \to -\infty$, but any time satisfying $|k\eta_i| \gg 1$ would work. At this time, all modes were deep inside the horizon and satisfy the equation of a harmonic oscillator with a fixed frequency $\omega_k \to k$, cf. (8.25). For each Fourier mode, we then choose the natural vacuum state of the simple harmonic oscillator. Above we derived the preferred mode function corresponding to a harmonic oscillator in the ground state, cf. (8.53). We use this mode function as the initial condition for each Fourier mode:

$$\lim_{k\eta \to -\infty} f_k(\eta) = \frac{1}{\sqrt{2k}} e^{-ik\eta}. \tag{8.63}$$

Solving the Mukhanov–Sasaki equation (8.55) with this boundary condition leads to a unique mode function called the **Bunch–Davies mode function**:

$$\boxed{f_k(\eta) = \frac{1}{\sqrt{2k}} \left(1 - \frac{i}{k\eta}\right) e^{-ik\eta}.} \tag{8.64}$$

The corresponding state is the **Bunch–Davies vacuum**.

Figure 8.3 shows the evolution of the imaginary part of the comoving curvature perturbation, $\mathcal{R} = f/z$, given the solution (8.64). As expected, the solution oscillates at early times and freezes after horizon crossing. The initial amplitude of the oscillations is set by quantum zero-point fluctuations.

Exercise 8.6 Show that (8.64) implies

$$\mathrm{Re}[\mathcal{R}] = A(\sin x - x\cos x), \tag{8.65}$$
$$\mathrm{Im}[\mathcal{R}] = A(\cos x + x\sin x), \tag{8.66}$$
$$|\mathcal{R}| = A\sqrt{1 + x^2}, \tag{8.67}$$

where $x \equiv k\eta$.

Zero-point fluctuations

Finally, we can predict the quantum statistics of the operator

$$\hat{f}(\eta, \mathbf{x}) = \int \frac{\mathrm{d}^3 k}{(2\pi)^3} \left[f_k(\eta)\,\hat{a}_{\mathbf{k}} + f_k^*(\eta)\,\hat{a}_{-\mathbf{k}}^\dagger\right] e^{i\mathbf{k}\cdot\mathbf{x}}. \tag{8.68}$$

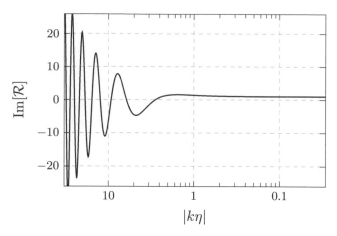

Fig. 8.3 Evolution of the imaginary part of the comoving curvature perturbation. As expected, the perturbation oscillates on small scales (or early times), $|k\eta| \gg 1$, and freezes on large scales (or late times), $|k\eta| \ll 1$.

The expectation value of $\hat{f}$ vanishes, i.e. $\langle \hat{f} \rangle \equiv \langle 0|\hat{f}|0 \rangle = 0$. However, the variance of inflaton fluctuations receives nonzero quantum fluctuations:

$$\langle |\hat{f}|^2 \rangle \equiv \langle 0|\hat{f}(\eta, \mathbf{0})\hat{f}(\eta, \mathbf{0})|0 \rangle$$

$$= \int \frac{d^3 k}{(2\pi)^3} \int \frac{d^3 k'}{(2\pi)^3} \, \langle 0|\big(f_k^*(\eta)\hat{a}_{-\mathbf{k}}^\dagger + f_k(\eta)\hat{a}_{\mathbf{k}}\big)\big(f_{k'}(\eta)\hat{a}_{\mathbf{k}'} + f_{k'}^*(\eta)\hat{a}_{-\mathbf{k}'}^\dagger\big)|0 \rangle$$

$$= \int \frac{d^3 k}{(2\pi)^3} \int \frac{d^3 k'}{(2\pi)^3} \, f_k(\eta)f_{k'}^*(\eta) \, \langle 0|[\hat{a}_{\mathbf{k}}, \hat{a}_{-\mathbf{k}'}^\dagger]|0 \rangle$$

$$= \int \frac{d^3 k}{(2\pi)^3} |f_k(\eta)|^2$$

$$= \int d\ln k \, \frac{k^3}{2\pi^2} |f_k(\eta)|^2 \,. \tag{8.69}$$

We see that the variance of the quantum fluctuations is determined by the (dimensionless) power spectrum of the classical solution:

$$\Delta_f^2(k, \eta) \equiv \frac{k^3}{2\pi^2} |f_k(\eta)|^2 \,. \tag{8.70}$$

Substituting the Bunch–Davies mode function (8.64) for $f = a\delta\phi$, we find

$$\Delta_{\delta\phi}^2(k, \eta) = \frac{\Delta_f^2(k, \eta)}{a^2(\eta)} = \left(\frac{H}{2\pi}\right)^2 \big[1 + (k\eta)^2\big] \xrightarrow{k\eta \to 0} \left(\frac{H}{2\pi}\right)^2 \,. \tag{8.71}$$

Note that in the superhorizon limit, $k\eta \to 0$, the dimensionless power spectrum $\Delta_{\delta\phi}^2$ approaches the same constant for all momenta, which is the characteristic of a *scale-invariant* spectrum.

Although the time dependence of the Hubble rate during inflation is small, it is crucial that it is nonzero. Otherwise, there wouldn't be any evolution and inflation would never end. Moreover, curvature perturbations would be ill-defined since $\mathcal{R} \propto \dot{H}^{-1} \to \infty$. To incorporate the effect of a varying $H(t)$, we evaluate the power spectrum (8.71) at horizon crossing, $k = aH$, which, for each Fourier mode, corresponds to a different moment in time. Since H is evolving, this leads to a slight scale dependence of the spectrum

$$\Delta_{\delta\phi}^2(k) \approx \left(\frac{H(t)}{2\pi}\right)^2 \Bigg|_{k=aH(t)}. \tag{8.72}$$

Because $H(t)$ decreases during inflation, the amplitude of fluctuations will be slightly larger for long-wavelength fluctuations which exit the horizon at the beginning of inflation.

From quantum to classical

An important consequence of the stretching of fluctuations from microscopic to macroscopic scales is the fact that the initial quantum fluctuations become classical fluctuations after horizon crossing. To see this, note from (8.64) that the mode function $f_k(\eta)$ becomes purely imaginary on superhorizon scales:

$$\begin{aligned} \mathrm{Re}[f_k] &\to 0 \,, \\ \mathrm{Im}[f_k] &\propto a(\eta) \,. \end{aligned} \tag{8.73}$$

The field operator and its conjugate momentum then become proportional to each other:

$$\hat{f}_{\mathbf{k}}(\eta) \xrightarrow{k\eta \to 0} f_k(\eta)\big(\hat{a}_{\mathbf{k}} - \hat{a}_{-\mathbf{k}}^\dagger\big)\,, \tag{8.74}$$

$$\hat{\pi}_{\mathbf{k}}(\eta) \xrightarrow{k\eta \to 0} f_k'(\eta)\big(\hat{a}_{\mathbf{k}} - \hat{a}_{-\mathbf{k}}^\dagger\big) \approx \frac{f_k'(\eta)}{f_k(\eta)}\,\hat{f}_{\mathbf{k}}(\eta)\,. \tag{8.75}$$

While both $\hat{f} \propto a(\eta)$ and $\hat{\pi} \propto a^2(\eta)$ grow outside the horizon, the commutator $[\hat{f}, \hat{\pi}]$ stays constant. Since the uncertainty $\Delta\hat{f}\Delta\hat{\pi}$ is proportional to $[\hat{f}, \hat{\pi}]$, a state $|f_k\rangle$ with a definite field value has an almost definite momentum. Using (8.64), we can compute the ratio

$$R \equiv \frac{\Delta\hat{f}\Delta\hat{\pi}}{|\hat{f}||\hat{\pi}|} = \frac{\hbar/2}{\hbar/2\sqrt{1+(k\eta)^{-6}}} = \frac{1}{\sqrt{1+(k\eta)^{-6}}} \xrightarrow{k\eta \to 0} (k\eta)^3 \ll 1. \tag{8.76}$$

We see that the quantum uncertainty is large ($R \sim 1$) at early times, but becomes very small ($R \ll 1$) at late times and on large scales.

8.3 Primordial Power Spectra

Our final task is to relate the fluctuations in the inflaton field to the observable fluctuations after inflation. As we described in Section 6.2.4, the bridge between

inflation and the late universe is provided by the primordial curvature perturbations. In this section, we will map the spectrum of inflaton fluctuations $\delta\phi$ to
the spectrum of the comoving curvature perturbation $\mathcal{R}$. We will also derive the
spectrum of primordial tensor fluctuations.

8.3.1 Curvature Perturbations

Recall that, in spatially flat gauge, $\mathcal{R}$ and $\delta\phi$ are related by a simple conversion
factor, cf. (8.29). The power spectrum of $\mathcal{R}$ can therefore be written as

$$\Delta_{\mathcal{R}}^2 = \left(\frac{H}{\dot{\bar\phi}}\right)^2 \Delta_{\delta\phi}\,, \tag{8.77}$$

and, using (8.72), we get

$$\boxed{\Delta_{\mathcal{R}}^2(k) = \left(\frac{H^2}{2\pi\dot{\bar\phi}^2}\right)^2 \Bigg|_{k=aH} = \frac{1}{8\pi^2\varepsilon}\frac{H^2}{M_{\mathrm{Pl}}^2}\Bigg|_{k=aH}}\,. \tag{8.78}$$

The time dependence of the conversion factor—captured by the slow-roll parameter
$\varepsilon(t)$—introduces an additional source of scale dependence. The scalar spectral index
is defined as

$$n_{\mathrm{s}} - 1 \equiv \frac{\mathrm{d}\ln\Delta_{\mathcal{R}}^2(k)}{\mathrm{d}\ln k}\,, \tag{8.79}$$

where the shift by -1 is a historical accident, cf. Section 5.3.3. Using $k = aH$, we
get

$$n_{\mathrm{s}} - 1 = \frac{\mathrm{d}\ln\Delta_{\mathcal{R}}^2}{\mathrm{d}\ln(aH)} \tag{8.80}$$

$$\approx \frac{\mathrm{d}\ln\Delta_{\mathcal{R}}^2}{\mathrm{d}\ln a} = \frac{\mathrm{d}\ln\Delta_{\mathcal{R}}^2}{H\mathrm{d}t} = \frac{2\,\mathrm{d}\ln H}{H\mathrm{d}t} - \frac{\mathrm{d}\ln\varepsilon}{H\mathrm{d}t} = 2\frac{\dot{H}}{H^2} - \frac{\dot\varepsilon}{H\varepsilon}\,, \tag{8.81}$$

which in terms of the Hubble slow-roll parameters defined in (4.34) and (4.36)
becomes

$$\boxed{n_{\mathrm{s}} - 1 = -2\varepsilon - \kappa}\,. \tag{8.82}$$

In the following insert, I will show how the same result can also be derived from a
systematic slow-roll expansion of the Mukhanov–Sasaki equation.

Slow-roll expansion The derivation above involved elements that didn't look
completely rigorous. To make up for this, let me now show how the same result
can be obtained from a systematic slow-roll expansion of the equation of motion.
The algebra is slightly more involved, but the physics is exactly the same.

Let us begin with the Mukhanov–Sasaki equation (8.18):

$$f'' + \left(k^2 - \frac{z''}{z}\right)f = 0\,, \quad \text{where} \quad z \equiv \frac{a\bar\phi'}{\mathcal{H}} = a\sqrt{2\varepsilon}\,M_{\mathrm{Pl}}\,. \tag{8.83}$$

It is easy to show that

$$\frac{z'}{z} = \mathcal{H}\left[1 + \frac{1}{2}\kappa\right],$$
(8.84)

$$\frac{z''}{z} \approx \mathcal{H}^2\left[2 - \varepsilon + \frac{3}{2}\kappa\right],$$
(8.85)

where (8.84) is exact and (8.85) is valid to first order in slow-roll parameters. Moreover, integrating $\varepsilon = 1 - \mathcal{H}'/\mathcal{H}^2 \approx$ const, we get

$$\mathcal{H} = -\frac{1}{\eta}(1 + \varepsilon),$$
(8.86)

and (8.85) becomes

$$\frac{z''}{z} = \frac{1}{\eta^2}\left[2 + 3\left(\varepsilon + \frac{1}{2}\kappa\right)\right].$$
(8.87)

Equation (8.83) can then be written as

$$f'' + \left(k^2 - \frac{\nu^2 - 1/4}{\eta^2}\right)f = 0, \quad \text{where} \quad \nu \equiv \frac{3}{2} + \varepsilon + \frac{1}{2}\kappa.$$
(8.88)

Imposing the normalization (8.60) and the Bunch–Davies initial condition (8.63), we get the following solution:

$$f_k(\eta) = \frac{\sqrt{\pi}}{2}\sqrt{-\eta}\,H_\nu^{(1)}(-k\eta),$$
(8.89)

where $H_\nu^{(1)}$ is a Hankel function of the first kind (see Appendix D). This has the correct early-time limit because

$$\lim_{k\eta \to -\infty} |H_\nu^{(1,2)}(-k\eta)| = \sqrt{\frac{2}{\pi}}\frac{1}{\sqrt{-k\eta}}\,e^{\mp ik\eta}.$$
(8.90)

To compute the power spectrum of $\mathcal{R} = f/z$, we need $z(\eta)$. Substituting (8.86) into (8.84), and integrating the resulting expression, we find

$$z(\eta) = z_*(\eta/\eta_*)^{\frac{1}{2} - \nu},$$
(8.91)

where η_* is some reference time. It will be convenient to take $\eta_* = -k_*^{-1}$, i.e. the moment of horizon crossing of a mode of wavenumber k_*. The power spectrum of $\mathcal{R}$ then is

$$\Delta_\mathcal{R}^2(k) = \frac{k^3}{2\pi^2}\frac{1}{z^2(\eta)}|f_k(\eta)|^2$$

$$= \frac{k^3}{2\pi^2}\frac{1}{2\varepsilon_* M_{\text{Pl}}^2 a_*^2}(-k_*\eta)^{2\nu - 1}\frac{\pi}{4}(-\eta)|H_\nu^{(1)}(-k\eta)|^2.$$
(8.92)

Using $a_* = k_*/H_*$ and the late-time limit of the Hankel function,

$$\lim_{k\eta \to 0} |H_\nu^{(1)}(-k\eta)|^2 = \frac{2^{2\nu}\Gamma(\nu)^2}{\pi^2}(-k\eta)^{-2\nu} \approx \frac{2}{\pi}(-k\eta)^{-2\nu},$$
(8.93)

we get

$$\Delta_{\mathcal{R}}^2(k) = \frac{1}{8\pi^2\varepsilon_*} \frac{H_*^2}{M_{\mathrm{Pl}}^2} (k/k_*)^{3-2\nu} \, , \tag{8.94}$$

which is equivalent to our previous result (8.78). Note that the time dependence in (8.92) has canceled, as it had to because $\mathcal{R}$ is conserved on superhorizon scales. The exponent in (8.94) is precisely the spectral index we found before

$$n_{\mathrm{s}} - 1 \equiv 3 - 2\nu = -2\varepsilon - \kappa \, , \tag{8.95}$$

where we used the definition of ν in (8.88).

In summary, we have found that the power spectrum of the curvature perturbation $\mathcal{R}$ takes a power law form:

$$\Delta_{\mathcal{R}}^2(k) = A_{\mathrm{s}} \left(\frac{k}{k_*} \right)^{n_{\mathrm{s}}-1} \, , \tag{8.96}$$

where the amplitude and the spectral index are

$$A_{\mathrm{s}} \equiv \frac{1}{8\pi^2} \frac{1}{\varepsilon_*} \frac{H_*^2}{M_{\mathrm{pl}}^2} \, , \tag{8.97}$$

$$n_{\mathrm{s}} \equiv 1 - 2\varepsilon_* - \kappa_* \, . \tag{8.98}$$

All quantities on the right-hand side are evaluated at the time when the reference scale k_* exited the horizon. The observational constraints on the scalar amplitude and spectral index are $A_{\mathrm{s}} = (2.098 \pm 0.023) \times 10^{-9}$ and $n_{\mathrm{s}} = 0.965 \pm 0.004$ (for $k_* = 0.05 \, \mathrm{Mpc}^{-1}$) [3]. The observed percent-level deviation from the scale-invariant value, $n_{\mathrm{s}} = 1$, is the first direct measurement of the time dependence during inflation.

8.3.2 Gravitational Waves

Arguably the most robust and model-independent prediction of inflation is a stochastic background of gravitational waves (GWs), or tensor metric perturbations $\delta g_{ij} = a^2 h_{ij}$. In Section 6.5, we reviewed the mathematical properties of cosmological GWs, and, in Chapter 7, we discussed their effect on the CMB anisotropies. We will now complete the story by deriving the tensor power spectrum predicted by inflation, which we define in such a way that $\langle h_{ij}^2 \rangle = \int \mathrm{d} \ln k \, \Delta_h^2(k)$.

In Problem 6.4 of Chapter 6, you are asked to show that tensor fluctuations in an FRW background satisfy the following wave equation:

$$h_{ij}'' + 2\mathcal{H} h_{ij}' - \nabla^2 h_{ij} = 0 \, . \tag{8.99}$$

This equation arises from the action

$$S_2 = \frac{N}{2} \int \mathrm{d}\eta \, \mathrm{d}^3 x \, a^2 \left[(h_{ij}')^2 - (\nabla h_{ij})^2 \right] \, . \tag{8.100}$$

This time, I don't know of a simple trick to determine the overall normalization.

However, expanding the Einstein–Hilbert action to second order in the tensor fluctuation isn't terribly complicated and leads to $N = M_{\rm Pl}^2/4$.

Derivation We first note that the Einstein–Hilbert action can be written as [4]

$$S = \frac{M_{\rm Pl}^2}{2} \int {\rm d}^4x \, \sqrt{-g} \, R \tag{8.101}$$

$$= \frac{M_{\rm Pl}^2}{2} \int {\rm d}^4x \, \sqrt{-g} \, g^{\mu\nu} \Gamma^{\alpha}_{\mu[\beta} \Gamma^{\beta}_{\nu]\alpha} + \text{boundary term}, \tag{8.102}$$

where the square brackets denote antisymmetrization of the indices. The first term in the second line is called the *Schrödinger form* of the action. A nice feature of writing the action in this way is that it contains only first derivatives of the metric. To find the overall normalization of the quadratic action for tensors, we evaluate (8.102) in the subhorizon limit. In this limit, we can ignore the time dependence of the scale factor and the nonzero Christoffel symbols are

$$\Gamma^0_{ij} = \Gamma^i_{0j} = \Gamma^i_{j0} = \frac{1}{2} h'_{ij}, \quad \Gamma^k_{ij} = \frac{1}{2} \left(\partial_i h^k{}_j + \partial_j h^k{}_i - \partial^k h_{ij} \right). \tag{8.103}$$

Substituting these into (8.102), we find

$$S_2^{(k \gg \mathcal{H})} = \frac{M_{\rm Pl}^2}{8} \int {\rm d}\eta {\rm d}^3x \, a^2 \left[(h'_{ij})^2 - (\nabla h_{ij})^2 \right], \tag{8.104}$$

where we have used that h_{ij} is transverse and traceless, and dropped the total derivative term $2\partial_k h_{ij} \partial_i h_{kj} = 2\partial_k(h_{ij} \partial_i h_{kj})$. Although (8.104) was derived in the subhorizon limit, we see from (8.99) that it is, in fact, valid on all scales. Comparison with (8.100) shows that $N = M_{\rm Pl}^2/4$.

It is convenient to use rotational symmetry to align the z-axis of the coordinate system with the momentum of the mode, i.e. $\mathbf{k} \equiv (0, 0, k)$, and write

$$\frac{M_{\rm Pl}}{\sqrt{2}} a \, h_{ij} \equiv \begin{pmatrix} f_+ & f_\times & 0 \\ f_\times & -f_+ & 0 \\ 0 & 0 & 0 \end{pmatrix}, \tag{8.105}$$

where f_+ and $f_\times$ describe the two polarization modes of the gravitational wave. Note that $h_{ij}^2 = (4/M_{\rm Pl}^2)(f_+^2 + f_\times^2)/a^2$. The action (8.100) then becomes

$$S_2 = \frac{1}{2} \sum_{\lambda=+,\times} \int {\rm d}\eta \, {\rm d}^3x \left[(f'_\lambda)^2 - (\nabla f_\lambda)^2 + \frac{a''}{a} f_\lambda^2 \right], \tag{8.106}$$

which is just two copies of the action of a massless scalar field, cf. (8.21). The power spectrum of tensor fluctuations is therefore simply a rescaling of the result for the inflaton fluctuations

$$\Delta_h^2(k) = 2 \times \frac{4}{M_{\rm Pl}^2} \times \Delta_{\delta\phi}^2(k, \eta) \bigg|_{k=aH}, \tag{8.107}$$

where the factor of 2 accounts for the sum over the two polarization modes. Using (8.72), we get

$$\Delta_h^2(k) = \frac{2}{\pi^2} \left(\frac{H}{M_{\mathrm{Pl}}} \right)^2 \Bigg|_{k=aH}. \tag{8.108}$$

Notice that the tensor power spectrum is a direct measure of the expansion rate H during inflation, while the scalar power spectrum depends on both H and ε. The time dependence of the expansion rate $H(t)$ determines the tensor spectral index:

$$n_t \equiv \frac{\mathrm{d}\ln\Delta_h^2}{\mathrm{d}\ln k} = \frac{\mathrm{d}\ln\Delta_h^2}{\mathrm{d}\ln(aH)}$$

$$\approx \frac{\mathrm{d}\ln\Delta_h^2}{\mathrm{d}\ln a} = \frac{\mathrm{d}\ln\Delta_h^2}{H\mathrm{d}t} = 2\frac{\dot{H}}{H^2} = -2\varepsilon. \tag{8.109}$$

In the following insert, I will show how this result can also be derived from a systematic slow-roll expansion.

Slow-roll expansion The equation of motion for each polarization mode is

$$f_k'' + \left(k^2 - \frac{a''}{a} \right) f_k = 0, \tag{8.110}$$

where the effective mass can be written as

$$\frac{a''}{a} = \frac{\nu^2 - 1/4}{\eta^2}, \quad \text{with} \quad \nu \approx \frac{3}{2} + \varepsilon. \tag{8.111}$$

The Bunch–Davies mode function is again given by (8.89). The superhorizon limit of the power spectrum of tensor fluctuations then is

$$\Delta_h^2(k) = 2 \times \left(\frac{2}{aM_{\mathrm{Pl}}} \right)^2 \lim_{k\eta\to 0} \frac{k^3}{2\pi^2} |f_k(\eta)|^2 = \frac{2}{\pi^2} \frac{H_*^2}{M_{\mathrm{Pl}}^2} (k/k_*)^{3-2\nu}, \tag{8.112}$$

where H_* is the Hubble parameter evaluated at the time when the reference scale k_* exited the horizon. This is equivalent to our previous result (8.108). In particular, the tensor spectral index is the same as before

$$n_t \equiv 3 - 2\nu = -2\varepsilon, \tag{8.113}$$

where we have used the definition of ν in (8.111).

In summary, we have found that the tensor spectrum has the following power law form:

$$\Delta_h^2(k) = A_t \left(\frac{k}{k_*} \right)^{n_t}, \tag{8.114}$$

where the amplitude and the spectral index are

$$A_{\rm t} \equiv \frac{2}{\pi^2} \frac{H_*^2}{M_{\rm Pl}^2} ,$$ (8.115)

$$n_{\rm t} \equiv -2\varepsilon_* .$$ (8.116)

Observational constraints on the tensor amplitude are usually expressed in terms of the **tensor-to-scalar ratio**,

$$\boxed{r \equiv \frac{A_{\rm t}}{A_{\rm s}} = 16\varepsilon_* } .$$ (8.117)

Since the amplitude of scalar fluctuations has been measured, the tensor-to-scalar ratio quantifies the size of the tensor fluctuations. Tensor modes will be observable by the next round of CMB polarization experiments if $r > 0.01$ (see Section 8.4.2).

Energy scale of inflation

If primordial gravitational waves were detected, it would tell us the energy at which inflation occurred. Using (8.115), we can write

$$\frac{H}{M_{\rm Pl}} = \pi \sqrt{A_{\rm s}} \sqrt{\frac{r}{2}} \approx 10^{-5} \left(\frac{r}{0.01}\right)^{1/2} ,$$ (8.118)

where we used $A_{\rm s} \approx 2.1 \times 10^{-9}$ for the scalar amplitude. Detecting inflationary tensor perturbations at the level $r \gtrsim 0.01$ would therefore imply that the expansion rate during inflation was about $10^{-5} M_{\rm Pl}$. This can also be expressed in terms of the **energy scale of inflation**,

$$E_{\rm inf} \equiv (3H^2 M_{\rm Pl}^2)^{1/4} = 5 \times 10^{-3} \left(\frac{r}{0.01}\right)^{1/4} M_{\rm Pl} .$$ (8.119)

Note that the tensor-to-scalar ratio depends sensitively on the energy scale of inflation, $r \propto E_{\rm inf}^4$. Decreasing $E_{\rm inf}$ by a factor of 10 reduces the tensor-to-scalar ratio by a factor of 10^4. Gravitational waves from inflation are therefore only observable if inflation occurred near the GUT scale, $E_{\rm inf} \sim 10^{16}$ GeV.

The Lyth bound

Another interesting feature of inflationary models with observable gravitational waves is the fact that the field moves over a super-Planckian distance, $\Delta\phi \equiv |\phi_* - \phi_e| > M_{\rm Pl}$. The derivation of this result is simple, but the theoretical consequences are profound.

Using the definition of the slow-roll parameter ε in (8.8), the tensor-to-scalar ratio (8.117) can be written as

$$r = \frac{8}{M_{\rm Pl}^2} \left(\frac{{\rm d}\phi}{{\rm d}N}\right)^2 ,$$ (8.120)

where $dN = H dt$. We see that the tensor-to-scalar ratio is proportional to the inflaton speed measured in terms of the e-folding time. The perturbations which eventually created the CMB fluctuations exited the horizon at early times, about $N_* = 50$ e-folds before the end of inflation. The total distance that the field moved between this time and the end of inflation can then be written as

$$\frac{\Delta\phi}{M_{\rm Pl}} = \int_0^{N_*} dN \sqrt{\frac{r}{8}} \approx \sqrt{\frac{r_*}{0.01}} \int_0^{N_*} \frac{dN}{50} \sqrt{\frac{r}{r_*}}, \tag{8.121}$$

where $r_* \equiv r(N_*)$ is the tensor-to-scalar ratio on CMB scales. During the slow-roll evolution, $r(N)$ doesn't change much and we obtain the so-called **Lyth bound** [5]:

$$\frac{\Delta\phi}{M_{\rm Pl}} = O(1) \times \sqrt{\frac{r_*}{0.01}}. \tag{8.122}$$

Large values of the tensor-to-scalar ratio, $r > 0.01$, therefore correlate with **super-Planckian field excursions**, $\Delta\phi > M_{\rm Pl}$. It is quite a remarkable coincidence that the level of tensors that is experimentally accessible in the near future is tied to the fundamental scale of quantum gravity, $M_{\rm Pl} = 2.4 \times 10^{18}\,{\rm GeV}$.

Ultraviolet sensitivity

How scared should we be of these super-Planckian field excursions? From (8.119) we see that the super-Planckian field values are not a problem of the energy density becoming larger than the Planck scale. It is also not a problem of radiative stability of the potential. A theory with only the inflaton and the graviton is radiatively stable, since quantum corrections are proportional to the parameters that break a shift symmetry for the inflaton. The small values of these parameters are therefore technically natural even if the field moves over a super-Planckian range. The essence of the problem of **large-field inflation** is instead that gravity requires an ultraviolet completion, and couplings of the inflaton to the additional degrees of freedom that provide this ultraviolet completion do not necessarily respect the symmetry structures needed to protect the inflaton potential in the low-energy theory.

From our general discussion of effective field theory in Chapter 4, we know that integrating out fields of mass Λ with order-unity couplings to the inflaton ϕ will lead to corrections to the inflaton potential of the form

$$\Delta V(\phi) = \sum_{n=1}^{\infty} \left(\frac{c_n}{\Lambda^{2n}} \phi^{4+2n} + \cdots \right), \tag{8.123}$$

where c_n are dimensionless Wilson coefficients that are typically of order one. We therefore expect that the functional form of the potential will change when the field moves a distance of order Λ: in other words, there is "structure" in the potential on scales of order Λ (see Fig. 8.4). Even under the optimistic assumption that $\Lambda \approx M_{\rm Pl}$, the potential (8.123) will not support large-field inflation unless one effectively fine-tunes the infinite set of Wilson coefficients c_n.

The leading idea for implementing large-field inflation is to use a symmetry to suppress the dangerous higher-dimension contributions in (8.123). For example, an unbroken **shift symmetry**,

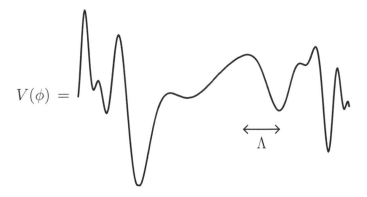

$$V(\phi) =$$

Fig. 8.4 In the absence of symmetries, effective field theory predicts that generic potentials have structure on sub-Planckian scales, $\Lambda < M_{\rm Pl}$. Observable tensor modes, on the other hand, require a smooth inflaton potential over a super-Planckian range.

$$\phi \mapsto \phi + {\rm const}\,, \tag{8.124}$$

forbids all non-derivative interactions, including the desirable parts of the inflaton potential, while a suitable weakly broken shift symmetry[2] can give rise to a radiatively stable model of large-field inflation. Whether such a shift symmetry can be UV-completed is a subtle and important question for a Planck-scale theory like string theory. This is discussed in more detail in my book with Liam McAllister [1] and explicit examples of large-field inflation in string theory were developed in [7, 8].

8.3.3 Slow-Roll Predictions

So far, we have expressed the inflationary predictions in terms of the Hubble rate during inflation, evaluated at the time when the modes of interest exited the horizon. We would like to relate this to the shape of the potential in slow-roll inflation. Using the relation between the Hubble and potential slow-roll parameters derived in (4.66), we can write the amplitude of the scalar spectrum (8.97) and its tilt (8.98) as

$$A_{\rm s} = \frac{1}{24\pi^2} \frac{1}{\varepsilon_{V,*}} \frac{V_*}{M_{\rm Pl}^4}\,, \tag{8.125}$$

$$n_{\rm s} - 1 = -6\varepsilon_{V,*} + 2\eta_{V,*}\,, \tag{8.126}$$

---

[2] Note that to realize an approximate shift symmetry in the low-energy theory, it would suffice for the inflaton to have weak couplings $g \ll 1$ to all the degrees of freedom of the UV completion; the Wilson coefficients in (8.123) would then be suppressed by powers of g. Equivalently, the effective cutoff scale would become $M_{\rm Pl}/g \gg M_{\rm Pl}$ and the coupling of the inflaton to any additional degrees of freedom would be weaker than gravitational [6].

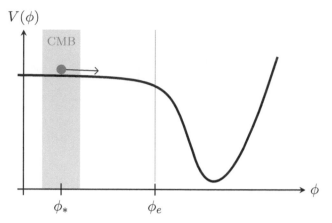

Fig. 8.5 Fluctuations of different wavelengths are produced at different times in the inflationary evolution and therefore probe different parts of the inflationary potential. The scales observed in the CMB probe the conditions roughly 50 e-folds before the end of inflation.

where ε_V and η_V were defined in (4.65). Similarly, the amplitude of the tensor spectrum (8.115) and its tilt (8.116) become

$$A_t = \frac{2}{3\pi^2} \frac{V_*}{M_{\text{Pl}}^4},$$
(8.127)

$$n_t = -2\varepsilon_{V,*}.$$
(8.128)

Observations near the reference scale k_* then probe the shape of the inflaton potential around $\phi_* \equiv \phi(t_*)$, where t_* is the moment of horizon crossing of the fluctuation with wavenumber k_* (see Fig. 8.5). The largest observable fluctuations are probed in the CMB and exit the horizon at early times. Short-wavelength fluctuations exit the horizon later as the field has rolled further down its potential.

It is often useful to define the moment of horizon exit by the number of e-folds remaining until the end of inflation

$$N_* \equiv \int_{\phi_*}^{\phi_e} \frac{H}{\dot{\phi}} \, d\phi \approx \int_{\phi_*}^{\phi_e} \frac{1}{\sqrt{2\varepsilon_V}} \frac{|d\phi|}{M_{\text{Pl}}},$$
(8.129)

where ϕ_e is the field value at which $\varepsilon_V = 1$. CMB observations probe a range of about 7 e-folds centered around $N_* \sim 50$ e-folds before the end of inflation.

Case study: quadratic inflation

Let me illustrate the slow-roll predictions for $m^2\phi^2$ inflation. In Chapter 4, we showed that the slow-roll parameters of this model are

$$\varepsilon_V(\phi) = \eta_V(\phi) = \frac{2M_{\text{Pl}}^2}{\phi^2},$$
(8.130)

and the number of e-folds before the end of inflation is

$$N(\phi) = \frac{\phi^2}{4M_{\text{Pl}}^2} - \frac{1}{2} \approx \frac{\phi^2}{4M_{\text{Pl}}^2} \,. \tag{8.131}$$

At the time when the fluctuations which eventually created the CMB anisotropies crossed the horizon, we have

$$\varepsilon_{V,*} = \eta_{V,*} \approx \frac{1}{2N_*} \approx 0.01 \,, \tag{8.132}$$

where the final equality is for a fiducial value of $N_* \approx 50$. The spectral tilt and the tensor-to-scalar ratio therefore are

$$n_{\text{s}} \equiv 1 - 6\varepsilon_{V,*} + 2\eta_{V,*} = 1 - \frac{2}{N_*} \approx 0.96 \,, \tag{8.133}$$

$$r \equiv 16\varepsilon_{V,*} = \frac{8}{N_*} \approx 0.16 \,. \tag{8.134}$$

As we will see in Section 8.4, the relatively large value of the tensor-to-scalar ratio is inconsistent with the data, so the model is ruled out. You can explore the predictions of other slow-roll models in the problems at the end of the chapter.

8.4 Observational Constraints*

> It doesn't matter how beautiful your theory is, it doesn't matter how smart you are or what your name is. If it doesn't agree with experiment, it's wrong.
>
> *Richard Feynman*

In the previous two sections, I showed that inflation contains a built-in mechanism to produce a nearly scale-invariant spectrum of primordial curvature perturbations. In this section, I will describe the current observational evidence for inflation and discuss future observational tests.

8.4.1 Current Results

It is remarkable that we now have observations that probe the physical conditions just fractions of a second after the Big Bang, when the energies were many orders of magnitude higher than the highest energies accessible to particle colliders. In the following, I will explain what we have learned from these observations.

Scale invariance

As we have seen in Chapter 7, the main features of the CMB spectrum arise from the evolution of coupled photon–baryon fluctuations on subhorizon scales. We found that a nearly featureless spectrum of initial fluctuations evolves into the peaks and troughs of the CMB anisotropy spectrum (see Fig. 8.6). Since the evolution of the

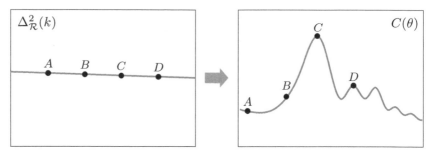

Fig. 8.6   The primordial curvature perturbations (*left*) excite sound waves in the photon–baryon plasma, leading to oscillatory features in the spectrum of CMB anisotropies (*right*). Since this evolution is well understood, observations of the CMB can be used to constrain the primordial spectrum.

fluctuations after inflation is well understood, we can use the CMB observations to put constraints on the primordial power spectrum and hence on the physics of inflation.

A key prediction of inflation is the fact that the primordial correlations are approximately scale invariant, corresponding to the approximate time-translation invariance of the inflationary dynamics. Inflation also predicts a percent-level deviation from perfect scale invariance arising from the small time dependence that is required because inflation has to end. This breaking of perfect scale invariance was measured by the WMAP and Planck satellites. The latest measurement of the scalar spectral index is [3]

$$n_\mathrm{s} = 0.9652 \pm 0.0042 \,, \tag{8.135}$$

which deviates at a statistically significant level from $n_\mathrm{s} = 1$. The sign of the deviation, $n_\mathrm{s} < 1$, is also what would be expected from a decreasing energy density during inflation.

Coherent phases

> You might think then that the shape of the power spectrum can be measured in observations and that this is what convinces us that inflation is right. It is true that we can measure the power spectrum and that the observations agree with the theory. But this is not what tingles our spines when we look at the data. Rather, the truly striking aspect of the perturbations generated during inflation is that all Fourier modes have the same phase.   *Scott Dodelson*

Possibly the most remarkable evidence for inflation comes from the **phase coherence** of the primordial fluctuations. If the perturbations in the photon–baryon fluid didn't have coherent phases, then the sound waves in the plasma wouldn't interfere constructively at recombination and the CMB power spectrum would not

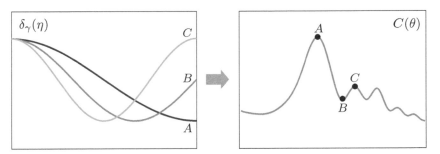

Fig. 8.7   Cartoon illustrating how sound waves in the primordial plasma (*left*) are associated to the peak structure of the CMB power spectrum (*right*). If all waves start with the same initial phase, then their amplitudes at photon decoupling will depend on their wavelengths (or oscillation frequencies). Waves that are captured at an extremum (*A* or *C*) produce the peaks in the CMB spectrum, while waves that have zero amplitude at decoupling (*B*) produce the troughs.

have its famous peak structure. I will now explain why this phase coherence arises naturally in inflationary models. A more detailed version of the same argument has appeared in a nice article by Dodelson [9].

The CMB is a snapshot of sound waves at the time of recombination (or, more precisely, photon decoupling). Each wave is captured at a certain phase in its evolution (see Fig. 8.7). Since the oscillation frequency is set by the wavelength of the wave, $\omega = c_s k$, waves with different wavelengths are observed at different moments in their evolution. This is the reason why the angular power spectrum of the temperature fluctuations goes up and down as a function of scale. While this cartoon is essentially correct, it doesn't highlight the fact that the initial phases must be synchronized. Even a single Fourier mode should be thought of as an ensemble of many waves with fixed wavelength, but amplitudes that are drawn from a Gaussian probability distribution. These waves need to be added to get the real space density distribution and hence the CMB anisotropies. The modes will only have the same phase at recombination if their initial phases are the same (see Fig. 8.8). This coherence of the phases is necessary in order for the superposition of many modes to lead to the peak structure seen in the CMB spectrum. If the initial phases were random, then the peaks of the spectrum would be washed out.

In inflation, the initial phases are set by the fact that the modes are frozen outside of the horizon. When the modes re-enter the horizon, the time derivative of the amplitude, $\mathcal{R}'$, is small. All modes with the same wavenumber k, but distinct wavevectors $\mathbf{k}$, therefore start their evolution at the same time. Thinking of the Fourier modes as a combination of sine and cosine solutions, inflation only excites the cosine solution. What is remarkable is that the coherence of the phases extends to superhorizon scales at recombination. This is seen most convincingly in the cross correlation between fluctuations in the CMB temperature and E-mode polarization. Since polarization is generated only by the scattering of the photons

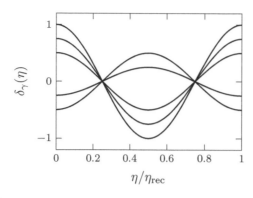 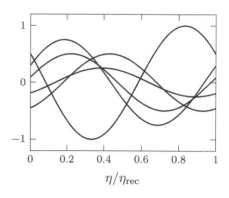

Fig. 8.8 Evolution of different Fourier modes with the same wavelength. The amplitudes are drawn from a Gaussian distribution with zero mean. The modes on the left start with the same initial phase and therefore are captured with the same phase at recombination. The modes on the right have random initial phases.

off the free electrons just before decoupling, this signal cannot be created after recombination. Without inflation (or something very similar), such a signal would violate causality [10]. The large-scale TE correlation signal was detected for the first time by the WMAP satellite [11].

Adiabaticity

In Section 6.2, we showed that the most general initial conditions for the fluid components of the hot Big Bang may include isocurvature contributions. However, these isocurvature fluctuations are *not* a natural outcome of the simplest inflationary models. Instead, single-field inflation has only one fluctuating scalar degree of freedom, which induces purely adiabatic initial conditions. Moreover, thermal equilibrium after inflation tends to wash out any isocurvature modes.

The peak structure of the CMB spectrum is evidence that the dominant contribution to the primordial perturbations is adiabatic. As we have seen, adiabatic initial conditions produce a cosine oscillation in the photon–baryon plasma. Isocurvature initial conditions, on the other hand, would create a sine oscillation which is not consistent with the observed peaks of the CMB. The possibility of a dominant isocurvature perturbation is therefore ruled out and even a subdominant level of isocurvature fluctuations is highly constrained.

Observational constraints on isocurvature modes depend on whether they are correlated or uncorrelated with the adiabatic mode. In the totally correlated case, all fluctuations are still proportional to $\mathcal{R}$ and we can write the matter isocurvature mode as

$$S_m = \sqrt{\alpha}\,\mathcal{R}\,. \tag{8.136}$$

The Planck limit on the proportionality constant is [12]

$$\alpha < 0.0003^{+0.0016}_{-0.0012}\,. \tag{8.137}$$

A more detailed analysis for CDM, neutrino density and neutrino velocity isocurvature modes gives similar results. There is, hence, no evidence for any isocurvature contribution to the initial perturbations, in agreement with the expectation from single-field inflation.

Gaussianity

So far, we have only described constraints arising from the two-point function of the primordial correlations. This two-point function, in fact, contains all the information about the initial conditions if the perturbations are drawn from a Gaussian probability distribution. The simplest inflationary models indeed predict Gaussianity to be a good approximation. It is easy to see why. Slow-roll inflation only occurs on the flat part of the potential where the self-interactions of the field are small. The linearized equation of motion is then a good approximation for the evolution of the inflaton fluctuations. This equation took the form of a harmonic oscillator equation, which we used to compute the expectation value of quantum fluctuations in the ground state. Since the ground state wave function of the harmonic oscillator is a Gaussian, we expect the initial perturbations created by inflation also to obey Gaussian statistics. All odd N-point functions then vanish, while all even N-point functions are related to the power spectrum (which therefore contains all the information).

Small amounts of non-Gaussianity may nevertheless still exist and would teach us a lot about the physics of inflation. In particular, extensions of slow-roll models can produce non-Gaussian fluctuations from enhanced self-interactions of the inflaton or couplings to additional fields. Below I will describe the current constraints and future tests of these types of non-Gaussianity.

8.4.2 Future Tests

> Extraordinary claims require extraordinary evidence.   *Carl Sagan*

Although the evidence for inflation is intriguing, we cannot yet claim that it is part of the standard model of cosmology at the same level as, for example, BBN or recombination are. This isn't surprising, since, unlike BBN and recombination, inflation requires physics beyond the Standard Model of particle physics. To gain more confidence in the theory, we need further observational tests. In this section, I will mention two avenues that I consider particularly promising: tensor modes and non-Gaussianity.

Tensor modes

A major goal of current efforts in observational cosmology is to detect the tensor component of the primordial fluctuations. As we have seen, the tensor amplitude depends on the energy scale of inflation and is therefore not predicted (i.e. it varies between models). While this makes the search for primordial tensor modes difficult, it is also what makes it exciting. Detecting tensors would reveal the energy scale

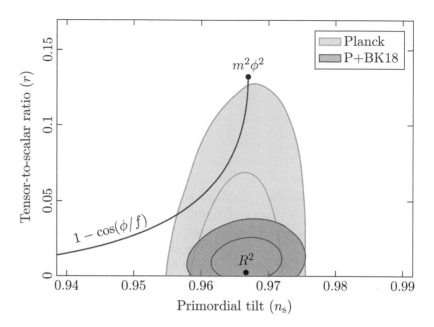

Fig. 8.9　Current constraints on the spectral tilt n_s and the tensor-to-scalar ratio r from CMB measurements of Planck and BICEP/Keck [13]. Shown also are the predictions of a few popular slow-roll models.

at which inflation occurred, providing an important clue about the physics driving the inflationary expansion. As we discussed in Section 8.3.2, the tensor signal is also unusually sensitive to high-energy corrections and therefore probes important aspects of the microphysics of inflation and the nature of quantum gravity [1].

Most searches for tensor modes focus on their imprint on the polarization of the CMB. As we saw in Chapter 7, polarization is generated through the scattering of the anisotropic radiation field off the free electrons just before decoupling. The presence of a gravitational wave background creates an anisotropic stretching of the spacetime which induces a special type of polarization pattern, the so-called **B-mode** pattern (a pattern whose "curl" doesn't vanish). Such a pattern cannot be created by scalar fluctuations and is therefore a unique signature of primordial tensors. Because tensors are such a clean prediction of inflation, a B-mode detection would be a milestone towards establishing that inflation really occurred in the early universe.

The latest observational constraints are shown in Fig. 8.9, implying an upper bound on the tensor-to-scalar ratio of $r < 0.035$ [13]. A large number of ground-based, balloon and satellite experiments are searching for the B-mode signal predicted by inflation (see e.g. [14]). The main challenge for these experiments is to isolate the primordial signal from secondary B-modes created by astrophysical foregrounds. This can be done by using the frequency dependence of the signal (the primordial signal is a blackbody) and its spatial dependence (the primordial signal is statistically isotropic).

Non-Gaussianity

We have seen that inflation predicts that the initial correlations are highly Gaussian and therefore described well by the power spectrum. Nevertheless, in principle, a significant amount of information about the physics of inflation can be encoded in small levels of **non-Gaussianity**. In particular, the higher-order correlations associated with non-Gaussian initial conditions are sensitive to nonlinear interactions, while the power spectrum only probes the free theory. Measurements of primordial non-Gaussianity would therefore probe the particle content and the interactions during inflation.[3]

The leading signature of non-Gaussianity is a nonzero three-point correlation function, or its Fourier equivalent, the **bispectrum**:

$$\langle \mathcal{R}_{\mathbf{k}_1} \mathcal{R}_{\mathbf{k}_2} \mathcal{R}_{\mathbf{k}_3} \rangle = (2\pi)^3 \delta_{\mathrm{D}}(\mathbf{k}_1 + \mathbf{k}_2 + \mathbf{k}_3) \frac{(2\pi^2)^2}{(k_1 k_2 k_3)^2} B_{\mathcal{R}}(k_1, k_2, k_3) , \qquad (8.138)$$

where the delta function is again a consequence of the homogeneity of the background. The sum of the three wavevectors must therefore form a closed triangle and the strength of the signal will depend on the shape of the triangle. As indicated by the notation, the bispectrum is a function of the magnitudes of the wavevectors, $k_{n=1,2,3}$, as required by the isotropy of the background. The factor of $(k_1 k_2 k_3)^{-2}$ has been extracted in (8.138) to make the function $B_{\mathcal{R}}(k_1, k_2, k_3)$ dimensionless. The *amplitude* of the non-Gaussianity is then defined as the size of the bispectrum in the equilateral configuration:

$$f_{\mathrm{NL}}(k) \equiv -\frac{5}{18} \frac{B_{\mathcal{R}}(k, k, k)}{\Delta_{\mathcal{R}}^4(k)} , \qquad (8.139)$$

where the numerical factor is a historical accident and the overall minus sign is a consequence of my convention for the sign of $\mathcal{R}$ (which is different from the convention used in the Planck and WMAP analysis). In general, the parameter f_{NL} can depend on the overall wavenumber (or the size of the triangle), but for scale-invariant initial conditions it would be a constant. The bispectrum can then be written as

$$B_{\mathcal{R}}(k_1, k_2, k_3) \equiv -\frac{18}{5} f_{\mathrm{NL}} \times S(x_2, x_3) \times \Delta_{\mathcal{R}}^4 , \qquad (8.140)$$

where $x_2 \equiv k_2/k_1$ and $x_3 \equiv k_3/k_1$.[4] The shape function $S(x_2, x_3)$ is normalized so that $S(1, 1) \equiv 1$. As we will see below, the *shape* of the non-Gaussianity contains

---

[3] An analogy with particle physics may be informative: the measurements of the power spectrum are the analog of measuring the propagators of particles. While these propagators are important features of long-lived particles, their form is completely fixed by Lorentz symmetry and therefore doesn't contain dynamical information. To learn about the dynamics of the theory we collide particles and study the resulting interactions. Measurements of higher-order correlations (or non-Gaussianity) are the analog of measuring collisions in particle physics. The study of non-Gaussianity therefore also goes by the name of **cosmological collider physics** [15].

[4] We will order the wavenumbers such that $x_3 \leq x_2 \leq 1$. Since the wavenumbers are side lengths of a triangle, we have the constraint $x_2 + x_3 > 1$.

a lot of information about the microphysics of inflation. This is to be contrasted with the power spectrum, which is described by just two numbers, A_s and n_s, and not a whole function.

In the following, I will present a mostly qualitative discussion of popular forms of non-Gaussianity. Further details and explicit computations can be found in [16, 17].

- **Local non-Gaussianity**   One way to create a non-Gaussian field is to take a Gaussian one and add to it the square of itself [18, 19]:[5]

$$\mathcal{R}(\mathbf{x}) = \mathcal{R}_g(\mathbf{x}) - \frac{3}{5} f_\mathrm{NL}^\mathrm{local} \Big[\mathcal{R}_g^2(\mathbf{x}) - \langle \mathcal{R}_g^2 \rangle \Big] . \tag{8.141}$$

This local field redefinition creates the so-called **local non-Gaussianity**. The corresponding bispectrum, and the associated shape function, are

$$B_\mathcal{R}^\mathrm{local} \propto (k_1 k_2 k_3)^2 \Big(\mathcal{P}_1 \mathcal{P}_2 + \mathcal{P}_1 \mathcal{P}_3 + \mathcal{P}_2 \mathcal{P}_3 \Big) ,$$
$$S_\mathrm{local} = \frac{1}{3} \left(\frac{k_3^2}{k_1 k_2} + \frac{k_2^2}{k_1 k_3} + \frac{k_1^2}{k_2 k_3} \right) , \tag{8.142}$$

where $\mathcal{P}_n \equiv \mathcal{P}_\mathcal{R}(k_n)$ and the result for S_local is for a scale-invariant power spectrum, $\mathcal{P}_n \propto k_n^{-3}$. The shape of the local bispectrum is shown in Fig. 8.10. We see that the signal is peaked in the "squeezed limit," where one side of the triangle is much smaller than the other two (say $k_3 \ll k_2 \approx k_1$):

Interestingly, this feature cannot arise in single-field inflation because of the **single-field consistency relation** [20, 21]. This theorem states that, for any single-field model of inflation, the signal in the squeezed limit must satisfy

$$\lim_{k_3 \to 0} B_\mathcal{R}(k_1, k_2, k_3) = (n_\mathrm{s} - 1) \frac{(k_1 k_2 k_3)^2}{(2\pi^2)^2} \mathcal{P}_\mathcal{R}(k_3) \mathcal{P}_\mathcal{R}(k_1) , \tag{8.143}$$

which vanishes for scale-invariant fluctuations. Detecting a signal in the squeezed limit would therefore rule out *all* models of single-field inflation (with Bunch–Davies initial conditions), not just slow-roll models. Conversely, the signal in the squeezed limit can be used as a particular clean diagnostic for extra fields during inflation. Because inflation occurred at very high energies, even very massive particles ($m \lesssim H$) can be produced by the rapid expansion of the spacetime [15, 22]. When these particles decay into the inflaton, they can lead to a distinct signature in the squeezed limit of the bispectrum.

The current Planck constraint on local non-Gaussianity is [23]

$$f_\mathrm{NL}^\mathrm{local} = -0.9 \pm 5.1 . \tag{8.144}$$

---

[5] The factor of 3/5 appears because this field redefinition was originally introduced for the gravitational potential Φ, which during the matter era is related to $\mathcal{R}$ by a factor of 3/5. This also explains the normalization factor in (8.139).

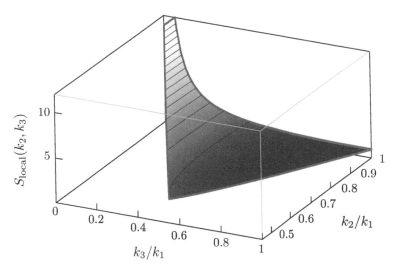

Fig. 8.10 Bispectrum of local non-Gaussianity. The signal is peaked for squeezed triangles with $k_3/k_1 \to 0$.

Future galaxy surveys may improve this bound by an order of magnitude [24], providing a sensitive probe of the particle spectrum during inflation. These measurements exploit the fact that local non-Gaussianity leads to a **scale-dependent bias** [25] on large scales, which can be measured in the galaxy power spectrum.

- **Equilateral non-Gaussianity** In slow-roll inflation, the flatness of the inflationary potential constrains the size of the inflaton self-interactions. However, interesting models of inflation have been suggested in which higher-derivative corrections—such as $(\partial\phi)^4$—play an important role during inflation [26, 27]. These inflaton interactions lead to cubic interactions of the curvature perturbations—such as $\dot{\mathcal{R}}^3$ and $\dot{\mathcal{R}}(\partial_i\mathcal{R})^2$—and hence a nonzero bispectrum.[6] A key characteristic of derivative interactions is that they are suppressed when any individual mode is far outside the horizon. This suggests that the bispectrum is small in the squeezed limit and maximal when all three modes have wavelengths equal to the horizon size. The bispectrum therefore has a shape that peaks for an equilateral triangle configuration ($k_1 \approx k_2 \approx k_3$):

The precise shape of this **equilateral non-Gaussianity** depends on the specific cubic interactions from which it arises; for example, the shapes coming from $\dot{\mathcal{R}}^3$

---

[6] A systematic way to classify these derivative interactions is in terms of an EFT for the inflationary fluctuations [2].

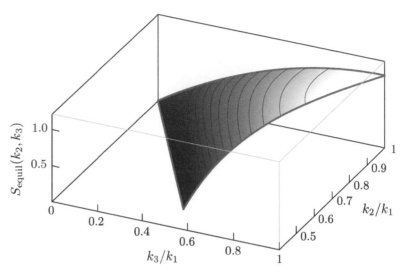

Fig. 8.11 Bispectrum for equilateral non-Gaussianity as described by the template shape (8.145). The signal is peaked for equilateral triangles with $k_3/k_1 \approx k_2/k_1 \approx 1$.

and $\dot{\mathcal{R}}(\partial_i \mathcal{R})^2$ are slightly different away from the equilateral limit. The CMB data is typically analyzed using the following template shape [28]:

$$S_{\text{equil}} = -\left(\frac{k_1^2}{k_2 k_3} + 2 \text{ perms} \right) + \left(\frac{k_1}{k_2} + 5 \text{ perms} \right) - 2 , \qquad (8.145)$$

which is plotted in Fig. 8.11. This template is a good approximation to all forms of equilateral non-Gaussianity near the equilateral limit, which is where most of the constraining power of the CMB data comes from. When using other data sets, however, it is important to remember that the template (8.145) is only an approximation. For example, the exact bispectrum coming from the interaction $\dot{\mathcal{R}}^3$ is [27]

$$S_{\dot{\mathcal{R}}^3} = \frac{27\, k_1 k_2 k_3}{(k_1 + k_2 + k_3)^3} , \qquad (8.146)$$

which agrees with (8.145) in the equilateral configuration, but differs away from it.

Using the template shape (8.145), the Planck collaboration derived the following constraint on equilateral non-Gaussianity [23]:

$$f_{\text{NL}}^{\text{equil}} = -26 \pm 47 . \qquad (8.147)$$

It will be hard (but not impossible) to improve on this constraint with future galaxy surveys, because gravitational nonlinearities also produce an equilateral signal. This makes it challenging to distinguish the primordial non-Gaussianity from the secondary non-Gaussianity produced by the late-time evolution of the perturbations.

- **Folded non-Gaussianity** The Gaussianity of slow-roll inflation also relies on the fact that we evaluated the quantum fluctuations in the Bunch–Davies vacuum

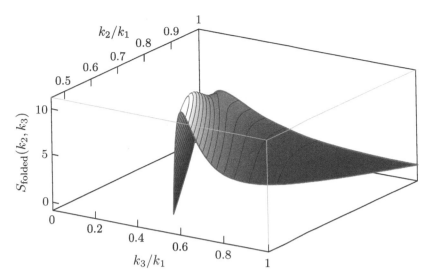

Fig. 8.12 Bispectrum for folded non-Gaussianity as described by the template shape (8.148), for $k_c/k_1 = 0.1$. The signal is peaked for folded triangles with $k_2 + k_3 \approx k_1$.

(corresponding to the ground state of the harmonic oscillator). In contrast, starting from an excited initial state would lead to non-Gaussianity. The detailed shape of this non-Gaussianity, and hence the constraint on f_{NL}, depends on the model for the excited initial state [27, 29]. A universal feature is that the correlations are enhanced for "folded" triangles where two of the wavevectors become colinear:

A template that captures the essential features of non-Gaussianity from excited initial states is [16]

$$S_{\mathrm{folded}} \propto k_1 k_2 k_3 \frac{-k_1 + k_2 + k_3}{(k_c - k_1 + k_2 + k_3)^4} + 2 \text{ perms}, \qquad (8.148)$$

where k_c is a cutoff that regulates the singularity in the folded limit, $k_2 + k_3 \to k_1$. This shape is plotted in Fig. 8.12. Ignoring the cutoff k_c, the function in (8.148) is almost the same as that in (8.146), except for a flipped sign in front of one of the wavenumbers ("energies"). This sign flip arises because an excited initial state contains an admixture of the negative-frequency solution $e^{+ik\eta}$. The alternating signs in (8.148) arise when correlating positive- and negative-frequency modes. It is these alternating signs that are the origin of the enhanced signal in the folded limit.

The signal in the folded configuration also provides an interesting test of the quantum origin of the fluctuations [30]. While classical fluctuations would generically have non-vanishing correlations in the folded limit, quantum fluctuations in the Bunch–Davies vacuum are characterized by the absence of such a signal.

8.5 Summary

In this chapter, we have studied the initial fluctuations created by inflation. We showed how quantum zero-point fluctuations get stretched outside the horizon during inflation, becoming the seeds for the large-scale structure of the universe.

We started by deriving the equation of motion for the inflaton fluctuations. In spatially flat gauge, the equation for $f \equiv a\delta\phi$ takes the form

$$f'' + \left(k^2 - \frac{z''}{z}\right)f = 0, \quad \text{where} \quad z \equiv \frac{a\bar{\phi}'}{\mathcal{H}}. \tag{8.149}$$

The comoving curvature perturbation is $\mathcal{R} = f/z$. The field f and its conjugate momentum $\pi \equiv f'$ are promoted to quantum operators which satisfy the equal time commutation relation

$$[\hat{f}(\eta, \mathbf{x}), \hat{\pi}(\eta, \mathbf{y})] = i\,\delta_{\mathrm{D}}(\mathbf{x} - \mathbf{y}). \tag{8.150}$$

The field operator is then written as

$$\hat{f}(\eta, \mathbf{x}) = \int \frac{\mathrm{d}^3 k}{(2\pi)^3} \left[f_k(\eta)\, \hat{a}_{\mathbf{k}} e^{i\mathbf{k}\cdot\mathbf{x}} + f_k^*(\eta)\, \hat{a}_{\mathbf{k}}^\dagger e^{-i\mathbf{k}\cdot\mathbf{x}}\right], \tag{8.151}$$

where $\hat{a}_{\mathbf{k}}$ and $\hat{a}_{\mathbf{k}}^\dagger$ are annihilation and creation operators, which obey

$$[\hat{a}_{\mathbf{k}}, \hat{a}_{\mathbf{k}'}^\dagger] = (2\pi)^3 \delta_{\mathrm{D}}(\mathbf{k} - \mathbf{k}'). \tag{8.152}$$

The vacuum state $|0\rangle$ is the state annihilated by $\hat{a}_{\mathbf{k}}$. A preferred initial state—the Bunch–Davies vacuum—is defined by the ground state of a harmonic oscillator. With this boundary condition, and assuming a de Sitter background, the mode function in (8.151) becomes

$$f_k(\eta) = \frac{e^{-ik\eta}}{\sqrt{2k}}\left(1 - \frac{i}{k\eta}\right). \tag{8.153}$$

Using this, the dimensionless power spectrum of the inflaton fluctuations is

$$\Delta_{\delta\phi}^2(k, \eta) = \frac{k^3}{2\pi^2}\frac{|f_k(\eta)|^2}{a^2(\eta)} = \left(\frac{H}{2\pi}\right)^2\left[1 + (k\eta)^2\right] \xrightarrow{k\eta \to 0} \left(\frac{H}{2\pi}\right)^2. \tag{8.154}$$

The power spectrum of the curvature perturbation $\mathcal{R}$ takes a power law form

$$\Delta_{\mathcal{R}}^2(k) = A_{\mathrm{s}}\left(\frac{k}{k_*}\right)^{n_{\mathrm{s}}-1}, \tag{8.155}$$

where the amplitude and the spectral index are

$$A_s \equiv \frac{1}{8\pi^2} \frac{1}{\varepsilon_*} \frac{H_*^2}{M_{\rm Pl}^2}, \tag{8.156}$$

$$n_s \equiv 1 - 2\varepsilon_* - \kappa_* . \tag{8.157}$$

All quantities on the right-hand side are evaluated at the time when the reference scale k_* exited the horizon, $k_* = (aH)_*$. Computing the power spectrum at horizon crossing mitigates the error that is being made in (8.153) by assuming a de Sitter background. The measured value of the scalar spectral index, $n_s = 0.965 \pm 0.004$ [3], shows a deviation from a scale-invariant spectrum, as expected from the time evolution during inflation.

A similar calculation for tensor fluctuations during inflation gives

$$\Delta_h^2(k) = A_t \left(\frac{k}{k_*}\right)^{n_t}, \tag{8.158}$$

where the amplitude and the spectral index are

$$A_t \equiv \frac{2}{\pi^2} \frac{H_*^2}{M_{\rm Pl}^2}, \tag{8.159}$$

$$n_t \equiv -2\varepsilon_* . \tag{8.160}$$

Observational constraints on the tensor amplitude are usually expressed in terms of the tensor-to-scalar ratio,

$$r \equiv \frac{A_t}{A_s} = 16\varepsilon_* . \tag{8.161}$$

A large value of the tensor-to-scalar ratio, $r > 0.01$, corresponds to the inflaton field moving over a super-Planckian distance during inflation, $\Delta\phi > M_{\rm Pl}$. The current upper bound on the tensor-to-scalar ratio is $r < 0.035$ [13]. Future observations of CMB polarization will improve this bound by at least an order of magnitude.

Further Reading

The inflationary quantum fluctuations are treated in many lecture notes (e.g. [31–33]) and textbooks (e.g. [34, 35]). My own contributions have been in the form of two TASI lectures [17, 36] and a book [1]. Nice introductions to "quantum field theory in curved spacetimes" are the book by Mukhanov and Winitzki [37] and the lecture notes of Jacobson [38]. For more on non-Gaussianity, I recommend the classic papers by Maldacena [20] and Weinberg [39], as well as the reviews by Chen [16], Wang [40], Lim [41] and Komatsu [42]. Finally, I also highly recommend reading the original papers by the Planck [12] and WMAP collaborations [11] to see the amazing work that has been done on the observational front.

Problems

8.1 Massless scalar in de Sitter

Consider a massless scalar field ϕ in a fixed de Sitter background.

1. Determine the equation of motion satisfied by the field. Imposing Bunch–Davies initial conditions, show that the unique solution of the equation is

$$\phi_k(\eta) = \frac{iH}{\sqrt{2k^3}}(1 + ik\eta)e^{-ik\eta}\,.$$

2. Derive the power spectrum of quantum zero-point fluctuations in the Bunch–Davies vacuum. By summing over all Fourier modes, calculate the equal-time two-point function in real space. What is the physical meaning of the divergence in the limit $\eta \to 0$?

3. Photons are massless, but they are not produced during inflation. Why?

8.2 Massive scalar in de Sitter

Consider a scalar field of mass m in a fixed de Sitter background.

1. Derive the classical equation of motion of the field. Discuss the behavior inside and outside the Hubble radius as function of m/H.

2. Compute the quantum zero-point fluctuations of the field in the Bunch–Davies vacuum.

3. For $m < \frac{3}{2}H$, show that the spectral index of the power spectrum is

$$n = 3 - 2\sqrt{\frac{9}{4} - \frac{m^2}{H^2}} \quad \xrightarrow{m \ll H} \quad \frac{2}{3}\frac{m^2}{H^2}\,.$$

4. Discuss the amplitude of the fluctuations for $m > \frac{3}{2}H$.

8.3 Inflationary fluctuations in comoving gauge

In the comoving gauge, the inflaton fluctuations vanish, $\delta\phi = 0$, and the scalar fluctuations are captured by the comoving curvature perturbation $\mathcal{R}$. The quadratic action for $\mathcal{R}$ is

$$S_2 = \int \mathrm{d}\eta \mathrm{d}^3x\, a^2 M_{\mathrm{Pl}}^2 \varepsilon \left((\mathcal{R}')^2 - (\partial_i \mathcal{R})^2\right),$$

where $\varepsilon \equiv -\dot{H}/H^2$ is the slow-roll parameter.

1. Derive the equation of motion for $\mathcal{R}$ and discuss its solution on scales that are smaller and larger than the Hubble radius.

2. Determine the equation of motion for $f \equiv z\mathcal{R}$, where $z \equiv a M_{\mathrm{Pl}}\sqrt{2\varepsilon}$.

3. Derive the power spectrum of $\mathcal{R}$ for slow-roll inflation.

8.4 Quantum-to-classical transition

Consider a particle of mass m moving in the potential

$$V(x) = -\frac{1}{2}kx^2,$$

where $k > 0$. Following Guth and Pi [43], we use this inverted harmonic oscillator as a model of the inflationary fluctuations after horizon crossing. We will show that the initial quantum fluctuations become classical at late times.

1. The wave function of the particle satisfies the Schrödinger equation

$$i\hbar\frac{\partial\psi}{\partial t} = -\frac{\hbar^2}{2m}\frac{\partial^2\psi}{\partial x^2} - \frac{1}{2}kx^2\psi.$$

Consider the ansatz $\psi(x,t) = A(t)\exp[-B(t)x^2]$. Show that the functions $A(t)$ and $B(t)$ satisfy

$$i\hbar\dot{A} = \frac{\hbar^2}{m}AB,$$

$$i\hbar\dot{B} = \frac{1}{2}k + 2\frac{\hbar^2}{m}B^2.$$

Introducing the parameters $a^2 \equiv \hbar/\sqrt{mk}$ and $\omega^2 \equiv k/m$, show that the solutions to these equations are

$$A = \frac{\alpha}{\sqrt{\cos(\phi - i\omega t)}},$$

$$B = \frac{1}{2a^2}\tan(\phi - i\omega t),$$

where α and ϕ are constants. Determine α by requiring the wave function to be correctly normalized. Show that the solution at late times is a Gaussian with width

$$\langle x^2\rangle \xrightarrow{t\to\infty} \frac{a^2}{4\sin(2\phi)}e^{2\omega t}.$$

2. Show that

$$\hat{x}\hat{p}\,\psi = \hbar\frac{x^2}{a^2}\psi + O(e^{-2\omega t}),$$

$$\hat{p}\hat{x}\,\psi = \hbar\frac{x^2}{a^2}\psi - i\hbar\psi + O(e^{-2\omega t}),$$

where $\hat{p} = -i\hbar(\partial/\partial x)$, and interpret the result.

8.5 Models of slow-roll inflation

In this problem, you will study the predictions of popular slow-roll models of inflation.

1. Consider slow-roll inflation with a polynomial potential $V(\phi) = \mu^{4-p}\phi^p$, where $p > 0$ and μ is a parameter with the dimension of mass. Show that the spectral index n_s and the tensor-to-scalar ratio r, evaluated at a reference scale k_*, are

$$n_s - 1 = -\frac{2+p}{2N_*}, \quad r = \frac{4p}{N_*},$$

where N_* is the number of e-folds between the horizon exit of k_* and the end of inflation. Which values of p are consistent with observations?

2. In Problem 4.4, you studied axion inflation driven by a potential of the form $V(\phi) = \mu^4[1 - \cos(\phi/f)]$, where f is the axion decay constant. Show that the scalar spectral index and the tensor-to-scalar ratio are

$$n_s - 1 = -\alpha\frac{e^{N_*\alpha}(1+\alpha/2)+1}{e^{N_*\alpha}(1+\alpha/2)-1}, \quad r = 8\alpha\frac{1}{e^{N_*\alpha}(1+\alpha/2)-1},$$

where $\alpha \equiv M_{\rm Pl}^2/f^2$. Sketch this prediction in the n_s–r plane. Discuss the limit $\alpha \ll 1$.

3. In Problem 4.6, you studied the Starobinsky model:

$$S = \frac{M_{\rm Pl}^2}{2}\int d^4x\sqrt{-g}\left(R + \frac{\alpha}{2M_{\rm Pl}^2}R^2\right),$$

where α is a dimensionless parameter. You showed that this can be written as a standard slow-roll model with potential

$$V(\phi) = \frac{M_{\rm Pl}^4}{4\alpha}\left(1 - e^{-\sqrt{2/3}\phi/M_{\rm Pl}}\right)^2,$$

and that inflation occurs for $\phi > 0.94\,M_{\rm Pl}$. We would now like to determine the observational predictions of the model.

- Assuming $\phi \gg M_{\rm Pl}$, show that the slow-roll parameters are

$$\eta_V \approx -\frac{4}{3}e^{-\sqrt{2/3}\phi/M_{\rm Pl}}, \quad \varepsilon_V \approx \frac{3}{4}\eta_V^2.$$

- Show that the scalar spectral tilt and the tensor-to-scalar ratio are

$$n_s - 1 \approx -\frac{2}{N_*}, \quad r \approx \frac{12}{N_*^2}.$$

This value of the tensor-to-scalar ratio is within reach of the next generation of CMB experiments.

- Show that the normalization of the scalar amplitude, $A_s \approx 2 \times 10^{-9}$, requires that $\alpha \approx 2 \times 10^9$. Explaining the origin of this large number is a challenge.

4. In Problem 4.7, you studied inflation driven by a non-minimally coupled scalar field

$$S = \int d^4x \sqrt{-g} \left[\frac{M_{\mathrm{Pl}}^2}{2} \left(1 + \xi \frac{\varphi^2}{M_{\mathrm{Pl}}^2} \right) R - \frac{1}{2} g^{\mu\nu} \partial_\mu \varphi \partial_\nu \varphi - \frac{\lambda}{4!} \varphi^4 \right],$$

where ξ is a dimensionless parameter. Describe (without detailed calculation) the predictions of this model.

8.6 Higgs inflation

The only known fundamental scalar field is the Higgs field. It is tempting to ask if the Higgs field could also have been the scalar field that created the inflationary expansion.

1. The Higgs potential is

$$V(\phi) = \lambda \left(\phi^2 - v^2 \right)^2,$$

where $v = 246 \, \mathrm{GeV}$ and ϕ is the radial part of the complex Higgs field. Sketch this potential and indicate the regions where slow-roll inflation might occur. Compute the slow-roll parameters ε_V and η_V.

2. Consider first the region $0 < \phi < v$. Can both slow-roll conditions, $\varepsilon_V \ll 1$ and $\eta_V \ll 1$, be satisfied simultaneously?

3. Now, look at the regime $\phi \gg v$. Show that $\varepsilon_V(\phi)$ and $\eta_V(\phi)$ become independent of v. Determine the field values at the end of inflation (ϕ_e) and $N_* = 50$ e-folds before (ϕ_*).

Compute the amplitude of the power spectrum of scalar fluctuations at ϕ_*. Express your answer in terms of N_*, v and the Higgs boson mass m_H. What value of m_H is required to match the observed amplitude $A_{\mathrm{s}} = 2 \times 10^{-9}$? How does this compare to the measured value, $m_H = 125 \, \mathrm{GeV}$?

4. A new version of Higgs inflation has been proposed in [44]. Its key ingredient is a non-minimal coupling of the Higgs to gravity, $\xi \phi^2 R$, with $\xi \gg 1$. Discuss the observational predictions of this model. How large does the parameter ξ have to be?

9   Outlook

This book has been a first introduction to the basic principles of cosmology. We have learned how a combination of theoretical advances and precision observations have transformed cosmology into a quantitative science. Questions about the age of the universe, its composition and its evolution now have very precise answers. Although cosmology has become part of mainstream science, it is a special type of historical science. The cosmological experiment cannot be repeated, since the Big Bang only happened once. Nevertheless, cosmologists have been able to reconstruct much of the history of the universe from a limited amount of observational clues. This culminated in the ΛCDM model which describes the cosmological evolution from 1 second after the Big Bang until today, 13.8 billion years later. The details of cosmological structure formation, through the gravitational clustering of small density variations in the primordial universe, are well understood. There is now even evidence for the rather spectacular proposal that quantum fluctuations during a period of exponential expansion about 10^{-34} seconds after the Big Bang provided the seed fluctuations for the large-scale structure of the universe.

At the same time, many fundamental questions in cosmology remain unanswered. We do not know what drove the inflationary expansion, what kind of matter filled the universe after inflation, and how the asymmetry between matter and antimatter was created. The nature of dark matter remains unknown and how dark energy fits into quantum field theory is still a complete mystery. Finally, we do not know what happened at the Big Bang singularity, and if there was any time (and space) before that moment. We hope that future observations of the polarization of the CMB and of the large-scale structure of the universe will help to unlock these mysteries. In addition, we will need new theoretical ideas. Maybe one of these ideas will come from a reader of this book. That would be wonderful.

A Elements of General Relativity

The following is a brief introduction to general relativity (GR). A familiarity with special relativity and classical dynamics will be assumed, but otherwise we will start from first principles. The presentation will be rather schematic, so this is not a substitute for a proper course on the subject. My goal is to present the minimal theoretical background required for a course in cosmology.

A.1 Spacetime and Relativity

The theory of relativity is based on a simple, yet profound, observation: the speed of light is the same in all inertial (non-accelerating) reference frames and does not depend on the motion of the observer. From this fact, Einstein deduced far-reaching consequences about the nature of space and time. In this section, I will provide a brief reminder of the basic concepts of special relativity.

A.1.1 Lorentz Transformations

In order for the speed of light to be the same in all inertial reference frames, the coordinates in these frames must be related by a Lorentz transformation. Consider two inertial frames S and S'. From the point of view of S, the frame S' is moving with a constant velocity v in the x-direction. The coordinates in S' are then related to those in S by the following **Lorentz transformation**:

$$
\begin{aligned}
t' &= \gamma(t - vx/c^2)\,, \\
x' &= \gamma(x - vt)\,, \\
y' &= y\,, \\
z' &= z\,,
\end{aligned}
\tag{A.1}
$$

where $\gamma \equiv 1/\sqrt{1 - v^2/c^2}$ is the Lorentz factor. It is then easy to confirm that the speed of light is the same in both frames.

Note that time and space have been mixed by the Lorentz transformation. Something similar happens for ordinary spatial rotations. Consider three-dimensional Euclidean space with coordinates $\mathbf{x} = (x, y, z)$ as defined in a frame S. A second frame S' may have coordinates $\mathbf{x}' = (x', y', z')$, where $\mathbf{x}' = R\mathbf{x}$, for some rotation matrix R. The two coordinate systems share the same origin, but are rotated with respect to each other, so that the coordinates in S' have become a mixture of the coordinates in S. Similarly, Lorentz transformations can be thought of as rotations between time and space. This mixing of space and time has profound implications: (1) events that are simultaneous in one frame are not simultaneous in another; (2) moving clocks run slow ("time dilation"); and (3) moving rods are shortened ("length contraction").

A.1.2 Spacetime and Four-Vectors

Although a rotation changes the components of a vector $\Delta\mathbf{x}$ connecting two points in space, it will not change the distance $|\Delta\mathbf{x}|$ between the points. In other words, $|\Delta\mathbf{x}|^2 = \Delta x^2 + \Delta y^2 + \Delta z^2$ is an invariant. Similarly, although time and space are relative, all observers will agree on the **spacetime interval**:

$$\Delta s^2 = -c^2\Delta t^2 + \Delta x^2 + \Delta y^2 + \Delta z^2 \,. \tag{A.2}$$

We can demonstrate this explicitly for the specific transformation in (A.1). Ignoring Δy and Δz, which just come along for the ride, the spacetime interval evaluated in the frame S' is

$$\begin{aligned}
\Delta s^2 &= -c^2(\Delta t')^2 + (\Delta x')^2 \\
&= -\gamma^2\left(c\Delta t - v\Delta x/c\right)^2 + \gamma^2(\Delta x - v\Delta t)^2 \\
&= -\gamma^2(c^2 - v^2)(\Delta t)^2 + \gamma^2\left(1 - v^2/c^2\right)(\Delta x)^2 \\
&= -c^2\Delta t^2 + \Delta x^2 \,.
\end{aligned} \tag{A.3}$$

In general relativity, we will encounter the spacetime interval between points that are infinitesimally close to each other. We then write the interval as

$$\mathrm{d}s^2 = -c^2\mathrm{d}t^2 + \mathrm{d}x^2 + \mathrm{d}y^2 + \mathrm{d}z^2 \,, \tag{A.4}$$

and call it the **line element**.

Note that Δs^2 is not positive definite. Two events that are *timelike* separated have $\Delta s^2 < 0$; they are closer in space than in time. In contrast, events with $\Delta s^2 > 0$ are said to be *spacelike* separated. Finally, two events with $\Delta s^2 = 0$ are *lightlike* separated and can be connected by a light ray. The set of all points that are lightlike separated from a point p defines its **lightcone**. Points that are timelike separated from p lie inside this lightcone, while spacelike separated points are outside the lightcone. To respect causality a particle must travel on a timelike path through spacetime. We call this path the particle's **worldline**.

Given the intimate connection between time and space in relativity it makes sense to combine them into a **four-vector**

$$x^\mu = (ct, x, y, z) \,, \tag{A.5}$$

where the Greek index μ runs from 0 to 3, and the zeroth component is time. To make the symmetry between time and space even more manifest, we will from now on use units where the speed of light is set to unity, $c \equiv 1$. The line element (A.4) can then be written as

$$ds^2 = \eta_{\mu\nu}dx^\mu dx^\nu \, , \tag{A.6}$$

where $\eta_{\mu\nu}$ is the **Minkowski metric**

$$\eta_{\mu\nu} = \begin{pmatrix} -1 & 0 & 0 & 0 \\ 0 & 1 & 0 & 0 \\ 0 & 0 & 1 & 0 \\ 0 & 0 & 0 & 1 \end{pmatrix}. \tag{A.7}$$

In (A.6), we used Einstein's summation convention which declares repeated indices to be summed over.

Under a Lorentz transformation the spacetime four-vector transforms as

$$x'^\mu = \Lambda^\mu{}_\nu x^\nu \, , \tag{A.8}$$

where $\Lambda^\mu{}_\nu$ is a 4×4 matrix, which for the specific transformation in (A.1) is

$$\Lambda^\mu{}_\nu = \begin{pmatrix} \gamma & -\gamma v & 0 & 0 \\ -\gamma v & \gamma & 0 & 0 \\ 0 & 0 & 1 & 0 \\ 0 & 0 & 0 & 1 \end{pmatrix}. \tag{A.9}$$

In general, the invariance of the line element (A.6) requires that

$$\eta_{\rho\sigma} = \Lambda^\mu{}_\rho \Lambda^\nu{}_\sigma \eta_{\mu\nu} \, , \tag{A.10}$$

and the set of matrices satisfying this constraint defines the **Lorentz group**.

The metric can also be used to lower the index of the vector x^μ to produce the components of the dual **co-vector**

$$x_\mu = \eta_{\mu\nu}x^\nu = (-t, x, y, z) \, . \tag{A.11}$$

Sometimes x_μ is called a covariant vector, while x^μ is a contravariant vector.[1] To raise an index, we need the inverse metric $\eta^{\mu\nu}$, defined by $\eta^{\mu\rho}\eta_{\rho\nu} = \delta^\mu_\nu$, so that $x^\mu = \eta^{\mu\nu}x_\nu$. An important co-vector is the differential operator

$$\partial_\mu \equiv \frac{\partial}{\partial x^\mu} = (\partial_t, \partial_x, \partial_y, \partial_z) \, , \tag{A.12}$$

which appears frequently in relativistic equations of motion.

The inner product of a vector and a co-vector is

$$x^\mu x_\mu = -t^2 + \mathbf{x} \cdot \mathbf{x} \, . \tag{A.13}$$

In order for this inner product to be Lorentz invariant, the components of a co-vector must transform as

$$x'_\mu = (\Lambda^{-1})^\nu{}_\mu x_\nu \, , \tag{A.14}$$

where $(\Lambda^{-1})^\nu{}_\mu$ is the inverse of $\Lambda^\mu{}_\nu$.

---

[1] Here, we are using slightly imprecise language by referring to x^μ and x_μ as "vectors" rather than the "components" of abstract vectors without indices.

Natural generalizations of vectors and co-vectors are **tensors**. A tensor of rank (m, n) has m contravariant (upper) indices and n covariant (lower) indices:

$$T^{\mu_1 \ldots \mu_m}{}_{\nu_1 \ldots \nu_n}. \tag{A.15}$$

The transformation of such a tensor is what you would guess from its indices:

$$(T')^{\mu_1 \ldots \mu_m}{}_{\nu_1 \ldots \nu_n} = \Lambda^{\mu_1}{}_{\sigma_1} \cdots (\Lambda^{-1})^{\rho_1}{}_{\nu_1} \cdots T^{\sigma_1 \ldots \sigma_m}{}_{\rho_1 \ldots \rho_n}. \tag{A.16}$$

The most complicated tensors one encounters in special relativity are the electromagnetic field strength $F_{\mu\nu}$ and the energy-momentum tensor $T_{\mu\nu}$ (see below). In general relativity, the most complicated tensor is the Riemann tensor $R_{\mu\nu\rho\sigma}$.

Why are tensors important? If a physical law can be written in terms of spacetime tensors, it means that it holds in any reference frame. In other words, if the law is true in one inertial frame, it will be true in any Lorentz-transformed frame. Newton's laws cannot be written in terms of spacetime tensors and therefore are not consistent with relativity. Maxwell's equations, on the other hand, can be written in tensorial form and therefore are consistent with relativity. This is not an accident. Einstein was motivated by Maxwell's equations because they imply that the speed of light should be independent of the motion of the observer.

Exercise A.1 Consider the inhomogeneous Maxwell equations:

$$\mathbf{\nabla} \cdot \mathbf{E} = \rho, \tag{A.17}$$

$$\mathbf{\nabla} \times \mathbf{B} - \partial_t \mathbf{E} = \mathbf{J}. \tag{A.18}$$

Defining the four-vector current $J^\mu = (\rho, J^i)$ and the field-strength tensor,

$$F^{\mu\nu} = \begin{pmatrix} 0 & E_x & E_y & E_z \\ -E_x & 0 & B_z & -B_y \\ -E_y & -B_z & 0 & B_x \\ -Ez & By & -B_x & 0 \end{pmatrix}, \tag{A.19}$$

show that these equations can be written as

$$\partial_\nu F^{\mu\nu} = J^\mu, \tag{A.20}$$

where $\partial_\nu \equiv \partial/\partial x^\nu$.

Hint: Write the Maxwell equations in components, using $(\mathbf{E})^i = F^{0i}$ and $(\mathbf{\nabla} \times \mathbf{B})^i = \epsilon^{ijk} \partial_j B_k = \partial_j F^{ij}$.

A.1.3 Relativistic Kinematics

Consider a massive particle moving through spacetime. The trajectory of the particle is specified by the function $x^\mu(\lambda)$, where λ is a parameter labeling the points along the particle's worldline. What should we choose for the parameter λ? One option is to use the time experienced by the particle called the **proper time**. Going to the rest frame of the particle, where its spatial coordinates are constant, we have

$$d\tau^2 = -ds^2. \tag{A.21}$$

Note that $d\tau^2 > 0$ for a timelike trajectory. Just like the interval ds^2, the proper time is something that all inertial observers will agree on. In a general frame, the spatial position $\mathbf{x}$ of the particle will be a function of the time t. In terms of these coordinates, the differential of the proper time is

$$d\tau = \sqrt{dt^2 - d\mathbf{x}^2} = dt\sqrt{1 - \left(\frac{d\mathbf{x}}{dt}\right)^2} = dt\sqrt{1 - v^2} = \frac{dt}{\gamma}. \tag{A.22}$$

Integrating this gives the proper time along the trajectory in terms of the background coordinates.

Given the function $x^\mu(\tau)$, we can define the **four-velocity** of the particle

$$U^\mu \equiv \frac{dx^\mu}{d\tau}. \tag{A.23}$$

Since τ is a Lorentz invariant, U^μ transforms in the same way as x^μ and is therefore also a four-vector. In contrast, dx^μ/dt is *not* a four-vector, since both x^μ and t change under a Lorentz transformation. Since U^μ is a four-vector, the inner product $U^\mu U_\mu$ is a Lorentz invariant. In fact, it is easy to show that $U^\mu U_\mu = -1$. Finally, it follows from (A.22) that the four-velocity in a general frame is

$$U^\mu = \gamma(1, \mathbf{v}), \tag{A.24}$$

while in the rest frame of the particle it becomes $U^\mu = (1, 0, 0, 0)$.

Another important quantity is the **four-momentum**

$$P^\mu = mU^\mu, \tag{A.25}$$

where m is the mass of the particle. Given (A.24), we have $P^\mu = \gamma m(1, \mathbf{v})$. The spatial part gives the relativistic generalization of the three-momentum, $\mathbf{p} = \gamma m\mathbf{v}$, while the time component is the energy of the particle $E = \gamma m$. The inner product of the four-momentum is $P^\mu P_\mu = -m^2$, which for $P^\mu = (E, p^i)$ becomes

$$E^2 = \mathbf{p}^2 + m^2. \tag{A.26}$$

This is the generalization of the famous $E = mc^2$ to include kinetic energy.

So far, we have only described massive particles. What about massless particles like photons? Massless particles travel on lightlike trajectories with $ds^2 = 0$. The proper time therefore vanishes and our analysis above breaks down. However, the result in (A.26) still holds in the massless limit where it gives

$$E = \sqrt{\mathbf{p}^2 + m^2} \to |\mathbf{p}|. \tag{A.27}$$

The four-momentum of a massless particle then is $P^\mu = (|\mathbf{p}|, \mathbf{p})$, with $P^\mu P_\mu = 0$.

Exercise A.2 Recall from quantum mechanics that the energy of a photon is

$$E = \frac{hc}{\lambda}, \tag{A.28}$$

where h is Planck's constant and λ is the wavelength of the light. Consider an observer traveling with a velocity v towards a photon moving in the x-direction. By transforming to the frame of the observer, determine the shift in the wavelength of the photon. This is the *relativistic Doppler effect*.

A.1.4 Relativistic Dynamics

We are often interested not in the motion of individual particles, but in the coarse-grained dynamics of a large collection of particles. In other words, instead of tracking the positions of each particle individually, we want to follow the evolution of average quantities, such as the number density n, energy density ρ and pressure P. We will now discuss how these quantities are described in relativity.

Number density

Consider a box of volume V centered around a position $\mathbf{x}$. The box contains N particles, so the density of particles is $n = N/V$. Taking the box size to be small, we can think of this as the local density at the point $\mathbf{x}$. Clearly, this number density is *not* a relativistic invariant. To see this, consider a frame S' in which the box is moving with a velocity v. The dimension of the box will be Lorentz contracted along the direction of travel, so its volume now is $V' = V/\gamma$. Since the number of particles inside the box stays the same, the number density in this frame will be $n' = \gamma n$. Using (A.24), we may also write this as

$$n' = nU^0, \tag{A.29}$$

where n is the number density in the rest frame of the box and U^0 is the time component of the four-velocity of the box. This suggests that the number density is the time component of a four-vector called the **number current**:

$$\boxed{N^\mu \equiv nU^\mu}. \tag{A.30}$$

This four-vector has components $N^\mu = (n', \mathbf{n}')$, where we reserve n (without the prime) for the density in the rest frame. The spatial part of this four-vector is the number current density, $\mathbf{n}' = \gamma n \mathbf{v}$. Given an area element $\mathbf{A}$, the inner product $\mathbf{n}' \cdot \mathbf{A}$ describes the number of particles flowing across the area per unit time.

If particles are neither created nor destroyed, then the number density only changes when particles flow into or out of the volume. Locally, this is described by the following **continuity equation**:

$$\frac{\partial n'}{\partial t} = -\boldsymbol{\nabla} \cdot \hat{\mathbf{n}}'\,.$$
(A.31)

Using the number current four-vector, this equation can be written as

$$\boxed{\partial_\mu N^\mu = 0}\,,$$
(A.32)

where ∂_μ was defined in (A.12).

Energy-momentum tensor

Of particular importance in general relativity are the densities of energy and momentum, since these are the sources for the spacetime curvature.

As we have seen above, energy and momentum are closely related as the time and space components of the momentum four-vector P^μ. We would now like to write the energy and momentum *densities* as the time components of four-vector *currents*. We then combine these currents into a single object, $T^{0\mu}$, where T^{00} is the density of the energy and T^{0i} is the density of the momentum (in the direction x^i). As you may guess from the double index, we are building a new rank-2 tensor $T^{\mu\nu}$ called the **energy-momentum tensor**. The second index tells us whether we are talking about the energy ($\nu = 0$) or the momentum ($\nu = i$), while the first index tells us whether we are talking about the density ($\mu = 0$) or the flow ($\mu = i$). Hence, we have

$$T^{00} = \text{density of energy}\,, \quad T^{i0} = \text{flow of energy}\,,$$
$$T^{0i} = \text{density of momentum}\,, \quad T^{ij} = \text{flow of momentum}\,.$$

Note that each component of the momentum has its own flux. For example, T^{12} is the flow of the x^2-momentum along the x^1-direction. The flow of the momentum density creates a stress (= force per unit area) and T^{ij} is therefore often called the *stress tensor*. Its diagonal components are the *pressure* and the off-diagonal components are the *anisotropic stress*. Integrating the densities over space gives the total energy and momentum, or $P^\nu = \int \mathrm{d}^3x\, T^{0\nu}$. By analogy with (A.32), we write the following conservation equation for the energy-momentum tensor:

$$\boxed{\partial_\mu T^{\mu\nu} = 0}\,.$$
(A.33)

These are four equations: one for the energy density ($\nu = 0$) and three for the different components of the momentum density ($\nu = i$).

As a simple example, let us return to our particles in the box. Ignoring the kinetic energies of the individual particles, the total energy density in the rest frame is $\rho = mn$. In the boosted frame, the energy of the particles and their number density each increase by a factor of γ, so that $\rho' = \gamma^2 \rho$. Similarly, the momentum density becomes $\boldsymbol{\pi}' = \gamma^2 \rho \mathbf{v}$. Using (A.24), we can also write this as

$$\rho' = \rho U^0 U^0 \,, \tag{A.34}$$

$$\pi'^i = \rho U^0 U^i \,, \tag{A.35}$$

where ρ is the energy density in the rest frame. A natural guess for the energy-momentum tensor of the particles inside the box therefore is

$$T^{\mu\nu} = \rho U^\mu U^\nu \,, \tag{A.36}$$

where $T^{0\nu} = (\rho', \boldsymbol{\pi}')$.

If we include the random motion of the particles, then we get an extra contribution from the pressure P created by this motion. Since the pressure is isotropic, the energy-momentum tensor in the rest frame must be diagonal:

$$T^{\mu\nu} = \begin{pmatrix} \rho & 0 & 0 & 0 \\ 0 & P & 0 & 0 \\ 0 & 0 & P & 0 \\ 0 & 0 & 0 & P \end{pmatrix}. \tag{A.37}$$

In a general frame, this becomes

$$\boxed{T^{\mu\nu} = (\rho + P)\, U^\mu U^\nu + P \eta^{\mu\nu}}\,. \tag{A.38}$$

This is the energy-momentum tensor of a **perfect fluid**. It plays an important role in cosmology, since on large scales all matter can be modeled by perfect fluids.

Relativistic field theory

In modern physics, **fields** are fundamental and particles are a derived concept. Elementary particles are now understood to be quantum excitations of fields and the Standard Model of particle physics is a relativistic quantum field theory. Even in classical physics, fields—like the gravitational field and the electromagnetic field—play an important role. In the following, I will briefly comment on the dynamics of fields in special relativity.

For simplicity, let us consider a real scalar field $\phi(t, \mathbf{x})$. The field has a "kinetic energy" (density) $\frac{1}{2}\dot{\phi}^2$, a "gradient energy" $\frac{1}{2}(\nabla\phi)^2$ and a "potential energy" $V(\phi)$. The kinetic and gradient energies can be combined into a Lorentz-invariant "kinetic term":

$$-\frac{1}{2}\eta^{\mu\nu}\partial_\mu\phi\partial_\nu\phi = \frac{1}{2}\dot{\phi}^2 - \frac{1}{2}(\nabla\phi)^2 \,, \tag{A.39}$$

which is often abbreviated as $-\frac{1}{2}(\partial\phi)^2$. The **Lagrangian** of the theory takes the form of "kinetic minus potential energy":

$$L = \int \mathrm{d}^3x \left[-\frac{1}{2}\eta^{\mu\nu}\partial_\mu\phi\partial_\nu\phi - V(\phi) \right], \tag{A.40}$$

and the **action** is

$$S = \int_{t_1}^{t_2} dt\, L\,.$$ (A.41)

The evolution of the field configuration $\phi(t, \mathbf{x})$ between two times t_1 and t_2 follows from the **principle of least action**. Consider an infinitesimal change of the field, $\phi \to \phi + \delta\phi$. The corresponding variation of the action is

$$\delta S \equiv S[\phi + \delta\phi] - S[\phi]$$
$$= \int d^4x \left[-\eta^{\mu\nu} \partial_\mu \phi \partial_\nu \delta\phi - \frac{dV}{d\phi}\delta\phi \right]$$
$$= \int d^4x \left[\eta^{\mu\nu} \partial_\mu \partial_\nu \phi - \frac{dV}{d\phi} \right]\delta\phi\,,$$ (A.42)

where the first term in the second line has been integrated by parts. The resulting boundary term has been dropped because $\delta\phi(t_1, \mathbf{x}) = \delta\phi(t_2, \mathbf{x}) = 0$. In order for the variation δS to vanish for arbitrary $\delta\phi$, the bracket in (A.42) must vanish. This gives the **Klein–Gordon equation**

$$\boxed{\Box\phi = \frac{dV}{d\phi}}\,,$$ (A.43)

where $\Box \equiv \eta^{\mu\nu} \partial_\mu \partial_\nu$ is the d'Alembertian operator.

A.1.5 Gravity and Relativity

> That one body may act upon another at a distance through a vacuum, without the mediation of anything else, by and through which their action and force may be conveyed from one to another, is to me so great an absurdity that I believe no man who has in philosophical matters a competent faculty of thinking can ever fall into it.   *Isaac Newton*

It is easy to see that Newtonian gravity is *not* consistent with relativity. Consider a particle of mass m in a gravitational field $\Phi(\mathbf{x}, t)$. The force it experiences is given by $\mathbf{F} = -m\boldsymbol{\nabla}\Phi$, where the gravitational potential satisfies the **Poisson equation**:

$$\nabla^2\Phi = 4\pi G\rho\,.$$ (A.44)

The Green's function solution to the Poisson equation is

$$\Phi(\mathbf{x}, t) = -G \int d^3x' \frac{\rho(\mathbf{x}', t)}{|\mathbf{x} - \mathbf{x}'|}\,,$$ (A.45)

which describes how a matter distribution with mass density $\rho(\mathbf{x}, t)$ creates the potential. The problem with this expression is that any change in $\rho(\mathbf{x}, t)$ propagates instantaneously throughout space in obvious violation of relativity. A related problem is that the Poisson equation is not a tensorial equation, so it depends on the reference frame. Lorentz transformations mix up time and space coordinates. Hence, if we transform to another inertial frame then the resulting equation would involve time derivatives. The above equation therefore does not take the same form in every inertial frame, which is another way of seeing that Newtonian gravity is incompatible with special relativity.

A similar issue arises in **Coulomb's law** of electrostatics. In particular, the equation for the electric potential ϕ takes a very similar form, $\nabla^2 \phi = -\rho_e/\epsilon_0$, where $\rho_e(\mathbf{x}, t)$ is the charge density. A change in the charge density would therefore also be experienced instantaneously throughout space. Of course, in the case of electrostatics, we know that the resolution is Maxwell's equations of electrodynamics, which can be written in tensorial form using the vector potential $A^\mu = (\phi, \mathbf{A})$ and the vector current $J^\mu = (\rho_e, \mathbf{J}_e)$; see Exercise A.1. Our challenge will be to find the analog of Maxwell's equations for gravity.

A.2 Gravity Is Geometry

In the rest of this appendix, I will give a basic introduction to general relativity. I will begin, in this section, with the equivalence principle as a motivation for associating gravity with the geometry of spacetime.

A.2.1 The Equivalence Principle

> I was sitting in a chair in the patent office in Bern when all of a sudden a thought occurred to me: "If a person falls freely he will not feel his own weight." I was startled. This simple thought made a deep impression on me. It impelled me toward a theory of gravitation.   *Albert Einstein*

The origin of general relativity lies in the following simple question: Why do objects with different masses fall at the same rate? We think we know the answer: the mass of an object cancels in Newton's law

$$\cancel{m}\,\mathbf{a} = \cancel{m}\,\mathbf{g}, \tag{A.46}$$

where $\mathbf{g}$ is the local gravitational acceleration. However, the meaning of "mass" on the two sides of (A.46) is quite different. We should really distinguish between the two masses by giving them different names:

$$m_I\,\mathbf{a} = m_G\,\mathbf{g}. \tag{A.47}$$

The **gravitational mass**, m_G, is a source for the gravitational field (just like the charge q_e is a source for an electric field), while the **inertial mass**, m_I, characterizes the dynamical response to any forces. In the case of the electric force, you wouldn't be tempted to cancel q_e and m_I. It is therefore a nontrivial result that experiments find[2]

$$\frac{m_I}{m_G} = 1 \pm 10^{-13} \,. \tag{A.48}$$

In Newtonian gravity, this equality of inertial and gravitational mass has no explanation and appears to be an accident. In GR, on the other hand, the observation that $m_I = m_G$ is taken as a fundamental property of gravity called the **weak equivalence principle** (WEP).

There are two other forces which are also proportional to the inertial mass. These are

$$\begin{aligned} \text{Centrifugal force}: \quad & \mathbf{F} = -m_I\,\boldsymbol{\omega} \times (\boldsymbol{\omega} \times \mathbf{r})\,, \\ \text{Coriolis force}: \quad & \mathbf{F} = -2m_I\,\boldsymbol{\omega} \times \dot{\mathbf{r}}\,. \end{aligned} \tag{A.49}$$

In this case, we understand that these forces are proportional to the inertial mass because they are "fictitious forces" in a non-inertial frame (one that is rotating with a frequency $\boldsymbol{\omega}$). Could gravity also be a fictitious force, arising only because we are in a non-inertial reference frame?

An important consequence of the equivalence principle is that gravity is "universal," meaning that it acts in the same way on all objects. Consider a particle in a gravitational field $\mathbf{g}$. Using the WEP, the equation of motion of the particle is

$$\ddot{\mathbf{x}} = \mathbf{g}(\mathbf{x}(t), t)\,. \tag{A.50}$$

Solutions of this equation are uniquely determined by the initial position and velocity of the particle. Any two particles with the same initial position and velocity will follow the same trajectory. As we will soon see, this simple observation has far-reaching consequences.

Imagine being confined to a sealed box. Your challenge, if you choose to accept it, is to determine the physical conditions outside the box by performing experiments inside the box. Consider first the case where the box is sitting in an electric field. How could you tell? Easy, just study the motion of an electron and a proton. Because these particles have opposite charges they will experience forces in opposite directions (see Fig. A.1). However, the same does not work for gravity. Since the gravitational charge (i.e. mass) is the same for all objects, two test particles with different masses will fall in exactly the same way. But, the particles are still falling, so haven't we detected the gravitational field? This is where Einstein's genius comes in. He pointed out that the motion of the two particles would be exactly the same if instead of sitting in a gravitational field, the box was actually in empty space but accelerating at a constant rate $\mathbf{a} = -\mathbf{g}$ (see Fig. A.1). The two particles will fall to the ground as before, but this time not because of the gravitational force, but because the box is accelerating into them. We conclude that:

A uniform gravitational field is indistinguishable from uniform acceleration.

---

[2] Note that (A.47) defines both m_G and $\mathbf{g}$. For any given material, we can therefore define $m_G = m_I$ by the rescaling $\mathbf{g} \to \lambda\mathbf{g}$ and $m_G \to \lambda^{-1}m_G$. What is nontrivial is that (A.48) then holds for other bodies made of other materials.

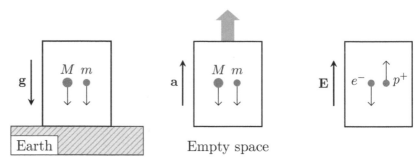

Illustration of Einstein's famous thought experiment showing that a uniform gravitational field (*left*) is indistinguishable from uniform acceleration (*middle*). This is to be contrasted with the case of an electric field (*right*) which acts differently on opposite charges and hence cannot be mimicked by acceleration.

A corollary of this observation is the fact that the effects of gravity can be removed by going to a *non-inertial* reference frame, like for the fictitious forces in (A.49). In particular, if the box is freely falling in the gravitational field (i.e. its acceleration is $\mathbf{a} = \mathbf{g}$) then the particles in the box will not fall to the ground. Einstein called this his "happiest thought": a freely falling observer doesn't feel a gravitational field (see Fig. A.2).

What about other experiments you could do (not just dropping test particles)? Could they discover the presence of a gravitational field? Einstein said no. There is no experiment—of any kind—that can distinguish uniform acceleration from a uniform gravitational field. This generalization of the WEP is called the **Einstein equivalence principle** (EEP). It implies that, in a small region of space (so that the gravitational field is approximately uniform), you can always find coordinates so that there is no acceleration. These coordinates correspond to a *local* inertial frame where the spacetime is approximately Minkowski space. Said differently:

In a small region, the laws of physics reduce to those of special relativity.

As we will see, the EEP suggests that the effects of gravity are associated with

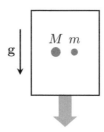

Fig. A.2 In a freely falling frame, objects do not experience the gravitational force.

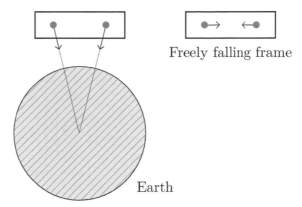

Illustration of tidal forces arising from the inhomogeneous gravitational field of the Earth. These tides cannot be removed by going to the freely falling "lab frame."

the curvature of spacetime which becomes relevant on larger scales where the field cannot be approximated as being uniform.

In arguing for the equivalence between gravity and acceleration it was essential that we restricted ourselves to uniform fields over small regions of space. But what if the gravitational field is not uniform? Consider a box that is freely falling towards the Earth (see Fig. A.3). We again drop two test particles. The gravitational attraction between the particles is minuscule and can therefore be neglected. Nevertheless, the two particles will accelerate towards each other because they each feel a force pointing towards the center of the Earth. This is an example of a **tidal force**, arising from the non-uniformity of the gravitational field. These tides are the real effects of gravity that cannot be canceled by going to an accelerating frame. Note that these tidal forces cause initially "parallel" trajectories to converge. As we will see, this violation of Euclidean geometry is a manifestation of the curvature of spacetime.

A.2.2 Gravity as Curved Spacetime

We have by now hinted several times at the fact that gravity should be interpreted as spacetime curvature. This is such an important feature of our modern understanding of gravity that it is worth belaboring the point. In the following, I will give a simple argument which will link the equivalence principle rather directly to the curvature of spacetime. For clarity, I will restore explicit factors of c in this section.

Let me begin by describing a famous observational consequence of the equivalence principle, the **gravitational redshift**. Consider Alice and Bob in a uniform gravitational field of strength g in the negative z-direction (see Fig. A.4). They are at heights $z_A = 0$ and $z_B = h$, respectively. Alice sends out a light signal with wavelength $\lambda_A = \lambda_0$. What is the wavelength λ_B received by Bob? By the equivalence principle, we should be able to obtain the result if we take Alice and

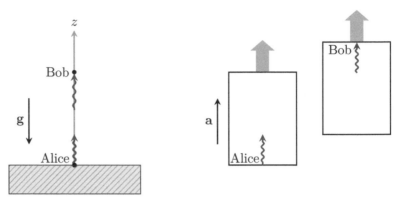

Left: Gravitational redshift in the Pound–Rebka experiment. Light emitted by Alice is received with a longer wavelength by Bob. *Right*: By the equivalence principle, the result can be derived by considering the Doppler shift of the light in an accelerating frame.

Bob to be moving with acceleration g in the positive z-direction in Minkowski spacetime (see Fig. A.4). Assuming $\Delta v/c$ to be small, the light reaches Bob after a time $\Delta t \approx h/c$. By this time, Bob's velocity has increased by $\Delta v = g\Delta t = gh/c$. Due to the Doppler effect, the received light will therefore have a slightly longer wavelength, $\lambda_B = \lambda_0 + \Delta\lambda$, with

$$\frac{\Delta\lambda}{\lambda_0} = \frac{\Delta v}{c} = \frac{gh}{c^2} . \tag{A.51}$$

By the equivalence principle, light emitted from the ground with wavelength λ_0 must therefore be "redshifted" by an amount

$$\boxed{\frac{\Delta\lambda}{\lambda_0} = \frac{\Delta\Phi}{c^2}} , \tag{A.52}$$

where $\Delta\Phi = gh$ is the change in the gravitational potential. This gravitational redshift was first measured by Pound and Rebka in 1959. Although we derived (A.52) for a uniform gravitational field, it holds for a non-uniform field if $\Delta\Phi$ is taken to be the integrated change in the gravitational potential between the two points in the spacetime.

We can also think of the gravitational redshift as an effect of **time dilation**. The period of the emitted light is $T_A = \lambda_A/c$ and that of the received light is $T_B = \lambda_B/c$. The result in (A.52) then implies that

$$\boxed{T_B = \left(1 + \frac{\Phi_B - \Phi_A}{c^2}\right) T_A} . \tag{A.53}$$

We conclude that time runs slower in a region of stronger gravity (smaller Φ). In the example above, we have $\Phi_A < \Phi_B$ (Alice feels a stronger gravitational field than

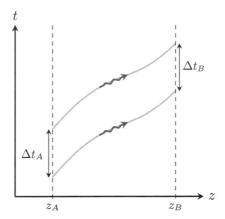

Fig. A.5   Spacetime diagram showing the wordlines of two light pulses. In a static spacetime, the worldlines must have identical shapes and hence $\Delta t_A = \Delta t_B$.

Bob), so that $T_A < T_B$ (time runs slower for Alice than for Bob). Although our thought experiment involved light signals, the result holds for any type of clock in a gravitational field. It therefore also applies to the heart rate of the observer. In our example this means that Alice will see Bob aging more rapidly. This "gravitational twin paradox" has been tested with atomic clocks on planes.[3]

Let us finally see why all of this implies that spacetime is curved. Consider the same setup as before. Alice now sends out two pulses of light, separated by a time interval Δt_A (as measured by her clock). Bob receives the signals spaced out by Δt_B (as measured by his clock). Figure A.5 shows the corresponding spacetime diagram. Since the gravitational field is static, the paths taken by the two pulses must have identical shapes (whatever that shape may be). But, this then seems to imply that $\Delta t_B = \Delta t_A$, in apparent contradiction to (A.53). What happened? When drawing the congruent wordlines in Fig. A.5 we implicitly assumed that the spacetime is flat. The resolution to the paradox is to accept that the spacetime is curved.

To see this more explicitly, consider a spacetime in which the interval between two nearby events is not given by $ds^2 = -dt^2 + d\mathbf{x}^2$, but by

$$ds^2 = -\left(1 + \frac{2\Phi(\mathbf{x})}{c^2}\right) dt^2 + d\mathbf{x}^2, \qquad (A.54)$$

---

[3] Accounting for time dilation effects is also essential for the successful operation of the Global Positioning System (GPS) [1]. The satellites used in GPS are about 20 000 km above the Earth where the gravitational field is four times weaker than that on the ground. Because of the gravitational time dilation, the clocks on the satellites tick faster by about 45 μs per day. Correcting for the relativistic time dilation due to the motion of the orbiting clocks (at about 14 000 km/hr), the net effect is 38 μs per day. This is a problem. To achieve a positional accuracy of 15 m, time throughout the GPS system must be known to an accuracy of 50 ns (the time required for light to travel 15 m). If we didn't correct for the effects of time dilation, the GPS would accumulate an error of about 10 km per day. Said differently, the accuracy we expect from the GPS would fail in less than two minutes.

with $\Phi \ll c^2$. In these coordinates, Alice sends signals at times t_A and $t_A + \Delta t$, and Bob receives them at t_B and $t_B + \Delta t$. Note that the spacetime diagram is still that shown in Fig. A.5, with two congruent worldlines. However, although the coordinate interval Δt is the same for Alice and Bob, their observed *proper times* are different. In particular, the proper time interval between the signals sent by Alice is

$$\Delta \tau_A = \sqrt{-g_{00}(\mathbf{x})} \, \Delta t = \sqrt{1 + \frac{2\Phi_A}{c^2}} \, \Delta t \approx \left(1 + \frac{\Phi_A}{c^2}\right) \Delta t \,, \qquad (\text{A.55})$$

where we have used that $\Delta \mathbf{x} = 0$ and expanded to first order in small $\Phi_A \equiv \Phi(\mathbf{x}_A)$. Similarly, the proper time between the signals received by Bob is

$$\Delta \tau_B \approx \left(1 + \frac{\Phi_B}{c^2}\right) \Delta t \,. \qquad (\text{A.56})$$

Combining (A.55) and (A.56), we find

$$\Delta \tau_B = \left(1 + \frac{\Phi_B}{c^2}\right) \left(1 + \frac{\Phi_A}{c^2}\right)^{-1} \Delta \tau_A \approx \left(1 + \frac{\Phi_B - \Phi_A}{c^2}\right) \Delta \tau_A \,, \qquad (\text{A.57})$$

which is the same as (A.53). The time dilation has therefore been explained by the geometry of spacetime.

A.3 Motion in Curved Spacetime

General relativity contains two key ideas: (1) "spacetime curvature tells matter how to move" (equivalence principle) and (2) "matter tells spacetime how to curve" (Einstein equations). In this section, we will develop the first idea a bit further. We will show that in the absence of any non-gravitational forces, particles move along special paths in the curved spacetime called **geodesics**. We will also derive an equation that determines these trajectories.

A.3.1 Relativistic Action

The action of a relativistic point particle is

$$S = -m \int d\tau \,, \qquad (\text{A.58})$$

where τ is the proper time along the worldline of the particle and m is its mass. It is not hard to understand why this is the correct action. The action must be a Lorentz scalar, so that all observers compute the same value for it. A natural candidate is the proper time, because all observers will agree on the amount of time that elapsed on a clock carried by the moving particle.

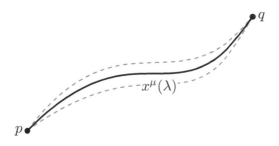

Illustration of a family of curves connecting two points in a spacetime. In order for a path to be a geodesic, its action must be an extremum, which implies that small variations of the path should not change the action.

As a useful consistency check, we evaluate the action (A.58) for a particular observer in Minkowski spacetime. Using (A.22), the action can be written as the following integral over time

$$S = -m \int \mathrm{d}t \sqrt{1 - v^2} \,, \qquad (A.59)$$

where $v^2 = \delta_{ij} \dot{x}^i \dot{x}^j$. For small velocities, $v \ll 1$, the integrand is $-m + \frac{1}{2} m v^2$. We see that the Lagrangian is simply the kinetic energy of the particle, plus a constant that doesn't affect the equations of motion.

Substituting the line element (A.54) into (A.58), we get

$$S = -m \int \mathrm{d}t \sqrt{(1 + 2\Phi) - v^2}$$
$$\approx \int \mathrm{d}t \left(-m + \frac{1}{2} m v^2 - m\Phi + \cdots \right), \qquad (A.60)$$

where, in the second line, we expanded the square root for small v and Φ. We see that the metric perturbation Φ indeed plays the role of the gravitational potential in Newtonian gravity. It is now also obvious why the inertial mass (appearing in the kinetic term $\frac{1}{2} m v^2$) is the same as the gravitational mass (appearing in the potential $m\Phi$).

A.3.2 Geodesic Equation

Let us now use the action (A.58) to study the motion of particles in a general curved spacetime with metric $g_{\mu\nu}(t, \mathbf{x})$. Consider an arbitrary timelike curve $\mathcal{C}$ connecting two points p and q (see Fig. A.6). A geodesic is the preferred curve for which the action is an extremum. We first introduce a parameter λ to label points along the curve, $x^\mu(\lambda)$, with $\mathcal{C}(0) = p$ and $\mathcal{C}(1) = q$. Notice that even the time coordinate x^0 is parameterized, which is necessary if we want to treat time and space on an equal footing. The action for the curve $\mathcal{C}$ can then be written as

$$S[x^\mu(\lambda)] = -m \int_0^1 \mathrm{d}\lambda \sqrt{-g_{\mu\nu} \frac{\mathrm{d}x^\mu}{\mathrm{d}\lambda} \frac{\mathrm{d}x^\nu}{\mathrm{d}\lambda}} \,. \qquad (A.61)$$

As we will show in the box below, this action has an extremum if the path satisfies the **geodesic equation**

$$\boxed{\frac{\mathrm{d}^2 x^\mu}{\mathrm{d}\tau^2} + \Gamma^\mu_{\alpha\beta} \frac{\mathrm{d}x^\alpha}{\mathrm{d}\tau} \frac{\mathrm{d}x^\beta}{\mathrm{d}\tau} = 0} \,, \tag{A.62}$$

where τ is the proper time along the curve and

$$\Gamma^\mu_{\alpha\beta} \equiv \frac{1}{2} g^{\mu\lambda}(\partial_\alpha g_{\beta\lambda} + \partial_\beta g_{\alpha\lambda} - \partial_\lambda g_{\alpha\beta}) \tag{A.63}$$

is the **Christoffel symbol** (or "connection coefficient").

Derivation Consider a small variation of the path $x^\mu \to x^\mu + \delta x^\mu$. The corresponding variation of the action is

$$\delta S \equiv S[x^\mu + \delta x^\mu] - S[x^\mu]$$
$$= m \int_0^1 \mathrm{d}\lambda \, \frac{1}{2L} \left(\delta g_{\mu\nu} \frac{\mathrm{d}x^\mu}{\mathrm{d}\lambda} \frac{\mathrm{d}x^\nu}{\mathrm{d}\lambda} + 2 g_{\mu\nu} \frac{\mathrm{d}x^\mu}{\mathrm{d}\lambda} \frac{\mathrm{d}\delta x^\nu}{\mathrm{d}\lambda} \right), \tag{A.64}$$

where we have defined $L^2 \equiv -g_{\mu\nu} \, \mathrm{d}x^\mu/\mathrm{d}\lambda \, \mathrm{d}x^\nu/\mathrm{d}\lambda$. Before continuing, it is convenient to switch from the general parameterization using λ to the parameterization using proper time τ. We could not have used τ from the beginning since the value of τ at the final point q is different for different curves, so that the range of integration would not have been fixed. Notice that

$$\left(\frac{\mathrm{d}\tau}{\mathrm{d}\lambda} \right)^2 = -g_{\mu\nu} \frac{\mathrm{d}x^\mu}{\mathrm{d}\lambda} \frac{\mathrm{d}x^\nu}{\mathrm{d}\lambda} = L^2 \,, \tag{A.65}$$

and hence $\mathrm{d}\tau/\mathrm{d}\lambda = L$. Equation (A.64) therefore becomes

$$\delta S = m \int \mathrm{d}\tau \left(\frac{\mathrm{d}\lambda}{\mathrm{d}\tau} \right) \frac{1}{2L} \left(\delta g_{\mu\nu} L^2 \dot{x}^\mu \dot{x}^\nu + 2L^2 g_{\mu\nu} \dot{x}^\mu \delta(\dot{x}^\nu) \right)$$
$$= m \int \mathrm{d}\tau \, \frac{1}{2} \left(\partial_\alpha g_{\mu\nu} \, \delta x^\alpha \, \dot{x}^\mu \dot{x}^\nu + 2 g_{\mu\nu} \dot{x}^\mu \delta(\dot{x}^\nu) \right), \tag{A.66}$$

where $\dot{x}^\mu \equiv \mathrm{d}x^\mu/\mathrm{d}\tau$. Integrating the second term of (A.66) by parts, we get

$$\delta S = m \int \mathrm{d}\tau \left(\frac{1}{2} \partial_\alpha g_{\mu\nu} \dot{x}^\mu \dot{x}^\nu \, \delta x^\alpha - g_{\mu\nu} \ddot{x}^\mu \delta x^\nu - \partial_\alpha g_{\mu\nu} \dot{x}^\alpha \dot{x}^\mu \delta x^\nu \right). \tag{A.67}$$

In the last term, we replaced $\partial_\alpha g_{\mu\nu}$ with $\frac{1}{2}(\partial_\alpha g_{\mu\nu} + \partial_\mu g_{\alpha\nu})$ because it is contracted with an object that is symmetric in α and μ. After a suitable relabeling of indices, we then get

$$\delta S = -m \int \mathrm{d}\tau \left(g_{\mu\nu} \ddot{x}^\mu + \frac{1}{2} (\partial_\alpha g_{\beta\nu} + \partial_\beta g_{\alpha\nu} - \partial_\nu g_{\alpha\beta}) \dot{x}^\alpha \dot{x}^\beta \right) \delta x^\nu \,. \tag{A.68}$$

Factoring out the metric $g_{\mu\nu}$, we can write this as

$$\delta S = -m \int \mathrm{d}\tau \, g_{\mu\nu} \left(\ddot{x}^\mu + \Gamma^\mu_{\alpha\beta} \dot{x}^\alpha \dot{x}^\beta \right) \delta x^\nu \,, \tag{A.69}$$

where $\Gamma^{\mu}_{\alpha\beta}$ is the Christoffel symbol defined in (A.63). In order for δS to vanish for arbitrary δx^{ν}, the bracket in (A.69) must vanish and we obtain the geodesic equation (A.62).

We see that the simple action (A.58) has given rise to a relatively complex equation of motion. The amount of index juggling involved in the above derivation might be disheartening, but fortunately the geodesic equation is relatively easy to apply in the cosmological context (see Section 2.2.1).

It is also useful to write the geodesic equation in terms of the four-velocity $U^{\mu} = \mathrm{d}x^{\mu}/\mathrm{d}\tau$. Using the chain rule to write

$$\frac{\mathrm{d}}{\mathrm{d}\tau} U^{\mu}(x^{\alpha}(\tau)) = \frac{\mathrm{d}x^{\alpha}}{\mathrm{d}\tau} \frac{\partial U^{\mu}}{\partial x^{\alpha}} = U^{\alpha} \frac{\partial U^{\mu}}{\partial x^{\alpha}} \,, \tag{A.70}$$

equation (A.62) becomes

$$U^{\alpha} \left(\frac{\partial U^{\mu}}{\partial x^{\alpha}} + \Gamma^{\mu}_{\alpha\beta} U^{\beta} \right) = 0 \,. \tag{A.71}$$

The term in brackets is the so-called **covariant derivative** of the four-vector U^{μ}, which we write as

$$\boxed{\nabla_{\alpha} U^{\mu} \equiv \partial_{\alpha} U^{\mu} + \Gamma^{\mu}_{\alpha\beta} U^{\beta}} \,. \tag{A.72}$$

The geodesic equation can then be written in the following compact way:

$$U^{\alpha} \nabla_{\alpha} U^{\mu} = 0 \,. \tag{A.73}$$

As we will see below, this form of the geodesic equation can also be derived directly by considering the concept of a "parallel transport" of the four-velocity.

Newtonian limit

In Newtonian gravity, the equation of motion for a test particle in a gravitational field is

$$\frac{\mathrm{d}^2 x^i}{\mathrm{d}t^2} = -\partial^i \Phi \,. \tag{A.74}$$

Let us see how to recover this result from the Newtonian limit of the geodesic equation (A.62). The Newtonian approximation assumes that (1) particles are moving slowly (relative to the speed of light), (2) the gravitational field is weak (and can therefore be treated as a perturbation of Minkowski space), and (3) the field is also static (i.e. has no time dependence). The first condition means that

$$\frac{\mathrm{d}x^i}{\mathrm{d}\tau} \ll \frac{\mathrm{d}t}{\mathrm{d}\tau} \,, \tag{A.75}$$

so that (A.62) becomes

$$\frac{\mathrm{d}^2 x^{\mu}}{\mathrm{d}\tau^2} + \Gamma^{\mu}_{00} \left(\frac{\mathrm{d}t}{\mathrm{d}\tau} \right)^2 = 0 \,. \tag{A.76}$$

In the static, weak-field limit, we then write the metric (and its inverse) as

$$g_{\mu\nu} = \eta_{\mu\nu} + h_{\mu\nu} \,,$$
$$g^{\mu\nu} = \eta^{\mu\nu} - h^{\mu\nu} \,, \tag{A.77}$$

where the perturbation is small, $|h_{\mu\nu}| \ll 1$, and time independent. To first order in $h_{\mu\nu}$, the relevant component of the Christoffel symbol is

$$\Gamma^{\mu}_{00} = \frac{1}{2} g^{\mu\lambda} (\partial_0 g_{\lambda 0} + \partial_0 g_{\lambda 0} - \partial_\lambda g_{00})$$
$$= -\frac{1}{2} \eta^{\mu j} \partial_j h_{00} \,. \tag{A.78}$$

The $\mu = 0$ component of (A.76) then reads $d^2 t/d\tau^2 = 0$, so that $dt/d\tau$ is a constant, while the $\mu = i$ component becomes

$$\frac{d^2 x^i}{d\tau^2} = \frac{1}{2} \left(\frac{dt}{d\tau} \right)^2 \partial^i h_{00} \,. \tag{A.79}$$

Dividing both sides by $(dt/d\tau)^2$, we get

$$\boxed{\frac{d^2 x^i}{dt^2} = \frac{1}{2} \partial^i h_{00}} \,, \tag{A.80}$$

which matches (A.74) if

$$h_{00} = -2\Phi \,. \tag{A.81}$$

Note that this identification of the metric perturbation with the gravitational potential is consistent with what we inferred previously from the equivalence principle, cf. (A.54).

Massless particles

For massless particles, our derivation of the geodesic equation doesn't apply since the action (A.61) vanishes. In the end, the general form of the geodesic equation still holds

$$\boxed{\frac{d^2 x^{\mu}}{d\lambda^2} = -\Gamma^{\mu}_{\alpha\beta} \frac{dx^\alpha}{d\lambda} \frac{dx^\beta}{d\lambda}} \,, \tag{A.82}$$

but λ cannot be identified with proper time (which vanishes along the path of a massless particle). Instead, we choose λ such that $P^\mu = dx^\mu/d\lambda$, where P^μ is the four-momentum of the particle. The geodesic equation for a massless particle can then be written as $P^\alpha \nabla_\alpha P^\mu = 0$, with $P_\mu P^\mu = 0$.

A.3.3 A Mathematical Interlude*

So far, I have used a physics-first approach to introducing general relativity. In this section, I will fill in some of the mathematical details. This material is important if you want to get a full appreciation for the logical structure of the theory, but it can be skipped if your immediate goal is to develop just enough understanding of the theory to get started with your studies of cosmology.

Tensors in general relativity

I have explained that, in special relativity, all physical laws must be written in terms of tensors if we want them to be valid in every inertial reference frame. The same is true in general relativity, but the theory now deals with non-inertial frames. These frames are related by coordinate transformations that depend on spacetime position.

Consider a general coordinate transformation $x^\mu \to x'^\mu(x)$. The infinitesimal coordinate differentials transform as

$$\mathrm{d}x'^\mu = \frac{\partial x'^\mu}{\partial x^\nu}\,\mathrm{d}x^\nu \equiv S^\mu{}_\nu\,\mathrm{d}x^\nu\,, \tag{A.83}$$

where the transformation matrix $S^\mu{}_\nu(x)$ plays the same role as the matrix $\Lambda^\mu{}_\nu$ in a Lorentz transformation, but crucially it now depends on the position x^μ. A four-vector field $V^\mu(x)$ is an object that transforms in the same way as $\mathrm{d}x^\mu$:

$$V'^\mu = S^\mu{}_\nu V^\nu\,. \tag{A.84}$$

Note that although the coordinate differentials $\mathrm{d}x^\mu$ transform like a vector, the coordinates x^μ do *not*. Unlike in special relativity, the transformation $x^\mu \to x'^\mu(x)$ can be an arbitrary nonlinear function and only $\mathrm{d}x^\mu$ transforms linearly.

To determine the transformation of a covariant vector W_μ, we use that the inner product $W_\mu V^\mu$ must be an invariant. The transformation of W_μ therefore involves the inverse matrix

$$W'_\mu = (S^{-1})_\mu{}^\nu W_\nu = \frac{\partial x^\nu}{\partial x'^\mu}W_\nu\,. \tag{A.85}$$

By inspection of (A.84) and (A.85), it is now not hard to guess that an arbitrary tensor transforms as

$$(T')^{\mu_1\ldots\mu_m}{}_{\nu_1\ldots\nu_n} = S^{\mu_1}{}_{\sigma_1}\cdots(S^{-1})_{\nu_1}{}^{\rho_1}\cdots T^{\sigma_1\ldots\sigma_m}{}_{\rho_1\ldots\rho_n}\,. \tag{A.86}$$

There is simply no other way to contract the indices. Moreover, the expression in (A.86) is the natural generalization of the analogous result (A.16) in special relativity. An important tensor is the metric tensor $g_{\mu\nu}$. Its transformation can also be derived from the invariance of the line element:

$$\mathrm{d}s^2 = g'_{\mu\nu}(x')\,\mathrm{d}x'^\mu\mathrm{d}x'^\nu = g_{\rho\sigma}(x)\,\mathrm{d}x^\rho\mathrm{d}x^\sigma$$
$$= g_{\rho\sigma}(x)\frac{\partial x^\rho}{\partial x'^\mu}\frac{\partial x^\sigma}{\partial x'^\nu}\mathrm{d}x'^\mu\mathrm{d}x'^\nu\,, \tag{A.87}$$

so that

$$g'_{\mu\nu} = g_{\rho\sigma}\frac{\partial x^\rho}{\partial x'^\mu}\frac{\partial x^\sigma}{\partial x'^\nu} \equiv (S^{-1})_\mu{}^\rho(S^{-1})_\nu{}^\sigma g_{\rho\sigma}\,, \tag{A.88}$$

which is consistent with the general transformation law (A.86).

Derivatives of tensors

Next, let us see how derivatives of scalars, vectors and tensors transform under a general coordinate transformation. Consider first the partial derivative of a scalar field, $\partial_\mu \phi$. Since $\phi'(x') = \phi(x)$, we get

$$\partial'_\mu \phi'(x') = \frac{\partial \phi'(x')}{\partial x'^\mu} = \frac{\partial x^\nu}{\partial x'^\mu} \frac{\partial \phi(x)}{\partial x^\nu} \equiv (S^{-1})_\mu{}^\nu \partial_\nu \phi(x) \,. \tag{A.89}$$

Unsurprisingly, $\partial_\mu \phi$ transforms like a covariant vector.

Things get more interesting when we take the partial derivative of a vector, $\partial_\lambda V^\mu$. We might naively expect this to transform like a rank $(1,1)$ tensor. However, it is easy to see that this is not the case:

$$\partial'_\lambda V'^\mu(x') = \frac{\partial V'^\mu(x')}{\partial x'^\lambda} = \frac{\partial x^\sigma}{\partial x'^\lambda} \frac{\partial}{\partial x^\sigma} \left(S^\mu{}_\nu(x) V^\nu(x) \right) \tag{A.90}$$

$$= (S^{-1})_\lambda{}^\sigma S^\mu{}_\nu \, \partial_\sigma V^\nu + \left[(S^{-1})_\lambda{}^\sigma \partial_\sigma S^\mu{}_\nu \right] V^\nu \,. \tag{A.91}$$

The first term in (A.91) is what we would expect if the derivative were indeed a tensor, but the second term spoils the transformation law. The offending term arises from the partial derivative acting on the transformation matrix $S^\mu{}_\nu(x)$. We would like to define a new derivative $\nabla_\lambda V^\mu$ that does transform like a tensor:

$$\nabla'_\lambda V'^\mu = (S^{-1})_\lambda{}^\sigma S^\mu{}_\nu \, \nabla_\sigma V^\nu \,. \tag{A.92}$$

In the following exercise, you will show that the covariant derivative defined in (A.72) has precisely this property.[4]

Exercise A.3 Show that the Christoffel symbol defined in (A.63) transforms as

$$\Gamma'^\mu_{\lambda\nu} = \frac{\partial x'^\mu}{\partial x^\rho} \frac{\partial x^\sigma}{\partial x'^\lambda} \frac{\partial x^\eta}{\partial x'^\nu} \Gamma^\rho_{\sigma\eta} + \frac{\partial x'^\mu}{\partial x^\eta} \frac{\partial^2 x^\eta}{\partial x'^\lambda \partial x'^\nu} \tag{A.93}$$

$$= S^\mu{}_\rho (S^{-1})_\lambda{}^\sigma (S^{-1})_\nu{}^\eta \Gamma^\rho_{\sigma\eta} + S^\mu{}_\eta (S^{-1})_\lambda{}^\rho \partial_\rho (S^{-1})_\nu{}^\eta \,.$$

We see that the Christoffel symbol does *not* transform as a tensor. Show that the non-tensorial part of the transformation precisely cancels the non-tensorial part in (A.91), so that (A.92) holds.

Hint: You will need that the derivative of an inverse matrix satisfies $\partial S^{-1} = -S^{-1}(\partial S) S^{-1}$, which holds because $SS^{-1} = 1$.

So far, we have only encountered the covariant derivative of a contravariant vector:

$$\boxed{\nabla_\mu V^\nu = \partial_\mu V^\nu + \Gamma^\nu_{\mu\alpha} V^\alpha} \,. \tag{A.94}$$

---

[4] The Christoffel symbol defined in (A.63) is actually a special case of a connection called the **Levi-Civita connection**. In general, the connection is defined independently of the metric, but still satisfies the transformation law in (A.93).

To determine how the covariant derivative acts on a covariant vector, W_ν, we consider how it acts on the scalar $f \equiv W_\nu V^\nu$. Using that $\nabla_\mu f = \partial_\mu f$, it is easy to show that

$$\boxed{\nabla_\mu W_\nu = \partial_\mu W_\nu - \Gamma^\alpha_{\mu\nu} W_\alpha} . \qquad (A.95)$$

Notice the change of the sign of the second term relative to (A.94) and the placement of the dummy index.

Exercise A.4 Derive (A.95) from (A.94).

The covariant derivative of the mixed tensor $T^\mu{}_\nu$ can be derived similarly by considering $f \equiv T^\mu{}_\nu V^\nu W_\mu$. This gives

$$\boxed{\nabla_\sigma T^\mu{}_\nu = \partial_\sigma T^\mu{}_\nu + \Gamma^\mu_{\sigma\alpha} T^\alpha{}_\nu - \Gamma^\alpha_{\sigma\nu} T^\mu{}_\alpha} . \qquad (A.96)$$

Again, pay careful attention to the signs and the placement of the dummy indices. Staring at this expression for a little bit should reveal the pattern for arbitrary tensors, although it won't be needed in this book.

Drowning in a sea of indices

If you don't have experience with index manipulations, the last few pages might have seemed like a bad dream. I can offer the following points of consolation: First, this is as bad as it gets. After you understand the index kung-fu in the above examples, you will be prepared for anything else that GR might throw at you. Second, things typically become much simpler in practical applications. This is especially true in cosmology. So don't despair and don't let the proliferation of indices distract you from the inherent beauty of the physical concepts.

From flat to curved spacetime

We have just seen that the covariant derivative of a tensor transforms like a tensor, while the partial derivative does not. This means that relativistic equations must be constructed out of covariant derivatives, not partial derivatives. A simple prescription to upgrade equations from flat space to curved space is therefore to replace every partial derivative by a covariant derivative, $\partial_\mu \to \nabla_\mu$.[5] For example, in (A.20) we showed that the inhomogeneous Maxwell equations can be written as $\partial_\nu F^{\mu\nu} = J^\mu$. The generalization of this equation to curved spacetimes is simply

$$\nabla_\nu F^{\mu\nu} = J^\mu , \qquad (A.97)$$

[5] Since the Christoffel symbols depend only on a single derivative of the metric, it is possible to find coordinates—called "Riemann normal coordinates"—so that they vanish at a given point, $\Gamma^\mu_{\alpha\beta}(p) = 0$. At that point p, covariant derivatives reduce to partial derivatives and the physics becomes that of special relativity (as required by the equivalence principle).

where the dependence on the metric is encoded in the covariant derivative and the associated Christoffel symbols. This describes the dynamics of electromagnetic fields in general relativity.

Similarly, we have seen that the conservation of the energy-momentum tensor in special relativity implies $\partial_\nu T^{\mu\nu} = 0$. In general relativity, this becomes

$$\nabla_\nu T^{\mu\nu} = 0 \,. \tag{A.98}$$

Again, the covariant derivative depends on the metric and hence defines a coupling between the matter and the gravitational degrees of freedom.

Parallel transport and geodesics

Given the notion of a covariant derivative, we can define the concept of a **parallel transport** of a tensor. In flat spacetime, "parallel transport" simply means translating a vector along a curve while "keeping it constant." More concretely, a vector V^μ is constant along a curve $x^\mu(\lambda)$ if its components don't depend on the parameter λ:

$$\frac{\mathrm{d}V^\mu}{\mathrm{d}\lambda} = \frac{\mathrm{d}x^\nu}{\mathrm{d}\lambda}\partial_\nu V^\mu = 0 \qquad \text{(flat spacetime)}. \tag{A.99}$$

We generalize this to curved spacetimes by replacing the partial derivative in (A.99) by a covariant derivative. This gives the so-called **directional covariant derivative**. A vector is parallel transported in general relativity if the directional covariant derivative of the vector along a curve vanishes:

$$\frac{DV^\mu}{D\lambda} \equiv \frac{\mathrm{d}x^\nu}{\mathrm{d}\lambda}\nabla_\nu V^\mu = 0 \qquad \text{(curved spacetime)}. \tag{A.100}$$

Although we have only written the equation for a vector field, an analogous equation applies for arbitrary tensors. Writing out the covariant derivative, the equation of parallel transport becomes

$$\frac{\mathrm{d}V^\mu}{\mathrm{d}\lambda} + \Gamma^\mu_{\sigma\nu}\frac{\mathrm{d}x^\sigma}{\mathrm{d}\lambda}V^\nu = 0 \,, \tag{A.101}$$

which tells us that the components of the vector will now change along the curve and that this change is determined by the connection $\Gamma^\mu_{\sigma\nu}$.

Using parallel transport, we can give an alternative definition of a geodesic as the curve generated by a tangent vector $\mathrm{d}x^\mu/\mathrm{d}\lambda$ that is transported parallel to itself. This generalizes the notion of a straight line in flat space, which can also be thought of as the path that parallel transports its own tangent vector. Substituting $V^\mu = \mathrm{d}x^\mu/\mathrm{d}\lambda$ into (A.101), we get

$$\frac{\mathrm{d}^2 x^\mu}{\mathrm{d}\lambda^2} + \Gamma^\mu_{\sigma\nu}\frac{\mathrm{d}x^\sigma}{\mathrm{d}\lambda}\frac{\mathrm{d}x^\nu}{\mathrm{d}\lambda} = 0 \,, \tag{A.102}$$

which is indeed the same as the geodesic equation that we found before, if we write $\lambda = a\tau + b$ for a massive particle.

A.4 The Einstein Equation

So far, we have described how test particles move in arbitrary curved spacetimes. Next, we will discuss how the spacetime curvature is determined by the local matter distribution. We are in search of the following relationship:

$$\begin{pmatrix} \text{a measure of local} \\ \text{spacetime curvature} \end{pmatrix} = \begin{pmatrix} \text{a measure of local} \\ \text{stress-energy density} \end{pmatrix}. \tag{A.103}$$

We will start on the left-hand side.

A.4.1 Tidal Forces and Curvature

In Euclidean space, parallel lines never meet. Similarly, in Minkowski spacetime, initially parallel geodesics stay parallel forever. In a curved space(time), on the other hand, initially parallel geodesics do not stay parallel. This gives us a way to measure the curvature of the spacetime.[6] In this section, we will study the relative acceleration of two test particles, first in Newtonian gravity and then in GR. We will find that the motion is determined by the **Riemann tensor**, an object of fundamental importance in differential geometry.

Consider two particles with positions $\mathbf{x}(t)$ and $\mathbf{x}(t) + \mathbf{b}(t)$. In Newtonian gravity, the two particles satisfy

$$\frac{\mathrm{d}^2 x^i}{\mathrm{d}t^2} = -\partial^i \Phi(x^j), \tag{A.104}$$

$$\frac{\mathrm{d}^2(x^i + b^i)}{\mathrm{d}t^2} = -\partial^i \Phi(x^j + b^j). \tag{A.105}$$

Subtracting (A.104) from (A.105), and expanding the result to first order in the infinitesimal separation vector b^j, we get

$$\frac{\mathrm{d}^2 b^i}{\mathrm{d}t^2} = -\partial_j \partial^i \Phi \, b^j. \tag{A.106}$$

We see the relative acceleration of the particles is determined by the **tidal tensor**[7] $\partial_i \partial_j \Phi$. The Poisson equation relates the *trace* of this tidal tensor to the mass density

$$\nabla^2 \Phi = \partial_i \partial^i \Phi = 4\pi G \rho. \tag{A.107}$$

In the next section, we will use this connection between the tidal tensor and the Poisson equation as an inspiration to guess the Einstein equation for the gravitational field.

---

[6] Note that following the motion of a single test particle is not enough to measure spacetime curvature, since the particle remains at rest in a freely falling frame. The relative motion of at least two particles is therefore needed to detect curvature.

[7] This is called the tidal tensor because of the role it plays in explaining the tides on Earth.

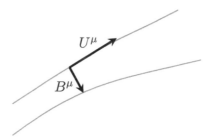

Fig. A.7   Evolution of two geodesics with separation B^μ in a curved spacetime. The relative acceleration of the geodesics depends on the Riemann tensor and is hence a measure of the spacetime curvature.

Let us now find the equivalent of (A.106) in GR where it is called the **geodesic deviation equation**. The algebra will be a bit more involved, but the physics is the same. The analog of the tidal tensor will give us a local measure of the spacetime curvature. Its trace will lead to the Einstein equation.

Consider two geodesics separated by an infinitesimal vector B^μ (see Fig. A.7). We define the "relative velocity" of the two geodesics as the directional covariant derivative of B^μ along one of the geodesics:

$$V^\mu \equiv \frac{DB^\mu}{D\tau} = U^\nu \nabla_\nu B^\mu = \frac{dB^\mu}{d\tau} + \Gamma^\mu_{\sigma\nu} U^\nu B^\sigma \,, \qquad (A.108)$$

where $U^\mu = dx^\mu/d\tau$. Similarly, the "relative acceleration" is

$$A^\mu \equiv \frac{D^2 B^\mu}{D\tau^2} = U^\nu \nabla_\nu V^\mu = \frac{dV^\mu}{d\tau} + \Gamma^\mu_{\sigma\nu} U^\nu V^\sigma \,. \qquad (A.109)$$

Using the geodesic equation and the definition of the covariant derivative, we can compute the relative acceleration. After some work (see the box below), we find

$$\boxed{\frac{D^2 B^\mu}{D\tau^2} = -R^\mu{}_{\nu\rho\sigma} U^\nu U^\sigma B^\rho} \,, \qquad (A.110)$$

where we have defined the **Riemann tensor**

$$\boxed{R^\mu{}_{\nu\rho\sigma} \equiv \partial_\rho \Gamma^\mu_{\nu\sigma} - \partial_\sigma \Gamma^\mu_{\nu\rho} + \Gamma^\mu_{\rho\lambda}\Gamma^\lambda_{\nu\sigma} - \Gamma^\mu_{\sigma\lambda}\Gamma^\lambda_{\nu\rho}} \,. \qquad (A.111)$$

This scary beast is the analog of the tidal tensor in Newtonian gravity.

Derivation Substituting (A.108) into (A.109), we get

$$\begin{aligned}
A^\alpha &= \frac{dV^\alpha}{d\tau} + \Gamma^\alpha_{\beta\gamma} U^\beta V^\gamma \\
&= \frac{d}{d\tau}\left(\frac{dB^\alpha}{d\tau} + \Gamma^\alpha_{\beta\gamma} U^\beta B^\gamma\right) + \Gamma^\alpha_{\beta\gamma} U^\beta \left(\frac{dB^\gamma}{d\tau} + \Gamma^\gamma_{\delta\epsilon} U^\delta B^\epsilon\right) \qquad (A.112) \\
&= \frac{d^2 B^\alpha}{d\tau^2} + \frac{d\Gamma^\alpha_{\beta\gamma}}{d\tau} U^\beta B^\gamma + \Gamma^\alpha_{\beta\gamma}\frac{dU^\beta}{d\tau} B^\gamma + 2\Gamma^\alpha_{\beta\gamma} U^\beta \frac{dB^\gamma}{d\tau} + \Gamma^\alpha_{\beta\gamma}\Gamma^\gamma_{\delta\epsilon} U^\beta U^\delta B^\epsilon \,.
\end{aligned}$$

The derivatives of the Christoffel symbol and the four-velocity can be written as

$$\frac{\mathrm{d}\Gamma^\alpha_{\beta\gamma}}{\mathrm{d}\tau} = U^\delta \partial_\delta \Gamma^\alpha_{\beta\gamma}, \tag{A.113}$$

$$\frac{\mathrm{d}U^\beta}{\mathrm{d}\tau} = -\Gamma^\beta_{\delta\epsilon} U^\delta U^\epsilon, \tag{A.114}$$

where (A.114) follows from the geodesic equation (A.71). We therefore get

$$A^\alpha = \frac{\mathrm{d}^2 B^\alpha}{\mathrm{d}\tau^2} + 2\Gamma^\alpha_{\beta\gamma} U^\beta \frac{\mathrm{d}B^\gamma}{\mathrm{d}\tau} + \left(\partial_\delta \Gamma^\alpha_{\beta\gamma} - \Gamma^\epsilon_{\delta\beta}\Gamma^\alpha_{\epsilon\gamma} + \Gamma^\alpha_{\beta\epsilon}\Gamma^\epsilon_{\delta\gamma}\right) U^\beta U^\delta B^\gamma, \tag{A.115}$$

where we have relabeled some dummy indices to extract the common factor $U^\beta U^\delta B^\gamma$ from three of the terms. To replace the derivatives of B^α, we note that $x^\alpha(\tau) + B^\alpha(\tau)$ obeys the geodesic equation

$$\frac{\mathrm{d}^2(x^\alpha + B^\alpha)}{\mathrm{d}\tau^2} + \Gamma^\alpha_{\beta\gamma}(x^\delta + B^\delta)\frac{\mathrm{d}(x^\beta + B^\beta)}{\mathrm{d}\tau}\frac{\mathrm{d}(x^\gamma + B^\gamma)}{\mathrm{d}\tau} = 0. \tag{A.116}$$

Subtracting the geodesic equation for $x^\alpha(\tau)$ and expanding the result to linear order in B^α, we get

$$\frac{\mathrm{d}^2 B^\alpha}{\mathrm{d}\tau^2} + 2\Gamma^\alpha_{\beta\gamma} U^\beta \frac{\mathrm{d}B^\gamma}{\mathrm{d}\tau} = -\partial_\delta \Gamma^\alpha_{\beta\gamma} B^\delta U^\beta U^\gamma$$

$$= -\partial_\gamma \Gamma^\alpha_{\beta\delta} U^\beta U^\delta B^\gamma, \tag{A.117}$$

where we relabeled some dummy indices in the second line. Substituting this into (A.115), we find

$$A^\alpha = -\underbrace{\left(\partial_\gamma \Gamma^\alpha_{\beta\delta} - \partial_\delta \Gamma^\alpha_{\beta\gamma} + \Gamma^\epsilon_{\delta\beta}\Gamma^\alpha_{c\gamma} - \Gamma^\alpha_{\beta\epsilon}\Gamma^\epsilon_{\delta\gamma}\right)}_{\equiv R^\alpha{}_{\beta\gamma\delta}} U^\beta U^\delta B^\gamma, \tag{A.118}$$

which confirms the result in (A.110).

In the local inertial frame of a freely falling observer, with four-velocity $U^\mu = (1,0,0,0)$, the geodesic deviation equation (A.110) becomes

$$\frac{\mathrm{d}^2 B^\mu}{\mathrm{d}\tau^2} = -R^\mu{}_{0\nu0} B^\nu. \tag{A.119}$$

For the static, weak-field metric (A.54), we have $R^i{}_{0j0} = \partial_j \Gamma^i_{00} = \partial^i \partial_j \Phi$ and (A.119) reduces to the Newtonian result (A.106).

A.4.2 The Mathematics of Curvature*

There is a lot of beautiful mathematics underlying the physics of spacetime curvature. We will only have time for a very brief snapshot.

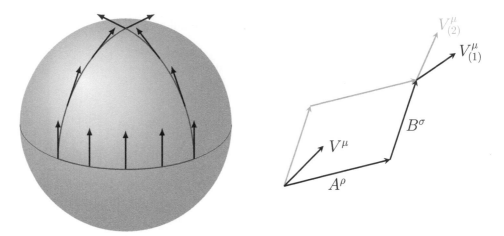

Fig. A.8 Path dependence of parallel transport. The example on the left shows the parallel transport of a vector on a two-sphere. Starting with a vector on the equator, pointing along a line of constant longitude, the direction of the vector at the North Pole clearly depends on the path along which it was transported. The diagram on the right defines an infinitesimal parallelogram in spacetime. If the spacetime is curved then the parallel transport along two different paths will not give the same vector.

Parallel transport and curvature

An important property of the parallel transport of a vector on a curved manifold is that it depends on the path along which the vector is transported. This is illustrated in Fig. A.8 for the case of a two-sphere. Consider a vector on the equator, pointing along a line of constant longitude. We wish to parallel transport this vector to the North Pole. We first do this along the line of constant longitude. Alternatively, we can first parallel transport the vector along the equator by an angle ϕ and then transport it to the North Pole along the new line of constant longitude. As you see from the figure, the two vectors at the North Pole are not the same, but point in different directions.

This path dependence of the parallel transport gives another way to diagnose whether the spacetime is curved. Consider a parallelogram spanned by the infinitesimal vectors A^ρ and B^σ (see Fig. A.8) and imagine parallel transporting a vector V^μ. From the equation of parallel transport (A.101), we have that the change of the vector along a side δx^ρ is

$$\delta V^\mu = -\Gamma^\mu_{\nu\rho} V^\nu \delta x^\rho . \tag{A.120}$$

On "path 1" we parallel transport the vector first along A^ρ and then along B^σ, while on "path 2" we reverse the order (giving the gray path in Fig. A.8). Using (A.120), we get

$$\delta V^{\mu}_{(1)} = -\Gamma^{\mu}_{\nu\rho}(x)V^{\nu}(x)A^{\rho} - \Gamma^{\mu}_{\nu\rho}(x+A)V^{\nu}(x+A)B^{\rho}\,,$$
$$\delta V^{\mu}_{(2)} = -\Gamma^{\mu}_{\nu\rho}(x)V^{\nu}(x)B^{\rho} - \Gamma^{\mu}_{\nu\rho}(x+B)V^{\nu}(x+B)A^{\rho}\,,$$

(A.121)

and the difference is

$$\delta V^{\mu} \equiv \delta V^{\mu}_{(1)} - \delta V^{\mu}_{(2)}$$
$$= \frac{\partial(\Gamma^{\mu}_{\nu\rho}V^{\nu})}{\partial x^{\sigma}}B^{\sigma}A^{\rho} - \frac{\partial(\Gamma^{\mu}_{\nu\rho}V^{\nu})}{\partial x^{\sigma}}A^{\sigma}B^{\rho}\,,$$

(A.122)

where we have Taylor expanded the arguments for small A^{ρ} and B^{ρ}. Swapping the dummy indices on the second term, $\rho \leftrightarrow \sigma$, and differentiating the products, we find

$$\delta V^{\mu} = (\partial_{\sigma}\Gamma^{\mu}_{\nu\rho}V^{\nu} + \Gamma^{\mu}_{\nu\rho}\partial_{\sigma}V^{\nu} - \partial_{\rho}\Gamma^{\mu}_{\nu\sigma}V^{\nu} - \Gamma^{\mu}_{\nu\sigma}\partial_{\rho}V^{\nu})A^{\rho}B^{\sigma}\,.$$

(A.123)

Using (A.101) again, we have $\partial_{\sigma}V^{\nu} = -\Gamma^{\nu}_{\sigma\lambda}V^{\lambda}$ and hence (A.123) becomes

$$\boxed{\delta V^{\mu} = R^{\mu}{}_{\nu\rho\sigma}V^{\nu}A^{\rho}B^{\sigma}}\,,$$

(A.124)

where $R^{\mu}{}_{\nu\rho\sigma}$ is the Riemann tensor as defined in (A.111).

Properties of the Riemann tensor

Only 20 of the $4^4 = 256$ components of $R^{\mu}{}_{\nu\rho\sigma}$ are independent. This is because the Riemann tensor has a lot of symmetries that relate its different components. These symmetries are easiest to present for the Riemann tensor with only lower indices $R_{\mu\nu\rho\sigma} = g_{\mu\lambda}R^{\lambda}{}_{\nu\rho\sigma}$. We then have

$$R_{\mu\nu\rho\sigma} = -R_{\nu\mu\rho\sigma}\,,$$

(A.125)

$$R_{\mu\nu\rho\sigma} = -R_{\mu\nu\sigma\rho}\,,$$

(A.126)

$$R_{\mu\nu\rho\sigma} = R_{\rho\sigma\mu\nu}\,,$$

(A.127)

$$R_{\mu\nu\rho\sigma} + R_{\mu\rho\sigma\nu} + R_{\mu\sigma\nu\rho} = 0\,.$$

(A.128)

In words: the Riemann tensor is antisymmetric in its first two indices [(A.125)] and its last two indices [(A.126)]. Moreover, it is symmetric under the exchange of the first two indices with the last two indices [(A.127)]. Finally, the sum of the cyclic permutations of the last three indices vanishes [(A.128)].

In addition to these algebraic symmetries, the Riemann tensor satisfies an important differential identity called the **Bianchi identity**. This identity states that the sum of the cyclic permutations of the first three indices of $\nabla_{\lambda}R_{\mu\nu\rho\sigma}$ vanishes:

$$\nabla_{\lambda}R_{\mu\nu\rho\sigma} + \nabla_{\nu}R_{\lambda\mu\rho\sigma} + \nabla_{\mu}R_{\nu\lambda\rho\sigma} = 0\,.$$

(A.129)

This is the analog of the homogeneous Maxwell equation $\partial_{\lambda}F_{\mu\nu} + \partial_{\nu}F_{\lambda\mu} + \partial_{\mu}F_{\nu\lambda} = 0$.

A.4.3 Guessing the Einstein Equation

We will determine the Einstein equation in two different ways. First, we will "guess" it. Then, we will construct an action for the metric and show that the corresponding equation of motion leads to the same Einstein equation.

We are searching for the relativistic generalization of the Poisson equation:

$$\nabla^2 \Phi = 4\pi G \rho. \tag{A.130}$$

We would like to write this equation in tensorial form, so that it is valid independently of the choice of coordinates. We know that in relativity the energy density is the temporal component of the energy-momentum tensor, $\rho = T_{00}$ (see Section A.1.4). This suggests that $T_{\mu\nu}$ should appear on the right-hand side of the Einstein equation. Moreover, we have also seen that the relativistic generalization of the gravitational potential Φ is the metric $g_{\mu\nu}$ (see Section A.2.2). On the left-hand side, we therefore expect an object with two derivatives acting on the metric. A naive guess would be to act with the d'Alembertian operator $\nabla^\sigma \nabla_\sigma$ on $g_{\mu\nu}$. However, this doesn't work because $\nabla_\sigma g_{\mu\nu} = 0$. To infer the correct quantity, we recall that the right-hand side of the Poisson equation is the trace of the tidal tensor, $\partial_i \partial_j \Phi$, and that the relativistic generalization of the tidal tensor is the Riemann tensor, $R^\mu{}_{\nu\rho\sigma}$ (see Section A.4.1). This suggests that the trace of the Riemann tensor would be an interesting object. Taking the trace means contracting the upper index with a lower index. The symmetries of the Riemann tensor imply that there is a unique way of doing so, which leads to the **Ricci tensor**:

$$\boxed{R_{\mu\nu} \equiv R^\lambda{}_{\mu\lambda\nu} = \partial_\lambda \Gamma^\lambda_{\mu\nu} - \partial_\nu \Gamma^\lambda_{\mu\lambda} + \Gamma^\lambda_{\lambda\rho}\Gamma^\rho_{\mu\nu} - \Gamma^\rho_{\mu\lambda}\Gamma^\lambda_{\nu\rho}} \ . \tag{A.131}$$

This has all the properties we want: it is a symmetric $(0, 2)$ tensor with second-order derivatives acting on the metric.

A first and second guess

Einstein's first guess for the field equation of GR therefore was

$$R_{\mu\nu} \overset{?}{=} \kappa T_{\mu\nu}, \tag{A.132}$$

where κ is a constant. However, this doesn't work because, in general, we can have $\nabla^\mu R_{\mu\nu} \neq 0$, which would be inconsistent with the conservation of the energy-momentum tensor, $\nabla^\mu T_{\mu\nu} = 0$. To see this, we consider the following double contraction of the Bianchi identity (A.129):

$$0 = g^{\sigma\lambda} g^{\mu\rho} \left(\nabla_\lambda R_{\mu\nu\rho\sigma} + \nabla_\nu R_{\lambda\mu\rho\sigma} + \nabla_\mu R_{\nu\lambda\rho\sigma} \right)$$
$$= \nabla^\sigma R_{\nu\sigma} - \nabla_\nu R + \nabla^\rho R_{\nu\rho}, \tag{A.133}$$

where $R = g^{\mu\nu} R_{\mu\nu}$ is the **Ricci scalar**. This implies that

$$\nabla^\mu R_{\mu\nu} = \frac{1}{2}\nabla_\nu R, \tag{A.134}$$

which doesn't vanish, except in the special case where R (and hence $T = g^{\mu\nu} T_{\mu\nu}$) is a constant.

The problem is easy to fix: we simply note that (A.134) can be written as

$$\nabla^{\mu} \left(R_{\mu\nu} - \frac{1}{2} g_{\mu\nu} R \right) = 0 \,. \tag{A.135}$$

This suggests an alternative measure of curvature, the so-called **Einstein tensor**:

$$\boxed{G_{\mu\nu} \equiv R_{\mu\nu} - \frac{1}{2} g_{\mu\nu} R} \,, \tag{A.136}$$

which is consistent with the conservation of the energy-momentum tensor. Our improved guess of the Einstein equation therefore is

$$G_{\mu\nu} \overset{?}{=} \kappa T_{\mu\nu} \,. \tag{A.137}$$

To show that this is the correct equation, we still have to verify that it reduces to the Poisson equation (A.130) in the Newtonian limit.

Newtonian limit

To save a few lines of algebra, it is convenient to first write the Einstein equation in a slightly different form. Contracting both sides of (A.137) gives

$$R = -\kappa T \,, \tag{A.138}$$

where we used that $\delta^{\mu}_{\mu} = 4$. Substituting this back into (A.137), we get the *trace-reversed* Einstein equation:

$$R_{\mu\nu} = \kappa \left(T_{\mu\nu} - \frac{1}{2} g_{\mu\nu} T \right) \,. \tag{A.139}$$

In the Newtonian limit, the energy-momentum tensor takes the form of a pressureless fluid, with $T_{00} = \rho$ and $T = g^{00} T_{00} \approx -T_{00} = -\rho$. Note that we have taken ρ to be small and could therefore use the unperturbed metric at leading order. The temporal component of (A.139) then is

$$R_{00} = \frac{1}{2} \kappa \rho \,. \tag{A.140}$$

We would like to evaluate R_{00} in the static, weak-field limit, where the metric can be written as $g_{\mu\nu} = \eta_{\mu\nu} + h_{\mu\nu}$, cf. (A.77). The temporal component of the Ricci tensor then is

$$\begin{aligned} R_{00} = R^{i}{}_{0i0} &= \partial_i \Gamma^{i}_{00} - \partial_0 \Gamma^{i}_{i0} + \Gamma^{i}_{j\lambda} \Gamma^{\lambda}_{00} - \Gamma^{i}_{0\lambda} \Gamma^{\lambda}_{j0} \\ &= \partial_i \Gamma^{i}_{00} \,. \end{aligned} \tag{A.141}$$

In the first line, we used that $R^{0}{}_{000} = 0$ and then wrote out the definition of the Riemann tensor (A.111). In the second line, we dropped the terms of the form Γ^2

which are second order in the metric perturbation, because the Christoffel symbols are first order. We also dropped $\partial_0\Gamma^i_{i0}$ because the metric perturbation is assumed to be time independent. The relevant component of the Christoffel symbol was computed in (A.78): $\Gamma^i_{00} = -\frac{1}{2}\partial^i h_{00}$. At first order in the metric perturbation, the temporal component of the Ricci tensor then is

$$R_{00} = -\frac{1}{2}\nabla^2 h_{00}\,, \tag{A.142}$$

and equation (A.140) becomes

$$\nabla^2 h_{00} = -\kappa\rho\,. \tag{A.143}$$

Recall that the Newtonian limit of the geodesic equation implied that $h_{00} = -2\Phi$. Equation (A.143) therefore reproduces the Poisson equation (A.130) if $\kappa = 8\pi G$.

The Einstein equation

The final form of the **Einstein equation** then is

$$\boxed{G_{\mu\nu} = 8\pi G\, T_{\mu\nu}}\,. \tag{A.144}$$

This is one of the most beautiful equations ever written down. It describes a wide range of phenomena, from falling apples and planetary orbits to the expansion of the universe and black holes.

Note that (A.144) are ten second-order partial differential equations for the metric. In fact, because the contracted Bianchi identity, $\nabla^\mu G_{\mu\nu} = 0$, imposes four constraints, we have only six independent equations. This counting makes sense since there are four coordinate transformations and hence the metric has only six independent components.

A.4.4 Einstein–Hilbert Action

An alternative way of deriving the Einstein equations is from an action principle. The action must be an integral over a scalar function. Moreover, this scalar function should be a measure of the local spacetime curvature and be at most second order in derivatives of the metric. The unique such object is the Ricci scalar[8] and the corresponding **Einstein–Hilbert action** is

$$\boxed{S = \int \mathrm{d}^4x\sqrt{-g}\,R}\,, \tag{A.145}$$

where $g \equiv \det g_{\mu\nu}$ is the determinant of the metric. The factor of $\sqrt{-g}$ was introduced so that the volume element $\mathrm{d}^4x\sqrt{-g}$ is invariant under a coordinate transformation.

---

[8] Gravity as an effective field theory also contains higher-order curvature terms such as R^2 or $R_{\mu\nu}R^{\mu\nu}$. These are only important at very short distances.

The Einstein equation then follows by varying the action with respect to the (inverse) metric. Writing the Ricci scalar as $R = g^{\mu\nu}R_{\mu\nu}$, we have

$$\delta S = \int d^4x \left((\delta\sqrt{-g})g^{\mu\nu}R_{\mu\nu} + \sqrt{-g}\,\delta g^{\mu\nu}R_{\mu\nu} + \sqrt{-g}g^{\mu\nu}\delta R_{\mu\nu} \right). \qquad (A.146)$$

With some effort (see the exercise below), it can be shown that the last term is a total derivative $g^{\mu\nu}\delta R_{\mu\nu} = \nabla_\mu X^\mu$, with $X^\mu \equiv g^{\rho\nu}\delta\Gamma^\mu_{\rho\nu} - g^{\mu\nu}\delta\Gamma^\rho_{\nu\rho}$, and can therefore be dropped without affecting the equation of motion. To evaluate the first term, we use that

$$\delta\sqrt{-g} = -\frac{1}{2}\sqrt{-g}\,g_{\mu\nu}\delta g^{\mu\nu}. \qquad (A.147)$$

To prove this identity, you have to use that any diagonalizable matrix M satisfies $\log(\det M) = \text{Tr}(\log M)$. Substituting (A.147) into (A.146), we find

$$\delta S = \int d^4x\sqrt{-g} \left(R_{\mu\nu} - \frac{1}{2}g_{\mu\nu}R \right)\delta g^{\mu\nu}. \qquad (A.148)$$

For the action to be an extremum, this variation must vanish for arbitrary $\delta g^{\mu\nu}$. This is only the case if $G_{\mu\nu} = 0$, which is the vacuum Einstein equation.

Exercise A.5 Show that the last term in (A.146) is a total derivative

$$g^{\mu\nu}\delta R_{\mu\nu} = \nabla_\mu X^\mu, \quad \text{with} \quad X^\mu = g^{\rho\nu}\delta\Gamma^\mu_{\rho\nu} - g^{\mu\nu}\delta\Gamma^\rho_{\nu\rho}. \qquad (A.149)$$

Hint: You may use that $\delta\Gamma^\mu_{\rho\nu}$ is a tensor, although $\Gamma^\mu_{\rho\nu}$ is not. Consider "normal coordinates" at a point p, for which $\Gamma^\mu_{\rho\nu}(p) = 0$. Compute $\delta R_{\mu\nu}$ in these coordinates. The result is a tensor equation and therefore holds in all coordinate systems.

A.4.5 Including Matter

To get the non-vacuum Einstein equation, we add an action for matter to the Einstein–Hilbert action. The complete action then is

$$S = \frac{1}{2\kappa}\int d^4x\sqrt{-g}R + S_M, \qquad (A.150)$$

where the constant κ allows for a difference in the relative normalization of the gravitational action and the matter action. Varying this action with respect to the metric gives

$$\delta S = \frac{1}{2}\int d^4x\sqrt{-g} \left(\frac{1}{\kappa}G_{\mu\nu} - T_{\mu\nu} \right)\delta g^{\mu\nu}, \qquad (A.151)$$

where we have defined the energy-momentum tensor as

$$\boxed{T_{\mu\nu} \equiv -\frac{2}{\sqrt{-g}}\frac{\delta S_M}{\delta g^{\mu\nu}}}. \qquad (A.152)$$

The action (A.150) therefore has an extremum when the metric satisfies (A.137): $G_{\mu\nu} = \kappa T_{\mu\nu}$. Fixing the constant κ in the same way as before then gives the Einstein equation (A.144).

A.4.6 The Cosmological Constant

There is one other term that can be added to the left-hand side of the Einstein equation which is consistent with the local conservation of $T_{\mu\nu}$, namely a term of the form $\Lambda g_{\mu\nu}$, for some constant Λ. Adding this term doesn't affect the conservation of the energy-momentum tensor, because the covariant derivative of the metric is zero, $\nabla^\mu g_{\mu\nu} = 0$. Einstein, in fact, did add such a term and called it the **cosmological constant**. The modified form of the Einstein equation is

$$\boxed{G_{\mu\nu} + \Lambda g_{\mu\nu} = 8\pi G\, T_{\mu\nu}}\,. \tag{A.153}$$

It has also become common practice to identify this cosmological constant with the stress-energy of the vacuum (if any) and include it on the right-hand side as a contribution to the energy-momentum tensor. The action leading to (A.153) is

$$S = \frac{1}{16\pi G} \int \mathrm{d}^4 x \sqrt{-g}(R - 2\Lambda) + S_M\,. \tag{A.154}$$

We see that the cosmological constant corresponds to a pure volume term in the action.

A.4.7 Some Simple Solutions

The Einstein equations are nonlinear functions of the metric, which makes solving them a complicated task. A few exact solutions nevertheless exist in situations with a large amount of symmetry. We will first consider the vacuum Einstein equation ($T_{\mu\nu} = 0$) with a cosmological constant. Contracting both sides of (A.153) with the metric, we get $R = 4\Lambda$ and hence

$$R_{\mu\nu} = \Lambda g_{\mu\nu}\,. \tag{A.155}$$

Let me mention a few famous solutions to this equation.

- First, we set $\Lambda = 0$. Reassuringly, the Minkowski line element,

$$\mathrm{d}s^2 = -\mathrm{d}t^2 + \mathrm{d}\mathbf{x}^2\,, \tag{A.156}$$

  satisfies the vacuum Einstein equation $R_{\mu\nu} = 0$. A more nontrivial solution to the same equation is the **Schwarzschild metric**:

$$\mathrm{d}s^2 = -\left(1 - \frac{2GM}{r}\right)\mathrm{d}t^2 + \left(1 - \frac{2GM}{r}\right)^{-1}\mathrm{d}r^2 + r^2\mathrm{d}\Omega_2^2\,, \tag{A.157}$$

  where $\mathrm{d}\Omega_2^2 \equiv \mathrm{d}\theta^2 + \sin^2\theta\,\mathrm{d}\phi^2$ is the metric on the unit two-sphere. This solution describes the spacetime around a spherically symmetric object of mass M.

- Next, we consider the case of a positive cosmological constant, $\Lambda > 0$. The corresponding solution to the Einstein equation is **de Sitter space** (in *static patch coordinates*):

$$ds^2 = -\left(1 - \frac{r^2}{R^2}\right)dt^2 + \left(1 - \frac{r^2}{R^2}\right)^{-1}dr^2 + r^2 d\Omega_2^2, \qquad (A.158)$$

where $R^2 \equiv 3/\Lambda$. The static patch coordinates cover only part of the de Sitter geometry, namely that accessible to a single observer which is bounded by the cosmological horizon at $r = R$. Alternative coordinates that cover the whole space are the so-called *global coordinates*

$$ds^2 = -dT^2 + R^2 \cosh^2(T/R) d\Omega_3^2, \qquad (A.159)$$

where $d\Omega_3^2 \equiv d\psi^2 + \sin^2\psi \, d\Omega_2^2$ is the metric on the unit three-sphere. In these coordinates, we can think of de Sitter space as an evolving three-sphere that starts infinitely large at $T \to -\infty$, shrinks to a minimal size at $T = 0$ and then expands to infinite size at $T \to +\infty$. In applications to inflation, we often use the *planar coordinates*

$$ds^2 = -d\hat{t}^2 + e^{2\hat{t}/R}(dr^2 + r^2 d\Omega_2^2), \qquad (A.160)$$

which cover half of the global geometry. This describes an exponentially expanding universe with flat spatial slices (although this time dependence only becomes physical when the time translation invariance of de Sitter space is broken by additional matter fields like the inflaton). More on the geometry of de Sitter space, and the different coordinates used to describe it, can be found in [2, 3].

- Finally, we can also take the cosmological constant to be negative, $\Lambda < 0$. The corresponding solution is **anti-de Sitter space**:

$$ds^2 = -\left(1 + \frac{r^2}{R^2}\right)dt^2 + \left(1 + \frac{r^2}{R^2}\right)^{-1}dr^2 + r^2 d\Omega_2^2, \qquad (A.161)$$

where $R^2 \equiv -3/\Lambda$.

Last, but not least, the **Robertson–Walker metric** for a homogeneous and isotropic universe is (see Chapter 2)

$$ds^2 = -dt^2 + a^2(t) \, d\mathbf{x}^2, \qquad (A.162)$$

where the scale factor $a(t)$ describes the expansion of the universe. This solves the Einstein equation if the energy-momentum tensor is that of a perfect fluid, $T_{\mu\nu} = \mathrm{diag}(\rho, P, P, P)$ (in the rest frame), and the scale factor satisfies the **Friedmann equations**:

$$\left(\frac{\dot{a}}{a}\right)^2 = \frac{8\pi G}{3}\rho, \qquad (A.163)$$

$$\frac{\ddot{a}}{a} = -\frac{4\pi G}{3}(\rho + 3P). \qquad (A.164)$$

These equations play an important role in cosmology.

A.5 Summary

In this appendix, we reviewed the fundamentals of general relativity. We started with the equivalence principle which states that a uniform gravitational field is indistinguishable from constant acceleration, and used it to motivate the modern view of gravity as the curvature of spacetime.

We then showed that freely falling particles move along geodesics in a curved spacetime. Starting from the relativistic action of a point particle, we derived the geodesic equation

$$\frac{\mathrm{d}^2 x^\mu}{\mathrm{d}\tau^2} = -\Gamma^\mu_{\alpha\beta} \frac{\mathrm{d}x^\alpha}{\mathrm{d}\tau} \frac{\mathrm{d}x^\beta}{\mathrm{d}\tau} \,, \tag{A.165}$$

where τ is the proper time along the trajectory and $\Gamma^\mu_{\alpha\beta}$ is the Christoffel symbol

$$\Gamma^\mu_{\alpha\beta} = \frac{1}{2} g^{\mu\lambda} (\partial_\alpha g_{\beta\lambda} + \partial_\beta g_{\alpha\lambda} - \partial_\lambda g_{\alpha\beta}) \,. \tag{A.166}$$

We explained that in order for relativistic equations to be valid in any reference frame they must be written in terms of spacetime tensors. This required the introduction of a new type of covariant derivative. The action of the covariant derivative on vectors with upper and lower indices is

$$\begin{aligned}
\nabla_\alpha V^\mu &= \partial_\alpha V^\mu + \Gamma^\mu_{\alpha\beta} V^\beta \,, \\
\nabla_\alpha W_\mu &= \partial_\alpha W_\mu - \Gamma^\beta_{\alpha\mu} W_\beta \,.
\end{aligned} \tag{A.167}$$

The action of the covariant derivative on arbitrary tensors is a natural generalization of these expressions.

We then computed the relative acceleration of two geodesics and showed that it is proportional to the Riemann tensor

$$R^\mu{}_{\nu\rho\sigma} = \partial_\rho \Gamma^\mu_{\nu\sigma} - \partial_\sigma \Gamma^\mu_{\nu\rho} + \Gamma^\mu_{\rho\lambda} \Gamma^\lambda_{\nu\sigma} - \Gamma^\mu_{\sigma\lambda} \Gamma^\lambda_{\nu\rho} \,. \tag{A.168}$$

We argued that the Riemann tensor plays the role of the tidal tensor, $\partial_i \partial_j \Phi$, in Newtonian gravity. Since the trace of the tidal tensor appears in the Poisson equation for the gravitational field, $\nabla^2 \Phi = \delta^{ij} \partial_i \partial_j \Phi = 4\pi G\rho$, we expect the trace of the Riemann tensor to play a role in Einstein's field equations for the metric. This trace is the Ricci tensor

$$\begin{aligned}
R_{\mu\nu} &\equiv R^\lambda{}_{\mu\lambda\nu} \\
&= \partial_\lambda \Gamma^\lambda_{\mu\nu} - \partial_\nu \Gamma^\lambda_{\mu\lambda} + \Gamma^\lambda_{\lambda\rho} \Gamma^\rho_{\mu\nu} - \Gamma^\rho_{\mu\lambda} \Gamma^\lambda_{\nu\rho} \,,
\end{aligned} \tag{A.169}$$

and the vacuum Einstein equation is indeed $R_{\mu\nu} = 0$. The non-vacuum Einstein equation is

$$R_{\mu\nu} - \frac{1}{2} g_{\mu\nu} R = 8\pi G \, T_{\mu\nu} \,, \tag{A.170}$$

where the term proportional to the Ricci scalar, $R \equiv g^{\mu\nu} R_{\mu\nu}$, had to be added in order for the left-hand side to be consistent with the conservation of the energy-momentum tensor, $\nabla^{\mu} T_{\mu\nu} = 0$.

We showed that the Newtonian limit of gravity is captured by the line element

$$\mathrm{d}s^2 = -(1 + 2\Phi)\mathrm{d}t^2 + (1 - 2\Phi)\delta_{ij}\mathrm{d}x^i\mathrm{d}x^j \,, \tag{A.171}$$

where $\Phi(x^i) \ll 1$. With this metric, the geodesic equation (A.165) and the Einstein equation (A.170) reduce to $\ddot{x}^i = -\partial^i\Phi$ and $\nabla^2\Phi = 4\pi G\rho$.

Further Reading

There is an abundance of excellent books and lecture notes on general relativity. A favorite of many students is Sean Carroll's book *Spacetime and Geometry* [4], and it is really excellent. I also have an affinity for D'Inverno's *Introducing Einstein's Relativity* [5] since I used it to teach myself the subject as an undergraduate. I remember its explicit style being particularly useful for self-study. Hartle's *Gravity* [6] uses a physics-first approach to general relativity that I used as an inspiration for parts of this appendix. In preparing this appendix, I have also consulted lectures notes by Reall [7], Tong [8] and Lim [9]. Finally, I must mention Misner, Thorne and Wheeler's classic text *Gravitation* [10]. It is full of insights, but its slightly idiosyncratic style (and its length of over 1300 pages) make it hard to use as a primary source.

B Details of the CMB Analysis

In this appendix, we will derive the CMB anisotropies more formally using the Boltzmann equation. This will provide a rigorous justification for some of the results used in Chapter 7. It will allow us to explain how the CMB fluctuations are computed with an accuracy that matches the incredible precision of the recent measurements of the microwave background.

This appendix follows closely the excellent lecture notes by Anthony Challinor [1], which I highly recommend for further details.

B.1 Boltzmann Equation

The hydrodynamic approximation used in Chapter 7 is only valid on scales larger than the mean free path of the photons. However, by definition, the mean free path becomes large when the photons decouple from the primordial plasma and corrections beyond the fluid description will be important. The rigorous way to describe the dynamics in this regime is the **Boltzmann equation**.

In Chapter 3, we introduced the concept of a distribution function f to describe a collection of particles in statistical mechanics. We showed that momentum integrals of the distribution function define the density and pressure of the coarse-grained system. We also derived the evolution equations for the fluid from energy and momentum conservation. To go beyond the fluid approximation, we take the distribution function as the fundamental object and study its time evolution in phase space. The relevant evolution equation is the Boltzmann equation, which schematically takes the form

$$\frac{\mathrm{d}f_a}{\mathrm{d}\eta} = C_a[\{f_b\}].\tag{B.1}$$

Note that there is a separate Boltzmann equation for each species $a = \gamma, \nu, e, b, c, \cdots$. The **collision term** on the right-hand side describes the interactions between the different species. As indicated by the notation, the collision term can depend on the distribution functions for several species, leading to a set of coupled equations.

We will focus on the evolution of the photon distribution function, f_γ, and drop the subscript, i.e. $f_\gamma \to f$. Photons interact most strongly with the electrons, so their Boltzmann equation becomes

$$\frac{\mathrm{d}f}{\mathrm{d}\eta} = C[\{f, f_e\}] . \tag{B.2}$$

Since the electrons are strongly coupled to the rest of the plasma, it makes no difference to think of the right-hand side of the Boltzmann equation as the photon–electron or photon–baryon coupling, and we will usually refer to the latter.

B.1.1 Without Scattering

In the absence of scattering, the right-hand side of (B.2) vanishes and we have

$$\frac{\mathrm{d}f}{\mathrm{d}\eta} = 0 . \tag{B.3}$$

This innocent looking equation expresses **Liouville's theorem**: the number of particles in a given element of phase space doesn't change with time. In a perturbed spacetime this contains nontrivial information since the phase space elements themselves evolve in a complicated way due to the perturbed metric.

In an inhomogeneous spacetime, the photon distribution function can depend on time η, position $\mathbf{x}$, the photon energy E and the photon direction of propagation $\hat{\mathbf{p}}$. It is convenient to introduce the *comoving energy* $\epsilon \equiv Ea$ which, in the absence of perturbations, would be constant along the photon path. The collisionless Boltzmann equation (B.3) can then be written as

$$\frac{\mathrm{d}}{\mathrm{d}\eta} f(\eta, \mathbf{x}, \epsilon, \hat{\mathbf{p}}) = \frac{\partial f}{\partial \eta} + \frac{\partial f}{\partial \mathbf{x}} \cdot \frac{\mathrm{d}\mathbf{x}}{\mathrm{d}\eta} + \frac{\partial f}{\partial \ln \epsilon} \frac{\mathrm{d}\ln \epsilon}{\mathrm{d}\eta} + \underbrace{\frac{\partial f}{\partial \hat{\mathbf{p}}} \cdot \frac{\mathrm{d}\hat{\mathbf{p}}}{\mathrm{d}\eta}}_{O(2)} = 0 . \tag{B.4}$$

As indicated, the last term is second order in perturbations and can therefore be dropped in a linearized analysis. (It describes the gravitational lensing of the CMB.) Moreover, since $\partial f/\partial \mathbf{x}$ and $\mathrm{d}\ln \epsilon/\mathrm{d}\eta$ are first order in perturbations, we can use the zeroth-order expressions $\mathrm{d}\mathbf{x}/\mathrm{d}\eta = \hat{\mathbf{p}}$ and $\partial f/\partial \ln \epsilon = \partial \bar{f}/\partial \ln \epsilon$ in the remaining terms. Equation (B.4) then becomes

$$\frac{\partial f}{\partial \eta} + \hat{\mathbf{p}} \cdot \nabla f + \frac{\partial \bar{f}}{\partial \ln \epsilon} \frac{\mathrm{d}\ln \epsilon}{\mathrm{d}\eta} = 0 . \tag{B.5}$$

At zeroth order, this implies $\partial \bar{f}/\partial \eta = 0$, so that the zeroth-order distribution function depends only on the comoving energy ϵ. This is consistent with the form of the equilibrium distribution function (3.4), as long as $T \propto a^{-1}$,

$$\bar{f}(\epsilon) = \left[\exp\left(\frac{E}{\bar{T}(\eta)} \right) - 1 \right]^{-1} = \left[\exp\left(\frac{\epsilon}{a\bar{T}(\eta)} \right) - 1 \right]^{-1}$$

$$= \left[\exp\left(\frac{\epsilon}{T_0} \right) - 1 \right]^{-1} , \tag{B.6}$$

where $T_0 = 2.7255\,\mathrm{K}$ is the average CMB temperature today. To study the effect of first-order perturbations, we write the distribution function as

$$f(\eta, \mathbf{x}, \epsilon, \hat{\mathbf{p}}) = \left[\exp\left(\frac{\epsilon}{a\bar{T}(\eta)[1 + \Theta(\eta, \mathbf{x}, \hat{\mathbf{p}})]} \right) - 1 \right]^{-1}, \tag{B.7}$$

where $\Theta \equiv \delta T / \bar{T}$ is the fractional temperature perturbation—sometimes called the **brightness function**. We assume that Θ does not depend on the energy of the photons, which is the case if the effect of perturbations is achromatic and does not lead to spectral distortions. We will have to check that this is a consistent assumption when we add scattering terms to the Boltzmann equation. Taking Θ to be a small perturbation, we get

$$f(\eta, \mathbf{x}, \epsilon, \hat{\mathbf{p}}) \approx \left[\exp\left(\frac{\epsilon[1 - \Theta(\eta, \mathbf{x}, \hat{\mathbf{p}})]}{a\bar{T}(\eta)} \right) - 1 \right]^{-1}$$

$$\approx \bar{f}(\epsilon) - \Theta(\eta, \mathbf{x}, \hat{\mathbf{p}}) \frac{\mathrm{d}\bar{f}}{\mathrm{d}\ln\epsilon}. \tag{B.8}$$

Equation (B.5) then becomes

$$-\frac{\mathrm{d}\bar{f}}{\mathrm{d}\ln\epsilon} \left(\frac{\partial\Theta}{\partial\eta} + \hat{\mathbf{p}} \cdot \boldsymbol{\nabla}\Theta - \frac{\mathrm{d}\ln\epsilon}{\mathrm{d}\eta} \right) = 0, \tag{B.9}$$

where we have kept the overall factor since we ultimately want to add nonzero collision terms on the right-hand side. At first order, the first two terms in the brackets combine into a total derivative,

$$\frac{\mathrm{d}}{\mathrm{d}\eta}\Theta(\eta, \mathbf{x}, \hat{\mathbf{p}}) = \frac{\partial\Theta}{\partial\eta} + \frac{\partial\Theta}{\partial\mathbf{x}} \cdot \frac{\mathrm{d}\mathbf{x}}{\mathrm{d}\eta} + \underbrace{\frac{\partial\Theta}{\partial\hat{\mathbf{p}}} \cdot \frac{\mathrm{d}\hat{\mathbf{p}}}{\mathrm{d}\eta}}_{O(2)} = \frac{\partial\Theta}{\partial\eta} + \hat{\mathbf{p}} \cdot \boldsymbol{\nabla}\Theta, \tag{B.10}$$

so we can write (B.9) as

$$-\frac{\mathrm{d}\bar{f}}{\mathrm{d}\ln\epsilon} \left(\frac{\mathrm{d}\Theta}{\mathrm{d}\eta} - \frac{\mathrm{d}\ln\epsilon}{\mathrm{d}\eta} \right) = 0 \quad \Rightarrow \quad \boxed{\frac{\mathrm{d}\Theta}{\mathrm{d}\eta} = \frac{\mathrm{d}\ln\epsilon}{\mathrm{d}\eta}}. \tag{B.11}$$

The evolution of the temperature perturbations is therefore directly related to the evolution of the comoving photon energy in the presence of metric fluctuations, $\mathrm{d}\ln\epsilon/\mathrm{d}\eta$. The latter follows from the geodesic equation (see Section 7.2.1):

$$\frac{\mathrm{d}\ln\epsilon}{\mathrm{d}\eta} = -\frac{\mathrm{d}\Psi}{\mathrm{d}\eta} + \Phi' + \Psi'. \tag{B.12}$$

Integrating this equation gives the gravitational redshifting of free-streaming photons in the presence of metric perturbations.

B.1.2 Including Scattering

Before recombination, the scattering between photons and electrons, $e + \gamma \leftrightarrow e + \gamma$, played an important role. In the following, I will show how to include this in the Boltzmann equation [1].

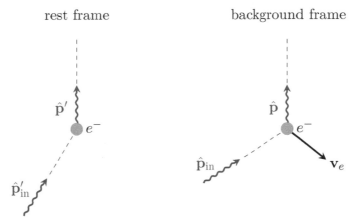

rest frame   background frame

Fig. B.1   Thomson scattering in the electron rest frame and in the background frame.

Electron rest frame

Let us start in the rest frame of a single electron. In this frame, an incoming photon has energy ϵ'_{in} and three-momentum $\mathbf{p}'_{in} = \epsilon'_{in}\hat{\mathbf{p}}'_{in}$ (see Fig. B.1). The scattered photon has energy $\epsilon' = \epsilon'_{in}$ and three-momentum $\mathbf{p}' = \epsilon'\hat{\mathbf{p}}'$, where we have used the fact that Thomson scattering doesn't change the energy of the photon (in the rest frame of the scatterer). We take the incoming radiation to be unpolarized and average over the polarization of the outgoing radiation. Ignoring the anisotropic nature of the scattering, the **differential cross section** is

$$\frac{d\sigma}{d\Omega} = \frac{\sigma_T}{4\pi} \, . \tag{B.13}$$

This describes the rate at which the electron scatters photons per solid angle and per unit incident photon flux. The parameter σ_T is the Thomson cross section

$$\sigma_T = \frac{8\pi}{3} \left(\frac{q_e^2}{4\pi\epsilon_0 m_e c^2} \right)^2 = 6.65 \times 10^{-29} \, \mathrm{m}^2 \, , \tag{B.14}$$

where q_e is the charge of the electron and m_e is its mass. Since (B.14) scales inversely with the mass of the scatterer, the scattering with the much heavier protons can be ignored.

The scattering rate with respect to the proper time τ' in the electron rest frame can then be written as

$$C'[f'(\epsilon', \hat{\mathbf{p}}')] \equiv \left. \frac{df'(\epsilon', \hat{\mathbf{p}}')}{d\tau'} \right|_{\mathrm{scatt.}} = n_e \int d\hat{\mathbf{p}}'_{in} \frac{d\sigma}{d\Omega} \, [\underset{\uparrow \atop \mathrm{in}}{f'(\epsilon', \hat{\mathbf{p}}'_{in})} - \underset{\uparrow \atop \mathrm{out}}{f'(\epsilon', \hat{\mathbf{p}}')}] \, , \tag{B.15}$$

where n_e is the proper number density of the electrons. The first term in the integrand captures the *in-scattering* of photons ($\hat{\mathbf{p}}'_{in} \to \hat{\mathbf{p}}'$), while the second term describes the *out-scattering* ($\hat{\mathbf{p}}' \to \hat{\mathbf{p}}'_{in}$). We see that the collision term vanishes for isotropic radiation, confirming that it starts at first order in perturbations. In the out-scattering term, we can pull the distribution function out of the integral and hence get

$$C'[f'(\epsilon', \hat{\mathbf{p}}')] = -n_e \sigma_T f'(\epsilon', \hat{\mathbf{p}}') + n_e \sigma_T \int \frac{\mathrm{d}\hat{\mathbf{p}}'_{\text{in}}}{4\pi} f'(\epsilon', \hat{\mathbf{p}}'_{\text{in}}) \,. \tag{B.16}$$

Next, we have to transform this result to the background frame in which the electrons are moving.

Background frame

To obtain the result in a general frame, we will perform a Lorentz boost which accounts for the bulk velocity of the electrons. At zeroth order, the proper time in the boosted frame is the same as that in the rest frame, $\tau = \tau' + O(1)$. Since the scattering rate starts at first order in perturbations, we don't need more accuracy than this. The scattering rate with respect to the conformal time in the boosted frame then is

$$C[f(\epsilon, \hat{\mathbf{p}})] \equiv \frac{\mathrm{d}f(\epsilon, \hat{\mathbf{p}})}{\mathrm{d}\eta}\bigg|_{\text{scatt.}} = a \frac{\mathrm{d}f'(\epsilon', \hat{\mathbf{p}}')}{\mathrm{d}\tau}\bigg|_{\text{scatt.}} + O(2)$$

$$= a\, C'[f'(\epsilon', \hat{\mathbf{p}}')] + O(2) \,, \tag{B.17}$$

where we have used that the distribution function is invariant under the boost, $f'(\epsilon', \hat{\mathbf{p}}') = f(\epsilon, \hat{\mathbf{p}})$. Equation (B.16) then leads to

$$C[f(\epsilon, \hat{\mathbf{p}})] = -\Gamma f(\epsilon, \hat{\mathbf{p}}) + \Gamma \int \frac{\mathrm{d}\hat{\mathbf{p}}'_{\text{in}}}{4\pi} f'(\epsilon', \hat{\mathbf{p}}'_{\text{in}}) \,, \tag{B.18}$$

where $\Gamma \equiv a \bar{n}_e \sigma_T$ is the scattering rate. (At first order, it is sufficient to use the background density $\bar{n}_e$.) Finally, we wish to rewrite the right-hand side in terms of quantities defined in the boosted frame.

While the energies of the incoming and outgoing photons were equal in the rest frame, in the boosted frame we have $\epsilon_{\text{in}} \neq \epsilon$. To derive the relation between ϵ_{in} and ϵ, we consider the Lorentz transformation

$$\epsilon' = \gamma \epsilon (1 - \hat{\mathbf{p}} \cdot \mathbf{v}_e) \,, \tag{B.19}$$

where $\gamma(v_e)$ is the Lorentz factor. Similarly, the inverse transform gives

$$\epsilon_{\text{in}} = \gamma \epsilon'_{\text{in}} (1 + \hat{\mathbf{p}}'_{\text{in}} \cdot \mathbf{v}_e)$$
$$= \gamma \epsilon' (1 + \hat{\mathbf{p}}'_{\text{in}} \cdot \mathbf{v}_e)$$
$$= \gamma^2 \epsilon (1 - \hat{\mathbf{p}} \cdot \mathbf{v}_e)(1 + \hat{\mathbf{p}}'_{\text{in}} \cdot \mathbf{v}_e)$$
$$\approx \epsilon \left(1 - (\hat{\mathbf{p}} - \hat{\mathbf{p}}'_{\text{in}}) \cdot \mathbf{v}_e \right) \,, \tag{B.20}$$

and the Lorentz invariance of the distribution function implies

$$f'(\epsilon', \hat{\mathbf{p}}'_{\rm in}) = f(\epsilon_{\rm in}, \hat{\mathbf{p}}_{\rm in})$$

$$= f[\epsilon(1 - (\hat{\mathbf{p}} - \hat{\mathbf{p}}'_{\rm in}) \cdot \mathbf{v}_e), \hat{\mathbf{p}}_{\rm in}]$$

$$= \bar{f}[\epsilon(1 - (\hat{\mathbf{p}} - \hat{\mathbf{p}}'_{\rm in}) \cdot \mathbf{v}_e)] - \frac{\mathrm{d}\bar{f}}{\mathrm{d}\ln\epsilon}\Theta(\hat{\mathbf{p}}_{\rm in})$$

$$= \bar{f}(\epsilon) - \frac{\mathrm{d}\bar{f}}{\mathrm{d}\ln\epsilon}(\hat{\mathbf{p}} - \hat{\mathbf{p}}'_{\rm in}) \cdot \mathbf{v}_e - \frac{\mathrm{d}\bar{f}}{\mathrm{d}\ln\epsilon}\Theta(\hat{\mathbf{p}}_{\rm in}) . \tag{B.21}$$

We substitute this into (B.18). Let us discuss the three terms of (B.21) in turn: The first term, $\bar{f}(\epsilon)$, integrates to $+\Gamma\bar{f}(\epsilon)$ and hence cancels the zeroth-order term in the out-scattering contribution, $-\Gamma f(\epsilon, \hat{\mathbf{p}}) \approx -\Gamma\bar{f}(\epsilon) + (\mathrm{d}\bar{f}/\mathrm{d}\ln\epsilon)\Theta(\hat{\mathbf{p}})$. In the second term, the $\hat{\mathbf{p}}'_{\rm in} \cdot \mathbf{v}_e$ part integrates to zero (by parity) and the $\hat{\mathbf{p}} \cdot \mathbf{v}_e$ part is independent of $\hat{\mathbf{p}}'_{\rm in}$ and can be pulled out of the integral. The last term is first order, so we can make the replacements $\mathrm{d}\hat{\mathbf{p}}'_{\rm in} \to \mathrm{d}\hat{\mathbf{p}}_{\rm in}$ and $\hat{\mathbf{p}}'_{\rm in} \cdot \hat{\mathbf{p}}' \to \hat{\mathbf{p}}_{\rm in} \cdot \hat{\mathbf{p}}$. With these simplifications, we get

$$C[f(\epsilon, \hat{\mathbf{p}})] = \frac{\mathrm{d}\bar{f}}{\mathrm{d}\ln\epsilon} \times \Gamma\left[\Theta(\hat{\mathbf{p}}) - \Theta_0 - \hat{\mathbf{p}} \cdot \mathbf{v}_e\right] , \tag{B.22}$$

where we have defined the **monopole** of the temperature anisotropy

$$\Theta_0 \equiv \int \frac{\mathrm{d}\hat{\mathbf{p}}_{\rm in}}{4\pi}\,\Theta(\hat{\mathbf{p}}_{\rm in}) . \tag{B.23}$$

As we will see, this monopole is proportional to the photon density fluctuations δ_γ. Since electrons and baryons are tightly coupled by Coulomb scattering, they have equal bulk velocities, so we can replace $\mathbf{v}_e$ by $\mathbf{v}_b$.

Adding (B.22) to the right-hand side of the collisionless Boltzmann equation (B.11), we get

$$\boxed{\frac{\mathrm{d}\Theta}{\mathrm{d}\eta} = \frac{\mathrm{d}\ln\epsilon}{\mathrm{d}\eta} - \Gamma\left[\Theta - \Theta_0 - \hat{\mathbf{p}} \cdot \mathbf{v}_b\right]} . \tag{B.24}$$

When the scattering rate is large, $\Gamma \gg \mathcal{H}$, the parenthesis on the right-hand side of (B.24) has to vanish, $\Theta \to \Theta_0 + \hat{\mathbf{p}} \cdot \mathbf{v}_b$. This is physically intuitive: scattering tends to make the photon distribution isotropic in the electron rest frame, which becomes $\Theta_0 + \hat{\mathbf{p}} \cdot \mathbf{v}_b$ in the boosted frame.

Anisotropic scattering

So far, our treatment has ignored the anisotropic nature of Thomson scattering. The more correct form of the differential cross section in the electron rest frame is

$$\frac{\mathrm{d}\sigma}{\mathrm{d}\Omega} = \frac{3\sigma_T}{16\pi}\left[1 + \cos^2\theta\right] , \tag{B.25}$$

where $\cos\theta \equiv \hat{\mathbf{p}}'_{\rm in} \cdot \hat{\mathbf{p}}'$ is the cosine of the scattering angle. The result we used in (B.13) is the angular average of (B.25). Repeating the analysis with the differential cross section (B.25) gives

$$C[f(\epsilon, \hat{\mathbf{p}})] = \frac{\mathrm{d}\bar{f}}{\mathrm{d}\ln\epsilon} \times \Gamma \left[\Theta(\hat{\mathbf{p}}) - \hat{\mathbf{p}} \cdot \mathbf{v}_b - \frac{3}{4} \int \frac{\mathrm{d}\hat{\mathbf{m}}}{4\pi} \Theta(\hat{\mathbf{m}}) \left[1 + (\hat{\mathbf{m}} \cdot \hat{\mathbf{p}})^2 \right] \right], \quad (B.26)$$

where I have defined $\hat{\mathbf{m}} \equiv \hat{\mathbf{p}}_{\text{in}}$ to spare us a subscript. Adding this to (B.11) then gives

$$\boxed{\frac{\mathrm{d}\Theta}{\mathrm{d}\eta} = \frac{\mathrm{d}\ln\epsilon}{\mathrm{d}\eta} - \Gamma \left[\Theta - \Theta_0 - \hat{\mathbf{p}} \cdot \mathbf{v}_b - \frac{3}{4} \int \frac{\mathrm{d}\hat{\mathbf{m}}}{4\pi} \Theta(\hat{\mathbf{m}}) \left((\hat{\mathbf{m}} \cdot \hat{\mathbf{p}})^2 - \frac{1}{3} \right) \right]}, \quad (B.27)$$

where we have extracted the monopole Θ_0 from the integral. As we will see below, when the photons are tightly coupled to the baryons, the photon monopole Θ_0 is much larger than the remaining integral contribution (which is a quadrupole moment), so that (B.27) reduces to (B.24).

B.2 Free Streaming and Projection

In Chapter 7, we showed that the relation between the primordial curvature perturbations $\mathcal{R}_i(\mathbf{k})$ and the observed temperature anisotropies $\delta T(\hat{\mathbf{n}})$ involves three elements:

$$\mathcal{R}_i(\mathbf{k}) \xrightarrow{\text{evolution (§B.3)}} \begin{pmatrix} \delta_\gamma \\ \Psi \\ v_b \end{pmatrix}_{\eta_*} \xrightarrow[\text{projection (§B.2.2)}]{\text{free streaming (§B.2.1)}} \delta T(\hat{\mathbf{n}}).$$

In the rest of this appendix, we will repeat this analysis using the Boltzmann formalism.

B.2.1 Line-of-Sight Solution

As in Chapter 7, we are interested in the line-of-sight solution of the CMB temperature perturbations, but this time we do not assume instantaneous recombination. Instead, we introduce the **optical depth** between times η and η_0 (see Section 3.1.5):

$$\tau(\eta) \equiv \int_\eta^{\eta_0} \Gamma(\tilde{\eta}) \, \mathrm{d}\tilde{\eta}. \quad (B.28)$$

Physically, this describes the opacity of the universe at a given time, when seen from today. (The probability of no scattering as a photon travels from η to η_0 is $e^{-\tau}$.) Mathematically, the function $e^{-\tau}$ will act as an integration factor for the Boltzmann equation. To see this, note that

$$\frac{\mathrm{d}\Theta}{\mathrm{d}\eta} + \Gamma\Theta = e^\tau \frac{\mathrm{d}}{\mathrm{d}\eta}(e^{-\tau}\Theta), \quad (B.29)$$

so that part of the Boltzmann equation turns into a total derivative after introducing the optical depth. Equation (B.24) then becomes

$$\frac{\mathrm{d}}{\mathrm{d}\eta}(e^{-\tau}\Theta) = e^{-\tau}\frac{\mathrm{d}\ln\epsilon}{\mathrm{d}\eta} + g\left[\Theta_0 - \hat{\mathbf{n}}\cdot\mathbf{v}_b\right],\tag{B.30}$$

where we have introduced the **visibility function**, $g(\eta) \equiv -\tau'(\eta)e^{-\tau(\eta)}$, and used that $\hat{\mathbf{n}} = -\hat{\mathbf{p}}$.

For scalar fluctuations, the geodesic equation (B.12) implies

$$e^{-\tau}\frac{\mathrm{d}\ln\epsilon}{\mathrm{d}\eta} = -e^{-\tau}\frac{\mathrm{d}\Psi}{\mathrm{d}\eta} + e^{-\tau}(\Phi' + \Psi')$$

$$= -\frac{\mathrm{d}}{\mathrm{d}\eta}(e^{-\tau}\Psi) + g\Psi + e^{-\tau}(\Phi' + \Psi'),\tag{B.31}$$

and the Boltzmann equation can be written as

$$\frac{\mathrm{d}}{\mathrm{d}\eta}\left[e^{-\tau}(\Theta + \Psi)\right] = S,\tag{B.32}$$

where the source term is

$$S \equiv g\left[\Theta_0 + \Psi - \hat{\mathbf{n}}\cdot\mathbf{v}_b\right] + e^{-\tau}(\Phi' + \Psi').\tag{B.33}$$

Integrating (B.32) along the line-of-sight, we find

$$\Theta(\hat{\mathbf{n}}) \equiv \Theta(\eta_0, \mathbf{x}_0, \hat{\mathbf{n}}) = \int_0^{\eta_0} \mathrm{d}\eta\, S(\eta, \mathbf{x}_0 + \chi(\eta)\hat{\mathbf{n}}, \hat{\mathbf{n}}),\tag{B.34}$$

where $\chi(\eta) = \eta_0 - \eta$ in a flat universe. To arrive at (B.34), we have used $\tau(\eta_0) = 0$ and $\tau(0) \approx \infty$, and dropped the unobservable monopole $\Psi(\eta_0, \mathbf{x}_0)$ on the left-hand side. Substituting the source term on the right-hand side, we get

$$\boxed{\Theta(\hat{\mathbf{n}}) = \int_0^{\eta_0} \mathrm{d}\eta\, \left(g\left[\Theta_0 + \Psi - \hat{\mathbf{n}}\cdot\mathbf{v}_b\right] + e^{-\tau}(\Phi' + \Psi')\right)}.\tag{B.35}$$

Notice that the terms which are proportional to the visibility function $g(\eta)$ are localized near the last-scattering surface, while terms proportional to $e^{-\tau(\eta)}$ have an integrated effect until today. In Chapter 7, we assumed that recombination occurred instantaneously. In that case, we have $g(\eta) \approx \delta_{\mathrm{D}}(\eta - \eta_*)$ and $e^{-\tau} \approx \theta(\eta - \eta_*)$, where θ is the Heaviside function. Using this in (B.35), we get

$$\Theta(\hat{\mathbf{n}}) \approx (\Theta_0 + \Psi)_* - (\hat{\mathbf{n}}\cdot\mathbf{v}_b)_* + \int_{\eta_*}^{\eta_0} \mathrm{d}\eta\, (\Phi' + \Psi'),\tag{B.36}$$

which is the same as the solution (7.29) found in Section 7.2.

B.2.2 Spatial-to-Angular Projection

The next step is the same as in Chapter 7. We want to extract the multipole moments Θ_l of the temperature anisotropies $\Theta(\hat{\mathbf{n}})$ and relate them to the Fourier modes of the inhomogeneities in the primordial plasma.

We first write the source term S in a Fourier decomposition:

$$S(\eta, \chi(\eta)\hat{\mathbf{n}}, \hat{\mathbf{n}}) = \int \frac{\mathrm{d}^3 k}{(2\pi)^3} S(\eta, \mathbf{k}, \hat{\mathbf{n}})\, e^{i\chi(\eta)\mathbf{k}\cdot\hat{\mathbf{n}}}$$

$$= \int \frac{\mathrm{d}^3 k}{(2\pi)^3} \left[F(\eta, \mathbf{k}) - i\hat{\mathbf{k}}\cdot\hat{\mathbf{n}}\, G(\eta, \mathbf{k}) \right] e^{i\chi(\eta)\mathbf{k}\cdot\hat{\mathbf{n}}}, \qquad (B.37)$$

where we have used that, for scalar fluctuations, $\mathbf{v}_b \equiv i\hat{\mathbf{k}}\, v_b$ and defined

$$F(\eta, \mathbf{k}) \equiv g(\Theta_0 + \Psi) + e^{-\tau}(\Phi' + \Psi'),$$
$$G(\eta, \mathbf{k}) \equiv g\, v_b. \qquad (B.38)$$

The corresponding transfer functions are

$$F(\eta, k) \equiv \frac{F(\eta, \mathbf{k})}{\mathcal{R}_i(\mathbf{k})} \quad \text{and} \quad G(\eta, k) \equiv \frac{G(\eta, \mathbf{k})}{\mathcal{R}_i(\mathbf{k})}. \qquad (B.39)$$

Using the Legendre expansion of the exponential,

$$e^{i\chi\mathbf{k}\cdot\hat{\mathbf{n}}} = \sum_l i^l (2l+1) j_l(k\chi) P_l(\hat{\mathbf{k}}\cdot\hat{\mathbf{n}}), \qquad (B.40)$$

we then get

$$\Theta(\hat{\mathbf{n}}) = \sum_l i^l (2l+1) \int \frac{\mathrm{d}^3 k}{(2\pi)^3}\, \Theta_l(k)\, \mathcal{R}_i(\mathbf{k}) P_l(\hat{\mathbf{k}}\cdot\hat{\mathbf{n}}), \qquad (B.41)$$

where

$$\Theta_l(k) \equiv \int_0^{\eta_0} \mathrm{d}\eta \left[F(\eta, k) j_l(k\chi) - G(\eta, k) j_l'(k\chi) \right]. \qquad (B.42)$$

Substituting the explicit forms of the transfer functions, this becomes

$$\Theta_l(k) = \int_0^{\eta_0} \mathrm{d}\eta \left[\left(g(\Theta_0 + \Psi) + e^{-\tau}(\Phi' + \Psi') \right) j_l(k\chi) - g\, v_b\, j_l'(k\chi) \right], \qquad (B.43)$$

which reduces to our previous result (7.43) if we assume instantaneous recombination. The CMB power spectrum is

$$C_l = 4\pi \int \mathrm{d}\ln k\, \Theta_l^2(k)\, \Delta_{\mathcal{R}}^2(k), \qquad (B.44)$$

as before.

B.3 Evolution Before Decoupling

In Chapter 7, we studied the evolution of fluctuations in the primordial plasma in a hydrodynamic approximation. We will now describe the same dynamics using the Boltzmann equation.

B.3.1 Photons

Recall the Boltzmann equation (B.24) for the evolution of the photon temperature fluctuations:

$$\frac{d\Theta}{d\eta} = \frac{d\ln\epsilon}{d\eta} - \Gamma\big[\Theta - \Theta_0 - \hat{\mathbf{p}}\cdot\mathbf{v}_b\big]\,, \tag{B.45}$$

where

$$\frac{d\Theta}{d\eta} = \Theta' + \hat{p}^i\partial_i\Theta\,, \tag{B.46}$$

$$\frac{d\ln\epsilon}{d\eta} = \Phi' - \hat{p}^i\partial_i\Psi\,. \tag{B.47}$$

In Fourier space, this becomes

$$\Theta' + ik\mu\Theta = \Phi' - ik\mu\Psi - \Gamma\big[\Theta - \Theta_0 - i\mu v_b\big]\,, \tag{B.48}$$

where we have used that $\mathbf{v}_b \equiv i\hat{\mathbf{k}}\,v_b$ and defined $\mu \equiv \hat{\mathbf{k}}\cdot\hat{\mathbf{p}}$ as the cosine of the angle between the wavevector of the inhomogeneity and the photon's direction of propagation. For $\mu = 1$, the photon moves along $\hat{\mathbf{k}}$, i.e. along the direction of changing temperature (or density), while for $\mu = 0$, the photon moves along the direction of constant temperature.

Multipole moments

Note that the equation of motion for Θ does not depend explicitly on the vectors $\mathbf{k}$ and $\hat{\mathbf{p}}$, but only on their relative orientation, $\mu \equiv \hat{\mathbf{k}}\cdot\hat{\mathbf{p}}$, and the wavenumber k. The initial conditions for Θ, on the other hand, depend on the wavevector $\mathbf{k}$ and $\mu \equiv \hat{\mathbf{k}}\cdot\hat{\mathbf{p}}$, but not $\hat{\mathbf{p}}$ (this is because the fluctuations are tightly coupled at early times, so that $\Theta \to \Theta_0 + \hat{\mathbf{p}}\cdot\mathbf{v}_b$). It is then useful to expand the fluctuations in terms of Legendre polynomials:

$$\Theta(\eta,\mathbf{k},\hat{\mathbf{p}}) \;\to\; \Theta(\eta,\mathbf{k},\mu) \equiv \sum_{l=0}^{\infty}(-i)^l(2l+1)\Theta_l(\eta,\mathbf{k})P_l(\mu)\,, \tag{B.49}$$

where Θ_l are the **multipole moments** of the distribution. The factor of $(-i)^l(2l+1)$ was introduced for later convenience.

To gain a physical understanding of the different multipole moments, we look at the energy-momentum tensor for the photons:

$$T^\mu{}_\nu = \int \frac{d^3p}{E(p)}\, f\, P^\mu P_\nu\,, \tag{B.50}$$

where the perturbed four-momentum was derived in the main text:

$$\begin{aligned} P^\mu &= (E/a)[1 - \Psi, (1+\Phi)\hat{\mathbf{p}}]\,, \\ P_\nu &= (aE)[-(1+\Psi), (1-\Phi)\hat{\mathbf{p}}]\,. \end{aligned} \tag{B.51}$$

At linear order in perturbations, the time–time component of the energy-momentum tensor becomes

$$
T^0{}_0 = \int \frac{d^3 p}{E(p)} f P^0 P_0 = -\int E \, dE \, d\hat{\mathbf{p}} \, f \left[a^{-1} E (1 - \Psi) \right] \left[a E (1 + \Psi) \right]
$$

$$
= -\frac{1}{a^4} \int d\hat{\mathbf{p}} \int d\epsilon \, \epsilon^3 \left(\bar{f} - \frac{d\bar{f}}{d \ln \epsilon} \Theta \right)
$$

$$
= -\frac{4\pi}{a^4} \int d\epsilon \, \bar{f} \epsilon^3 + \frac{1}{a^4} \int d\epsilon \, \frac{d\bar{f}}{d\epsilon} \epsilon^4 \int d\hat{\mathbf{p}} \, \Theta
$$

$$
= -\underbrace{\frac{4\pi}{a^4} \int d\epsilon \, \bar{f} \epsilon^3}_{\bar{\rho}_\gamma} \left(1 + \underbrace{4 \int \frac{d\hat{\mathbf{p}}}{4\pi} \Theta}_{\delta_\gamma} \right), \tag{B.52}
$$

where $\bar{\rho}_\gamma \propto a^{-4}$, as expected, and

$$
\boxed{\delta_\gamma \equiv 4\Theta_0} . \tag{B.53}
$$

The **monopole** of the temperature fluctuations is therefore given by the density fluctuations of the photon gas. The relation in (B.53) is consistent with the fact that $\rho_\gamma \propto T^4$.

Similar manipulations for the space–time component $T^i{}_0$ lead to

$$
\mathbf{v}_\gamma(\eta, \mathbf{k}) = 3 \int \frac{d\hat{\mathbf{p}}}{4\pi} \Theta(\eta, \mathbf{k}, \mu) \, \hat{\mathbf{p}}
$$

$$
= 3 \int \frac{d\hat{\mathbf{p}}}{4\pi} \sum_l (-i)^l (2l + 1) \Theta_l(\eta, \mathbf{k}) P_l(\hat{\mathbf{k}} \cdot \hat{\mathbf{p}}) \, \hat{\mathbf{p}} . \tag{B.54}
$$

By symmetry, the integral over $\hat{\mathbf{p}}$ must be proportional to $\hat{\mathbf{k}}$. The proportionality constant is determined as follows

$$
A_l \hat{\mathbf{k}} = \int \frac{d\hat{\mathbf{p}}}{4\pi} P_l(\hat{\mathbf{k}} \cdot \hat{\mathbf{p}}) \, \hat{\mathbf{p}} \quad \Rightarrow \quad A_l = \int \frac{d\hat{\mathbf{p}}}{4\pi} P_l(\hat{\mathbf{k}} \cdot \hat{\mathbf{p}}) \hat{\mathbf{k}} \cdot \hat{\mathbf{p}}
$$

$$
= \int_{-1}^{1} \frac{d\mu}{2} \mu P_l(\mu)
$$

$$
= \frac{1}{3} \delta_{l1} . \tag{B.55}
$$

Equation (B.54) then becomes $\mathbf{v}_\gamma = -3i\hat{\mathbf{k}} \, \Theta_1 \equiv i\hat{\mathbf{k}} \, v_\gamma$, so that

$$
\boxed{v_\gamma = -3\Theta_1} . \tag{B.56}
$$

The **dipole** of the temperature fluctuations is therefore related to the bulk velocity of the photon gas.

Finally, performing the same exercise for the $T^i{}_j$ component of the energy-momentum tensor gives

$$\Pi_\gamma^{ij}(\eta, \mathbf{k}) = 4\bar{\rho}_\gamma \int \frac{d\hat{\mathbf{p}}}{4\pi} \, \Theta(\eta, \mathbf{k}, \mu) \, \hat{p}^{\langle i}\hat{p}^{j\rangle}$$

$$= 4\bar{\rho}_\gamma \int \frac{d\hat{\mathbf{p}}}{4\pi} \sum_l (-i)^l (2l+1) \Theta_l(\eta, \mathbf{k}) P_l(\hat{\mathbf{k}} \cdot \hat{\mathbf{p}}) \, \hat{p}^{\langle i}\hat{p}^{j\rangle} \,, \tag{B.57}$$

where $\hat{p}^{\langle i}\hat{p}^{j\rangle} \equiv \hat{p}^i\hat{p}^j - \delta^{ij}/3$. The integral over $\hat{\mathbf{p}}$ must be a symmetric and trace-free tensor made from $\hat{\mathbf{k}}$, i.e.

$$B_l \hat{k}^{\langle i}\hat{k}^{j\rangle} = \int \frac{d\hat{\mathbf{p}}}{4\pi} P_l(\hat{\mathbf{k}} \cdot \hat{\mathbf{p}}) \, \hat{p}^{\langle i}\hat{p}^{j\rangle} \quad \Rightarrow \quad \frac{2}{3}B_l = \int \frac{d\hat{\mathbf{p}}}{4\pi} P_l(\hat{\mathbf{k}} \cdot \hat{\mathbf{p}}) \left((\hat{\mathbf{k}} \cdot \hat{\mathbf{p}})^2 - \frac{1}{3} \right)$$

$$= \frac{1}{3} \int_{-1}^{1} d\mu \, P_l(\mu) P_2(\mu)$$

$$= \frac{2}{15}\delta_{l2} \,. \tag{B.58}$$

Hence, we have

$$\Pi_\gamma^{ij}(\eta, \mathbf{k}) = -4\bar{\rho}_\gamma \Theta_2(\eta, \mathbf{k})\hat{k}^{\langle i}\hat{k}^{j\rangle} \,. \tag{B.59}$$

Defining the scalar part of the anisotropic stress via $\Pi_\gamma^{ij} = -(\bar{\rho}_\gamma + \bar{P}_\gamma)\hat{k}^{\langle i}\hat{k}^{j\rangle}\Pi_\gamma$, we get

$$\boxed{\Pi_\gamma = 3\Theta_2} \,. \tag{B.60}$$

This shows that the **quadrupole** of the temperature fluctuations is associated with the anisotropic stress of the photons.

Anisotropic scattering

The Boltzmann equation (B.45) did not account for anisotropic scattering. Including this effect, as in (B.27), gives

$$\Theta' + ik\mu\Theta = \Phi' - ik\mu\Psi$$
$$- \Gamma \left[\Theta - \Theta_0 - i\mu v_b - \frac{3}{16\pi}\hat{p}^i\hat{p}^j \int d\hat{\mathbf{m}} \, \Theta(\eta, \mathbf{k}, \hat{\mathbf{m}}) \, \hat{m}^{\langle i}\hat{m}^{j\rangle} \right], \tag{B.61}$$

where $\hat{m}^{\langle i}\hat{m}^{j\rangle} \equiv \hat{m}^i\hat{m}^j - \delta^{ij}/3$. Substituting (B.49) into the integral term leads to

$$\boxed{\Theta' + ik\mu\Theta = \Phi' - ik\mu\Psi - \Gamma \left[\Theta - \Theta_0 - i\mu v_b + \frac{1}{2}\Theta_2 P_2(\mu) \right]}, \tag{B.62}$$

where we have used the same trick as in (B.58) to perform the integral.

Boltzmann hierarchy

It is also useful to write the Boltzmann equation as an infinite set of coupled equations for each of the multipole moments. Substituting the multipole expansion (B.49) into (B.62), we get

$$\sum_l (-i)^l (2l+1) P_l(\mu) \left(\Theta_l' + ik\mu\Theta_l \right) = \sum_l (-i)^l (2l+1) P_l(\mu) \left(\delta_{l0}\Phi' + \delta_{l1}\frac{k\Psi}{3} \right.$$
$$\left. - \Gamma \left[(1 - \delta_{l0})\Theta_l + \delta_{l1}\frac{v_b}{3} - \frac{1}{10}\delta_{l2}\Theta_2 \right] \right), \quad \text{(B.63)}$$

where we have used that $P_0(\mu) = 1$ and $P_1(\mu) = \mu$. To deal with the $\mu P_l(\mu)$ term on the left-hand side, we use the recursion relation

$$(2l+1)\mu P_l(\mu) = (l+1)P_{l+1}(\mu) + l P_{l-1}(\mu), \quad \text{(B.64)}$$

which allows us to write

$$\sum_l (-i)^l (2l+1) P_l(\mu)\, ik\mu\Theta_l = \sum_l (-i)^l \left((l+1)P_{l+1}(\mu) + l P_{l-1}(\mu) \right) ik\Theta_l$$
$$= \sum_l (-i)^l\, k \left((l+1)\Theta_{l+1} - l\Theta_{l-1} \right) P_l(\mu). \quad \text{(B.65)}$$

Since the Legendre polynomials are an orthogonal basis, we can equate the coefficients in (B.63) for each l. This gives an infinite set of ordinary differential equations for the multipole moments called the **Boltzmann hierarchy**:

$$\Theta_l' + \frac{k}{2l+1} \left((l+1)\Theta_{l+1} - l\Theta_{l-1} \right)$$
$$= \delta_{l0}\Phi' + \delta_{l1}\frac{k\Psi}{3} - \Gamma \left[(1 - \delta_{l0})\Theta_l + \delta_{l1}\frac{v_b}{3} - \frac{1}{10}\delta_{l2}\Theta_2 \right]. \quad \text{(B.66)}$$

Writing the Boltzmann equation in the multipole expansion is convenient since all higher moments are suppressed in the tight-coupling limit.

Tight-coupling limit

To show that the higher moments are suppressed in the limit of large scattering rate, we consider the Boltzmann equation for $\Theta_{l>2}$:

$$\Theta_l' + \frac{k}{2l+1} \left((l+1)\Theta_{l+1} - l\Theta_{l-1} \right) = -\Gamma\Theta_l, \quad \text{(for } l > 2). \quad \text{(B.67)}$$

For $\Gamma \gg k$, we have $\Gamma\Theta_l \gg \Theta_l'$, so that we can drop the Θ_l' term on the left-hand side. Ignoring further the Θ_{l+1} term, we get

$$\Theta_l \sim \frac{k}{\Gamma}\Theta_{l-1} \ll \Theta_{l-1}, \quad \text{(B.68)}$$

which shows that all higher moments are indeed small in the tight-coupling regime. (Note that this justifies why we were allowed to drop the Θ_{l+1} term.) This makes intuitive sense: scattering tries to make the radiation field isotropic in the rest frame of the scatterer. The dominant multipole moments are therefore the monopole and the dipole (from boosting to the background frame), which evolve according to the $l = 0$ and $l = 1$ moments of the Boltzmann equation (B.66):

$$\Theta'_0 = -k\Theta_1 + \Phi', \tag{B.69}$$

$$\Theta'_1 = -\frac{k}{3}(2\Theta_2 - \Theta_0) + \frac{k}{3}\Psi - \Gamma\left(\Theta_1 + \frac{v_b}{3}\right). \tag{B.70}$$

These equations become more recognizable when we write them in terms of $\delta_\gamma = 4\Theta_0$, $v_\gamma = -3\Theta_1$ and $\Pi_\gamma = 3\Theta_2$. We get the continuity and Euler equations for the photons:

$$\delta'_\gamma - \frac{4}{3}(kv_\gamma + 3\Phi') = 0, \tag{B.71}$$

$$v'_\gamma + \frac{1}{4}k\delta_\gamma - \frac{2}{3}k\Pi_\gamma + k\Psi = -\Gamma(v_\gamma - v_b). \tag{B.72}$$

At leading order in the tight-coupling approximation, we can drop the photon anisotropic stress Π_γ. At next-to-leading order, however, it must be taken into account and leads to a damping of the fluctuations. The Boltzmann equation for the quadrupole moment Θ_2 implies

$$\Theta_2 = -\frac{10}{9}\frac{k}{\Gamma}\left(\frac{\Theta'_2}{k} + \frac{3}{5}\Theta_3 - \frac{2}{5}\Theta_1\right) \approx \frac{4}{9}\frac{k}{\Gamma}\Theta_1. \tag{B.73}$$

In the second equality, we have dropped the Θ_3 term and the time dependence of the quadrupole, since both are suppressed for $\Gamma \gg k$. The photon anisotropic stress and bulk velocity are therefore related as

$$\Pi_\gamma = -\frac{4}{9}\frac{k}{\Gamma}v_\gamma. \tag{B.74}$$

The factor of $4/9$ becomes $8/15$ when polarization corrections to the scattering are included.

B.3.2 Baryons

We see from (B.72) that the evolution of the photons depends on the baryon velocity v_b. To obtain the evolution equation for v_b, we apply momentum conservation to the coupled photon–baryon system:

$$q'_i + 4\mathcal{H}q_i = -\partial_i\delta P - \partial_j\Pi^j{}_i - (\bar{\rho} + \bar{P})\partial_i\Psi. \tag{B.75}$$

Ignoring the baryon pressure, we have

$$q_i \equiv i\hat{k}_i\left(\frac{4}{3}\bar{\rho}_\gamma v_\gamma + \bar{\rho}_b v_b\right), \quad \delta P \approx \frac{1}{3}\bar{\rho}_\gamma\delta_\gamma, \quad \Pi_{ij} = -\frac{4}{3}\bar{\rho}_\gamma\hat{k}_{\langle i}\hat{k}_{j\rangle}\Pi_\gamma, \tag{B.76}$$

and, eliminating v'_γ using (B.72), we find

$$v'_b + \mathcal{H}v_b + k\Psi = \frac{\Gamma}{R}(v_\gamma - v_b), \tag{B.77}$$

where $R \equiv (3/4)\bar{\rho}_b/\bar{\rho}_\gamma$. In Chapter 7, we combined the Euler equations (B.77) and (B.72) to study the evolution of the photon–baryon system as an expansion in $k/\Gamma \ll 1$.

B.3.3 Neutrinos

Neutrinos play an important role during the radiation-dominated era: about 40% of the background density is in neutrinos and perturbations in the cosmic neutrino background leave a subtle imprint in the CMB anisotropies. We can write the perturbed neutrino distribution function as

$$f_\nu \equiv \bar{f}_\nu - \frac{\mathrm{d}\bar{f}_\nu}{\mathrm{d}\ln\epsilon_\nu}\mathcal{N}\,, \tag{B.78}$$

where $\mathcal{N}$ plays the same role as Θ for photons. Below $1\,\mathrm{MeV}$, neutrinos are decoupled from the rest of the plasma and therefore satisfy the collisionless Boltzmann equation. The moments $\mathcal{N}_0$ and $\mathcal{N}_1$ then obey (B.69) and (B.70) without the collision term:

$$\mathcal{N}_0' = -k\mathcal{N}_1 + \Phi'\,, \tag{B.79}$$

$$\mathcal{N}_1' = \frac{k}{3}\left(\mathcal{N}_0 - 2\mathcal{N}_2 + \Psi\right)\,. \tag{B.80}$$

This describes the evolution of the neutrino density contrast $\delta_\nu = 4\mathcal{N}_0$ and the neutrino velocity $v_\nu = -3\mathcal{N}_1$. The perturbations in the neutrino distribution function are mediated to the rest of the plasma through the metric perturbations Φ and Ψ.

Because neutrinos are free streaming, they can develop significant anisotropic stress. The $l = 2$ moment of the Boltzmann equation is

$$\mathcal{N}_2' = \frac{k}{5}\left(2\mathcal{N}_1 - 3\mathcal{N}_3\right)\,. \tag{B.81}$$

Neglecting the $\mathcal{N}_3$ term—which is suppressed by a factor of $k\eta$—the evolution equation for the neutrino anisotropic stress $\Pi_\nu = 3\mathcal{N}_2$ becomes

$$\Pi_\nu' \approx -\frac{2}{5}kv_\nu\,. \tag{B.82}$$

This result was used in Problem 6.2.

B.3.4 Einstein Equations

To close the system of equations, we must specify the Einstein equations for the metric perturbations. The Poisson equation (6.111) is sourced by the density perturbations (or $l = 0$ moments of the distribution functions) for all species:

$$\nabla^2\Phi + 3\mathcal{H}(\Phi' + \mathcal{H}\Psi) = 4\pi G a^2\Big[\underbrace{\bar{\rho}_c\delta_c + \bar{\rho}_b\delta_b}_{\bar{\rho}_m\delta_m} + \underbrace{4\bar{\rho}_\gamma\Theta_0 + 4\bar{\rho}_\nu\mathcal{N}_0}_{\bar{\rho}_r\delta_r}\Big]\,, \tag{B.83}$$

while the constraint equation (6.113) depends on the $l = 2$ moments:

$$\nabla^2(\Phi - \Psi) = -32\pi Ga^2 \left[\bar{\rho}_\gamma \Theta_2 + \bar{\rho}_\nu \mathcal{N}_2\right]. \tag{B.84}$$

The two potentials are equal and opposite unless the photons and neutrinos have appreciable quadrupole moments. We have seen that the photon quadrupole Θ_2 is suppressed due to the tight coupling to the baryons. Neutrinos, on the other hand, are purely free streaming and can therefore have a large quadrupole moment $\mathcal{N}_2$. To determine the evolution of $\mathcal{N}_2$ requires solving the hierarchy of Boltzmann equations to higher order.

B.3.5 Boltzmann Codes

We have found that the moments of the photon temperature fluctuations satisfy a hierarchy of coupled Boltzmann equations:

$$\Theta_l' + \frac{k}{2l+1}\left((l+1)\Theta_{l+1} - l\Theta_{l-1}\right) - \delta_{l0}\Phi' - \delta_{l1}\frac{k\Psi}{3}$$
$$= -\Gamma\left[(1 - \delta_{l0})\Theta_l + \delta_{l1}\frac{v_b}{3} - \frac{1}{10}\delta_{l2}\Theta_2\right]. \tag{B.85}$$

These equations must be solved together with the Euler equation for the baryon velocity v_b and the Einstein equations for the metric potentials Φ and Ψ. The first CMB codes (e.g. COSMICS [2]) solved these coupled equations directly. Unfortunately, this is very inefficient and the codes took hours to run. Modern codes—such as CMBFAST [3], CAMB [4] and CLASS [5]—truncate the Boltzmann hierarchy at a relatively low order and solve for the low-l moments that are needed in the evolution of other fluid species and the metric perturbations. The higher moments of the photon distribution are then found from the following integral solution:

$$\Theta_l(\eta) = \int_0^\eta d\eta \left(\left[g(\Theta_0 + \Psi) + e^{-\tau}(\Phi' + \Psi')\right]j_l(k\chi) \right.$$
$$\left. + g\left[\Theta_1 j_l'(k\chi) - \frac{1}{20}\Theta_2(j_l + 3j_l'')(k\chi)\right]\right), \tag{B.86}$$

where $\chi(\eta) \equiv \eta_0 - \eta$. This turns out to be much more efficient than solving the full hierarchy and allows CMB codes now to run in seconds.

B.4 CMB Polarization

As we discussed in Section 7.6, an important property of the CMB is its polarization. The polarized signal can also be derived from the Boltzmann equation, but the analysis is a bit more involved than for the temperature anisotropies. I will therefore present the main results without detailed derivations, following closely the treatment in Dodelson's book [6]. Further details can be found in Challinor's lecture notes [1].

In Section 7.6.4, we showed that density perturbations only produce E-mode polarization. For a wavevector along the x-direction this corresponds to a non-vanishing Stokes Q parameter. We will denote the corresponding polarization strength by Θ_P. We would like to predict the signal from a Boltzmann equation for Θ_P.

Boltzmann equation

The left-hand side of the Boltzmann equation describes the free streaming of the photons and is therefore the same for Θ_P as for Θ, cf. (B.61). Dodelson gives a heuristic explanation for the right-hand side of the Boltzmann: First, we recall from (7.200) that the polarization strength generated by a plane wave perturbation is proportional to $(1-\mu^2)\Theta_2$, where μ is the cosine of the angle between the wavevector and the direction of the outgoing photon. The constant of proportionality must involve the number of scatterings per unit conformal time, i.e. it must contain the interaction rate Γ. The source term in the Boltzmann equation therefore is $\alpha\,\Gamma(1-\mu^2)\Theta_2$, where α is an order-one numerical factor. Without this source term, multiple scatterings would gradually unpolarize the radiation. This is captured by a loss term proportional to Θ_P. Combing the source and loss terms, the Boltzmann equation then reads

$$\Theta_P' + ik\mu\Theta_P = \Gamma\left[\alpha(1-\mu^2)\Theta_2 - \Theta_P\right]. \tag{B.87}$$

This simplified discussion has ignored the polarization of the incoming radiation. Taking this into account, Bond and Efstathiou derived the following form of the Boltzmann equation [7]:

$$\Theta_P' + ik\mu\Theta_P = \Gamma\left[\frac{3}{4}(1-\mu^2)\Sigma - \Theta_P\right], \tag{B.88}$$

where $\Sigma \equiv \Theta_2 + \Theta_{P0} + \Theta_{P2}$, with Θ_{P0} and Θ_{P2} being the $l = 0$ and $l = 2$ moments of Θ_P.

In the tight-coupling limit, the source term Σ can be written in terms of Θ_2 alone. To see this, we note that the terms on the right-hand side of (B.88) must cancel for large Γ, so that

$$\Theta_P = \frac{1}{2}[1 - P_2(\mu^2)](\Theta_2 + \Theta_{P0} + \Theta_{P2}). \tag{B.89}$$

Writing $\Theta_P(\mu)$ in a series of Legendre polynomials and equating the coefficients of P_0 and P_2 on both sides gives

$$\begin{aligned}2\Theta_{P0} &= \Theta_2 + \Theta_{P0} + \Theta_{P2} \\ 10\Theta_{P2} &= \Theta_2 + \Theta_{P0} + \Theta_{P2}\end{aligned} \quad \Rightarrow \quad \Theta_{P0} = 5\Theta_{P2} = 5\Theta_2/4. \tag{B.90}$$

Hence, we get $\Sigma = 5\Theta_2/2$.

Line-of-sight solution

Following the same procedure as in Section B.2.1, the formal solution of the Boltzmann equation (B.88) can be written as

$$\Theta_P(\mathbf{k}, \mu) = \frac{3}{4}(1 - \mu^2) \int_0^{\eta_0} d\eta \, g(\eta) \, e^{ik(\eta_0 - \eta)\mu} \, \Sigma(\eta, \mathbf{k}) \,, \tag{B.91}$$

where $g(\eta)$ is the visibility function and we have switched to $\mu \equiv \hat{\mathbf{k}} \cdot \hat{\mathbf{n}}$. Except for the rapidly changing visibility function, all terms in the integrand can be evaluated at the decoupling time η_*. These terms can then be pulled out of the integral and the remaining integral over $g(\eta)$ is unity. We therefore get

$$\Theta_P(\mathbf{k}, \mu) \approx \frac{3}{4}\Sigma(\eta_*, \mathbf{k})(1 - \mu^2)e^{ik\eta_0\mu} \,, \tag{B.92}$$

where we have used $\eta_0 - \eta_* \approx \eta_0$ in the exponential.

E-mode power spectrum

To derive the E-mode power spectrum, we have to extract the multipole moments from the line-of-sight solution and then integrate them against the power spectrum of the initial conditions. My treatment will be very schematic, but the details can be found in [8].

Recall that the line-of-sight solution for the temperature fluctuations can be written as an expansion in terms of the $m = 0$ modes of the spherical harmonics (or the Legendre polynomials):

$$\Theta(\mathbf{k}, \mu) = \sum_l i^l \sqrt{4\pi(2l + 1)} \, \Theta_l(\mathbf{k}) \, Y_{l,0}(\hat{\mathbf{n}})$$

$$= \sum i^l (2l + 1) \, \Theta_l(\mathbf{k}) \, P_l(\mu) \,. \tag{B.93}$$

The fluctuations in the polarization field must instead be expanded in terms of the $m = 0$ modes of the spin-2 spherical harmonics (or the associated Legendre polynomials) [8]:

$$\Theta_P(\mathbf{k}, \mu) = \sum_l i^l \sqrt{4\pi(2l + 1)} \, \Theta_{Pl}(\mathbf{k}) \, _{\pm 2}Y_{l,0}(\hat{\mathbf{n}})$$

$$= \sum i^l (2l + 1) \, \Theta_{Pl}(\mathbf{k}) \sqrt{\frac{(l - 2)!}{(l + 2)!}} P_l^2(\mu) \,. \tag{B.94}$$

To compare this to our solution (B.92), we note that

$$e^{ik\eta_0\mu} = -\frac{1}{(k\eta_0)^2} \frac{d^2}{d\mu^2} e^{ik\eta_0\mu} = -\sum i^l (2l + 1) \frac{j_l(k\eta_0)}{(k\eta_0)^2} \frac{d^2}{d\mu^2} P_l(\mu)$$

$$= -\sum_l i^l (2l + 1) \frac{j_l(k\eta_0)}{(k\eta_0)^2} \frac{P_l^2(\mu)}{1 - \mu^2} \,, \tag{B.95}$$

so that

$$\Theta_P(\mathbf{k}, \mu) \approx \frac{3}{4}\Sigma(\eta_*, \mathbf{k})(1 - \mu^2)e^{ik\eta_0\mu}$$

$$= -\frac{3}{4}\Sigma(\eta_*, \mathbf{k}) \sum_l i^l(2l+1) \frac{j_l(k\eta_0)}{(k\eta_0)^2} P_l^2(\mu). \qquad (\text{B.96})$$

Equation (B.94) then implies that

$$E_l(\mathbf{k}) \equiv -\Theta_{Pl}(\mathbf{k}) = \frac{3}{4}\Sigma(\eta_*, \mathbf{k})\sqrt{\frac{(l+2)!}{(l-2)!}} \frac{j_l(k\eta_0)}{(k\eta_0)^2}, \qquad (\text{B.97})$$

where we introduced the overall sign to be consistent with the definition of a_{lm}^E in (7.175). In the tight-coupling approximation, we have $\Sigma \approx 5\Theta_2/2$, where the quadrupole moment is related to the dipole Θ_1 by (B.73). Combining these relations, we get the E-mode transfer function

$$\Theta_{E,l}(k) \equiv -\frac{E_l(\mathbf{k})}{\mathcal{R}_i(\mathbf{k})} \approx \frac{5}{6}\frac{k}{\Gamma(\eta_*)}\frac{\Theta_1(\eta_*, \mathbf{k})}{\mathcal{R}_i(\mathbf{k})}\frac{l^2}{(k\eta_0)^2} j_l(k\eta_0). \qquad (\text{B.98})$$

The E-mode spectrum and its cross correlation with the temperature anisotropies then are

$$C_l^{EE} = 4\pi \int_0^\infty \mathrm{d}\ln k \; \Theta_{E,l}^2(k) \, \Delta_{\mathcal{R}}^2(k),$$

$$C_l^{TE} = 4\pi \int_0^\infty \mathrm{d}\ln k \; \Theta_l(k)\Theta_{E,l}(k) \, \Delta_{\mathcal{R}}^2(k). \qquad (\text{B.99})$$

Measurements of these power spectra are shown in Chapter 7 (see Fig. 7.30).

C Useful Quantities and Relations

In this appendix, I collect important quantities and relations that are used frequently in cosmology. Unless stated otherwise, the values of all measured parameters are those presented by the Particle Data Group. I will use the notation that figures in parentheses give the 1σ-uncertainties in the last digit(s) of a measured quantity. For example, $6.674\,30(15)$ is to be read as $6.674\,30 \pm 0.000\,15$.

C.1 Units and Conversions

All physical quantities are described by three fundamental units, which we usually take to be the units of mass, length and time. This defines the International System of Units (SI). All other units are derived from these. For example, the unit of force is $\mathrm{kg\,m\,s^{-2}}$ (remember: $F = ma$). We sometimes gives these derived units new names, like the "Newton" for a force, but this doesn't change the essential fact that everything is a combination of mass, length and time. There have been a few unlucky historical accidents: temperature was given its own unit ("Kelvin") and then Boltzmann's constant k_B was introduced to relate temperature to energy. In hindsight, it really makes sense to set $k_\mathrm{B} = 1$ and measure temperature in units of energy. Similarly, electric charge has its own unit ("Coulomb") and the electric permittivity ε_0 was introduced to relate charge to force or energy. It makes sense to set $\varepsilon_0 \equiv 1$ and measure charge in terms of the "electrostatic unit" (esu) which has dimensions of $[\mathrm{mass}]^{1/2}[\mathrm{length}]^{3/2}[\mathrm{time}]^{-1}$.

Meters, kilograms and seconds are convenient units in everyday life, but they stop being a natural set of units in cosmology and particle physics. I showed in Chapter 1 that the enormous length scales relevant to the cosmos are best described in "light-years" or "parsec," and that the unit of choice in cosmology is Mpc.

The microscopic realm is governed by relativity and quantum mechanics. The inadequacy of everyday units in this situation is illustrated by the fact that the fundamental constants of these theories—the speed of light c and Planck's constant h—appear with very large exponents:

$$c = 2.997\,924\,58 \times 10^8 \,\mathrm{m\,s^{-1}},$$
$$h = 6.626\,070\,15 \times 10^{-34} \,\mathrm{J\,s}. \tag{C.1}$$

To describe microscopic phenomena, it is therefore more natural to measure velocities in units of the speed of light c and actions in units of the (reduced) Planck constant:

$$\hbar \equiv \frac{h}{2\pi} = 1.054\,571\,817\ldots \times 10^{-34}\,\mathrm{J\,s}\,. \tag{C.2}$$

Sometimes, this is expressed by setting the speed of light and Planck's constant to unity:

$$c = \hbar \equiv 1\,, \tag{C.3}$$

but it really just means extracting factors of c and $\hbar$ from all quantities and then not writing them. The third independent unit is usually taken to be the unit of energy, which in high-energy physics is chosen to be the "electron volt":

$$1\,\mathrm{eV} = 1.602\,177\,33(49) \times 10^{-19}\,\mathrm{J}\,. \tag{C.4}$$

We use eV in atomic physics, MeV in nuclear physics and GeV in particle physics.

Instead of the units of mass, length and time, we now use velocity, action and energy as the building blocks for the system of units. The resulting units are called **natural units** because that's what they are. A quantity with the SI units $\mathrm{kg}^\alpha \mathrm{m}^\beta \mathrm{s}^\gamma$ in natural units becomes

$$\left(\frac{E}{c^2}\right)^\alpha \left(\frac{\hbar c}{E}\right)^\beta \left(\frac{\hbar}{E}\right)^\gamma = E^{\alpha-\beta-\gamma}\,\hbar^{\beta+\gamma}\,c^{\beta-2\alpha}\,. \tag{C.5}$$

Suppressing the factors of c and $\hbar$ means that everything is measured in units of energy to some power. A few examples are:

$$
\begin{aligned}
[\text{mass}] &= E\,c^{-2} \\
[\text{time}] &= E^{-1}\,\hbar \\
[\text{length}] &= E^{-1}\,\hbar c \\
[\text{momentum}] &= E\,c^{-1} \\
[\text{force}] &= E^2\,(\hbar c)^{-1} \\
[\text{pressure}] &= E^4\,(\hbar c)^{-3} \\
[\text{energy density}] &= E^4\,(\hbar c)^{-3} \\
[\text{charge}] &= (\hbar c)^{1/2} \\
[\text{magnetic field}] &= E^2\,(\hbar c)^{-3/2}
\end{aligned}
\tag{C.6}
$$

To perform conversions from SI units to natural units, it is helpful to know that

$$
\begin{aligned}
1\,\mathrm{kg} &= 5.609\,586\,16(17) \times 10^{35}\,\mathrm{eV}\,c^{-2}\,, \\
1\,\mathrm{m} &= 5.067\,728\,86(15) \times 10^{6}\,\mathrm{eV}^{-1}\,(\hbar c)\,, \\
1\,\mathrm{s} &= 1.519\,266\,89(46) \times 10^{15}\,\mathrm{eV}^{-1}\,\hbar\,.
\end{aligned}
\tag{C.7}
$$

Other useful conversion factors are

$$\hbar c = 197.327\,053(59)\,\mathrm{MeV\,fm}\,,$$
$$\hbar = 6.582\,122\,0(20) \times 10^{-22}\,\mathrm{MeV\,s}\,, \qquad (\mathrm{C.8})$$
$$e^2 = [137.035\,999\,084(21)]^{-1}\,(\hbar c)\,,$$

where $\mathrm{fm} = 10^{-15}\,\mathrm{m}$.

The strength of the gravitational force is determined by Newton's constant:

$$
\begin{aligned}
G &= 6.674\,30(15) \times 10^{-11}\,\mathrm{m^3\,kg^{-1}\,s^{-2}} \\
&= 6.708\,72(10) \times 10^{-39}\,\mathrm{GeV^{-2}}\,\hbar c^5\,.
\end{aligned}
\qquad (\mathrm{C.9})
$$

This can be used to define the Planck scale as the characteristic energy scale at which the effects of quantum gravity are expected to become important:

$$
\begin{aligned}
E_{\mathrm{Pl}} \equiv \sqrt{\frac{\hbar c^5}{G}} &= 1.220\,890(14) \times 10^{19}\,\mathrm{GeV} \\
&= 1.956\,082(22) \times 10^{9}\,\mathrm{J}\,.
\end{aligned}
\qquad (\mathrm{C.10})
$$

Measuring all energies in units of the Planck energy—or equivalently setting $G \equiv 1$—reduces all physical quantities to dimensionless numbers. These **Planck units** are often used in inflationary calculations (see Chapters 4 and 8).

C.2 Constants and Parameters

In this section, I collect some physical constants and parameters that play an important role in cosmology.

C.2.1 Physical Constants

The table below presents the values of important physical constants. Note that the speed of light c, Planck's constant h, Boltzmann's constant k_{B} and the elementary electric charge e have defined values and are therefore exact. For the remaining constants, the figures in parentheses give the 1σ-uncertainties in the last digit(s).

Quantity	Symbol	Value
speed of light in vacuum	c	$299\,792\,458\,\mathrm{m\,s^{-1}}$
Planck's constant	h	$6.626\,070\,15 \times 10^{-34}\,\mathrm{J\,s}$
Planck's constant, reduced	$\hbar \equiv h/2\pi$	$1.054\,571\,817\ldots \times 10^{-34}\,\mathrm{J\,s}$
Boltzmann's constant	k_{B}	$1.380\,649 \times 10^{-23}\,\mathrm{J\,K^{-1}}$
Avogadro's constant	N_{A}	$6.022\,140\,76 \times 10^{23}\,\mathrm{mol^{-1}}$
Newton's constant	G	$6.674\,30(15) \times 10^{-11}\,\mathrm{m^3 kg^{-1} s^{-2}}$
Planck mass	$m_{\mathrm{Pl}} \equiv \sqrt{\hbar c/G}$	$1.220\,890(14) \times 10^{19}\,\mathrm{GeV}$
Planck mass, reduced	$M_{\mathrm{Pl}} \equiv \sqrt{\hbar c/8\pi G}$	$2.435\,323(48) \times 10^{18}\,\mathrm{GeV}$
Planck length	$l_{\mathrm{Pl}} \equiv \sqrt{\hbar G/c^3}$	$1.616\,255(18) \times 10^{-35}\,\mathrm{m}$

Planck time	$t_{\mathrm{Pl}} \equiv \sqrt{\hbar G/c^5}$	$5.391\,246(27) \times 10^{-44}\,\mathrm{s}$
elementary charge	e	$1.602\,176\,634 \times 10^{-19}\,\mathrm{C}$
electric permittivity	ε_0	$8.854\,187\,812\,8(13) \times 10^{-12}\,\mathrm{Fm}^{-1}$
fine-structure constant	$\alpha \equiv e^2/(4\pi\varepsilon_0\,\hbar c)$	$1/137.035\,999\,084(21)$
strong coupling constant	$\alpha_s(m_Z)$	$0.1179(10)$
classical electron radius	$r_e \equiv e^2/(4\pi\varepsilon_0 m_e c^2)$	$2.817\,940\,326\,2(13) \times 10^{-15}\,\mathrm{m}$
Bohr radius	$a_0 \equiv r_e\alpha^{-2}$	$0.529\,177\,210\,903(80) \times 10^{-10}\,\mathrm{m}$

C.2.2 Particle Physics

Quantity	Symbol	Value
electron mass	m_e	$0.510\,998\,950\,00(15)\,\mathrm{MeV}$
muon mass	m_μ	$105.658\,375\,5(23)\,\mathrm{MeV}$
proton mass	m_p	$938.272\,088\,16(29)\,\mathrm{MeV}$
neutron mass	m_n	$939.565\,420\,52(54)\,\mathrm{MeV}$
proton–neutron mass difference	Q	$1.293\,332\,36(46)\,\mathrm{MeV}$
neutron lifetime	τ_n	$879.4(6)\,\mathrm{s}$
deuteron mass	m_{D}	$1875.612\,942\,57(57)\,\mathrm{MeV}$
$W^\pm$ boson mass	m_W	$80.379(12)\,\mathrm{GeV}$
Z^0 boson mass	m_Z	$91.187\,6(21)\,\mathrm{GeV}$
Higgs boson mass	m_H	$125.10(14)\,\mathrm{GeV}$
Thomson cross section	σ_T	$0.665\,245\,873\,21(60) \times 10^{-28}\,\mathrm{m}^2$
Fermi's constant	G_F	$1.116\,378\,7(6) \times 10^{-5}\,\mathrm{GeV}^{-2}$

C.2.3 Astrophysics and Cosmology

The table below lists important astrophysical and cosmological parameters, as well as their measured values. Note that the precise values for the cosmological parameters depend on the data sets that are being used. I will cite the values adopted by the Particle Data Group, which are derived from a 6-parameter ΛCDM cosmology fit to the Planck 2018 temperature, polarization and lensing data.

Quantity	Symbol	Value
sidereal year	yr	$31\,558\,149.8\,\mathrm{s} \approx \pi \times 10^7\,\mathrm{s}$
astronomical unit	au	$149\,597\,870\,700\,\mathrm{m}$
parsec	pc	$3.085\,677\,581\,49 \times 10^{16}\,\mathrm{m} = 3.261\,56\ldots\,\mathrm{ly}$
light-year	ly	$0.306\,601\ldots\,\mathrm{pc} = 0.946\,073\ldots \times 10^{16}\,\mathrm{m}$
Solar mass	$M_\odot$	$1.988\,41(4) \times 10^{30}\,\mathrm{kg}$
CMB temperature, today	T_0	$2.7255(6)\,\mathrm{K} \approx 0.235\,\mathrm{meV}$
CMB dipole amplitude, today	d_0	$3.3621(10)\,\mathrm{mK}$
Solar velocity with respect to CMB	$v_\odot$	$369.82(11)\,\mathrm{km\,s}^{-1}$
Local Group velocity	v_{LG}	$620(15)\,\mathrm{km\,s}^{-1}$
number density of CMB photons	n_γ	$410.7(3)\,\mathrm{cm}^{-3}$
energy density of CMB photons	ρ_γ	$4.645(4) \times 10^{-34}\,\mathrm{g\,cm}^{-3} \approx 0.260\,\mathrm{eV\,cm}^{-3}$
dimensionless CMB density	Ω_γ	$2.473 \times 10^{-5}h^{-2} = 5.38(15) \times 10^{-5}$

Hubble expansion rate, today	H_0	$100\,h\,\mathrm{km\,s^{-1}\,Mpc^{-1}}$
		$= h \times (9.777\,\mathrm{Gyr})^{-1}$
		$= h \times 2.133 \times 10^{-33}\,\mathrm{eV}$
Hubble parameter, dimensionless	h	$0.674(5)$
Hubble length	c/H_0	$1.372(10) \times 10^{26}\,\mathrm{m} = 4\,446(32)\,\mathrm{Mpc}$
Hubble time	$1/H_0$	$14.507(23)\,\mathrm{Gyr}$
critical density of the universe	ρ_{crit}	$8.532\,85(2) \times 10^{-30}\,\mathrm{g\,cm^{-3}}$
		$= 4.786\,58(11) \times 10^{-5}\,\mathrm{(GeV}/c^2)\,\mathrm{cm^{-3}}$
		$= 1.260\,780 \times 10^{11}\,M_\odot\,\mathrm{Mpc^{-3}}$
baryon-to-photon ratio	η	$5.8 \times 10^{-10} \le \eta \le 6.5 \times 10^{-10}$
number density of baryons	n_b	$2.515(17) \times 10^{-7}\,\mathrm{cm^{-3}}$
baryon density of the universe	Ω_b	$0.0493(6)$
dark matter density parameter	Ω_c	$0.265(7)$
matter density parameter	Ω_m	$0.315(7)$
dark energy density parameter	Ω_Λ	$0.685(7)$
energy density of dark energy	ρ_Λ	$5.83(16) \times 10^{-30}\,\mathrm{g\,cm^{-3}}$
cosmological constant	Λ	$1.088(30) \times 10^{-56}\,\mathrm{cm^{-2}}$
dark energy equation of state	w	$-1.028(31)$
sum of neutrino masses	$\sum m_\nu$	$0.06\,\mathrm{eV} < \sum m_\nu < 0.12\,\mathrm{eV}$
effective number of neutrinos	N_{eff}	$2.99(17)$
neutrino density of the universe	Ω_ν	$0.0012 < \Omega_\nu < 0.003$
curvature parameter	Ω_k	$|\Omega_k| < 0.005$
primordial helium fraction	Y_P	$0.2453(34)$
reionization optical depth	τ	$0.054(7)$
fluctuation amplitude at $8h^{-1}\mathrm{Mpc}$	σ_8	$0.811(6)$
amplitude of scalar fluctuations	A_s	$2.098(23) \times 10^{-9}$
scalar spectral index	n_s	$0.965(4)$
tensor-to-scalar ratio	r	< 0.035
redshift of matter–radiation equality	z_{eq}	$3\,402(26)$
redshift of photon decoupling	z_*	$1089.92(25)$
redshift at half reionization	z_{re}	$7.7(7)$
redshift when acceleration was zero	z_q	$0.636(18)$
age at matter–radiation equality	t_{eq}	$51.1(8)\,\mathrm{kyr}$
age at photon decoupling	t_*	$372.9(10)\,\mathrm{kyr}$
age at half reionization	t_{re}	$690(90)\,\mathrm{Myr}$
age when acceleration was zero	t_q	$7.70(10)\,\mathrm{Gyr}$
age of the universe today	t_0	$13.797(23)\,\mathrm{Gyr}$

C.3 Important Relations

In the following, I collect a few relations that appear frequently in cosmological calculations. The presentation was inspired by an appendix in Wayne Hu's thesis.

C.3.1 Timescales

Four different quantities are used to measure time in cosmology: the scale factor a, the redshift z, coordinate time t and conformal time η.

The scale factor and redshift at matter–radiation equality are

$$\frac{a_{\rm eq}}{a_0} = \frac{\Omega_r}{\Omega_m - \Omega_r} \approx 2.95 \times 10^{-4} \left(\frac{\Omega_m}{0.32}\right)^{-1}, \tag{C.11}$$

$$z_{\rm eq} = \frac{a_0}{a_{\rm eq}} - 1 \approx 3395 \left(\frac{\Omega_m}{0.32}\right), \tag{C.12}$$

while the equality of matter and dark energy is at

$$\frac{a_{\Lambda m}}{a_0} = \left(\frac{\Omega_m}{\Omega_\Lambda}\right)^{1/3} \approx 0.77 \left(\frac{\Omega_m}{0.32}\right)^{1/3} \left(\frac{0.68}{\Omega_\Lambda}\right)^{1/3}, \tag{C.13}$$

$$z_{\Lambda m} = \left(\frac{\Omega_\Lambda}{\Omega_m}\right)^{1/3} - 1 \approx 1.30 \left(\frac{0.32}{\Omega_m}\right)^{1/3} \left(\frac{\Omega_\Lambda}{0.68}\right)^{1/3} - 1. \tag{C.14}$$

During the matter and radiation eras, we have

$$\eta = 2H_0^{-1}\Omega_m^{-1/2}(a_{\rm eq}/a_0) \left(\sqrt{1 + a/a_{\rm eq}} - 1\right), \tag{C.15}$$

$$t = \frac{2}{3}H_0^{-1}\Omega_m^{-1/2}(a_{\rm eq}/a_0)^{3/2} \left(2 - \sqrt{1 + a/a_{\rm eq}}\,(2 - a/a_{\rm eq})\right), \tag{C.16}$$

which in the appropriate limits become

$$\eta = H_0^{-1} \begin{cases} \Omega_r^{-1/2}(a/a_0) & {\rm RD}, \\ 2\Omega_m^{-1/2}(a/a_0)^{1/2} & {\rm MD}, \end{cases} \tag{C.17}$$

$$t = H_0^{-1} \begin{cases} \frac{1}{2}\Omega_r^{-1/2}(a/a_0)^2 & {\rm RD}, \\ \frac{2}{3}\Omega_m^{-1/2}(a/a_0)^{3/2} & {\rm MD}. \end{cases} \tag{C.18}$$

Assuming spatial flatness, the age of the universe is

$$\eta_0 \approx 2H_0^{-1}\Omega_m^{-1/2}(1 + \ln \Omega_m^{0.085}), \tag{C.19}$$

$$t_0 = \frac{2}{3}H_0^{-1}(1 - \Omega_m)^{-1/2} \ln \left(\frac{1 + \sqrt{1 - \Omega_m}}{\sqrt{\Omega_m}}\right), \tag{C.20}$$

where $\Omega_m + \Omega_\Lambda = 1$.

C.3.2 Length Scales

Throughout this book, special attention has been drawn to the important role played by the Hubble radius H^{-1}, or $(aH)^{-1}$ in comoving coordinates. The wavenumber corresponding to the Hubble scale is

$$
k_H \equiv aH = a_0 H_0 \begin{cases} \sqrt{\Omega_r}(a_0/a) & \text{RD}, \\ \sqrt{\Omega_m}(a_0/a)^{1/2} & \text{MD}, \\ \sqrt{\Omega_\Lambda}(a/a_0) & \Lambda\text{D}, \end{cases} \tag{C.21}
$$

with the critical wavenumber at matter–radiation equality being

$$
k_{\rm eq} \equiv (aH)_{\rm eq} = a_0 H_0 \sqrt{2\Omega_m}(a_0/a_{\rm eq})^{1/2}
$$
$$
= 0.01 \left(\frac{\Omega_m}{0.32} \right) \text{Mpc}^{-1}. \tag{C.22}
$$

The latter is an important scale in structure formation, since modes with $k < k_{\rm eq}$ behave differently from those with $k > k_{\rm eq}$. The measured value of the equality scale is $k_{\rm eq} = 0.010\,384(81)\,\text{Mpc}^{-1}$.

The sound horizon before recombination is

$$
r_s(\eta) = \frac{2}{3k_{\rm eq}} \sqrt{\frac{6}{R_{\rm eq}}} \ln \left(\frac{\sqrt{1+R(\eta)} + \sqrt{R(\eta)+R_{\rm eq}}}{1 + \sqrt{R_{\rm eq}}} \right), \tag{C.23}
$$

where

$$
R \equiv \frac{3}{4} \frac{\rho_b}{\rho_\gamma} = 0.6120 \left(\frac{\Omega_b}{0.049} \right) \frac{1090}{1+z}. \tag{C.24}
$$

The sound horizon at decoupling is the characteristic scale imprinted in the anisotropies of the CMB. Its measured value is $r_s(\eta_*) = 144.43(26)\,\text{Mpc}$.

The mean free path of photons before recombination is

$$
\ell_\gamma = (X_e n_e \sigma_T a/a_0)^{-1}
$$
$$
= 18.9 \left(\frac{X_e}{0.1} \right)^{-1} \left(\frac{1-\frac{1}{2}Y_P}{0.878} \right)^{-1} \left(\frac{\Omega_b}{0.049} \right)^{-1} \left(\frac{1+z}{1090} \right)^{-2} \text{Mpc}, \tag{C.25}
$$

where X_e is the free-electron fraction (see Chapter 3). The diffusion length is roughly the geometric mean of ℓ_γ and the horizon η. More precisely, the Silk damping scale is

$$
k_S^{-2}(\eta) \equiv \int_0^\eta \frac{d\tilde{\eta}}{6(1+R)\Gamma(\tilde{\eta})} \left(\frac{16}{15} + \frac{R^2}{1+R} \right), \tag{C.26}
$$

where $\Gamma(\eta) \equiv n_e \sigma_T a$ is the interaction rate. The effective damping scale arises from a combination of Silk and Landau damping, $k_D^{-2} = k_S^{-2} + k_L^{-2}$, with the latter being a consequence of the finite thickness of the surface of last-scattering. The fluctuations in the photon–baryon fluid are exponentially suppressed below the damping scale (see Chapter 7). The measured damping scale is $k_D = 0.140\,87(30)\,\text{Mpc}^{-1}$.

D Special Functions

A number of special functions make an appearance in cosmology. In this appendix, I will collect them and describe their essential properties.

D.1 Fourier Transforms

Since the universe is statistically homogeneous and isotropic, it is convenient to expand small fluctuations in terms of eigenfunctions of the spatial Laplacian ∇^2. Our convention for these Fourier transforms is

$$f(\mathbf{x}) = \int \frac{\mathrm{d}^3 k}{(2\pi)^3} f(\mathbf{k}) \, e^{i\mathbf{k}\cdot\mathbf{x}} \,, \tag{D.1}$$

$$f(\mathbf{k}) = \int \mathrm{d}^3 x \, f(\mathbf{x}) \, e^{-i\mathbf{k}\cdot\mathbf{x}} \,. \tag{D.2}$$

If the field $f(\mathbf{x})$ is real, then its Fourier components must satisfy $f^*(\mathbf{k}) = f(-\mathbf{k})$ and the power spectrum is

$$\langle f(\mathbf{k}) f^*(\mathbf{k}') \rangle = (2\pi)^3 \delta_{\mathrm{D}}(\mathbf{k} - \mathbf{k}') \, \mathcal{P}_f(k) \,. \tag{D.3}$$

The inverse Fourier transform of the power spectrum is the two-point correlation function in position space:

$$\langle f(\mathbf{x}) f(\mathbf{x}') \rangle = \int \frac{\mathrm{d}k}{k} \frac{k^3}{2\pi^2} \mathcal{P}_f(k) \, j_0(k|\mathbf{x} - \mathbf{x}'|) \,, \tag{D.4}$$

where $j_0(x) \equiv \sin x / x$.

D.2 Spherical Harmonics

Spherical harmonics are eigenfunctions of the spherical Laplacian ∇^2 and of ∂_ϕ:

$$\left[\frac{\partial^2}{\partial\theta^2} + \cot\theta \frac{\partial}{\partial\theta} + \frac{1}{\sin^2\theta} \frac{\partial^2}{\partial\phi^2} \right] Y_{lm} = -l(l+1) Y_{lm} \,,$$
$$\frac{\partial}{\partial\phi} Y_{lm} = im Y_{lm} \,, \tag{D.5}$$

where l is a positive integer and m is an integer with $|m| \leq l$. The lowest harmonics are

$$Y_{00} = \frac{1}{\sqrt{4\pi}}\,, \quad Y_{10} = -\sqrt{\frac{3}{4\pi}}\cos\theta\,, \qquad Y_{20} = \sqrt{\frac{5}{16\pi}}(1 - 3\cos^2\theta)\,,$$

$$Y_{1,\pm1} = \mp i\sqrt{\frac{3}{8\pi}}\sin\theta e^{\pm i\phi}\,, \quad Y_{2,\pm1} = \pm i\sqrt{\frac{15}{8\pi}}\cos\theta\sin\theta e^{\pm i\phi}\,, \quad \text{(D.6)}$$

$$Y_{2,\pm2} = -\sqrt{\frac{15}{32\pi}}\sin^2\theta e^{\pm 2i\phi}\,.$$

The spherical harmonics are an orthonormal basis in the sense that

$$\int d\hat{n}\, Y_{lm}^*(\hat{n})Y_{l'm'}(\hat{n}) = \delta_{ll'}\delta_{mm'}\,. \tag{D.7}$$

Any function on a sphere can therefore be written as

$$f(\hat{n}) = \sum_{lm} f_{lm}Y_{lm}(\hat{n})\,, \tag{D.8}$$

and its multipole moments are

$$f_{lm} = \int d\hat{n}\, f(\hat{n})\, Y_{lm}^*\,. \tag{D.9}$$

If $f(\hat{n})$ is real, then the multipole moments must satisfy $f_{lm}^* = (-1)^m f_{l,-m}$, where we have used that $Y_{lm}^* = (-1)^m Y_{l,-m}$ in our phase convention. The angular power spectrum is defined as

$$\langle f_{lm}f_{l'm'}^* \rangle = C_l\,\delta_{ll'}\delta_{mm'}\,, \tag{D.10}$$

and the two-point correlation function is

$$\langle f(\hat{n})f(\hat{n}') \rangle = \sum_l \frac{2l+1}{4\pi}C_l P_l(\hat{n}\cdot\hat{n}')\,, \tag{D.11}$$

where P_l is the Legendre polynomial.

D.3 Legendre Polynomials

Legendre polynomials are solutions of

$$\left[(1 - \mu^2)\frac{d^2}{d\mu^2} - 2\mu\frac{d}{d\mu}\right]P_l = -l(l+1)P_l\,. \tag{D.12}$$

As the name suggests, the solutions of this differential equation are polynomials in μ. The lowest order ones are

$$P_0(\mu) = 1\,, \quad P_1(\mu) = \mu\,, \quad P_2(\mu) = \frac{1}{2}(3\mu^2 - 1)\,. \tag{D.13}$$

Higher-order solutions can be obtained from the following recursion relation:

$$(l+1)P_{l+1} = (2l+1)\mu P_l - lP_{l-1}.\tag{D.14}$$

The Legendre polynomials are orthogonal on the interval $[-1, 1]$:

$$\int_{-1}^{1} \mathrm{d}\mu\, P_l(\mu)P_{l'}(\mu) = \frac{2}{2+l}\delta_{ll'}.\tag{D.15}$$

Any function defined on the interval can therefore be written as

$$f(\mu) = \sum_l \frac{2l+1}{2} f_l P_l(\mu).\tag{D.16}$$

An important special case is the Rayleigh plane wave expansion:

$$e^{i\mathbf{k}\cdot\mathbf{x}} = \sum_l i^l(2l+1)j_l(kx)P_l(\hat{\mathbf{k}}\cdot\hat{\mathbf{x}}),\tag{D.17}$$

where j_l are spherical Bessel functions.

D.4 Spherical Bessel Functions

Spherical Bessel functions play an important role in the description of the CMB anisotropies where they arise when three-dimensional inhomogeneities are projected onto the two-dimensional surface of last-scattering (see Chapter 7). They satisfy the equation

$$\frac{\mathrm{d}^2 j_l}{\mathrm{d}x^2} + \frac{2}{x}\frac{\mathrm{d}j_l}{\mathrm{d}x} + \left(1 - \frac{l(l+1)}{x^2}\right)j_l = 0.\tag{D.18}$$

The two lowest-order solutions are

$$j_0(x) = \frac{\sin x}{x}, \quad j_1(x) = \frac{\sin x - x\cos x}{x^2},\tag{D.19}$$

and higher-order ones can be obtained by the recursion relation

$$j_{l+1} = \frac{2l+1}{x}j_l - j_{l-1}.\tag{D.20}$$

An important relation between the spherical Bessel functions and Legendre polynomials is

$$\int_{-1}^{1} \mathrm{d}\mu\, P_l(\mu)e^{ix\mu} = \frac{2}{(-i)^l}\, j_l(x).\tag{D.21}$$

The spherical Bessel functions can also be written as

$$j_l(x) = \sqrt{\frac{\pi}{2x}}J_{l+1/2}(x),\tag{D.22}$$

where $J_l(x)$ are the ordinary Bessel functions.

D.5 Bessel and Hankel Functions

Ordinary Bessel functions satisfy

$$\frac{\mathrm{d}^2 J_\nu}{\mathrm{d}z^2} + \frac{1}{z}\frac{\mathrm{d}J_\nu}{\mathrm{d}z} + \left(1 - \frac{\nu^2}{z^2}\right) J_\nu = 0\,. \tag{D.23}$$

A series solution of this equation is

$$J_\nu(z) = \left(\frac{z}{2}\right)^{-\nu} \sum_{n=0}^{\infty} \frac{(-1)^n}{n!\,\Gamma(\nu + n + 1)} \left(\frac{z}{2}\right)^{2n}\,. \tag{D.24}$$

As long as ν is not an integer, then a second linearly independent solution is $J_{-\nu}(z)$. The functions $J_\nu(z)$ and $J_{-\nu}(z)$ are called *Bessel functions of the first kind*. If ν is an integer n, then $J_n(z)$ and $J_{-n}(z)$ are not independent, but related by

$$J_{-n}(z) = (-1)^n J_n(z)\,. \tag{D.25}$$

A function that remains linearly independent of $J_\nu(z)$, even for integer ν, is the *Bessel function of the second kind* (sometimes called the *Neumann function*):

$$Y_\nu(z) = \frac{J_\nu(z)\cos(\nu\pi) - J_{-\nu}(z)}{\sin(\nu\pi)}\,. \tag{D.26}$$

For integer ν, this definition must be understood as the limit $\nu \to n$. The Bessel functions have the following limiting behavior for $z \to 0$:

$$\begin{aligned} J_\nu(z) &\xrightarrow{z\to 0} \frac{1}{\Gamma(\nu+1)}\left(\frac{z}{2}\right)^{\nu}, \\ Y_\nu(z) &\xrightarrow{z\to 0} -\frac{1}{\pi}\Gamma(\nu)\left(\frac{z}{2}\right)^{-\nu}\,. \end{aligned} \tag{D.27}$$

These limits make manifest that the Bessel equation (D.23) has a scaling symmetry for $z \to 0$. We see that $J_\nu(z)$ is regular for small argument, while $Y_\nu(z)$ diverges.

In the limit $z \to \infty$, the Bessel equation exhibits a translation symmetry and should have solutions that scale as $e^{\pm iz}$. This is made more manifest by defining the *Hankel functions* (or *Bessel functions of the third kind*):

$$\begin{aligned} H_\nu^{(1)}(z) &\equiv J_\nu(z) + iY_\nu(z)\,, \\ H_\nu^{(2)}(z) &\equiv J_\nu(z) - iY_\nu(z)\,. \end{aligned} \tag{D.28}$$

We encountered Hankel functions in Chapter 8 when we derived the mode functions for the inflationary fluctuations. The asymptotic limits of the Hankel functions for large arguments are

$$\begin{aligned} H_\nu^{(1)}(z) &\xrightarrow{z\to\infty} \sqrt{\frac{2}{\pi z}}\, e^{i(z - \frac{1}{2}\nu\pi - \frac{1}{4}\pi)}\,, \\ H_\nu^{(2)}(z) &\xrightarrow{z\to\infty} \sqrt{\frac{2}{\pi z}}\, e^{-i(z - \frac{1}{2}\nu\pi - \frac{1}{4}\pi)}\,. \end{aligned} \tag{D.29}$$

These limits were used to define the initial conditions of the inflationary fluctuations. The scaling for small arguments follows from (D.27):

$$
\begin{aligned}
H_\nu^{(1)}(z) \xrightarrow{z \to 0} -\frac{i}{\pi}\Gamma(\nu)\left(\frac{z}{2}\right)^{-\nu}, \\
H_\nu^{(2)}(z) \xrightarrow{z \to 0} +\frac{i}{\pi}\Gamma(\nu)\left(\frac{z}{2}\right)^{-\nu},
\end{aligned}
\tag{D.30}
$$

which was used in Section 8.3 to determine the superhorizon behavior of the inflationary fluctuations.

D.6 Gamma and Zeta Functions

The Euler gamma function is

$$
\Gamma(z) = \int_0^\infty \mathrm{d}t\, e^{-t} t^{z-1},
\tag{D.31}
$$

where $\mathrm{Re}(z) > 0$. It is an analytic function with simple poles at $z = -p$, for integer p. A useful relation is

$$
\Gamma(z+1) = z\Gamma(z),
\tag{D.32}
$$

which implies that for integer arguments the gamma function reduces to factorials, $\Gamma(n+1) = n!$. In Chapter 3, we encountered gamma functions with half integer arguments. These can be evaluated by first showing that

$$
\Gamma(1/2) = \sqrt{\pi},
\tag{D.33}
$$

and then using (D.32) to infer that $\Gamma(3/2) = 1/2\,\Gamma(1/2) = \sqrt{\pi}/2$ and so on.

The Riemann zeta function arises in certain integrals in statistical mechanics and is defined as

$$
\zeta(s) = \frac{1}{\Gamma(s)} \int \mathrm{d}x\, \frac{x^{s-1}}{e^x - 1}.
\tag{D.34}
$$

Important special cases are

$$
\zeta(2) = \frac{\pi^2}{6}, \quad \zeta(3) = 1.20205\ldots, \quad \zeta(4) = \frac{\pi^4}{90}.
\tag{D.35}
$$

When the real part of s is greater than 1, then the zeta function can be written as a convergent series

$$
\zeta(s) = \sum_{n=1}^\infty \frac{1}{n^s},
\tag{D.36}
$$

which appeared in Section 3.1.2.

References

The references in this book are collected separately for each individual chapter. Preprint versions of many articles can be found at https://arxiv.org/. Another useful resource is the INSPIRE database at https://inspirehep.net/.

Chapter 1

[1] F. Zwicky *Helv. Phys. Acta* **6** (1933) 110–127.

[2] V. Rubin and W. Ford *Astrophys. J.* **159** (1970) 379.

[3] G. Bertone and D. Hooper *Rev. Mod. Phys.* **90** no. 4 (2018) 045002, `arXiv:1605.04909 [astro-ph.CO]`.

[4] J. de Swart, G. Bertone, and J. van Dongen *Nature Astron.* **1** (2017) 0059, `arXiv:1703.00013 [astro-ph.CO]`.

[5] N. Aghanim *et al.* [Planck Collaboration] *Astron. Astrophys.* **641** (2020) A8, `arXiv:1807.06210 [astro-ph.CO]`.

[6] S. Maddox, G. Efstathiou, W. Sutherland, and J. Loveday *Mon. Not. Roy. Astron. Soc.* **243** (1990) 692–712.

[7] G. Efstathiou, W. Sutherland, and S. Maddox *Nature* **348** (1990) 705–707.

[8] A. Guth *Phys. Rev. D* **23** (1981) 347–356.

[9] L. Krauss and M. Turner *Gen. Rel. Grav.* **27** (1995) 1137–1144, `arXiv:astro-ph/9504003`.

[10] J. Ostriker and P. Steinhardt *Nature* **377** (1995) 600–602.

[11] S. Perlmutter *et al.* [Supernova Cosmology Project Collaboration] *Astrophys. J.* **517** (1999) 565–586, `arXiv:astro-ph/9812133`.

[12] A. Riess *et al.* [Supernova Search Team] *Astron. J.* **116** (1998) 1009–1038, `arXiv:astro-ph/9805201`.

[13] P. de Bernardis *et al.* [Boomerang Collaboration] *Nature* **404** (2000) 955–959, `arXiv:astro-ph/0004404`.

[14] S. Hanany *et al.* [MAXIMA Collaboration] *Astrophys. J. Lett.* **545** (2000) L5, `arXiv:astro-ph/0005123`.

[15] S. Weinberg *Rev. Mod. Phys.* **61** (1989) 1–23.

[16] E. Hubble *Proc. Natl. Acad. Sci.* **15** no. 3 (1929) 168–173.

[17] S. Chatrchyan *et al.* [CMS Collaboration, CERN] *Phys. Lett. B* **716** (2012) 30–61, `arXiv:1207.7235 [hep-ex]`.

[18] G. Aad *et al.* [ATLAS Collaboration, CERN] *Phys. Lett. B* **716** (2012) 1–29, `arXiv:1207.7214 [hep-ex]`.

[19] G. Gamow *Phys. Rev.* **70** (1946) 572–573.

[20] R. Alpher and R. Herman *Nature* **162** no. 4124 (1948) 774–775.

[21] R. Alpher, J. Follin, and R. Herman *Phys. Rev.* **92** (1953) 1347–1361.

[22] A. Penzias and R. Wilson *Astrophys. J.* **142** (1965) 419–421.

[23] A. Linde *Phys. Lett. B* **108** (1982) 389–393.

[24] A. Albrecht and P. Steinhardt *Phys. Rev. Lett.* **48** (1982) 1220–1223.

[25] V. Mukhanov and G. Chibisov *JETP Lett.* **33** (1981) 532–535.
[26] J. Bardeen, P. Steinhardt, and M. Turner *Phys. Rev. D* **28** (1983) 679.
[27] S. Hawking *Phys. Lett. B* **115** (1982) 295.
[28] A. Starobinsky *Phys. Lett. B* **117** (1982) 175–178.
[29] A. Guth and S.-Y. Pi *Phys. Rev. Lett.* **49** (1982) 1110–1113.

Chapter 2

[1] E. Hubble *Proc. Natl. Acad. Sci.* **15** no. 3 (1929) 168–173.
[2] E. Hubble and M. Humason *Astrophys. J.* **74** (1931) 43–80.
[3] B. Schutz *Nature* **323** no. 6086 (1986) 310–311.
[4] A. Riess, S. Casertano, W. Yuan, L. Macri, and D. Scolnic *Astrophys. J.* **876** no. 1 (2019) 85, arXiv:1903.07603 [astro-ph.CO].
[5] N. Aghanim *et al.* [Planck Collaboration] *Astron. Astrophys.* **641** (2020) A6, arXiv:1807.06209 [astro-ph.CO].
[6] E. Corbelli and P. Salucci *Mon. Not. Roy. Astron. Soc.* **311** no. 2 (2000) 441–447, arXiv:astro-ph/9909252 [astro-ph].
[7] G. Bertone, D. Hooper, and J. Silk *Phys. Rept.* **405** (2005) 279–390, arXiv:hep-ph/0404175.
[8] C. Enz and A. Thellung *Helv. Phys. Acta* **33** (1960) 839–848.
[9] T.Y. Cao [Editor], *Conceptual Foundations of Quantum Field Theory.* Cambridge University Press, 1999.
[10] S. Weinberg *Phy. Rev. Lett.* **59** no. 22 (1987) 2607.
[11] A. Einstein and W. de Sitter *Proc. Natl. Acad. Sci.* **18** no. 3 (1932) 213–214.
[12] S. Hawking and R. Penrose *Proc. Roy. Soc. Lond. A* **314** (1970) 529–548.
[13] J. Maldacena *Int. J. Theor. Phys.* **38** (1999) 1113–1133, arXiv:hep-th/9711200.
[14] D. Spergel, *private communication.*
[15] D. Fixsen *et al.* [COBE Collaboration] *Astrophys. J.* **473** (1996) 576, arXiv:astro-ph/9605054.
[16] D. Scolnic *et al. Astrophys. J.* **859** no. 2 (2018) 101, arXiv:1710.00845 [astro-ph.CO].
[17] S. Weinberg, *Cosmology.* Oxford University Press, 2008.
[18] D. Tong, *Cosmology (Lecture Notes).*
[19] B. Ryden, *Introduction to Cosmology.* Cambridge University Press, 2016.
[20] D. Hogg arXiv:astro-ph/9905116.
[21] T. Davis and C. Lineweaver *Publ. Astron. Soc. Australia* **21** no. 1 (2004) 97–109.
[22] S. Weinberg *Rev. Mod. Phys.* **61** (1989) 1–23.
[23] J. Polchinski arXiv:hep-th/0603249.
[24] S. Carroll *Living Rev. Rel.* **4** (2001) 1, arXiv:astro-ph/0004075.
[25] R. Bousso *Gen. Rel. Grav.* **40** (2008) 607–637, arXiv:0708.4231 [hep-th].
[26] C. Burgess arXiv:1309.4133 [hep-th].
[27] A. Padilla arXiv:1502.05296 [hep-th].
[28] H. Kragh, *Cosmology and Controversy.* Princeton University Press, 1996.
[29] P.J.E. Peebles, *Cosmology's Century: An Inside History of Our Modern Understanding of the Universe.* Princeton University Press, 2020.
[30] S. Carroll, *Spacetime and Geometry.* Cambridge University Press, 2019.
[31] J. Hartle, *Gravity.* Cambridge University Press, 2021.
[32] B. Schutz, *A First Course in General Relativity.* Cambridge University Press, 2009.
[33] D. Tong, *General Relativity (Lecture Notes).*
[34] H. Reall, *General Relativity (Lecture Notes).*
[35] E. Lim, *General Relativity (Lecture Notes).*

Chapter 3

[1] N. Aghanim *et al.* [Planck Collaboration] *Astron. Astrophys.* **641** (2020) A6, arXiv:1807.06209 [astro-ph.CO].

[2] M. Gonzalez-Garcia and M. Yokoyama [Particle Data Group], *Neutrino Masses, Mixing, and Oscillations*, 2019, https://pdg.lbl.gov/.

[3] K. Abazajian *et al.* [CMB-S4 Collaboration] arXiv:1610.02743 [astro-ph.CO].

[4] D. Fixsen *et al.* [COBE Collaboration] *Astrophys. J.* **473** (1996) 576, arXiv:astro-ph/9605054.

[5] A. Sakharov *Sov. Phys. Usp.* **34** no. 5 (1991) 392–393.

[6] C. Burgess and G. Moore, *The Standard Model: A Primer*. Cambridge University Press, 2006.

[7] N. Manton *Phys. Rev. D* **28** (1983) 2019.

[8] F. Klinkhamer and N. Manton *Phys. Rev. D* **30** (Nov, 1984) 2212–2220.

[9] H. Cliff, *How to Make an Apple Pie from Scratch: In Search of the Recipe for Our Universe*. Picador, 2021.

[10] M. Schwartz, *Quantum Field Theory and the Standard Model*. Cambridge University Press, 2014.

[11] A. Riotto, *Theories of Baryogenesis*. arXiv:hep-ph/9807454.

[12] S. Dimopoulos and L. Susskind *Phys. Rev. D* **18** (1978) 4500–4509.

[13] J. Ellis, M. Gaillard, and D. Nanopoulos *Phys. Lett. B* **80** (1979) 360.

[14] M. Yoshimura *Phys. Rev. Lett.* **41** (1978) 281–284.

[15] D. Toussaint, S. Treiman, F. Wilczek, and A. Zee *Phys. Rev. D* **19** (1979) 1036–1045.

[16] S. Weinberg *Phys. Rev. Lett.* **42** (1979) 850–853.

[17] S. Barr, G. Segre, and H. Weldon *Phys. Rev. D* **20** (1979) 2494.

[18] D. Nanopoulos and S. Weinberg *Phys. Rev. D* **20** (1979) 2484.

[19] K. Kajantic, M. Laine, K. Rummukainen, and M. Shaposhnikov *Nucl. Phys. B* **466** (1996) 189–258, arXiv:hep-lat/9510020.

[20] M. Gavela, M. Lozano, J. Orloff, and O. Pene *Nucl. Phys. B* **430** (1994) 345–381, arXiv:hep-ph/9406288.

[21] M. Gavela, P. Hernandez, J. Orloff, O. Pene, and C. Quimbay *Nucl. Phys. B* **430** (1994) 382–426, arXiv:hep-ph/9406289.

[22] P. Huet and E. Sather *Phys. Rev. D* **51** (1995) 379–394, arXiv:hep-ph/9404302.

[23] V. Kuzmin, V. Rubakov, and M. Shaposhnikov *Phys. Lett. B* **155** (1985) 36.

[24] M. Shaposhnikov *JETP Lett.* **44** (1986) 465–468.

[25] M. Shaposhnikov *Nucl. Phys. B* **287** (1987) 757–775.

[26] A. Cohen, D. Kaplan, and A. Nelson *Ann. Rev. Nucl. Part. Sci.* **43** (1993) 27–70, arXiv:hep-ph/9302210.

[27] M. Trodden *Rev. Mod. Phys.* **71** (1999) 1463–1500.

[28] D. Morrissey and M. Ramsey-Musolf *New J. Phys.* **14** (2012) 125003, arXiv:1206.2942 [hep-ph].

[29] M. Fukugita and T. Yanagida *Phys. Lett. B* **174** (1986) 45–47.

[30] K. Abe *et al.* [T2K Collaboration] *Nature* **580** no. 7803 (2020) 339–344, arXiv:1910.03887 [hep-ex].

[31] C. Pitrou, A. Coc, J.-P. Uzan, and E. Vangioni *Phys. Rept.* **754** (2018) 1–66, arXiv:1801.08023 [astro-ph.CO].

[32] S. Dodelson, *Modern Cosmology*. Academic Press, 2003.

[33] B. Fields, P. Molaro, and S. Sarkar [Particle Data Group], *Big Bang Nucleosynthesis*, 2017, https://pdg.lbl.gov/.

[34] M. Valerdi, A. Peimbert, M. Peimbert, and A. Sixtos *Astrophys. J.* **876** no. 2 (2019) 98, arXiv:1904.01594 [astro-ph.GA].

[35] E. Aver, K. Olive, and E. Skillman *JCAP* **07** (2015) 011, arXiv:1503.08146 [astro-ph.CO].

[36] Y. Izotov, T. Thuan, and N. Guseva *Mon. Not. Roy. Astron. Soc.* **445** no. 1 (2014) 778–793, arXiv:1408.6953 [astro-ph.CO].

[37] M. Pettini and R. Cooke *Mon. Not. Roy. Astron. Soc.* **425** (2012) 2477–2486, arXiv:1205.3785 [astro-ph.CO].

[38] P. Bonifacio and P. Molaro *Mon. Not. Roy. Astron. Soc.* **285** (1997) 847–861, arXiv:astro-ph/9611043.

[39] P. Bonifacio *et al. Astron. Astrophys.* **390** (2002) 91, arXiv:astro-ph/0204332.

[40] J. Melendez *et al. Astron. Astrophys.* **515** (2010) L3, arXiv:1005.2944 [astro-ph.SR].

[41] L. Sbordone *et al. Astron. Astrophys.* **522** (2010) A26, arXiv:1003.4510 [astro-ph.GA].

[42] P.J.E. Peebles *Astrophys. J.* **153** (1968) 1.

[43] Y. Zeldovich, V. Kurt, and R. Sunyaev *Sov. Phys. JETP* **28** (1969) 146.

[44] C.-P. Ma and E. Bertschinger *Astrophys. J.* **455** (1995) 7–25, arXiv:astro-ph/9506072.

[45] S. Weinberg, *Cosmology.* Oxford University Press, 2008.

[46] S. Seager, D. Sasselov, and D. Scott *Astrophys. J.* **523** (1999) L1–L5, arXiv:astro-ph/9909275.

[47] Y. Ali-Haimoud and C. Hirata *Phys. Rev. D* **83** (2011) 043513, arXiv:1011.3758 [astro-ph.CO].

[48] J. Chluba, J. Rubino-Martin, and R. Sunyaev *Mon. Not. Roy. Astron. Soc.* **374** (2007) 1310–1320, arXiv:astro-ph/0608242.

[49] Y. Ali-Haimoud and C. Hirata *Phys. Rev. D* **82** (2010) 063521, arXiv:1006.1355 [astro-ph.CO].

[50] J. Chluba and R. Thomas *Mon. Not. Roy. Astron. Soc.* **412** (2011) 748, arXiv:1010.3631 [astro-ph.CO].

[51] J. Chluba and R. Sunyaev *Astron. Astrophys.* **446** (2006) 39–42, arXiv:astro-ph/0508144.

[52] C. Hirata *Phys. Rev. D* **78** (2008) 023001, arXiv:0803.0808 [astro-ph].

[53] J. Chluba and R. Sunyaev *Astron. Astrophys.* **475** (2007) 109, arXiv:astro-ph/0702531.

[54] E. Kholupenko, A. Ivanchik, and D. Varshalovich *Phys. Rev. D* **81** no. 8 (2010) 083004, arXiv:0912.5454 [astro-ph.CO].

[55] J. Chluba and R. Sunyaev *Astron. Astrophys.* **496** (2009) 619, arXiv:0810.1045 [astro-ph].

[56] C. Hirata and J. Forbes *Phys. Rev. D* **80** no. 2 (2009) 023001, arXiv:0903.4925 [astro-ph.CO].

[57] J. Chluba and R. Sunyaev *Astron. Astrophys.* **503** no. 2 (2009) 345–355, arXiv:0904.2220 [astro-ph.CO].

[58] V. Dubrovich and S. Grachev arXiv:astro-ph/0501672.

[59] J. Chluba and R. Sunyaev *Astron. Astrophys.* **480** (2008) 629, arXiv:0705.3033 [astro-ph].

[60] S. Grachev and V. Dubrovich *Astron. Lett.* **34** (2008) 439, arXiv:0801.3347 [astro-ph].

[61] E. Kolb and M. Turner, *The Early Universe.* CRC Press, 1990.

[62] C. Hirata, *Cosmology (Lecture Notes).*

[63] B. Fields and S. Sarkar arXiv:astro-ph/0601514.

[64] J. Cline, *Les Houches Lectures on Baryogenesis.* arXiv:hep-ph/0609145.

[65] N. Straumann, *Cosmological Phase Transitions.* arXiv:astro-ph/0409042.

[66] A. Mazumdar and G. White *Rept. Prog. Phys.* **82** no. 7 (2019) 076901, arXiv:1811.01948 [hep-ph].

[67] M. Kardar, *Statistical Physics of Particles.* Cambridge University Press, 2007.

[68] F. Mandl, *Statistical Physics.* John Wiley & Sons, 2008.

[69] R. Bowley and M. Sanchez, *Introductory Statistical Mechanics.* Clarendon Press, 2000.

[70] D. Tong, *Statistical Physics (Lecture Notes).*

Chapter 4

[1] D. Tong, *Cosmology (Lecture Notes)*.

[2] B. Bassett, S. Tsujikawa, and D. Wands *Rev. Mod. Phys.* **78** (2006) 537–589, `arXiv:astro-ph/0507632`.

[3] R. Allahverdi, R. Brandenberger, F.-Y. Cyr-Racine, and A. Mazumdar *Ann. Rev. Nucl. Part. Sci.* **60** (2010) 27–51, `arXiv:1001.2600 [hep-th]`.

[4] M. Amin, M. Hertzberg, D. Kaiser, and J. Karouby *Int. J. Mod. Phys. D* **24** (2014) 1530003, `arXiv:1410.3808 [hep-ph]`.

[5] K. Lozanov `arXiv:1907.04402 [astro-ph.CO]`.

[6] G. 't Hooft *NATO Adv. Study Inst. Ser. B Phys.* **59** (1980) 135.

[7] E. Silverstein and A. Westphal *Phys. Rev. D* **78** (2008) 106003, `arXiv:0803.3085 [hep-th]`.

[8] L. McAllister, E. Silverstein, and A. Westphal *Phys. Rev. D* **82** (2010) 046003, `arXiv:0808.0706 [hep-th]`.

[9] D. Baumann and L. McAllister, *Inflation and String Theory*. Cambridge University Press, 2015.

[10] E. Silverstein, *TASI Lectures on Cosmological Observables and String Theory*. `arXiv:1606.03640 [hep-th]`.

[11] E. Silverstein, *Les Houches Lectures on Inflationary Observables and String Theory*. `arXiv:1311.2312 [hep-th]`.

[12] E. Silverstein, *TASI Lectures on Moduli and Microphysics*. `arXiv:hep-th/0405068`.

[13] C. Burgess, *Lectures on Cosmic Inflation and Its Potential Stringy Realizations*. `arXiv:0708.2865 [hep-th]`.

[14] D. Baumann and L. McAllister *Ann. Rev. Nucl. Part. Sci.* **59** (2009) 67–94, `arXiv:0901.0265 [hep-th]`.

[15] J. Hartle and S. Hawking *Adv. Ser. Astrophys. Cosmol.* **3** (1987) 174–189.

[16] D. Goldwirth and T. Piran *Phys. Rev. Lett.* **64** (1990) 2852–2855.

[17] D. Goldwirth and T. Piran *Phys. Rev. D* **40** (1989) 3263–3279.

[18] W. East, M. Kleban, A. Linde, and L. Senatore *JCAP* **09** (2016) 010, `arXiv:1511.05143 [hep-th]`.

[19] K. Clough, E. Lim, B. DiNunno, W. Fischler, R. Flauger, and S. Paban *JCAP* **09** (2017) 025, `arXiv:1608.04408 [hep-th]`.

[20] A. Vilenkin *Phys. Rev. Lett.* **74** (1995) 846–849, `arXiv:gr-qc/9406010 [gr-qc]`.

[21] J. Garriga, D. Schwartz-Perlov, A. Vilenkin, and S. Winitzki *JCAP* **0601** (2006) 017, `arXiv:hep-th/0509184 [hep-th]`.

[22] R. Bousso, B. Freivogel, and I.-S. Yang *Phys. Rev. D* **79** (2009) 063513, `arXiv:0808.3770 [hep-th]`.

[23] R. Bousso *Phys. Rev. Lett.* **97** (2006) 191302, `arXiv:hep-th/0605263 [hep-th]`.

[24] A. De Simone *et al. Phys. Rev. D* **82** (2010) 063520, `arXiv:0808.3778 [hep-th]`.

[25] G. Gibbons and N. Turok *Phys. Rev. D* **77** (2008) 063516, `arXiv:hep-th/0609095 [hep-th]`.

[26] J. Khoury, B. Ovrut, P. Steinhardt, and N. Turok *Phys. Rev. D* **64** (2001) 123522, `arXiv:hep-th/0103239`.

[27] P. Steinhardt and N. Turok *Phys. Rev. D* **65** (2002) 126003, `arXiv:hep-th/0111098`.

[28] W. Kinney, *TASI Lectures on Inflation*. `arXiv:0902.1529 [astro-ph.CO]`.

[29] L. Senatore, *TASI Lectures on Inflation*.

[30] D. Lyth and A. Riotto *Phys. Rept.* **314** (1999) 1–146, `arXiv:hep-ph/9807278`.

[31] C. Lineweaver, *Inflation and the Cosmic Microwave Background*. `arXiv:astro-ph/0305179`.

[32] D. Baumann, *TASI Lectures on Inflation*. `arXiv:0907.5424 [hep-th]`.

[33] D. Baumann, *TASI Lectures on Primordial Cosmology*. `arXiv:1807.03098 [hep-th]`.

[34] A. Manohar, *Les Houches Lectures on Effective Field Theories.* `arXiv:1804.05863 [hep-ph]`.

[35] C. Burgess *Ann. Rev. Nucl. Part. Sci.* **57** (2007) 329–362, `arXiv:hep-th/0701053`.

[36] C. Cheung, P. Creminelli, A. L. Fitzpatrick, J. Kaplan, and L. Senatore *JHEP* **03** (2008) 014, `arXiv:0709.0293 [hep-th]`.

[37] S. Weinberg *Phys. Rev. D* **77** (2008) 123541, `arXiv:0804.4291 [hep-th]`.

[38] A. Guth *J. Phys. A* **40** (2007) 6811–6826, `arXiv:hep-th/0702178`.

[39] L. Kofman, A. Linde, and A. Starobinsky *Phys. Rev. D* **56** (1997) 3258–3295, `arXiv:hep-ph/9704452`.

[40] A. Guth *Phys. Rev. D* **23** (1981) 347–356.

[41] F. Lucchin and S. Matarrese *Phys. Rev. D* **32** (1985) 1316.

[42] K. Freese, J. Frieman, and A. Olinto *Phys. Rev. Lett.* **65** (1990) 3233–3236.

[43] C. Armendariz-Picon, T. Damour, and V. Mukhanov *Phys. Lett. B* **458** (1999) 209–218, `arXiv:hep-th/9904075`.

[44] A. Starobinsky *Phys. Lett. B* **91** (1980) 99–102.

[45] T. Futamase and K.-i. Maeda *Phys. Rev. D* **39** (1989) 399–404.

[46] R. Fakir and W. Unruh *Phys. Rev. D* **41** (1990) 1783–1791.

[47] F. Bezrukov and M. Shaposhnikov *Phys. Lett. B* **659** (2008) 703–706, `arXiv:0710.3755 [hep-th]`.

Chapter 5

[1] S. Weinberg *Astrophys. J.* **581** (2002) 810–816, arXiv:astro-ph/0207375.

[2] E. Harrison *Phys. Rev. D* **1** (1970) 2726–2730.

[3] P.J.E. Peebles and J. Yu *Astrophys. J.* **162** (1970) 815–836.

[4] Y. Zeldovich *Mon. Not. Roy. Astron. Soc.* **160** (1972) 1P–3P.

[5] N. Aghanim *et al.* [Planck Collaboration] *Astron. Astrophys.* **641** (2020) A6, arXiv:1807.06209 [astro-ph.CO].

[6] S. Chabanier, M. Millea, and N. Palanque-Delabrouille *Mon. Not. Roy. Astron. Soc.* **489** no. 2 (2019) 2247–2253, arXiv:1905.08103 [astro-ph.CO].

[7] J. Gunn and J. Gott *Astrophys. J.* **176** (1972) 1–19.

[8] C. Knobel, *An Introduction to the Theory of Cosmological Structure Formation.* arXiv:1208.5931 [astro-ph.CO].

[9] S. Weinberg *Phys. Rev. Lett.* **59** (1987) 2607.

[10] W. Press and P. Schechter *Astrophys. J.* **187** (1974) 425–438.

[11] A. Zentner *Int. J. Mod. Phys. D* **16** (2007) 763–816, arXiv:astro-ph/0611454.

[12] J. Bond, S. Cole, G. Efstathiou, and N. Kaiser *Astrophys. J.* **379** (1991) 440.

[13] A. Cooray and R. Sheth *Phys. Rept.* **372** (2002) 1–129, arXiv:astro-ph/0206508.

[14] R. Sheth and G. Tormen *Mon. Not. Roy. Astron. Soc.* **308** no. 1 (09, 1999) 119–126.

[15] N. Kaiser *Astrophys. J. Lett.* **284** (1984) L9–L12.

[16] J. Bardeen, J. Bond, N. Kaiser, and A. Szalay *Astrophys. J.* **304** (1986) 15–61.

[17] S. Cole and N. Kaiser *Mon. Not. Roy. Astron. Soc.* **237** (1989) 1127–1146.

[18] H. Mo and S. White *Mon. Not. Roy. Astron. Soc.* **282** (1996) 347, arXiv:astro-ph/9512127.

[19] V. Desjacques, D. Jeong, and F. Schmidt *Phys. Rept.* **733** (2018) 1–193, arXiv:1611.09787 [astro-ph.CO].

[20] H. Mo, F. van den Bosch, and S. White, *Galaxy Formation and Evolution.* Cambridge University Press, 2010.

[21] P.J.E. Peebles, *The Large-Scale Structure of the Universe.* Princeton University Press, 2020.

[22] D. Tong, *Cosmology (Lecture Notes).*

[23] F. van den Bosch, *Theory of Galaxy Formation (Lecture Notes).*

Chapter 6

[1] J. Bardeen *Phys. Rev. D* **22** (1980) 1882–1905.

[2] C.-P. Ma and E. Bertschinger *Astrophys. J.* **455** (1995) 7–25, `arXiv:astro-ph/9506072`.

[3] J. Lesgourgues, `arXiv:1104.2932 [astro-ph.IM]`.

[4] E. Lifshitz *J. Phys. (USSR)* **10** no. 2 (1946) 116.

[5] J. Maldacena *JHEP* **05** (2003) 013, `arXiv:astro-ph/0210603`.

[6] D. Baumann, A. Nicolis, L. Senatore, and M. Zaldarriaga *JCAP* **07** (2012) 051, `arXiv:1004.2488 [astro-ph.CO]`.

[7] P. Creminelli, M. Luty, A. Nicolis, and L. Senatore *JHEP* **12** (2006) 080, `arXiv:hep-th/0606090`.

[8] M. Bucher, K. Moodley, and N. Turok *Phys. Rev. D* **62** (2000) 083508, `arXiv:astro-ph/9904231`.

[9] M. Bucher, K. Moodley, and N. Turok *Phys. Rev. D* **66** (2002) 023528, `arXiv:astro-ph/0007360`.

[10] S. Weinberg, *Cosmology*. Oxford University Press, 2008.

[11] M. Amin, *Cosmology (Lecture Notes)*.

[12] W. Hu and N. Sugiyama *Astrophys. J.* **471** (1996) 542–570, `arXiv:astro-ph/9510117`.

[13] N. Aghanim *et al.* [Planck Collaboration] *Astron. Astrophys.* **641** (2020) A1, `arXiv:1807.06205 [astro-ph.CO]`.

[14] F. Beutler *et al.* [BOSS Collaboration] *Mon. Not. Roy. Astron. Soc.* **464** no. 3 (2017) 3409–3430, `arXiv:1607.03149 [astro-ph.CO]`.

[15] C. Caprini and D. Figueroa *Class. Quant. Grav.* **35** no. 16 (2018) 163001, `arXiv.1801.04268 [astro-ph.CO]`.

[16] H. Kodama and M. Sasaki *Prog. Theor. Phys. Suppl.* **78** (1984) 1–166.

[17] V. Mukhanov, H. Feldman, and R. Brandenberger *Phys. Rept.* **215** (1992) 203–333.

[18] K. Malik and D. Wands *Phys. Rept.* **475** (2009) 1–51, `arXiv:0809.4944 [astro-ph]`.

[19] J. Lesgourgues, *TASI Lectures on Cosmological Perturbations*. `arXiv:1302.4640 [astro-ph.CO]`.

[20] R. Brandenberger *Lect. Notes Phys.* **646** (2004) 127–167, `arXiv:hep-th/0306071`.

[21] O. Piattella, *Lecture Notes in Cosmology*. `arXiv:1803.00070 [astro-ph.CO]`.

[22] W. Hu *Lect. Notes Phys.* **470** (1996) 207, `arXiv:astro-ph/9511130`.

[23] W. Hu `arXiv:0802.3688 [astro-ph]`.

Chapter 7

[1] C. Bennett *et al.* [WMAP Collaboration] *Astrophys. J.* **208** no. 2 (2013) 20, `arXiv:1212.5225 [astro-ph.CO]`.

[2] N. Aghanim *et al.* [Planck Collaboration] *Astron. Astrophys.* **641** (2020) A1, `arXiv:1807.06205 [astro-ph.CO]`.

[3] K. Story *et al.* [SPT Collaboration] *Astrophys. J.* **779** no. 1 (2013) 86, `arXiv:1210.7231 [astro-ph.CO]`.

[4] S. Aiola *et al.* [ACT Collaboration] *JCAP* **12** (2020) 047, `arXiv:2007.07288 [astro-ph.CO]`.

[5] A. Yahil, G. Tammann, and A. Sandage *Astrophys. J.* **217** (1977) 903–915.

[6] A. Sandage, B. Reindl, and G. Tammann *Astrophys. J.* **714** no. 2 (2010) 1441–1459, `arXiv:0911.4925 [astro-ph.CO]`.

[7] A. Nusser and M. Davis *Astrophys. J.* **736** (2011) 93, `arXiv:1101.1650` `[astro-ph.CO]`.

[8] R. Sachs and A. Wolfe *Astrophys. J.* **147** (1967) 73–90.

[9] S. Weinberg, *Cosmology.* Oxford University Press, 2008.

[10] J. Lesgourgues, `arXiv:1104.2932 [astro-ph.IM]`.

[11] Z. Pan, L. Knox, B. Mulroe, and A. Narimani *Mon. Not. Roy. Astron. Soc.* **459** no. 3 (2016) 2513–2524, `arXiv:1603.03091 [astro-ph.CO]`.

[12] W. Hu and N. Sugiyama *Astrophys. J.* **444** (1995) 489–506, `arXiv:astro-ph/9407093`.

[13] S. Bashinsky and U. Seljak *Phys. Rev. D* **69** (2004) 083002, `arXiv:astro-ph/0310198`.

[14] D. Baumann, D. Green, J. Meyers, and B. Wallisch *JCAP* **01** (2016) 007, `arXiv:1508.06342 [astro-ph.CO]`.

[15] J. Silk *Astrophys. J.* **151** (1968) 459.

[16] M. Zaldarriaga and D. Harari *Phys. Rev. D* **52** (1995) 3276–3287, `arXiv:astro-ph/9504085`.

[17] N. Aghanim *et al.* [Planck Collaboration] *Astron. Astrophys.* **641** (2020) A6, `arXiv:1807.06209 [astro-ph.CO]`.

[18] A. Riess, S. Casertano, W. Yuan, L. Macri, and D. Scolnic *Astrophys. J.* **876** no. 1 (2019) 85, `arXiv:1903.07603 [astro-ph.CO]`.

[19] K. Abazajian *et al.* [CMB-S4 Collaboration] `arXiv:1610.02743 [astro-ph.CO]`.

[20] M. Zaldarriaga and U. Seljak *Phys. Rev. D* **55** (1997) 1830–1840, `arXiv:astro-ph/9609170`.

[21] M. Kamionkowski, A. Kosowsky, and A. Stebbins *Phys. Rev. D* **55** (1997) 7368–7388, `arXiv:astro-ph/9611125`.

[22] E. Leitch *et al. Nature* **420** no. 6917 (2002) 763–771.

[23] S. Choi *et al.* [ACT Collaboration] *JCAP* **12** (2020) 045, `arXiv:2007.07289 [astro-ph.CO]`.

[24] J. Henning *et al.* [SPT Collaboration] *Astrophys. J.* **852** no. 2 (2018) 97, `arXiv:1707.09353 [astro-ph.CO]`.

[25] S. Dodelson and F. Schmidt, *Modern Cosmology.* Academic Press, 2020.

[26] S. Dodelson, *Modern Cosmology.* Academic Press, 2003.

[27] M. Zaldarriaga, *An Introduction to CMB Anisotropies (ICTP Lectures).*

[28] P. Ade *et al.* [BICEP2/Keck Array Collaboration] *Phys. Rev. Lett.* **121** (2018) 221301, `arXiv:1810.05216 [astro-ph.CO]`.

[29] J. Sayre *et al.* [SPT Collaboration] *Phys. Rev. D* **101** no. 12 (2020) 122003, `arXiv:1910.05748 [astro-ph.CO]`.

[30] P. Ade *et al.* [POLARBEAR Collaboration] *Astrophys. J.* **848** no. 2 (2017) 121, `arXiv:1705.02907 [astro-ph.CO]`.

[31] J. Pritchard and M. Kamionkowski *Annals Phys.* **318** (2005) 2–36, `arXiv:astro-ph/0412581`.

[32] P. Ade *et al.* [BICEP/Keck Collaboration] `arXiv:2110.00483 [astro-ph.CO]`.

[33] J. Lesgourgues, *TASI Lectures on Cosmological Perturbations.* `arXiv:1302.4640 [astro-ph.CO]`.

[34] W. Hu *Lect. Notes Phys.* **470** (1996) 207, `arXiv:astro-ph/9511130`.

[35] A. Challinor and H. Peiris *AIP Conf. Proc.* **1132** no. 1, (2009) 86–140, `arXiv:0903.5158 [astro-ph.CO]`.

[36] W. Hu `arXiv:0802.3688 [astro-ph]`.

[37] A. Challinor, *Advanced Cosmology (Lecture Notes).*

[38] E. Komatsu, *Cosmic Microwave Background (Lecture Notes).*

[39] W. Hu, *Wandering in the Background (PhD Thesis).* `arXiv:astro-ph/9508126`.

[40] W. Hu and M. White *New Astron.* **2** (1997) 323, `arXiv:astro-ph/9706147`.

[41] A. Kosowsky *New Astron. Rev.* **43** (1999) 157, `arXiv:astro-ph/9904102`.

[42] P. Cabella and M. Kamionkowski, *Theory of Cosmic Microwave Background Polarization.* `arXiv:astro-ph/0403392`.

[43] Y.-T. Lin and B. Wandelt *Astropart. Phys.* **25** (2006) 151–166, `arXiv:astro-ph/0409734`.

[44] A. Lewis and A. Challinor *Phys. Rept.* **429** (2006) 1–65, `arXiv:astro-ph/0601594`.

Chapter 8

[1] D. Baumann and L. McAllister, *Inflation and String Theory.* Cambridge University Press, 2015.

[2] C. Cheung, P. Creminelli, A. Fitzpatrick, J. Kaplan, and L. Senatore *JHEP* **03** (2008) 014, `arXiv:0709.0293 [hep-th]`.

[3] N. Aghanim *et al.* [Planck Collaboration] *Astron. Astrophys.* **641** (2020) A6, `arXiv:1807.06209 [astro-ph.CO]`.

[4] P. Dirac, *General Theory of Relativity.* Princeton University Press, 2016.

[5] D. Lyth *Phys. Rev. Lett.* **78** (1997) 1861–1863, `arXiv:hep-ph/9606387`.

[6] N. Arkani-Hamed, L. Motl, A. Nicolis, and C. Vafa *JHEP* **06** (2007) 060, `arXiv:hep-th/0601001`.

[7] E. Silverstein and A. Westphal *Phys. Rev. D* **78** (2008) 106003, `arXiv:0803.3085 [hep-th]`.

[8] L. McAllister, E. Silverstein, and A. Westphal *Phys. Rev. D* **82** (2010) 046003, `arXiv:0808.0706 [hep-th]`.

[9] S. Dodelson *AIP Conf. Proc.* **689** no. 1 (2003) 184–196, `arXiv:hep-ph/0309057`.

[10] D. Spergel and M. Zaldarriaga *Phys. Rev. Lett.* **79** (1997) 2180–2183, `arXiv:astro-ph/9705182`.

[11] H. Peiris *et al.* [WMAP Collaboration] *Astrophys. J. Suppl.* **148** (2003) 213–231, `arXiv:astro-ph/0302225`.

[12] Y. Akrami *et al.* [Planck Collaboration] *Astron. Astrophys.* **641** (2020) A10, `arXiv:1807.06211 [astro-ph.CO]`.

[13] P. Ade *et al.* [BICEP/Keck Collaboration] `arXiv:2110.00483 [astro-ph.CO]`.

[14] K. Abazajian *et al.* [CMB-S4 Collaboration] `arXiv:1610.02743 [astro-ph.CO]`.

[15] N. Arkani-Hamed and J. Maldacena `arXiv:1503.08043 [hep-th]`.

[16] X. Chen *Adv. Astron.* **2010** (2010) 638979, `arXiv:1002.1416 [astro-ph.CO]`.

[17] D. Baumann, *TASI Lectures on Primordial Cosmology.* `arXiv:1807.03098 [hep-th]`.

[18] A. Gangui, F. Lucchin, S. Matarrese, and S. Mollerach *Astrophys. J.* **430** (1994) 447–457, `arXiv:astro-ph/9312033`.

[19] E. Komatsu and D. Spergel *Phys. Rev. D* **63** (2001) 063002, `arXiv:astro-ph/0005036`.

[20] J. Maldacena *JHEP* **05** (2003) 013, `arXiv:astro-ph/0210603`.

[21] P. Creminelli and M. Zaldarriaga *JCAP* **10** (2004) 006, `arXiv:astro-ph/0407059`.

[22] N. Arkani-Hamed, D. Baumann, H. Lee, and G. Pimentel *JHEP* **04** (2020) 105, `arXiv:1811.00024 [hep-th]`.

[23] Y. Akrami *et al.* [Planck Collaboration] *Astron. Astrophys.* **641** (2020) A9, `arXiv:1905.05697 [astro-ph.CO]`.

[24] O. Doré *et al.* [SPHEREx Science Team], `arXiv:1412.4872 [astro-ph.CO]`.

[25] N. Dalal, O. Doré, D. Huterer, and A. Shirokov *Phys. Rev. D* **77** (2008) 123514, `arXiv:0710.4560 [astro-ph]`.

[26] E. Silverstein and D. Tong *Phys. Rev. D* **70** (2004) 103505, `arXiv:hep-th/0310221`.

[27] X. Chen, M.-x. Huang, S. Kachru, and G. Shiu *JCAP* **01** (2007) 002, `arXiv:hep-th/0605045`.

[28] P. Creminelli, A. Nicolis, L. Senatore, M. Tegmark, and M. Zaldarriaga *JCAP* **05** (2006) 004, `arXiv:astro-ph/0509029`.

[29] R. Flauger, D. Green, and R. Porto *JCAP* **08** (2013) 032, `arXiv:1303.1430 [hep-th]`.

[30] D. Green and R. Porto *Phys. Rev. Lett.* **124** no. 25 (2020) 251302, `arXiv:2001.09149 [hep-th]`.

[31] A. Riotto, *Inflation and the Theory of Cosmological Perturbations.* `arXiv:hep-ph/0210162`.

[32] W. Kinney, *TASI Lectures on Inflation.* `arXiv:0902.1529 [astro-ph.CO]`.

[33] L. Senatore, *TASI Lectures on Inflation.*

[34] S. Dodelson, *Modern Cosmology.* Academic Press, 2003.

[35] V. Mukhanov, *Physical Foundations of Cosmology.* Cambridge University Press, 2005.

[36] D. Baumann, *TASI Lectures on Inflation.* `arXiv:0907.5424 [hep-th]`.

[37] V. Mukhanov and S. Winitzki, *Introduction to Quantum Effects in Gravity.* Cambridge University Press, 2007.

[38] T. Jacobson, *Introduction to Quantum Fields in Curved Spacetimes.* `arXiv:gr-qc/0308048`.

[39] S. Weinberg *Phys. Rev. D* **72** (2005) 043514, `arXiv:hep-th/0506236 [hep-th]`.

[40] Y. Wang *Commun. Theor. Phys.* **62** (2014) 109–166, `arXiv:1303.1523 [hep-th]`.

[41] E. Lim, *Advanced Cosmology: Primordial Non-Gaussianities (Lecture Notes).*

[42] E. Komatsu *Class. Quant. Grav.* **27** (2010) 124010, `arXiv:1003.6097 [astro-ph.CO]`.

[43] A. Guth and S.-Y. Pi *Phys. Rev. D* **32** (1985) 1899–1920.

[44] F. Bezrukov and M. Shaposhnikov *Phys. Lett. B* **659** (2008) 703–706, `arXiv:0710.3755 [hep-th]`.

Appendix A

[1] C. Will and N. Yunes, *Is Einstein Still Right?* Oxford University Press, 2020.

[2] M. Spradlin, A. Strominger, and A. Volovich, *Les Houches Lectures on de Sitter Space*, 2001. `arXiv:hep-th/0110007`.

[3] D. Anninos *Int. J. Mod. Phys. A* **27** (2012) 1230013, `arXiv:1205.3855 [hep-th]`.

[4] S. Carroll, *Spacetime and Geometry.* Cambridge University Press, 2019.

[5] R. d'Inverno, *Introducing Einstein's Relativity.* Oxford University Press, 1992.

[6] J. Hartle, *Gravity.* Cambridge University Press, 2021.

[7] H. Reall, *General Relativity (Lecture Notes).*

[8] D. Tong, *General Relativity (Lecture Notes).*

[9] E. Lim, *General Relativity (Lecture Notes).*

[10] C. Misner, K. Thorne, and J. Wheeler, *Gravitation.* Freeman, 2000.

Appendix B

[1] A. Challinor, *Advanced Cosmology (Lecture Notes).*

[2] E. Bertschinger `arXiv:astro-ph/9506070`.

[3] U. Seljak and M. Zaldarriaga *Astrophys. J.* **469** (1996) 437–444, `arXiv:astro-ph/9603033`.

[4] A. Lewis, A. Challinor, and A. Lasenby *Astrophys. J.* **538** (2000) 473–476, `arXiv:astro-ph/9911177 [astro-ph]`.

[5] J. Lesgourgues, `arXiv:1104.2932 [astro-ph.IM]`.

[6] S. Dodelson, *Modern Cosmology*. Academic Press, 2003.

[7] J. Bond and G. Efstathiou *Mon. Not. Roy. Astron. Soc.* **226** (1987) 655–687.

[8] M. Zaldarriaga and U. Seljak *Phys. Rev. D* **55** (1997) 1830–1840, `arXiv:astro-ph/9609170`.

Index